Methods in Enzymology

Volume 62
VITAMINS AND COENZYMES
Part D

METHODS IN ENZYMOLOGY

EDITORS-IN-CHIEF

Sidney P. Colowick   Nathan O. Kaplan

Methods in Enzymology

Volume 62

Vitamins and Coenzymes

Part D

EDITED BY

Donald B. McCormick and Lemuel D. Wright

DIVISION OF NUTRITIONAL SCIENCES AND
THE SECTION OF BIOCHEMISTRY, MOLECULAR AND CELL BIOLOGY
CORNELL UNIVERSITY
ITHACA, NEW YORK

ACADEMIC PRESS   New York   San Francisco   London   1979
A Subsidiary of Harcourt Brace Jovanovich, Publishers

ACADEMIC PRESS, INC.
111 Fifth Avenue, New York, New York 10003

United Kingdom Edition published by
ACADEMIC PRESS, INC. (LONDON) LTD.
24/28 Oval Road, London NW1 7DX

Library of Congress Cataloging in Publication Data

Main entry under title:

Vitamins and coenzymes, part A[B]

(Methods in enzymology, v. 18, 62)
Includes bibliographical references.
1. Vitamins. 2. Coenzymes. I. McCormick,
Donald Bruce, ed. II. Wright, Lemuel D., Date
ed. [DNLM: 1. Coenzymes. 2. Vitamins. W1 ME9615K
v. 18]
QP601.M49 [QP771] vol. 18 etc. 574.1'925'08s
ISBN 0–12–181962–0 [599'.01'926] 72–26903
(vol. 62)

PRINTED IN THE UNITED STATES OF AMERICA

79 80 81 82   9 8 7 6 5 4 3 2 1

Table of Contents

Section I. Ascorbic Acid

Section II. Thiamine: Phosphates and Analogs

Section III. Lipoic Acid and Derivatives

Section IV. Pantothenic Acid, Coenzyme A, and Derivatives

Section V. Biotin and Derivatives

Section VI. Pyridoxine, Pyridoxamine, and Pyridoxal: Analogs and Derivatives

Contributors to Volume 62

Article numbers are in parentheses following the names of contributors.
Affiliations listed are current.

ELIJAH ADAMS (69, 81), *Department of Biological Chemistry, University of Maryland School of Medicine, Baltimore, Maryland 21201*

R. KALERVO AIRAS (48), *Department of Biochemistry, University of Turku, SF 20500 Turku 50, Finland*

L. ALLAN (50), *Department of Biochemistry, Faculty of Medicine, University of Manitoba, Winnipeg, Manitoba R3E 0W3, Canada*

TERUO AMACHI (43), *Central Research Institute, Sunroy Limited, 1-1, Wakayamadai, Shimamoto-cho, Mishima-gun, Osaka, Japan*

BARBARA B. ANDERSON (85), *Department of Haematology, St. Bartholomew's Hospital, London EC1, England*

A. ARNONE (80), *Department of Biochemistry, The University of Iowa, Iowa City, Iowa 52240*

L. DAVID ARSCOTT (37), *Veterans Administration Hospital, Ann Arbor, Michigan 48105*

ROBERT L. BARCHI (25), *Departments of Neurology and of Biochemistry and Biophysics, University of Pennsylvania School of Medicine, Philadelphia, Pennsylvania 19104*

EDWARD A. BAYER (55, 63), *Department of Biophysics, Weizmann Institute of Science, Rehovot, Israel*

MELVIN BERGER (57), *National Institute of Allergy and Infectious Diseases, National Institutes of Health, Bethesda, Maryland 20014*

HUBERT E. BLUM (20), *Department of Internal Medicine, University Hospital, D-7800 Freiburg, West Germany*

DAVID N. BURTON (46), *Department of Microbiology, University of Manitoba, Winnipeg, Manitoba R3T 2N2, Canada*

ALICE DEL CAMPILLO-CAMPBELL (64), *Department of Biological Sciences, Stanford University, Stanford, California 94305*

CORALIE A. C. CARRAWAY (16), *Department of Biochemistry, Oklahoma State University, Stillwater, Oklahoma 74074*

JEAN-PAUL CARREAU (32), *Laboratoire de Physiologie Cellulaire, Université Pierre et Marie Curie (Jussieu), Paris, France*

DORIANO CAVALLINI (47), *Istituto di Chimica Biologica, Università di Roma, 00185 Roma, Italy*

HSIEN-HSIN CHANG (27), *Research Laboratories, Pillsbury Company, 311 Second Street, S.E., Minneapolis, Minnesota 55414*

M. S. CHAUHAN (68), *Department of Biochemistry, Faculty of Medicine, University of Manitoba, Winnipeg, Manitoba R3E 0W3, Canada*

COLIN F. CHIGNELL (54), *Laboratory of Environmental Biophysics, National Institute of Environmental Health Sciences, P. O. Box 12233, Research Triangle Park, North Carolina 27709*

P. PATRICK CLEARY (64), *Department of Microbiology, University of Minnesota Medical School, Minneapolis, Minnesota 55455*

CARLO COLOMBINI (70), *R7 Box 244, Charlottesville, Virginia 22901*

VALERIO CONSALVI (45), *Istituto di Chimica Biologica, Università di Roma, 00185 Roma, Italy*

JOSEPH CUPANO (6), *Chemical Research Department, Hoffmann–La Roche Inc., Nutley, New Jersey 07110*

K. DAKSHINAMURTI (50, 67, 68), *Department of Biochemistry, Faculty of Medicine, University of Manitoba, Winnipeg, Manitoba R3E 0W3, Canada*

CHARLES R. DAWSON (5), *Department of Chemistry, Columbia University, New York, New York 10027*

BRUNO DEUS (20), *Central Laboratory,*

University Hospital, D-7800 Freiburg, West Germany

SILVESTRO DUPRÈ (47), *Centro di Biologia Molecolare del Consiglio Nazionale della Ricerche, c/o Istituto di Chimica Biologica, Università di Roma, 00185 Roma, Italy*

ROBERT E. DYAR (40), *Laboratory for Experimental Pathology, William S. Middleton Memorial Veterans Medical Center, Madison, Wisconsin 53705*

DANIEL DYKHUIZEN (64), *Department of Biological Sciences, Purdue University, West Lafayette, Indiana 47907*

E. E. EDWIN (8, 15, 24), *Biochemical Department, Central Veterinary Laboratory, Ministry of Agriculture, Fisheries and Food, Weybridge, Surrey, England*

MAX A. EISENBERG (59, 60, 61), *Department of Biochemistry, Columbia University College of Physicians and Surgeons, New York, New York 10032*

HEIKKI A. ELO (52, 53), *Department of Biomedical Sciences, University of Tampere, Tampere, Finland*

R. RAY FALL (66), *Department of Chemistry, University of Colorado, Boulder, Colorado 80309*

SABURO FUKUI (78), *Laboratory of Industrial Biochemistry, Department of Industrial Chemistry, Faculty of Engineering, Kyoto University, Yoshida, Sakyo-Ku, Kyoto, Japan*

HAROLD C. FURR (27), *Department of Biochemistry and Biophysics, Iowa State University, Ames, Iowa 50010*

A. GIARTOSIO (77), *Istituto di Chimica Biologica, Facoltà di Medicina, Università di Roma, 00185 Roma, Italy*

P. M. GILLEVET (67), *Department of Biochemistry, Faculty of Medicine, University of Manitoba, Winnipeg, Manitoba R3E 0W3, Canada*

MARIAN GORECKI (31), *Department of Organic Chemistry, Weizmann Institute of Science, Rehovot, Israel*

RALPH GREEN (85), *Department of Clinical Research, Scripps Clinic and Research Foundation, 10666 North Torrey Pines Road, La Jolla, California 92037*

C. J. GUBLER (9, 12, 19), *Graduate Section*

of Biochemistry, Brigham Young University, Provo, Utah 84602

JAN A. GUTOWSKI (26), *Plasma Unit, Connaught Laboratories Ltd., 1755 Steeles Avenue West, Willowdale, Ontario M2N 5T8, Canada*

EARL H. HARRISON (27), *The Rockefeller University, New York, New York 10021*

RYOJI HAYASHI (17, 18), *Department of Microbiology, Yamaguchi University Medical School, Ube, Yamaguchi-Ken, 755 Japan*

IDA K. HEGNA (38), *Institute of Pharmacy, University of Oslo, Blindern, Oslo 3, Norway*

B. C. HEMMING (12), *Department of Plant Pathology, Montana State University, Bozeman, Montana 59717*

ROSS L. HOOD (49), *Commonwealth Scientific & Industrial Research Organisation, Division of Food Research, North Ryde, New South Wales 2113, Australia*

KIHACHIRO HORIIKE (83), *Department of Biochemistry, Osaka University Medical School, 33 Joancho, Kita-ku, Osaka 530, Japan*

SEI-ICHIRO IKEDA (78), *Department of Biochemistry, The Ohio State University, Columbus, Ohio 43210*

WHA BIN IM (65), *Department of Physiology, University of North Carolina School of Medicine, Chapel Hill, North Carolina 27514*

SHOJI IMAMOTO (43), *Central Research Institute, Suntory Limited, 1-1-1, Wakayama-dai, Shimamoto-cho, Mishima-gun, Osaka, Japan*

KIMIKAZU IWAMI (22), *Laboratory of Nutritional Chemistry, Faculty of Agriculture, Kyoto University, Sakyo-ku, Kyoto 606, Japan*

AKIO IWASHIMA (21), *Department of Biochemistry, Kyoto Prefectural University of Medicine, Kyoto, 602 Japan*

YOSHIKAZU IZUMI (58), *Department of Agricultural Chemistry, Kyoto University, Kyoto, Japan*

ROBERT J. JAWORSKI (73), *Department of Chemistry, University of Wisconsin, Madison, Wisconsin 53706*

ROBERT A. JENIK (46), *Lipid Metabolism*

Laboratory, William S. Middleton Memorial Veterans Center and the Department of Physiological Chemistry, University of Wisconsin, Madison, Wisconsin 53706

SOHAN L. JINDAL (29), Oswal Vanaspati Allied Industries, Ludhiana, India

FUSAKO KAWAI (42), Department of Chemistry, Kobe University of Commerce, Kobe 655, Japan

TAKASHI KAWASAKI (14), Department of Biochemistry, Hiroshima University School of Medicine, 1-2-3 Kasumi, Hiroshima, Japan

MICHAEL N. KAZARINOFF (83), Division of Nutritional Sciences and the Section of Biochemistry, Molecular and Cell Biology, Savage Hall, Cornell University, Ithaca, New York 14853

ASKAR G. KHALMURADOV (11), Department of Biochemistry of Coenzymes of A. V. Palladin Institute of Biochemistry, Academy of Sciences of the Ukrainian SSR, 9, Leontovicha Str., Kiev 252030, USSR

HARLEY L. KING, JR. (40), Department of Biochemistry, University of Utah Medical Center, Salt Lake City, Utah 84132

JACK F. KIRSCH (51), Department of Biochemistry, University of California at Berkeley, Berkeley, California 94720

PETER T. KISSINGER (3), Department of Chemistry, Purdue University, West Lafayette, Indiana 47907

SHOZABURO KITAOKA (41), Department of Agricultural Chemistry, University of Osaka Prefecture, Sakai, Osaka 591, Japan

HAROLD J. KLOSTERMAN (75), Department of Biochemistry, North Dakota State University, Fargo, North Dakota 58105

W. KORYTNYK (74), Department of Experimental Therapeutics, Roswell Park Memorial Institute, 666 Elm Street, Buffalo, New York 14263

KENNETH KRELL (61), Bureau of Radiological Health, Food and Drug Administration, 5600 Fisher Lane, Rockville, Maryland 20852

MARKKU S. KULOMAA (53), Department of Biomedical Sciences, University of Tampere, Tampere, Finland

M. G. KUNITANI (62), Department of Biochemistry and Biophysics, University of California at San Francisco, San Francisco, California 94143

FRANKLIN R. LEACH (16, 33), Department of Biochemistry, Oklahoma State University, Stillwater, Oklahoma 74074

MEN HUI LEE (5), Department of Pharmacology, Yale University School of Medicine, New Haven, Connecticut 06510

TING-KAI LI (84), Departments of Medicine and Biochemistry, Indiana University School of Medicine, and Veterans Administration Hospital, Indianapolis Indiana 46202

ARNOLD A. LIEBMAN (6), Chemical Research Department, Hoffmann–La Roche Inc., Nutley, New Jersey 07110

HENRY J. LIN (51), Department of Biochemistry, University of California at Berkeley, Berkeley, California 94720

LAWRENCE LUMENG (84), Departments of Medicine and Biochemistry, Indiana University School of Medicine, and Veterans Administration Hospital, Indianapolis, Indiana 46202

DONALD B. MCCORMICK (27, 65, 83), Division of Nutritional Sciences and the Section of Biochemistry, Molecular and Cell Biology, Savage Hall, Cornell University, Ithaca, New York 14853

ERNEST E. MCCOY (70), Department of Pediatrics, University of Alberta, Edmonton, Alberta T6G 2G3, Canada

HENRY H. MANTSCH (72), Division of Chemistry, National Research Council, Ottawa K1A OR6, Canada

RONALD MARKEZICH (6), Chemical Research Department, Hoffmann–La Roche Inc., Nutley. New Jersey 07110

DON S. MARTIN (80), Department of Chemistry, Iowa State University, Ames, Iowa 50010

ROWENA G. MATTHEWS (37), Department of Biological Chemistry, University of Michigan, Ann Arbor, Michigan 48109

ALFRED H. MERRILL (83), Section of Biochemistry, Molecular and Cell Biology,

Savage Hall, Cornell University, Ithaca, New York 14853

HARRY W. MESLAR (56), E.I. du Pont de Nemours and Company, Automatic Clinical Analysis Division, Wilmington, Delaware 19898

CAROL M. METZLER (80), Department of Biochemistry and Biophysics, Iowa State University, Ames, Iowa 50011

DAVID E. METZLER (79, 80), Department of Biochemistry and Biophysics, Iowa State University, Ames, Iowa 50011

A. YU. MISHARIN (76), Institute of Molecular Biology, USSR Academy of Sciences, Vavilov Str. 32, Moscow 117313, USSR

HISATERU MITSUDA (22), Laboratory of Food Science and Technology, Research Institute for Production Development, Simogamo Morimotocho, Sakyo-ku, Kyoto 606, Japan

KAZUTAKA MIYATAKE (41), Department of Agricultural Chemistry, University of Osaka Prefecture, Sakai, Osaka 591, Japan

MARIO MORIGGI (45), Istituto di Chimica Biologica, Università di Roma, 00185 Roma, Italy

RICHARD R. MUCCINO (6), Chemical Research Department, Hoffmann–La Roche Inc., Nutley, New Jersey 07110

D. S. MURDOCK (9), Utah State Bureau of Laboratories, 44 Medical Drive, Salt Lake City, Utah 84113

ROBERT W. MURRAY (29), Department of Chemistry, University of Missouri-St. Louis, St. Louis, Missouri 63121

N. KRISHNA MURTY (2), Department of Chemistry, Andhra University, Waltair 530003, India

KENJI NAKAJIMA (22), Laboratory of Nutritional Chemistry, Faculty of Agriculture, Kyoto University, Sakyo-ku, Kyoto 606, Japan

YOSHIHISA NAKANO (41), Department of Agricultural Chemistry, University of Osaka Prefecture, Sakai, Osaka 591, Japan

HIDEO NAKAYAMA (18), Department of Microbiology, Yamaguchi University Medical School, Ube, Yamaguchi-Ken, 755 Japan

MORIMITSU NISHIKIMI (4), Department of Medical Chemistry, Kochi Medical School, Nangoku-shi 781-51, Japan

TAKAHIRO NISHIMUNE (17), Department of Microbiology, Yamaguchi University Medical School, Ube, Yamaguchi-Ken, 755 Japan

YOSHITSUGU NOSE (21), Department of Biochemistry, Kyoto Prefectural University of Medicine, Kyoto, 602 Japan

KOICHI OGATA* (42, 44, 58), Department of Agricultural Chemistry, Kyoto University, Kyoto 606, Japan

MARION H. O'LEARY (73), Department of Chemistry, University of Wisconsin, Madison, Wisconsin 53706

STANLEY T. OMAYE (1), Biochemistry Division, Department of Nutrition, Letterman Army Institute of Research, Presidio of San Francisco, San Francisco, California 94129

LAWRENCE A. PACHLA (3), McNeil Laboratories, Biochemical Research, Camp Hill Road, Fort Washington, Pennsylvania 19034

JULIA M. PARKHOMENKO (11), Department of Biochemistry of Coenzymes of A.V. Palladin Institute of Biochemistry, Academy of Sciences of the Ukrainian SSR, 9, Leontovicha Str., Kiev 252030, USSR

RONALD J. PARRY (62), Department of Chemistry, Rice University, Houston, Texas 77001

ABRAHAM PATCHORNIK (31), Department of Organic Chemistry, Weizmann Institute of Science, Rehovot, Israel

H. K. PENTTINEN (10, 13, 23), Department of Medical Chemistry, University of Helsinki, Helsinki 17, Finland

CLARK W. PERRY (6), Chemical Research Department, Hoffmann–La Roche Inc., Nutley, New Jersey 07110

LAURA POLITI (45), Istituto di Chimica Biologica, Università di Roma, 00185 Roma, Italy

O. L. POLYANOVSKY (76), Institute of Molecular Biology, USSR Academy of Sciences, Vavilov Str. 32, Moscow 117312, USSR

* Deceased.

JOHN W. PORTER (46), *Lipid Metabolism Laboratory, William S. Middleton Memorial Veterans Center and the Department of Physiological Chemistry, University of Wisconsin, Madison, Wisconsin 53706*

SUE GLENN POWERS (39), *Department of Biochemistry, Cornell University Medical College, New York, New York 10021*

K. RAMA RAO (2), *Department of Chemistry, Andhra University, Waltair 530003, India*

F. RIVA (77), *Istituto di Chimica Biologica, Facoltà di Scienze, Università di Cagliari, 09100 Cagliari, Italy*

PAUL H. ROGERS (80), *Department of Biochemistry, Iowa State University, Ames, Iowa 50011*

A. B. ROY (7), *Department of Physical Biochemistry, The John Curtin School of Medical Research, The Australian National University, P.O. Box 334, Canberra City, A.C.T. 2601, Australia*

ALLA A. RYBINA (11), *Department of Biochemistry of Coenzymes of A.V. Palladin Institute of Biochemistry, Academy of Sciences of the Ukrainian SSR, 9, Leontovicha Str., Kiev 252030, USSR*

HOWERDE E. SAUBERLICH (1), *Department of Nutrition, Letterman Army Institute of Research, Presidio of San Francisco, San Francisco, California 94129*

ROBERTO SCANDURRA (45), *Istituto di Chimica Biologica, Università di Roma, 00185 Roma, Italy*

JASON C. H. SHIH (27, 34), *Laboratory of Applied Biochemistry, Department of Poultry Science, North Carolina State University, Raleigh, North Carolina 27650*

SAKAYU SHIMIZU (44), *Department of Agricultural Chemistry, Kyoto University, Kyoto 606, Japan*

MARVIN SILVER (28), *Canada Centre for Mineral and Energy Technology, 555 Booth Street, Ottawa, Ontario K1A OG1, Canada*

BIRANDRA K. SINHA (54), *Laboratory of Environmental Biophysics, National Institute of Environmental Health Sciences,* P.O. Box 12233, Research Triangle Park, North Carolina 27709

EHUD SKUTELSKY (55), *Section of Biological Ultrastructure, Weizmann Institute of Science, Rehovot, Israel*

IAN C. P. SMITH (72), *Division of Biological Sciences, National Research Council, Ottawa, Ontario K1A OR6, Canada*

ESMOND E. SNELL (39, 82), *Departments of Microbiology and Chemistry, University of Texas at Austin, Austin, Texas 78712*

CAMELLIA SOBHY (46), *Department of Biochemistry, University of Wisconsin, Madison, Wisconsin 53706*

OTTO SOLBERG (38), *National Institute of Public Health, Geitmyrsveien 75, Oslo 4, Norway*

JOSEPH T. SPENCE (27), *McArdle Laboratory for Cancer Research, University of Wisconsin Medical Center, Madison, Wisconsin 53706*

FRANK E. STARY (29), *Department of Chemistry, Maryville College, St. Louis, Missouri 63141*

GERALD L. STONER (60), *Armaeur Hansen Research Institute, P.O. Box 1005, Addis Abba, Ethiopia*

KEN STRYNADKA (70), *Department of Pediatrics, University of Alberta, Edmonton, Alberta T6G 2G3, Canada*

CLARENCE H. SUELTER (82), *Department of Biochemistry, Michigan State University, East Lansing, Michigan 48824*

MAKOTO TAKAGI (30), *Department of Organic Synthesis, Faculty of Engineering, Kyushu University, Fukuoka 812, Japan*

YUKIO TAKII (22), *Laboratory of Nutritional Chemistry, Faculty of Agriculture, Kyoto University, Sakyo-ku, Kyoto 606, Japan*

YOSHIKI TANI (44, 58), *Department of Agricultural Chemistry, Kyoto University, Kyoto 606, Japan*

COLIN THORPE (37), *Department of Chemistry, University of Delaware, Newark, Delaware 19711*

V. P. TIMOFEEV (76), *Institute of Molecular Biology, USSR Academy of Sciences, Vavilov Str. 32, Moscow 117312, USSR*

HARUHITO TSUGE (83), *Department of Agri-*

/cultural Chemistry, Gifu University, Kagamigahara 504, Gifu, Japan

PENTTI J. TUOHIMAA (52, 53), Department of Biomedical Sciences, University of Tampere, Tampere, Finland

C. TURANO (77), Istituto di Chimica Biologica, Facoltà di Farmacia, Università di Roma, 00185 Roma, Italy

J. DAVID TURNBULL (1), Biochemistry Division, Department of Nutrition, Letterman Army Institute of Research, Presidio of San Francisco, San Francisco, California 94129

GERALD G. VERNICE (6), Airco/Ohio Medical Products, Murray Hill, New Jersey 07974

C. BORRI VOLTATTORNI (77), Istituto di Chimica Biologica, Facoltà di Farmacia, Università di Perugia, 06100 Perugia, Italy

YASUO WAKABAYASHI (21), Department of Biochemistry, Kyoto Prefectural University of Medicine, Kyoto, 602 Japan

HAROLD B. WHITE, III (56), Department of Chemistry, University of Delaware, Newark, Delaware 19711

MEIR WILCHEK (55, 63), Department of Biophysics, Weizmann Institute of Science, Rehovot, Israel

DAVID R. WILKEN (40), Laboratory for Experimental Pathology, William S. Middleton Memorial Veterans Medical Center. and Department of Physiological Chemistry, University of Wisconsin Medical School, Madison, Wisconsin 53705

KEITH D. WILKINSON (37), Institute for Cancer Research, Fox Chase Center, Philadelphia, Pennsylvania 19111

A. K. WILLIAMS (71), Russell Research Center, U. S. Department of Agriculture, Athens, Georgia 30604

CHARLES H. WILLIAMS, JR. (37), Veterans Administration Hospital, and Department of Biological Chemistry, University of Michigan, Ann Arbor, Michigan 48105

LEMUEL D. WRIGHT (27, 65), Division of Nutritional Sciences and the Section of Biochemistry, Molecular and Cell Biology, Savage Hall, Cornell University, Ithaca, New York 14853

HIDEAKI YAMADA (42), Department of Agricultural Chemistry, Kyoto University, Kyoto 606, Japan

HIROSHI YANAGAWA (35, 36), Mitsubishi-Kasei Institute of Life Sciences, Machida-shi, Tokyo 194, Japan

BOB IN-YU YANG (79), Department of Chemistry, University of Missouri, Kansas City, Missouri 64110

KYODEN YASUMOTO (22), Laboratory of Nutritional Chemistry, Faculty of Agriculture, Kyoto University, Sakyo-ku, Kyoto 606, Japan

HAJIME YOSHIZUMI (43), Central Research Institute, Suntory Limited, 1-1-1, Wakayama-dai, Shimamoto-cho, Mishima-gun, Osaka, Japan

Preface

Since 1970–1971, when the earlier volumes (XVIII, A, B, and C) on "Vitamins and Coenzymes" were published as part of the *Methods in Enzymology* series, there has been a considerable expansion of techniques and methodology attendant to the assay, isolation, and characterization of the vitamins and those systems responsible for their biosynthesis, transport, and metabolism. In part, this has been generated by an increasing awareness of the diversity of such vitaminic forms as comprise essential moieties of coenzymes and also through recognition of the function of some derived metabolites as hormones, regulators, and even antioxidants.

As a consequence of this new body of information and its expected impact in the stimulation of further research on vitamins and coenzymes, we have sought to provide investigators with the more current modifications of "tried and true" methods as well as those which have only now become available. Volume 62 is the first of three volumes resulting from our efforts in soliciting contributions from numerous, active experimentalists who have published most of their findings in the usual, refereed research journals. The amount of material which appeared to warrant coverage necessitated a division into three parts, each comprising a volume: D, which covers the vitamin and coenzyme forms of ascorbate, thiamine, lipoate, pantothenate, biotin, and pyridoxine; E, nicotinate, flavins, and pteridines; and F, the B_{12} group and those classically considered as "fat-soluble."

We should like to express our gratitude to the contributors for their willingness to supply the information requested and, in some instances, their tolerance of editorial emendations. There has been an attempt to allow such overlap as would offer flexibility in the choice of method, such as modification of an assay procedure. Where some omissions seemingly occur, these may, in some cases, be attributed to the inadvertent oversight of the editors; however, in other cases it was felt that the topics were adequately covered in the earlier volumes on this subject or in other volumes in the *Methods of Enzymology* series.

Finally, we again wish to thank Mrs. Patricia MacIntyre for her excellent secretarial assistance and the numerous persons at Academic Press for their efficient and kind guidance.

DONALD B. MCCORMICK
LEMUEL D. WRIGHT

METHODS IN ENZYMOLOGY

EDITED BY

Sidney P. Colowick and Nathan O. Kaplan

VANDERBILT UNIVERSITY
SCHOOL OF MEDICINE
NASHVILLE, TENNESSEE

DEPARTMENT OF CHEMISTRY
UNIVERSITY OF CALIFORNIA
AT SAN DIEGO
LA JOLLA, CALIFORNIA

I. Preparation and Assay of Enzymes
II. Preparation and Assay of Enzymes
III. Preparation and Assay of Substrates
IV. Special Techniques for the Enzymologist
V. Preparation and Assay of Enzymes
VI. Preparation and Assay of Enzymes *(Continued)*
Preparation and Assay of Substrates
Special Techniques
VII. Cumulative Subject Index

METHODS IN ENZYMOLOGY

EDITORS-IN-CHIEF

Sidney P. Colowick   Nathan O. Kaplan

Methods in Enzymology

Volume 62
VITAMINS AND COENZYMES
Part D

Section I

Ascorbic Acid

[1] Selected Methods for the Determination of Ascorbic Acid in Animal Cells, Tissues, and Fluids[1]

By STANLEY T. OMAYE, J. DAVID TURNBULL, and
HOWERDE E. SAUBERLICH

Vitamin C is the enolic form of an α-ketolactone. The endiol groups at the second and third carbon atoms are sensitive to oxidation and can easily convert into a diketo group, L-dehydroascorbic acid. This oxidized form of the vitamin is just as effective against scurvy as the reduced substance. Aqueous solutions of the nutrient L-dehydroascorbic acid are air-oxidized further both rapidly and irreversibly to L-diketogulonic acid and other biological inactive substances.

Research on vitamin C has been exhaustive; however, the precise biochemical function of ascorbic acid remains obscure. This limited knowledge concerning the metabolic functions of ascorbic acid has hindered efforts to develop fully satisfactory procedures to identify vitamin C deficiency or to assess nutritional status. Vitamin C is the only water-soluble vitamin for which there is no microbial assay.

Methods for determining ascorbic acid are numerous. In general, chemical analyses for the vitamin may be divided into two groups; the determination of the reduced form and the determination of the oxidized form. The former group of analyses are usually based upon the oxidation–reduction properties of ascorbic acid. These are widely used as the fundamental reactions in the measurement of vitamin C. The latter group of analyses is usually based upon the oxidation of the ascorbic acid and the subsequent formation of a hydrazone or a fluorophor. Best results are obtained if samples, especially plasma, are quickly stabilized with either trichloroacetic acid or metaphosphoric acid and immediately analyzed. Prompt stabilization is especially important in the case of plasma or serum. The greater stability of ascorbic acid in acid solution is due to the decreased tendency for hydrolysis of the lactone ring with decreasing pH. In alkaline solution the hydrolysis is fairly rapid, and such solutions lose vitamin activity in a short period of time. Also the oxidation of ascorbate in solution is catalyzed by various metal ions. If necessary, acidified samples may be stored frozen at $-70°$.[2]

[1] The opinions or assertions contained herein are the private views of the authors and are not to be construed as official or as reflecting the views of the Department of the Army or the Department of Defense.

[2] D. W. Bradley, G. Emery, and J. E. Maynard, *Clin. Chim. Acta* **44**, 47 (1973).

METHODS IN ENZYMOLOGY, VOL. 62

Despite some limitations, plasma ascorbic acid has served as the index for the biochemical evaluation of vitamin C status. Low plasma levels of ascorbic acid do not necessarily indicate scurvy, although in clinical cases of scurvy the patients invariably have low or no serum ascorbic acid. Continued low levels of plasma ascorbic acid of less than 0.10 mg/100 ml would eventually lead to symptoms of scurvy.[3,4]

Plasma ascorbate levels of 0.4 to 1.4 mg/100 ml reflect a daily ascorbate intake of 40 mg or more in the adult. When tissues are saturated with vitamin C, the plasma ascorbate concentrations are between 0.8 and 1.5 mg/100 ml, the whole blood levels are between 1.0 and 1.5 mg/100 ml, and the buffy-coat ascorbic acid levels are between 25 and 35 mg/100 ml. Higher plasma ascorbic acid concentration can be attained temporarily following the ingestion of a large dose of the vitamin. Consequently, serum ascorbate values tend to reflect, in general, recent dietary intakes of vitamin C.

Whole-blood ascorbic acid values may be a less sensitive indicator of vitamin C nutriture than serum or plasma levels of the vitamin. This is because the ascorbic acid levels in erythrocytes never fall to the low levels encounted in serum.

Leukocyte ascorbate concentrations have been reported to be more closely related to tissue stores of the vitamin. A direct proportionality between leukocyte and total body ascorbic acid has been shown in the guinea pig.[5] Normal human values of ascorbate in leukocytes range from 20 to 53 $\mu g/10^8$ white blood cells (WBC's). In general, men have been reported to have lower leukocyte ascorbic acid values than women. Besides scurvy there also appears to be a decline of leukocyte ascorbic acid levels with increasing age. Peptic ulcer, gastroduodenal disorders, and postoperative surgery have been associated with decreased leukocyte ascorbic acid of less than 12.7 $\mu g/10^8$ WBC's compared to control values of more than 20.8 $\mu g/10^8$ WBC's. Unfortunately, the determination of ascorbic acid in WBC's is technically difficult and requires large blood samples, making the procedure impractical for routine use. Caution must be used in the isolation of WBC's for ascorbate measurement, as improper handling of cells will cause a loss of ascorbic acid resulting in false low values.

In general, individuals with low intakes of ascorbic acid excrete less of a given dose than those on an adequate or saturating intake. Urinary ascorbic acid excretion rapidly declines to undetectable levels during vitamin C depletion, however metabolites of ascorbate continue to be ex-

[3] H. E. Sauberlich, R. P. Dowdy, and J. H. Skala, in "Laboratory Test For the Assessment of Nutritional Status," p. 13. Chem. Rubber Publ. Co., Cleveland, Ohio, 1974.

[4] H. E. Sauberlich, Ann. N. Y. Acad. Sci. 258, 438 (1975).

[5] H. B. Burch, Ann. N. Y. Acad. Sci. 19, 268 (1961).

TABLE I
ASCORBIC ACID CONTENT OF ADULT HUMAN TISSUES

Tissue	Ascorbic acid (mg/100 g wet tissue)
Adrenal glands	30–40
Pituitary gland	40–50
Liver	10–16
Spleen	10–15
Lungs	7
Kidneys	5–15
Testes	3
Thyroid	2
Heart muscle	5–15
Skeletal muscle	3–4
Brain	13–15
Pancreas	10–15
Eye lens	25–31
Plasma	0.4–10
Saliva	0.07–0.09

creted. Urinary levels of ascorbic acid are prone to reflect recent dietary intakes and are subject to analytical problems, such as stability. Consequently measurements of urinary ascorbate levels are seldom used. Nevertheless, in the scorbutic patient, the urinary excretion of vitamin C would be expected to be near zero and therefore could provide supportive diagnostic information.

A large number of publications have appeared containing data on the tissue content of ascorbic acid in humans and animals. Table I is a partial reproduction of ascorbic acid values for human tissues as collected and compiled from the literature.[6] Highest ascorbate values were found in adrenal glands, pituitary gland, and eye lens with small but detectable concentrations in saliva.

The following four procedures described represent vitamin C assays frequently used at the present time. The original procedures were adapted or modified and tested for use in our own laboratory. No attempt at this time has been made to evaluate some of the more recent published assays. Although subject to controversy, it is generally accepted that the level of ascorbic acid in buffy coat (leukocyte-platelet fraction) provides a reliable index of vitamin C nutrition and tissue concentration. Therefore we have ended this paper on ascorbic acid analysis with a description of a procedure for the isolation of buffy-coat leukocytes. This procedure, which is in use in our laboratory, is a modification of existing methods.[7]

[6] D. Hornig, *Ann. N. Y. Acad. Sci.* **258**, 103 (1975).
[7] H. S. Loh, and C. W. M. Wilson, *Int. J. Vitam. Nutr. Res.* **41**, 90 (1971).

Ascorbic Acid Analysis I: Reduction of 2,6-Dichlorophenolindophenol

Principle. Ascorbic acid reduces the dye 2,6-dichlorophenolindo-phenol and causes a decrease in the absorption of dye at 520 nm. The reaction must be carried out within a pH range of 3.0–4.5. A blank value for each sample is obtained by adding a few crystals of ascorbic acid, completely reducing the dye and rendering it colorless. Nonspecific reduction of the dye by thiol compounds present in the sample is blocked by *p*-chloromercuribenzoate, which is included in the buffer.[8,9]

Reagents

    Metaphosphoric acid (HPO_3), 10% and 5% solutions in glass-distilled water

    Citrate/acetate buffer containing *p*-chloromercuribenzoate (pCMB): 22 g of trisodium citrate dihydrate is placed in about 40 ml of water, the pH is adjusted to 4.15 with glacial acetic acid, and the total volume is brought to 100 ml with water. Two hundred milligrams of pCMB are added, and the solution is centrifuged to remove any excess pCMB.

    2,6-Dichlorophenolindophenol (DCIP), sodium salt, 0.1 mg/ml in water.

Procedure. Serum, plasma, or urine is deproteinized by addition of 1 ml of sample to 1 ml of ice-cold 10% HPO_3 and, after thorough mixing, centrifuged for 20 min at 3500 *g*. Liver is homogenized with a Potter-Elvehjem homogenizer in 9 ml of ice-cold 5% HPO_3 per gram of tissue, then centrifuged in the same way. In each case, a 0.6-ml aliquot of the supernatant constitutes the sample for a single analysis.

Three-tenths milliliter of citrate/acetate buffer is added to each sample, and any turbid sample is centrifuged. From this point the samples are analyzed individually. Three-tenths milliliter of DCIP solution is added to the sample and, after exactly 30 sec, it is read against distilled water at 520 nm. A few crystals of ascorbic acid are added to bleach the dye by reducing it completely, and the sample is read again. This value serves as a blank for the sample. A standard curve, including a reagent blank, is constructed with standards ranging between 0 and 20 μg of ascorbic acid per milliliter of 5% HPO_3. The change in absorbance (ΔA) due to reduction of the dye by ascorbic acid in the sample is calculated from the following equation:

$$\Delta A = (RB - RB_b) - (S - S_b) \tag{1}$$

[8] J. A. Owen and B. Iggo, *Biochem. J.* **62**, 675 (1956).
[9] R. J. Henry, "Clinical Chemistry Principles and Technics," p. 715. Harper & Row, New York, 1964.

where RB is the absorbance of the reagent blank; RB_b is the absorbance of the reagent blank after bleaching with ascorbic acid; S is the absorbance of the sample; and S_b is the absorbance of the sample after bleaching with ascorbic acid.

ΔA is linearly related to ascorbic acid concentration, and the concentration in a sample is obtained by comparison of ΔA with the standard curve.

Comment. pCMB may be omitted from the buffer in the analysis of plasma or serum samples.

Ascorbic Acid Analysis II. Determination after Derivatization with 2,4-Dinitrophenylhydrazine

Principle. Ascorbic acid is oxidized by copper to form dehydroascorbic acid and diketogulonic acid. These products are treated with 2,4-dinitrophenylhydrazine to form the derivative bis-2,4-dinitrophenylhydrazone. This compound, in strong sulfuric acid, undergoes a rearrangement to form a product with an absorption band that is measured at 520 nm. The reaction is run in the presence of thiourea to provide a mildly reducing medium, which helps to prevent interference from nonascorbic acid chromogens.[10-12]

Reagents. All are stable for at least one week.

Trichloroacetic acid (TCA), 5% and 10% solutions in distilled water

2,4-Dinitrophenylhydrazine/thiourea/copper (DTC) solution, add 0.4 g thiourea, 0.05 g $CuSO_4 \cdot 5H_2O$, and 3.0 g 2,4-dinitrophenylhydrazine and bring to a total volume of 100 ml with 9 N H_2SO_4

65% H_2SO_4

Norit, acid washed (available from Sigma Chemical Company)

Procedure. One milliliter of whole blood, serum, plasma, or urine is added to 1.0 ml of ice-cold 10% TCA, mixed thoroughly, and centrifuged for 20 min at 3500 g. Liver is homogenized with a Potter-Elvehjem homogenizer in 9 ml of ice-cold 5% TCA per gram tissue and centrifuged for 20 min at 3500 g. Urinary supernatants are treated with Norit to remove interfering chromogens. To do this, approximately 0.2 g of Norit is added to the urinary supernatant and the resulting suspension is shaken vigorously for 10 min, then centrifuged to remove the Norit. From this point on, treatment of the various sample types is identical. In order to

[10] J. H. Roe and C. A. Kuether, *J. Biol. Chem.* **147,** 399 (1943).

[11] J. H. Roe *in* "Methods of Biochemical Analysis" (D. Glick, ed.), p. 115. John Wiley, New York, 1954.

[12] O. A. Bessey, O. H. Lowry, and M. J. Brock, *J. Biol. Chem.* **168,** 197 (1947).

form the bis-2,4-dinitrophenylhydrazone, 0.5 ml of supernatant is mixed with 0.1 ml of DTC and incubated for 3 hr at 37°. To convert this to the rearranged product, which will be measured spectrophotometrically, 0.75 ml of ice-cold 65% H_2SO_4 is added and mixed well, and the solutions are allowed to stand at room temperature for an additional 30 min. Absorbances are determined at 520 nm. Standards should be made in 5% TCA and range from 0 to 20 $\mu g/ml$.

Comments. This assay may be scaled down, with the use of microcuvettes. For example, a similar procedure has been used to analyze as little as 10 μl of serum.[13]

The above conditions for formation of bis-2,4-dinitrophenylhydrazone (3 hr at 37°) are recommended rather than a higher temperature for a shorter time in order to avoid possible interference by ascorbic acid-2-sulfate.[14]

Ascorbic Acid Analysis III: Fluorometric Determination

Assay Method

Principle. Ascorbic acid is oxidized to dehydroascorbic in the presence of iodine. The excess iodine is destroyed by adding sodium thiosulfate.[15] Dehydroascorbic acid is then condensed with *o*-phenylenediamine to form the fluorophor quinoxaline. A blank value is obtained by inhibiting the formation of the dehydroascorbic acid fluorophor formation with boric acid. With proper microcuvettes 0.05 μg of ascorbic acid per milliliter of plasma can be detected.[16]

Reagents

Metaphosphoric acid, 5% in glass-distilled water

Sodium acetate ($CH_3COONa \cdot 3H_2O$) solution, 25% in glass-distilled water

Sodium acetate–borate solution, 20 g of H_3BO_3 in 200 ml of 25% sodium acetate solution

o-Phenylenediamine solution, 0.1% in glass-distilled water; prepared fresh prior to use

Iodine solution, 0.1 N. Weigh out 12 g of iodate-free KI and dissolve in 100 ml of glass-distilled water. Add 6.5 g of iodine to the KI solu-

[13] O. H. Lowry, J. A. Lopez, and O. A. Bessey, *J. Biol. Chem.* **196**, 609 (1945).

[14] E. M. Baker, C. C. Hammer, J. E. Kennedy, and B. M. Tolbert, *Anal. Biochem.* **55**, 641 (1973).

[15] G. Brubacher and J. P. Vuilleumier, *in* "Clinical Biochemistry" (H. C. Curtius and M. Roth, eds.), Vol. II, p. 989. de Gruyter, Berlin, 1974.

[16] M. J. Deutsch and C. E. Weeks, *J. Assoc. Off. Anal. Chem.* **48**, 1248 (1965).

tion; stir until it is all dissolved. Dilute to final volume of 500 ml with glass-distilled water.

Sodium thiosulfate solution, 0.5 M in glass-distilled water

Procedure. One milliliter of plasma or urine is mixed with 18 ml of 5% metaphosphoric acid. The mixture is agitated vigorously for 30 sec and centrifuged. Tissue homogenates (10–20%) are stabilized with 5% metaphosphoric acid and centrifuged. To 2 ml of the clear supernatant an excess of iodine is added, i.e., 10 μl of 0.1 N iodine solution; after 10 sec the excess iodine is reduced by adding 5 μl of 0.5 M sodium thiosulfate solution. A blank and a test solution are prepared from the same oxidized sample. To 500 μl of oxidized sample, 500 μl of sodium acetate solution or 500 μl of sodium acetate–borate solution are added, respectively, to the test or blank solutions. The solutions are allowed to stand 15 min in the dark at room temperature, followed by the addition of 200 μl of o-phenylenediamine solution and another 15 min incubation in the dark at room temperature. Fluorescence should be measured within 5 min with excitation at 348 nm and emission 423 nm. A standard curve using ascorbic acid in 5% metaphosphoric acid is constructed in range of 0.1 to 10 μl/ml.

Comments. Trichloroacetic acid (TCA) is not a suitable stabilizing agent for fluorometric ascorbic acid determinations.

Ascorbic Acid Analysis IV: Dipyridyl Reaction (Reduction of Fe^{3+} and the Subsequent Determination of the $Fe^{2+}-\alpha,\alpha'$-Dipyridyl Complex)

Assay Method

Principle. This chemical test was developed[17] and modified[18–20] to determine quantitatively the amount of ascorbic acid in plasma, animal tissues, and urine. Determination of ascorbic acid depends on the reduction of ferric ion to ferrous ion by ascorbic acid and determining the ferrous ion as the red-orange, α, α'-dipyridyl complex. In the presence of orthophosphoric acid at pH 1–2, other reducing or interfering material, e.g., reductone, glucosone, reductic acid, α-tocopherol, glutathione, cysteine, acetol, methyl glyoxal, or creatinine, are inhibited. With this simple method it is possible to run a large number of samples in less than 2 hr,

[17] M. X. Sullivan and H. C. N. Clarke, *J. Assoc. Off. Agric. Chem.* **38,** 514 (1955).
[18] R. P. Maickel, *Anal. Biochem.* **1,** 498 (1960).
[19] V. Zannoni, M. Lynch, S. Goldstein, and P. Sato, *Biochem. Med.* **11,** 41 (1974).
[20] B. I. Sicki, E. G. Mimnough, and T. E. Gram, *Biochem. Pharmacol.* **26,** 2037 (1977).

and with proper microequipment it is possible to measure as low as 0.1 μg of ascorbic acid in as little as 0.01 ml sample size.[19]

Reagents

   α,α'-Dipyridyl, aqueous 0.5%, fresh (requires heat to dissolve and is stable at room temperature)

   Orthophosphoric acid, 85%

   Reagent-grade ferric chloride, aqueous 1%, reagent grade; made fresh each 3 days

   Trichloroacetic acid (TCA), 5% aqueous; made fresh for tissue and urine protein stabilization/precipitation

Procedure. To 0.5 ml of tissue homogenate (10–20%) or urine add 1.0 ml of 5% TCA. To 1 ml of plasma add 2 ml of 5% TCA. Centrifuge the samples at 4° at 14,000 g for 20 min. The following reagents are added in sequence to 270 μl of the supernatant: 80 μl of 85% H_3PO_4, 1.37 ml of 0.5% α,α'-dipyridyl, and 280 μl of 1% ferric chloride. The samples are allowed to stand at room temperature for 60 min for the ferrous-dipyridyl chromophore to develop. Samples are centrifuged and then read at 525 nm in a spectrophotometer with microcuvettes. Standards using ascorbic acid in 5% TCA are analyzed in a range of 0.5 μg/ml to 5 μg/ml.

Comments. Ascorbic acid analysis may be done on whole blood or red blood cells if the samples are first treated with carbon monoxide.[8]

Measurement of Buffy-Coat Ascorbic Acid

Principle. Blood is layered on a solution containing methylcellulose and either sodium metrizoate or sodium diatrizoate. Erythrocytes tend to clump at the interface and settle to the bottom of the tube, whereas the majority of leukocytes and platelets remain in the plasma. The upper layer is removed and centrifuged. The pellet is taken up in TCA, homogenized, and analyzed by the DNPH procedure described above.

Reagents

   Methylcellulose (15 centipoise, Fisher Scientific, Pittsburgh, Pa.), 2%, dissolved in 0.9% saline. This is best dissolved by placing in warm, but not boiling, saline, stirring, then chilling to 5°. This solution is stable at room temperature.

   Sodium diatrizoate (Winthrop Laboratories, 90 Park Ave., New York) or sodium metrizoate (Accurate Chemical and Scientific Corp., 28 Tee St., Hicksville, New York), 33.9% dissolved in water.

   Leukocyte preparation solution is made by mixing 16 parts of the methyl cellulose solution with 10 parts of the sodium metrizoate or sodium diatrizoate solution.

Procedure. Four milliliters of whole blood are carefully layered on 4 ml of the leukocyte separation mixture. This is allowed to stand at room temperature until the erythrocytes have separated from the upper layer (about 20–30 min). The upper layer, which contains leukocytes and platelets, is removed and the volume recorded. An aliquot of known volume is removed, and its leukocyte concentration is determined. The remainder is centrifuged for 10 min at 2000 g. The pellet is taken up in 0.75 ml of 5% TCA and homogenized in a small Potter-Elvehjem homogenizer. A 0.5-ml aliquot of the homogenate is analyzed for ascorbic acid by the DNPH method described above. Calculations are performed as follows:

$$\mu\text{g ascorbic acid}/10^8 \text{ leukocytes} = (A \times 0.75 \times 10^8)/[B \times (C - D)] \quad (2)$$

where A is micrograms of ascorbic acid per milliliter of homogenate; B is leukocytes per milliliter of plasma after erythrocyte separation; C is volume of plasma after erythrocytes have been separated; D is milliliter of plasma layer removed for leukocyte count.

Comments. Although 4 ml of blood is suggested as a convenient volume for a single analysis, this assay may be scaled up for greater sensitivity or down when less sample is available.

Sodium metrizoate and sodium diatrizoate are about equally effective in the leukocyte separation mixture, although to our knowledge only the use of metrizoate has been described in the literature.

The erythrocyte contamination for this leukocyte isolation procedure is about 1–2 erythrocytes per leukocyte.

TABLE II

RECOMMENDATIONS ON SUITABILITY OF ASSAYS FOR VARIOUS SAMPLE TYPES

	Tissue or body fluid			
Assay	Plasma	Whole blood	Liver	Urine
I. DCIP	Yes	No[a]	Yes	Yes
II. DNPH	Yes	Yes	Yes	Yes
III. Fluorometric	Yes	No[a]	Yes	Yes
IV. Dipyridyl	Yes	No[a]	Yes	Yes

[a] These methods may be adapted for use with whole blood by bubbling CO through the blood prior to addition of the deproteinizing agent [J. A. Owen and B. Iggo, *Biochem. J.* **62,** 675 (1956)].

[2] Vitamin C (Ascorbic Acid)

By N. Krishna Murty and K. Rama Rao

Determination with Iodine, Potassium Iodate, Potassium Bromate and Iodine Monochloride

The assay of ascorbic acid with iodine,[1] potassium iodate,[2] potassium bromate,[3] and iodine monochloride[4] using starch as indicator is reported by different workers, but Erdey and Bodor[5] and Gopala Rao and Narayana Rao[6] were of the opinion that starch cannot be used as indicator because it decreases the reaction rate between ascorbic acid and iodine. Variamine blue is proposed as the indicator in the place of starch for the assay of ascorbic acid with iodine,[7] potassium iodate,[7] and iodine monochloride.[4] Carbon tetrachloride[8] or chloroform in the presence of mercuric chloride[9] is also used for the detection of the equivalence point by extractive end point. Schulek, Kovaes, and Rozsa[10] used *p*-ethoxychrysoidine as indicator with potassium bromate in the presence of potassium bromide, but they stated that the endpoint is not sharp. We have developed a simple and rapid titrimetric procedure for the assay of ascorbic acid with iodine, potassium iodate, potassium bromate, or iodine monochloride using naphthol blue black, amaranth, or Brilliant Ponceau 5R as indicator.[11]

Reagents

Iodine, 50 mM

Potassium iodate, $M/60$

Potassium bromate, $M/60$

Iodine monochloride, 50 mM

Ascorbic acid, 50 mM

Potassium iodide, 1%

[1] A. J. Lorenz, R. W. Reynolds, and S. J. W. Stevans, *Symp. Vitamin., Am. Chem. Soc., 92nd Meeting,* 1936.

[2] R. Ballentine, *Ind. Eng. Chem., Anal. Ed.* **13,** 89 (1941).

[3] L. Szekeres, E. Sugar, and E. Pop, *Z. Anal. Chem.* **121,** 17 (1951).

[4] J. Cihalik, Doctoral thesis, University of Chemical Technology, Prague, 1957.

[5] L. Erdey and E. Bodor, *Anal. Chem.* **24,** 418 (1952); cf. *Z. Anal. Chem.* **137,** 293 (1952/1953).

[6] G. Gopala Rao and V. Narayana Rao, *Z. Anal. Chem.* **147,** 338 (1955).

[7] L. Erdey and L. Kaplar, *Z. Anal. Chem.* **162,** 180 (1958).

[8] G. S. Deshmukh and M. G. Bapat, *Z. Anal. Chem.* **145,** 254 (1955).

[9] B. Singh, G. P. Kashyap, and S. S. Sahota, *Z. Anal. Chem.* **162,** 357 (1958).

[10] E. Schulek, J. Kovaes, and R. Rozsa, *Z. Anal. Chem.* **121,** 17 (1951).

[11] N. Krishna Murty and K. Rama Rao, *J. Indian Chem. Soc.* **53,** 532 (1976).

Naphthol blue black, Brilliant Ponceau 5R, and amaranth, aqueous (0.2% solutions)

Procedure. An aliquot of ascorbic acid is taken in a 150-ml titration vessel, and the required amount of hydrochloric acid is added to give an overall acid concentration of 0.1–2.0 M for potassium iodate, 0.1–1.0 M for iodine, 0.2–0.3 M for potassium bromate, and 1.0–2.5 M for iodine monochloride titration. To the above mixture, about 0.1 ml of the indicator solution is added; the titration is carried out against a standard solution of the respective oxidant. In the assay of ascorbic acid with potassium iodate or with potassium bromate, 2–10 ml of 1% potassium iodide solution are added before the titration commences. The color changes are from blue to green for naphthol blue black, or from rose pink to brownish yellow for both amaranth and Brilliant Ponceau 5R. The color changes are fairly reversible, and the indicator correction is negligible.

Determination with Potassium Dichromate

The iodometric procedure of Lorenz, Reynolds, and Stevans[1] underwent numerous modifications to overcome interferences present along with ascorbic acid in the various vitamin sources. In view of the advantages of potassium dichromate as a primary standard oxidimetric reagent, the usual difficulty encountered in the assay of ascorbic acid in colored solutions by visual methods, and the problem of stability of iodine in very dilute solutions, the authors extended the use of potassium dichromate through a potentiometric end point.[12] A visual method is also proposed using naphthol blue black or amaranth or Brilliant Ponceau 5R as indicator.[12] The principle involved in this procedure is that potassium dichromate, when added to the titration mixture, first reacts with potassium iodide, added in the beginning, liberating iodine, which in turn reacts with ascorbic acid that is reduced to iodide. Oxalic acid is used as the catalyst to accelerate the reaction between potassium dichromate and potassium iodide.

Materials

Potassium dichromate, $M/60$

Ascorbic acid, 50 mM

Oxalic acid, 1 M

Potassium iodide, 1%

Naphthol blue black, amaranth, and Brilliant Ponceau 5R

Aqueous (0.2%) solutions

---

[12] K. Rama Rao, Ph.D. thesis, Andhra University, Waltair, 1977. This research was supported by the U.G.C. (India) Junior Research Fellowship.

The potentiometric titration assembly consists of a saturated calomel reference electrode, a bright platinum rod as indicator electrode, and a salt bridge with porous plates filled with saturated potassium chloride solution.

Procedure. A suitable volume of 50 mM ascorbic acid solution is taken in a 250-ml titration vessel and treated with enough dilute sulfuric acid (5–15 ml of 50 mM acid) to give an overall acid concentration of 0.05–0.15 M when diluted to 50 ml. About 5–15 ml of 1 M oxalic acid and 2–5 ml of 1% potassium iodide solutions are added to the above solution, and the latter is diluted to 50 ml with distilled water. The resulting mixture is titrated with $M/60$ potassium dichromate either potentiometrically or visually, using 0.1 ml of the naphthol blue black, amaranth, or Brilliant Ponceau 5R as indicator. In the potentiometric titration, stable potentials can be recorded immediately after the addition of each portion of the titrant. A potential break of about 130–140 mV per 0.04 ml of $M/60$ potassium dichromate is observed. In the visual titrations, the color changes are from pinkish blue to pinkish green for naphthol blue black, or from rose pink to brownish yellow for amaranth or Brilliant Ponceau 5R.

During the visual method of titration, the solution acquires a pink color due to the formation of chromium(III)–oxalate complex. Hence, the detection of the endpoint becomes difficult if the ascorbic acid content exceeds 80 mg where a potentiometric method can be used. This procedure is applied for the assay of ascorbic acid in juices of such fruits as citrus fruits (*Citrus aurantifolia* and *Citrus sinensis*), tomato (*Lycoperiscon esculentum*), and emblic myrobalam (*Phyllanthus emblica* or *Emblica officinalis*) and in pharmaceutical preparations containing vitamin C alone and in the presence of other vitamins. The extraction of vitamin C from fruit juices is carried out using oxalic acid as the extractant as recommended by Goldblith and Harris.[13] Pharmaceutical preparations containing vitamin C are directly assayed as given in the procedure either potentiometrically or visually. In the assay of pharmaceutical preparations containing vitamin C, compounds are separated using light petroleum ether (boiling range 40°–60°C) as recommended by Hashmi,[14] and then the determination is carried as described above. Potentiometric procedure is used for the pharmaceuticals containing coloring materials.

[13] S. A. Goldblith and R. S. Harris, *Anal. Chem.* **20,** 649 (1948).
[14] M. H. Hashmi, "Assay of Vitamins in Pharmaceutical Preparations," p. 236. Wiley, New York, 1973.

[3] Analysis of Ascorbic Acid by Liquid Chromatography with Amperometric Detection

By LAWRENCE A. PACHLA and PETER T. KISSINGER

Direct electrochemical methods for ascorbic acid have been available for many years.[1] These methods have not been popular owing to the inconvenience of the dropping mercury electrode, the tendency of electrodes to become fouled by adsorbed materials in biological samples, and the relatively poor resolution of electrochemical techniques in distinguishing between easily oxidized substances. All three of these difficulties can be overcome by the use of hydrodynamic thin-layer electrochemistry as a means of detection in high-performance liquid chromatography. The combined technique (LCEC) provides extraordinary sensitivity and selectivity while maintaining advantages in both cost and convenience.

Principles

The electrooxidation of ascorbic acid at carbon electrodes follows an irreversible EC type of electrode mechanism.[2] The electrochemical behavior of ascorbic acid (AA) and other easily oxidized compounds is most conveniently ascertained by linear sweep voltammetry. For example, Fig. 1 illustrates the typical behavior of ascorbic acid and its metabolite ascorbic acid 2-sulfate in acetate buffer at pH 5.25. An initial positive potential scan for an ascorbic acid solution yields an anodic peak, $E_{p,a}$ at $+0.45$ V. The product of the electrochemical oxidation of AA is dehydroascorbic acid. Dehydroascorbic acid is then very rapidly hydrated to yield an electroinactive product. The equilibrium constant favors the hydrated form to the extent that for all practical purposes the reaction is irreversible. The electrooxidation of ascorbic acid 2-sulfate also follows a similar pathway. The sulfated analog is oxidized via a $2e^-$ process at a much higher potential ($E_{p,a} = +0.88$ V) and, therefore, unlike ascorbic acid, is a very poor reducing agent. Again, the hydrated form of dehydroascorbic acid is the principal product. An overview of these electrode processes is given in Figs. 2 and 3.

As can be seen from the linear-sweep voltammograms, it would be possible to quantitate ascorbic acid in the presence of ascorbic acid 2-sulfate using an electroanalytical technique, because their oxidation

[1] J. Heyrovsky and P. Zuman, "Practical Polarography," p. 102. Academic Press, New York, 1968.
[2] S. P. Perone and W. J. Keetlow, *Anal. Chem.* **38**, 1760 (1966).

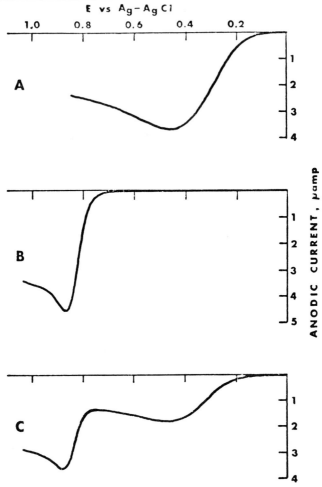

FIG. 1. Linear sweep voltammetry (0.03 V/sec) at the carbon paste electrode ($d_w = 3$ mm) in 1.0 M acetate buffer (pH 5.25): (A) 1 mM ascorbic acid; (B) 1 mM ascorbic acid 2-sulfate; (C) 0.5 mM ascorbic acid, 0.5 mM ascorbic acid 2-sulfate.

potentials are distinctly different. If a sample to be assayed contains only compounds that oxidize at much greater potentials than ascorbic acid, then direct electroanalysis provides the desired specificity. Unfortunately, many "real" samples contain a number of components whose oxidation potentials either overlap or are very close to that of ascorbic acid. Direct electroanalysis of such a sample results in positively biased values for the AA determination.

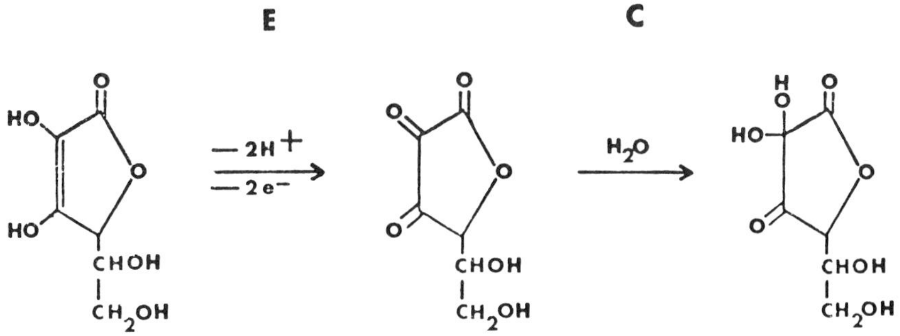

FIG. 2. Mechanism for the electrochemical oxidation of ascorbic acid.

One means of circumventing the poor specificity of electroanalytical techniques is to combine electrochemistry with high-performance liquid chromatography. Liquid chromatography with electrochemical detection (LCEC) has the resolving power of modern liquid chromatography but maintains the sensitivity of thin-layer electrochemistry.[3,4] The specificity of LCEC is primarily due to the chromatographic separation, although considerable selectivity also results from the electrochemistry. Most components of a sample can be categorized into two classes. Those compounds that oxidize at much greater potentials than the analyte (e.g., ascorbic acid 2-sulfate) or compounds whose oxidation potential is lower or very close to that for the analyte of interest. The selectivity associated with the electrochemistry is apparent in the first case. Operation of the chromatographic detector in a potential region that would allow ascorbic acid to be oxidized, but would prevent the electrooxidation of these compounds, precludes concern for their separability by the chromatographic column. For those compounds that fall into the second class, the specificity of an assay for ascorbic acid would be entirely dependent on the

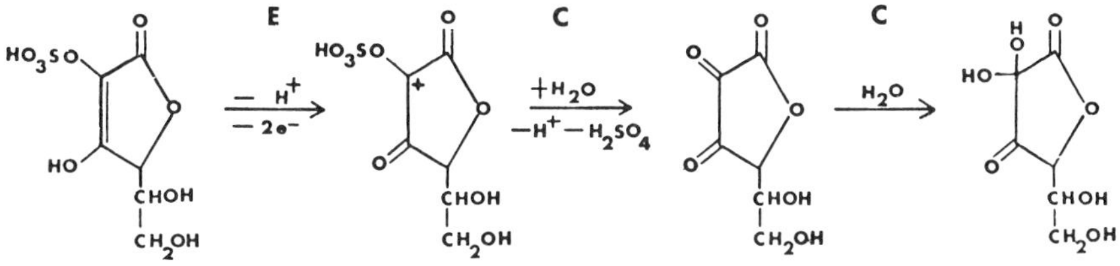

FIG. 3. Mechanism for the electrochemical oxidation of ascorbic acid 2-sulfate.

[3] P. T. Kissinger, *Anal. Chem.* **49**, 447A (1977).
[4] P. T. Kissinger, C. S. Bruntlett, G. C. Davis, L. J. Felice, R. M. Riggin, and R. E. Shoup, *Clin. Chem.* **23**, 1449 (1977).

selectivity of the chromatographic separation because each compound would be detected.

Ideal mobile phases for LCEC are aqueous buffers with or without a solvent (typ-methanol or acetonitrile). Because of this requirement, ion-exchange or reverse-phase packing materials have been found to be the most compatible with the technique. Each class of stationary phase can be further subdivided according to structure and mean diameter. Pellicular packings have an average diameter between 20 and 60 μm whereas the more efficient microparticulates are in the range of 5–10 μm. The authors' original work[5] involved the use of pellicular anion-exchange packing materials where the alkyl cationic sites were either chemically bonded to an inert core (e.g., Vydac SAX) or were mechanically coated onto a glass bead (e.g., Zipax SAX). The use of these pellicular packings provided an assay procedure with adequate selectivity while maintaining minimal cost. Since ascorbic acid exists as an anion in mildly acidic solution, it is readily retained by anion-exchange packing materials. Zipax SAX was found to be an excellent choice, since ascorbic acid was eluted in ca. 4 min; this gave adequate resolution from the void volume peak. This stationary phase is especially useful in the analysis of food products and biological fluids because of the probability of large void volume peaks due to nonretained oxidizable substances. A useful but less selective stationary phase is Vydac SAX, which was found to be useful for samples (e.g., multivitamin preparations), where the nonretained components are generally not detectable. In this case, the ascorbic acid is eluted shortly after the void volume, and therefore a higher sampling rate is achieved.

The microparticulate reverse-phase packings are an important alternative to the pellicular anion-exchange packing materials. These stationary phases consist of chemically modified silica gel particles having alkyl side chains extending out into the mobile phase. The separation process for these packings primarily involves a partitioning of the analyte between the nonpolar stationary phase and the polar mobile phase due to hydrophobic forces.[6] The primary advantage of these stationary phases is the improved efficiency. In addition, microparticle reverse-phase packings are very versatile. Not only can nonpolar molecules be retained, but also cationic and anionic species if they are sufficiently hydrophobic and/or if the mobile phase composition is modified to include an ion-pair reagent. The use of a suitable ion-pair reagent enables the analyte to form an ion-pair complex, which can then partition onto the stationary phase. In addition to the partitioning process, the ion-pair reagent modifies the

[5] L. A. Pachla and P. T. Kissinger, *Anal. Chem.* **48**, 364 (1976).
[6] C. Horvath, W. Melander, and I. Molnar, *Anal. Chem.* **49**, 142 (1977).

packing material so that it appears to act as an ion exchanger. In most cases published to date, the latter process dominates. Ascorbic acid is not readily retained by reverse-phase packings, therefore a cationic ion-pair reagent must be added to the mobile phase. Careful selection of the column modifying reagent allows the development of a specific method for ascorbic acid. The relative retention time of AA and other components can be changed by the judicious choice of the ion-pair reagent, pH, and ionic strength. For example, AA is more strongly retained on the column if tridecylamine is used instead of a *tert*-butyl ammonium salt.[7] Therefore, proper selection of the ion-pair reagent allows the chromatographic separation to be optimized for different sample types without changing the column.

Materials and Equipment

Reagents

Acetate buffer, 1.0 M, pH 4.00. Prepare by diluting 83.5 ml glacial acetic acid and 44.5 g of anhydrous sodium acetate to a total volume of 2 liters.

Actetate buffer, 1.0 M, pH 4.75. Prepare by diluting 58.0 ml of glacial acetic acid and 82.05 g of anhydrous sodium acetate to a final volume of 2 liters.

Acetate buffer, 1.0 M, pH 5.25. Prepare by diluting 27.9 ml of glacial acetic acid and 124.05 g of anhydrous sodium acetate to 2 liters.

Acetate buffer, 80 mM, plus 1 mM tridecylamine, and 15% methanol. This solution is prepared by appropriately mixing 1.0 M acetate buffer, pH 4.00, 100 mM tridecylamine and methanol to obtain the desired concentrations. The final pH of this solution is 4.5.

Metaphosphoric acid, 3%; acetic acid, 8%

Tridecylamine, 100 mM, in methanol

Perchloric acid, 50 mM

Acetate buffer, 70 mM, pH 4.75

Acetate buffer, 70 mM, pH 5.25

Instrumentation. All liquid chromatographic data were obtained using commercially available components and an amperometric detector (Bioanalytical Systems Inc., West Lafayette, Indiana, Model LC-50). Glass columns, 50 cm × 2 mm i.d., were dry-packed with the following pellicular high-performance, anion-exchange packing materials: Vydac SAX (The Separations Group, Hesperia, California, No. 301) and Zipax SAX (E. I. Dupont de Nemours and Co., Inc., Instrument Products Divi-

[7] S. P. Sood, L. E. Sartori, D. P. Wittmer, and W. G. Haney, *Anal. Chem.* **48,** 796 (1976).

sion, Wilmington, Delaware, No. 820960005). A stainless steel column, 15 cm × 4.6 mm i.d., was slurry-packed with microparticulate C_{18} reverse-phase packing material (EM Laboratories Inc., Elmsford, New York, Type Lichrosorb RP-18). The potential of the chromatographic detector was set at 0.70 V versus a Ag/AgCl reference. An injection volume of 20 μl is recommended for all solutions. To prevent clogging of the analytical column with particulate matter, a short plexiglas precolumn was placed in between the injection valve and the pellicular columns.[8] Samples to be analyzed on the microparticulate reverse-phase column were filtered using a centrifugal filter assembly (Bioanalytical Systems Inc., Model MF-1).

Preparation of Solutions

A stock 1 mg/ml solution in either the conventional extracting solution (3% metaphosphoric acid–8% acetic acid) or 50 mM $HClO_4$ is used to prepare calibration standards containing 0.1–5.0 mg/ml by diluting the stock solution with cold 50 mM $HClO_4$. These concentrations should be modified if an injection volume other than 20 μl is to be used. The amount of ascorbic acid to be introduced onto the chromatographic column should be within the range 2–100 ng.

Tablets. Five tablets are accurately weighed and pulverized into a fine powder. A known amount of the total mass corresponding to 1 mg of ascorbic acid is transferred into a 100-ml volumetric flask and diluted to volume with 50 mM $HClO_4$ just prior to chromatographic analysis.

Liquids. Liquid formulations are diluted with cold 50 mM $HClO_4$ to obtain a final assay solution with a nominal concentration of 1–3 μg/ml.

Capsules. Five capsules are mechanically ruptured in the presence of the extracting solution, and the supernatant is quantitatively transferred into a 100-ml volumetric flask. This solution should then be diluted to yield an ascorbate concentration between 1 and 3 μg/ml.

Milk. Whole milk is diluted 20-fold with cold 50 mM $HClO_4$. The heterogeneous mixture is filtered using a centrifugal filter, and a 20-μl aliquot of the clear supernatant is used for chromatographic analysis.

Powdered milk samples are weighed and diluted with cold water to obtain a concentration of 20 μg/ml. The reconstituted sample is then worked up in the same manner as fresh milk samples. Powdered water-soluble food products, in appropriate amounts, are weighed into and transferred into dry 100-ml volumetric flasks. Each sample is then diluted to volume with cold 50 mM $HClO_4$, just prior to analysis, to yield a nominal 1–3 μg/ml solution.

[8] L. A. Pachla and P. T. Kissinger, *Anal. Chem.* **48,** 237 (1976).

Food Products. Heterogeneous mixtures (e.g., fruits, baby foods) are centrifuged at 17,000 g and 5° for 10–15 min. The supernatant is decanted into a volumetric flask, and the residue is extracted using the extracting solution. This mixture is centrifuged, and the supernatants are pooled. The extraction procedure should be repeated twice. The supernatants are then diluted to volume. The final assay solution is prepared by diluting the above solution to a known volume with cold 50 mM HClO$_4$.

Biological Samples. Aliquots (0.5 ml) of urine samples are diluted between 1 : 10 and 1 : 100 with cold water or 50 mM perchloric acid and analyzed immediately. Much smaller aliquots can be used for studies with tissue homogenates or fluids from laboratory animals. For these types of sample, the precolumn filter becomes an essential component of the chromatographic system. Replacement of the precolumn is generally needed after 100–1000 samples have been processed.

Chromatographic Assay

Acetate buffer (70 mM, pH 4.7) is employed in the assay using Zipax SAX columns, whereas acetate buffer (70 mM, pH 5.2) is used for columns packed with Vydac SAX. In each case the flow rate should be about 0.3 ml/min. When the C$_{18}$ reverse-phase column is to be used, acetate buffer (80 mM, pH 4.0), 1 mM tridecylamine, and methanol (15%) is the mobile phase of choice. The flow rate for this column should be set at 1.26 ml/min. Twenty-microliter aliquots of calibration standards or samples with ascorbic acid concentrations in the range of 0.1 to 5.0 μg/ml are injected onto the chromatographic column. The concentration of AA in the samples are determined by comparing the area or peak height to that of an aqueous calibration standard or by using the standard addition technique. Because of the high acidity and ionic strength, it is not desirable to directly inject solutions prepared with the classical extracting solution (3% metaphosphoric acid–8% acetic acid). Samples prepared with this solution should be diluted with cold 50 mM perchloric acid prior to analysis.

Discussion

The retention times are such that assay of ascorbic acid can be performed at a sampling rate of 12–20 per hour depending on the chromatographic packing material chosen. Multivitamin formulations do not present any particular problems for iodimetric titration methods or LCEC except in those preparations containing ferrous sulfate. For these sample types, the chromatographic approach is superior. The selectivity and sensitivity of the LCEC approach becomes more apparent when it is applied to food products. Titrimetric methodology for food products yield positively biased values for such samples as limes. In several sample types, such as fortified cereals and milk products, titrimetric methods are unable to provide a distinct end point. To date, LCEC has been successfully applied to the analysis of AA in baby foods, fruit juices, artificial fruit drinks, fortified cereals, whole fruits (including limes), and milk products. The values obtained are lower than those obtained by nonspecific redox methods.[5] The precision of this chromatographic approach is ca. ±2% with a sampling rate of 15 per hour. The sensitivity of LCEC is clearly demonstrated in Fig. 4, which illustrates chromatograms obtained from homogenized milk diluted 1:20. In this case, quantitation was achieved by standard addition.

The present scheme was originally developed for biological fluids. In samples from individuals ingesting popular dosages of ascorbic acid (0.2–1.0 g/day), it is usually possible to simply dilute the sample 100-fold with 50 mM perchloric acid and inject it onto the chromatographic column.

Several well known stabilizing solutions were examined; these included (1) acetate buffer, pH 4.7; (2) 50 mM perchloric acid, 2 mM thiourea, and 1 mM EDTA; (3) 3% metaphosphoric acid–8% acetic acid; and (4) 50 mM perchloric acid. Standards prepared with acetate buffer exhibited significant decomposition within a period of hours and could not be used in the assay. The second solution containing EDTA proved to be a good candidate for stabilizing ascorbate solutions. However, a large void volume peak and a broad peak eluting just past AA hindered its usefulness. The classical extracting solution, 3% metaphosphoric acid–8% acetic acid, afforded excellent stabilizing properties. When samples and standards were prepared with this solution, a noticeable decrease in retention time was observed. This effect was due to the plug of high ionic strength that passed through the column. Samples prepared with the extracting solution were diluted prior to injection on the anion-exchange column. Cold 50 mM perchloric acid was found to be the ideal medium to inhibit the oxidation of ascorbic acid. Samples and standards were frequently diluted with this solution and stored refrigerated for several hours. Approximately 99% of the ascorbic acid remained after 12 hr when prepared in cold dilute perchloric acid.

When powdered samples are to be analyzed, the samples should be weighed into dry volumetric flasks and then diluted just prior to their analysis. Drying the flasks with a vacuum oven is necessary to remove extraneous moisture; otherwise decomposition of the vitamin will occur. By weighing all the samples at the same time, it is possible to prepare a set of samples that could be assayed the following day. This approach improves

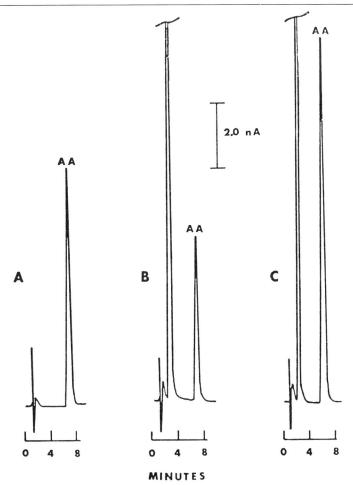

FIG. 4. Chromatogram for ascorbic acid (AA) in homogenized milk diluted 1:20. (A) Ten-nanogram standard injected. (B) Milk sample containing 6.8 μg/ml (6.8 ng of AA injected). (C) Sample following standard addition of 10 ng (16.8 ng of AA injected). Merck C_{18} reverse phase, 15 cm × 4.6 mm, stainless steel column. Mobile phase: 80 mM acetate buffer, 1 mM tridecylamine, 15% methanol (pH of final solution = 4.5) at a flow rate of 1.26 ml/min. Applied potential = 0.7 V vs Ag/Ag/AgCl.

the overall rate of analysis. In some cases (e.g., blood) the stability of AA is particularly poor, and sample handling prior to work-up should be carried out in the shortest possible time. Another problem encountered is that when the pellicular anion-exchange column has not been used for some time (ca. 15–30 min), the first sample injected always gives a value that is too low by 5–15%. Once the column has been treated with the sample or standard, quantitation of subsequent injections are both accurate and precise.

Conclusion

The LCEC approach affords convenience of sample preparation, sensitivity, and selectivity equal or superior to any method for ascorbic acid published to date. The sensitivity of LCEC is superior by two orders of magnitude when compared to liquid chromatography with ultraviolet detection. In addition, the selectivity associated with electrochemistry is a decided advantage. Although the use of pellicular, high-performance, anion-exchange packing materials is adequate for most sample types, increased selectivity is obtainable when microparticulte reverse-phase packings and different ion-pairing reagents are employed. A decided advantage of the latter approach is that an assay procedure can be optimized for maximum sample rate when repetitive analysis of a single sample type is desired.

[4] L-Gulono-γ-lactone Oxidase (Rat and Goat Liver)

By Morimitsu Nishikimi

$$\text{L-Gulono-}\gamma\text{-lactone} + O_2 \rightarrow \text{L-ascorbic acid} + H_2O_2$$

L-Gulono-γ-lactone oxidase catalyzes the last step in the biosynthesis of L-ascorbic acid in animals. Humans, primates, and guinea pigs are unable to synthesize this vitamin because they lack this enzyme. The primary product of the above reaction is thought to be 2-keto-L-gulono-γ-lactone, which isomerizes spontaneously to give L-ascorbic acid.

Assay Methods

Principle. The L-ascorbic acid formed can be determined colorimetrically by the 2,4-dinitrophenylhydrazine method[1] or by reduction of Fe^{3+} to Fe^{2+}, followed by coupling of the Fe^{2+} with α,α'-dipyridyl.[2] These methods can be used to assay crude tissue preparations provided that corrections are made for endogenous L-ascorbic acid. The Fe^{2+}-α,α'-dipyridyl method was used in the purification of the enzyme because it is rapid and simple. The dehydrogenase activity of the enzyme can be mea-

[1] J. H. Roe and C. A. Kuether, *J. Biol. Chem.* **147,** 399 (1943).
[2] M. X. Sullivan and H. C. N. Clarke, *J. Assoc. Off. Agric. Chem.* **38,** 514 (1955).

sured spectrophotometrically by reduction of 2,6-dichlorophenolindo-phenol in the presence of phenazine methosulfate.[3]

Reagents

Potassium phosphate buffer, 0.1 M, pH 7.5, containing EDTA, 2 mM
L-Gulono-γ-lactone, 25 mM, freshly prepared
Trichloroacetic acid, 50%
Orthophosphoric acid, 85%
α,α'-Dipyridyl, 0.5%, prepared by dissolving in hot water
Ferric chloride, 1%

Procedure. The assay mixture contains 0.5 ml of phosphate buffer, 0.1 ml of L-gulono-γ-lactone, and enzyme, in a final volume of 1 ml. The reaction is initiated by the addition of either L-gulono-γ-lactone or enzyme, and the mixture is incubated with shaking in air for 15 min at 37°. The reaction is stopped by adding 0.1 ml of trichloroacetic acid, and the precipitated protein is removed by centrifugation. To 0.2 ml of the supernatant solution are added 75 μl of orthophosphoric acid, 1 ml of α,α'-dipyridyl, and 0.2 ml of ferric chloride.[4] After the mixture has stood for 15 min at room temperature, absorbance at 525 nm is measured in a cuvette having a 1-cm light path. L-Ascorbic acid autoxidizes partially during the incubation, and the quantity of L-ascorbic acid formed should be corrected for this.

Definition of Unit and Specific Activity. One unit of enzyme is the quantity that catalyzes the formation of 1 μmol of L-ascorbic acid in 1 min under the conditions described above. Specific activity is expressed as units per milligram of protein. Protein concentration is determined by the Lowry method[5] or by the fluorescamine method,[6] using bovine serum albumin as the standard.

Purification of the Enzyme from Rat Liver

The following procedure, based on the method of Nakagawa and Asano,[7] was developed by Nishikimi *et al.*[8]

Step 1. Preparation of Microsomes. Livers are obtained from male albino rats after decapitation and are stored at $-20°$. All steps in the puri-

[3] C. Bublitz, *Biochim. Biophys. Acta* **48**, 61 (1961); see also this series, Vol. 6 [42].

[4] The sensitivity of the assay of L-ascorbic acid can be increased with some modifications [V. Zannoni, M. Lynch, S. Goldstein, and P. Sato, *Biochem. Med.* **11**, 41 (1974)].

[5] H. Lowry, N. J. Rosebrough, A. L. Farr, and R. J. Randall, *J. Biol. Chem.* **193**, 265 (1951); see also this series, Vol. 3 [73].

[6] N. Nakai, C. Y. Lai, and B. L. Horecker, *Anal. Biochem.* **58**, 563 (1974).

[7] H. Nakagawa and A. Asano, *J. Biochem.* (*Tokyo*) **68**, 737 (1970).

[8] M. Nishikimi, B. M. Tolbert, and S. Udenfriend, *Arch. Biochem. Biophys.* **175**, 427 (1976).

fication are carried out at 0–4°. The livers are homogenized in 4 volumes of 0.25 M sucrose by use of a Polytron homogenizer (Kinematica GmbH, Lucerne, Switzerland) and the homogenate is centrifuged for 15 min at 10,000 g. The supernatant solution is centrifuged for 60 min at 100,000 g, and the sedimented microsomes are suspended in 1.15% KCl containing 10 mM EDTA (pH 7.4) and stored at −20°.

Step 2. Tryptic Digestion. The microsomes (2–3 g of protein) are washed with 1.15% KCl, then suspended at a protein concentration of 10 mg/ml in 20 mM Tris-acetate buffer, pH 8.0, containing trypsin (0.3 mg/ml) and 1 mM EDTA, and are stirred gently overnight at 4° under nitrogen.

Step 3. Solubilization with Tween 20. The tryptic digest is centrifuged for 90 min at 100,000 g, and the precipitate is suspended at a protein concentration of 10 mg/ml in a solution containing 1.5% Tween 20, 20 mM Tris-acetate buffer, pH 8.0, and 1 mM EDTA. After standing for 30 min, the suspension is centrifuged for 90 min at 100,000 g, and the clear supernatant layer is collected.

Step 4. Ammonium Sulfate Fractionation. Solid ammonium sulfate (161 mg/ml) is added with stirring. After standing for 15 min, the mixture is centrifuged for 10 min at 10,000 g, and the pellicle formed is removed. Ammonium sulfate (148 mg/ml) is then added to the solution. The resulting precipitate is dissolved in a minimum volume (7.5 ml) of 20 mM Tris-acetate buffer, pH 8.0, containing 10 mM KCl, 1 mM EDTA, and 0.4% Brij 35.

Step 5. Gel Filtration. The solution is passed through a column of Sephadex G-150 (2 × 55 cm) which has previously been equilibrated with the above Tris-acetate buffer. The protein is then eluted with the same buffer, and collected in 2.5-ml fractions. Those fractions containing enzyme activity are combined.

Step 6. DEAE-Sephadex A-50 Chromatography. DEAE-Sephadex A-50 is swollen in an appropriate volume of 1 M sodium acetate, washed with distilled water, and packed into a column. About 10 bed volumes of the above Tris-acetate buffer are then passed through the column. The pooled fractions (step 5) are placed on the column of DEAE-Sephadex A-50 (2.1 × 10 cm). The column is washed with the same buffer and then eluted with a 200-ml linear gradient of KCl (10–300 mM) containing 20 mM Tris-acetate buffer, pH 8.0, 1 mM EDTA, and 0.4% Brij 35. Fractions of 2.5 ml are collected, and those containing activity greater than 0.06 unit/mg of protein are combined and dialyzed against 2 liters of 10 mM potassium phosphate buffer, pH 7.0, containing 1 mM EDTA and 0.4% Brij 35.

TABLE I
PURIFICATION OF RAT L-GULONO-γ-LACTONE OXIDASE[a]

Step	Total protein[b] (mg)	Total activity (units)	Specific activity (units/mg protein)	Yield (%)
1. Microsomes	2940	11.6	0.0039	100
2. Tryptic digestion	1375	10.1	0.0073	87
3. Tween 20	403	5.97	0.015	51
4. $(NH_4)_2SO_4$	148	4.60	0.031	40
5. Sephadex G-150	86	4.57	0.053	39
6. DEAE-Sephadex	16	1.38	0.086	12
7. Hydroxyapatite	1.2	0.649	0.54	5.6

[a] From M. Nishikimi, B. M. Tolbert, and S. Udenfriend, *Arch. Biochem. Biophys.* **175**, 427 (1976).

[b] Protein concentration was determined by the Lowry method, except for the preparation from hydroxyapatite column which was determined by the fluorescamine method.

Step 7. Hydroxyapatite Chromatography. The dialyzed solution is applied to a column of hydroxyapatite (BioGel HTP, Bio-Rad Laboratories, Richmond, California) (2.1 × 2.5 cm) which has previously been equilibrated with the above phosphate buffer. Enzyme is eluted with 30 mM potassium phosphate buffer, pH 7.0, containing 1 mM EDTA and 0.4% Brij 35. Fractions of 2.5 ml are collected, and those containing activity are combined and concentrated by ultrafiltration with a Diaflo membrane PM 30. A summary of the purification is given in Table I.

Purification of the Enzyme from Goat Liver

The purification procedure for the goat enzyme is similar to that described for the rat enzyme, but some modifications are required. Microsomes are prepared from frozen livers of young male goats (uncastrated), digested with trypsin, and treated with Tween 20, as described for the rat enzyme. The solubilized microsomal preparation is fractionated with ammonium sulfate. Solid ammonium sulfate (161 mg/ml) is added to the preparation. After removal of the pellicle formed by centrifugation, ammonium sulfate (181 mg/ml) is added to the solution. The resulting pellicle is dissolved and passed through a column of Sephadex G-150; the fractions containing activity are placed on a column of DEAE-Sephadex A-50, and the column is washed with the buffer as described for the rat enzyme. Most of the enzyme activity passes through the column. After dial-

TABLE II

PURIFICATION OF GOAT L-GULONO-γ-LACTONE OXIDASE[a]

Step	Total protein[b] (mg)	Total activity (units)	Specific activity (units/mg protein)	Yield (%)
1. Microsomes	1970	33.5	0.017	100
2. Tryptic digestion	902	24.3	0.027	73
3. Tween 20	367	13.3	0.036	40
4. $(NH_4)_2SO_4$	172	8.73	0.051	26
5. Sephadex G-150	99	8.73	0.088	26
6. DEAE-Sephadex	7.3	2.74	0.375	8.2
7. Hydroxyapatite	0.56	1.15	2.05	3.4

[a] From M. Nishikimi, B. M. Tolbert, and S. Udenfriend, *Arch. Biochem. Biophys.* **175,** 427 (1976).

[b] Protein concentration was determined by the Lowry method, except for the preparation from hydroxyapatite column, which was determined by the fluorescamine method.

ysis of this breakthrough fraction, the enzyme is adsorbed on a hydroxyapatite column, as described in the purification of the rat enzyme, and is eluted with 100 mM potassium phosphate buffer, pH 7.0, containing 1 mM EDTA and 0.4% Brij 35. The fractions containing activity are concentrated as described for the rat enzyme. A summary of the purification is shown in Table II.

Staining of Enzyme Activity on Gels[8]

Polyacrylamide gel electrophoresis in the presence of a nonionic detergent is performed as described by Dewald *et al.*[9] Tween 20, Triton X-100, and Brij 35 can be used as the detergent. The addition of the detergent is essential for the enzyme protein to migrate as a sharp band. Approximately 5 μg of the purified rat enzyme[10] is placed on a gel. After electrophoresis the gel is incubated in a mixture containing 2.5 mM L-gulono-γ-lactone, 0.33 mM phenazine methosulfate, 0.12 mM nitroblue tetrazolium, 1 mM EDTA, and 50 mM potassium phosphate buffer, pH 7.5. The incubation is carried out in the dark at room temperature (23°) until the color of blue formazan appears; the reaction is then stopped by placing the gel in 7% acetic acid. This method is also useful for the staining of precipitin lines of the enzyme in immunological studies.

[9] B. Dewald, J. T. Dulaney, and O. Touster, this series, Vol. 32 [8].

[10] Goat enzyme does not enter the gel under the electrophoresis conditions.

Properties[8]

Purity. Purified preparations of the rat and the goat enzymes appear nearly homogeneous, as judged by sodium dodecyl sulfate (SDS)–polyacrylamide gel electrophoresis. However, they might possibly contain some contaminating proteins of the same molecular weight as that of the enzyme. Monospecific antiserum against the goat enzyme can be raised in rabbits using the preparation of the final step.

Molecular Weights. The molecular weights of dissociated enzymes of both the rat and the goat are 51,000 as determined by SDS–polyacrylamide gel electrophoresis. In the native state, both enzymes demonstrate apparent molecular weights of 500,000, based on gel filtration, indicating that they occur as large aggregates.

Prosthetic Group. The absorption spectrum of the goat enzyme shows two peaks at 455 and 350 nm, which are characteristic of a flavin prosthetic group. The spectrum of the enzyme preparation of the rat has an additional peak at 415 nm, suggesting contamination by a heme protein. It has been shown that the flavin of the rat enzyme is covalently bound to the protein. The structure of the flavin, obtained by treatment with trypsin and chymotrypsin followed by acid hydrolysis, has been determined to be 8α-[N(1)-histidyl]riboflavin.[11] The enzyme can be localized on SDS–polyacrylamide gels by the fluorescence of the flavin.[12]

Specificity. In addition to L-gulono-γ-lactone, the enzyme oxidizes L-galactono-γ-lactone, D-mannono-γ-lactone, D-altrono-γ-lactone, and ethyl-D-idonate.[13] However, L-gulonic acid does not serve as substrate. It also attacks the semicarbazone, oxime, and cyanohydrine of D-glucurono-γ-lactone,[14] but not D-glucurono-γ-lactone itself.

Phenazine methosulfate acts as an electron acceptor in place of molecular oxygen.[3] Assay of the dehydrogenase activity of the enzyme and the staining of enzyme activity on polyacrylamide gels are based on this property.

Kinetic Constants. The apparent K_m of the rat enzyme for L-gulono-γ-lactone is 66 μM, and that of the goat enzyme for this substrate is 150 μM.

Inhibitors and Activators. A variety of strong chelating agents show almost no inhibitory effect at millimolar concentrations, suggesting no metal requirement. *p*-Chloromercuribenzoate, *p*-chloromercuribenzene

---

[11] W. C. Kenney, D. E. Edmondson, T. P. Singer, H. Nakagawa, A. Asano, and R. Sato, *Biochem. Biophys. Res. Commun.* **71,** 1194 (1976).

[12] M. Nishikimi, K. Kiuchi, and K. Yagi, *FEBS Lett.* **81,** 323 (1977).

[13] J. Kanfer, J. J. Burn, and G. Ashwell, *Biochem. Biophys. Acta* **31,** 556 (1959).

[14] P. Sato, M. Nishikimi, and S. Udenfriend, *Biochem. Biophys. Res. Commun.* **71,** 293 (1976).

sulfonate, and Hg^{2+} are strong inhibitors[7] and the purified enzyme is activated by dithiothreitol or 2-mercaptoethanol.[14] It appears that the enzyme has a sulfhydryl group, which is essential for activity.

Distribution.[15] Distribution of the enzyme in the animal kingdom is interesting from the phylogenetic standpoint. The enzyme is present in the kidney of amphibians and reptiles, and in the liver of mammals. It is missing in humans, primates, and guinea pigs. Primitive birds, such as the chicken, have the enzyme in the kidney, whereas the more highly evolved species have it in the liver.

[15] I. B. Chatterjee, *Science* **182**, 1271 (1973).

[5] Ascorbate Oxidase

By Men Hui Lee and Charles R. Dawson

$$\text{L-Ascorbic acid} + \tfrac{1}{2}O_2 \rightarrow \text{L-dehydroascorbic acid} + H_2O$$

Assay Method

Principle. Enzyme activity is assayed by measuring the initial rate of oxygen consumption during the oxidation of L-ascorbic acid either employing a manometric technique or by the use of an oxygen electrode.

Reagents and Procedure. The ascorbate oxidase activity is determined manometrically in a conventional Warburg apparatus at 25°. The reagents, optimal conditions of reaction volume, pH, substrate concentration, and assay procedure have been described by Dawson and Magee.[1] However, the positions of enzyme and substrate, during incubation in the Warburg flask, have been interchanged.[2] The final reaction volume can also be reduced to 2.5 ml employing respirometer flasks of about 20 ml volume.[3] In recent years the rate of oxygen uptake has been followed, under the same reaction conditions, using a 1.5-ml reaction volume containing a Clark oxygen electrode on a Gilson Model KM Oxygraph.[4-6]

Definition of Unit and Specific Activity. One unit of ascorbate oxidase

[1] C. R. Dawson and R. J. Magee, this series, Vol. 2, p. 831.
[2] R. J. Magee and C. R. Dawson, *Arch. Biochem. Biophys.* **99**, 338 (1962).
[3] K. Tokuyama, E. E. Clark, and C. R. Dawson, *Biochemistry* **4**, 1362 (1965).
[4] K. G. Strothkamp and C. R. Dawson, *Biochemistry* **13**, 434 (1974).
[5] R. E. Strothkamp and C. R. Dawson, *Biochemistry* **16**, 1926 (1977).
[6] K. G. Krul and C. R. Dawson, *Bioinorg. Chem.* **7**, 71 (1977).

activity has been defined as the amount of enzyme that causes an initial rate of oxygen uptake of 10 μl/min under the prescribed conditions. Specific activity is expressed in units per milligram of protein, or, alternatively, as units per microgram of copper. Protein is determined by the method of Lowry et al.[7] Copper is determined by the method of Stark and Dawson.[8]

Holoascorbate Oxidase

Purification Procedure

The enzyme has been purified from a variety of plant sources; mainly from cucumber[9] (*Cucumis sativus*), yellow summer crookneck squash[3,10] (*Cucurbita pepo condensa*), and green zucchini squash[3,11,12] (*Cucurbita pepo medullosa*). The following procedure[11] yields enzyme specimens of the highest specific activities and copper contents yet reported.

Raw Material. Green zucchini squash has been found to be the best source of the enzyme. The squash, usually 15–20 cm in length and obtained from commercial sources, is harvested during the months of June and July and brought immediately to the laboratory for processing. Because of this limited season of commercial availability of the fresh raw material, the first two steps in the purification process are developed for preparing a crude (yet stable) form of the enzyme that can be stored as a suitable starting material. Several 40-bushel batches of the zucchini squash are thus processed to provide crude ammonium sulfate precipitates (step 2), which are stored under deepfreeze conditions ($-15°$) until subsequently used in about 1-kg amounts for the purification operation described in steps 3, 4, etc. However, it should be mentioned that prolonged storage of the starting material under the deepfreeze conditions tends to allow deterioration of enzyme quality (i.e., specific activity and copper content values, etc.) of the purified specimens. Storage of the starting material under deepfreeze conditions for more than a year is not recommended. All the purification operations described below are carried

[7] O. H. Lowry, N. J. Rosebrough, A. L. Farr, and R. J. Randall, *J. Biol. Chem.* **193,** 265 (1951).
[8] G. R. Stark and C. R. Dawson, *Anal. Chem.* **30,** 191 (1958).
[9] T. Nakamura, N. Makino, and Y. Ogura, *J. Biochem.* (*Tokyo*) **64,** 189 (1968).
[10] F. J. Dunn and C. R. Dawson, *J. Biol. Chem.* **189,** 485 (1951).
[11] M. H. Lee and C. R. Dawson, *J. Biol. Chem.* **248,** 6596 (1973).
[12] L. Avigliano, P. Gerosa, G. Rotilio, A. Finazzi Agro, L. Calabrese, and B. Mondovi, *Ital. J. Biochem.* **21,** 248 (1972).

out at a temperature of 0°–4° in the cold room or the refrigerator except that the squash is peeled at room temperature.

Step 1. Preparation of Crude Juice. A 40-bushel batch of green squash is hand-peeled, and the pulp is discarded. The peelings (about ¼ inch thick) are minced with a power-driven meat grinder. A small amount of solid sodium borate ($Na_2B_4O_7 \cdot 10H_2O$) is periodically added to the fluid mince from the grinder in order to raise the pH of the crude juice (pH 5.9) to near neutrality (pH 6.8). The mince is placed in canvas bags and subjected to hydraulic pressure, with the use of a hand wine press. The juice is then immediately transferred to the cold room and treated with ammonium sulfate as described below. In the usual case, a 40-bushel batch of squash yields about 110 liters of juice possessing a total oxidase activity of 7 to 10×10^6 units.

Step 2. Ammonium Sulfate Precipitation. Solid ammonium sulfate is added slowly to the crude juice in amount corresponding to 65% saturation at 4°. The addition is carried out with continuous hand stirring for about 30 min after all the ammonium sulfate has dissolved. The proteinaceous greenish precipitate that develops is allowed to settle overnight, and most of the clear supernatant fluid is then removed by siphon and discarded. The green precipitate is then collected by batchwise gravity filtration through 12 to 15 fluted Eaton Dyckman No. 617 papers (40 cm in diameter). A total of about 4 kg of moist green precipitate is usually obtained possessing about 90% of the original enzyme activity.

Step 3. Fractionation with Acetone. A suitable quantity (usually about 1 kg of the frozen ammonium sulfate precipitate from step 2) is weighed and suspended in about 3 liters of ice-cold distilled water to extract the soluble enzyme. The suspension is stirred for about 30 min, and the insoluble residue is removed by centifugation (11,700 g for 10 min) in the cold. The precipitate is discarded, and the supernatant solution is divided into seven or eight 400-ml aliquots. To each aliquot is added solid NaCl (1.3 g/100 ml) and the solution is chilled to about −3° by immersion in a Dry-Ice acetone bath. By slowly revolving the beaker in the bath (or by stirring) the frozen material is deposited on the wall of the beaker to a thickness of about 0.5 inch. The beaker is then removed from the bath and the solid is allowed to thaw and soften until it can be broken up by a glass rod into a semisolid mass of snowlike consistency. A 0.9 volume of chilled acetone (−15°) is then added slowly in increments to each beaker to precipitate the enzyme while the suspension is stirred vigorously. The resulting precipitate is allowed to settle for about 10 min, and the clear supernatant fluid is decanted. The precipitates are collected separately on filter paper (Whatman No. 2) by suction filtration, combined, and then suspended in about 800 ml of ice-cold 2 mM, pH 7.6, phosphate buffer.

The material is then dialyzed against cold running tap water for 4–6 hr. Insoluble matter is removed by centrifugation and discarded.

Step 4. Phosphocellulose Chromatography. The dialyzed enzyme solution obtained in step 3 is further dialyzed against 3 volumes of 10 mM phosphate buffer, pH 5.5, for a period of 24 hr (the buffer is changed after each 8-hr period).[13] The small amount of nonenzymic protein material that precipitates is removed by centifugation and the enzyme solution is then directly added to the phosphocellulose column (3.3 × 11 cm) which has been treated with 0.5 N HCl and equilibrated with 10 mM phosphate buffer, pH 5.5. It is observed that a blue enzymically active component in the solution is absorbed in a band at the top of the column, while yellowish components (inactive) pass through. The column is then washed with 10 mM phosphate buffer, pH 5.5, until no further yellow color is observable in the eluent. With this concentration of phosphate there is essentially no movement of the blue band. The column is then washed with about 3 bed volumes of 30 mM phosphate buffer, pH 5.5, which removes colorless nonenzymic protein components. Finally the enzyme is eluted with 0.1 M phosphate buffer, pH 5.5, and fractions are collected by the use of a fraction collector. The elution of the enzyme is easily followed by observing the movement of the blue-green band down the column.

Step 5. Dialysis I. The blue enzyme solutions combined from several phosphocellulose columns (usually about 80 ml) are dialyzed against a 1-liter volume of 10 mM phosphate buffer, pH 5.5, for 24 hr in the refrigerator. The buffer is exchanged at the end of each 8-hr period. During the dialysis about one-half of the activity precipitates in the form of a blue flocculent sediment which is separated by centrifugation (17,300 g for 10 min at 4°).[14] The resulting blue pellet is then suspended in about 40 ml of 10 mM phosphate buffer, pH 7.6, to give an intensely blue solution. A small amount of colorless, inactive, and insoluble protein is removed by recentrifugation under the same conditions. The specific activity of the blue enzyme solution at this stage is usually about 3000 units/mg.

---

[13] Occasionally a blue mass of enzymic protein precipitates at this step of dialysis when the concentration is high. The blue precipitate should be dissolved in 10 mM phosphate buffer, pH 7.6, and combined with enzyme from step 5 for TEAE-cellulose chromatography described in step 6.

[14] The enzyme in the supernatant is subsequently purified following the same procedure as described below for the redissolved precipitate fraction. However, the final enzyme preparation of the supernatant fraction exhibits a maximum purity of only about 80% of that obtainable using the precipitate fraction. Recent experience has revealed that occasionally less than 50% of the activity precipitates, particularly when the protein concentration is low. Under these conditions, it is advantageous to discontinue the dialysis at pH 5.5 and equilibrate the enzyme against 10 mM phosphate buffer, pH 7.6, in preparation for the next step.

Step 6. TEAE-Cellulose Chromatography. The blue enzyme solution from step 5 is applied to a TEAE-cellulose column which has been previously equilibrated with 10 mM phosphate buffer, pH 7.6, under constant air pressure of 20 cm Hg. When the blue band has developed to about one-half of the length of the column, the application of enzyme is stopped. The column is then treated with an additional 5 volumes of 10 mM phosphate buffer, pH 7.6, and then with 2 volumes of 20 mM phosphate buffer, pH 7.6. During the latter treatment (20 mM phosphate) the blue band moves down the column somewhat. The blue enzyme is then eluted with 60 mM phosphate buffer, pH 7.6, and collected in small fractions with the use of a fraction collector. The movement of the blue band down and off the column shows the progress of the enzyme elution.

Step 7. Dialysis II. The fractions from the TEAE-cellulose column possessing specific activities higher than about 3400 units/mg are combined for further purification by precipitation dialysis. The blue enzyme solution is dialyzed for 24 hr in the refrigerator against a 600-ml volume of 2 mM phosphate buffer, pH 6.4. The buffer is changed twice during the dialysis as described earlier. Under these conditions of pH and ionic strength, about 90% of the activity is precipitated during the dialysis process. The blue precipitate is collected by centrifugation and then redissolved in 10 mM phosphate buffer, pH 7.6, as described in step 5.

The overall purification results achievable by the stepwise procedures described above are summarized in the table for a typical enzyme purification involving a 1.5-kg sample of frozen raw material (see step 2) dissolved in about 4 liters of distilled water.

Properties

Purity. The enzyme obtained from the above purification procedure is homogeneous by the criteria of ultracentrifugal analysis and electrophoresis on polyacrylamide gel.

Stability. Incubation of the purified enzyme (freshly prepared) at 80° for 10 min results in complete loss of activity. Enzyme solutions containing in the order of 10,000 units/ml or higher (3 mg or higher of enzyme protein per milliliter) retain about 80% of their activity after storage for 3 months at 4°.

Substrate Specificity. Ascorbate oxidase exhibits a high degree of specificity toward L-ascorbic acid and certain related compounds.[15] The structural requirements for an ascorbate oxidase substrate may include only (a) an enolic system, (b) existence primarily in an anionic

[15] G. R. Stark and C. R. Dawson, *in* "The Enzymes" (P. D. Boyer and K. Myrbäck, eds.), 2nd ed., Vol. 8, p. 297. Academic Press, New York, 1963.

PURIFICATION SUMMARY FOR ASCORBATE OXIDASE FROM GREEN ZUCCHINI SQUASH (*Cucurbita pepo medullosa*)

Purification steps	Total volume (ml)	Total activity (units)	Total protein (mg)	Specific activity (units/mg)	Enzyme yield (%)	Enzyme purification factor	Copper data		
							Total (µg)	Content (%)	Specific activity (units/µg Cu)
1. Crude juice	28,300	1,989,600	99,480	20	100	1	34,818	0.04	58
2. Ammonium sulfate precipitation	4,490	1,715,180	24,156	71	86	3	15,701	0.07	109
3. Fractionation with acetone	3,000	1,354,000	7,319	185	68	9	6,806	0.09	198
4. Phosphocellulose chromatography	183	688,400	385	1,790	34	89	1,015	0.26	676
5. Dialysis I	43	363,200	116	3,130	18	156	455	0.39	800
6. TEAE-cellulose chromatography	15.4	330,270	92.1	3,550	16	177	403	0.44	819
7. Dialysis II	9.7	284,510	66.9	4,250	14	213	310	0.46	918

form at physiological pH, and (c) the capacity for oxidation to a quinoid type of product via a free radical (semiquinoid) intermediate.[16]

Specific Activity and Copper Content. The typical ascorbate oxidase preparation obtained via the above purification procedure has a specific activity of 3800 ± 400 units per milligram of protein and a copper content of 0.46 ± 0.06%.[17] The copper specific activity value has been found[17] to average at 760 units per microgram of copper, and the maximum value achievable is approximately of 1000 units per microgram of copper. On the basis of a molecular weight of 140,000, the copper value of 0.46 ± 0.06% corresponds to 10 ± 2 atoms per enzyme molecule.

Sedimentation Constant, Molecular Weight, Protein Conformation, and Quaternary Structure. The sedimentation constant $s^{\circ}_{20,w}$ of the purified enzyme is 7.52. Studies employing sedimentation and diffusion indicate that ascorbate oxidase has a molecular weight of 140,000.[11] Conformational analysis based on circular dichroism reveals that the purified enzyme exists, in 10 mM phosphate buffer, pH 7.6, predominantly (about 65%) in the β-conformation.[18] No helical content is found. There is experimental evidence to support the view that the native enzyme is made up of two identical subunits. Each subunit has a molecular weight of about 70,000 and is believed to consist of two polypeptide chains of molecular weights approximating 30,000 and 40,000. The two identical subunits are linked by one or two disulfide bonds.[4]

Absorption Spectrum.[18] In the ultraviolet region, the purified enzyme exhibits an absorption maximum at 280 nm with a distinct shoulder absorption at 290 nm. The $E^{1\%}_{1cm}$ (280 nm) value is 21.7 ± 1.2. In the visible light region of the spectrum, purified ascorbate oxidase exhibits five absorption bands i.e., at 330, 460, 610, 770, and 880 nm. The extinction coefficients at each wavelength in 10 mM phosphate buffer, pH 7.6, are 2000, 390, 1300, 600, and 560 per gram atom copper per liter per centimeter, respectively. Under the same buffer conditions the molar extinction coefficient of the major blue absorption band at 610 nm is 9400 cm^{-1}. The ratio $A_{280\,nm} : A_{610\,nm}$ is found to be 25.6.

Electron Paramagnetic Resonance (EPR) Spectrum.[18] The EPR spectrum of ascorbate oxidase is characterized by extremely small low-field hyperfine splittings. The g_m, $g_{\parallel}$, and 1A1 values are found to be 2.074, 2.244, and 0.006 cm^{-1}, respectively. The spectrum also reveals the presence of two different types of EPR detectable copper.

Inhibition and Inactivation. Azide and fluoride are competitive inhibitors with respect to substrate and are noncompetitive inhibitors with

---

[16] J. Dayan and C. R. Dawson, *Biochem. Biophys. Res. Commun.* **73**, 451 (1976).
[17] M. H. Lee and C. R. Dawson, *Arch. Biochem. Biophys.* **191**, 119 (1978).
[18] M. H. Lee and C. R. Dawson, *J. Biol. Chem.* **248**, 6603 (1973).

respect to oxygen.[5] Inhibition by cyanide is complicated by significant inactivation of the enzyme. Small amounts of H_2O_2 added to the enzyme results in reversible formation of a complex that may be similar to a catalytic intermediate, but a large excess of H_2O_2 results in complete inactivation by a mechanism not yet elucidated.[19]

Apo- and Reconstituted Ascorbate Oxidase

Preparation

Dialysis of ascorbate oxidase against aqueous cyanide solution results in the formation of an inactive, copper-free protein (apoascorbate oxidase).[20] A successful procedure for the restoration of copper and activity to reconstitute the blue oxidase was first demonstrated by Penton.[20] The procedure involving addition of cuprous copper solution to the apoprotein solution results in a restoration of about 80% of the oxidase activity and 83.5% of the copper. The incomplete restoration of activity and copper was suspected to be due to the generation of H_2O_2 during the reconstitution process. Such H_2O_2 would cause inactivation of the reconstituted enzyme,[21] or oxidative modification of the apoenzyme. Consequently a more recent procedure for the restoration of the enzyme introduces catalase to destroy any accumulated H_2O_2, and this modification results in full restoration of both the activity and the copper content to the ascorbate oxidase.[22]

Reagents

MacIlvaine's buffer, 0.2 M, pH 5.6

NaCN, 0.1 N in 0.2 M MacIlvaine's buffer, pH 5.6

Catalase (Calbiochem), 47,000 IU/mg protein

Cuprous chloride solution. Prepared by reducing the cupric chloride solution with L-ascorbic acid under anaerobic conditions. A 0.2-ml sample of the cupric solution is placed in a Thunberg tube and a 5 M excess of crystalline L-ascorbic acid is placed in the side arm. The solution is then frozen by an acetone–Dry-Ice bath and attached to a high-vacuum system for evacuation at 1×10^{-4} to

[19] R. E. Strothkamp and C. R. Dawson, *Biochem. Biophys. Res. Commun.*, submitted.

[20] Z. G. Penton and C. R. Dawson, *in* "Oxidases and Related Redox Systems" (T. E. King, H. S. Mason, and M. Morrison, eds.), p. 222. Wiley, New York, 1965.

[21] C. R. Dawson, *in* "The Biochemistry of Copper" (J. Peisach, P. Aisen, and W. E. Blumberg, eds.), p. 305. Academic Press, New York, 1966.

[22] H. T. Chang, Ph.D. Dissertation, Columbia University, 1969.

1×10^{-5} mm Hg pressure. Evacuation of the system is accomplished by repeated evacuation of the frozen solution and thawing several times. Finally, the solution and side-arm contents are mixed for the reduction reaction to form the colorless stable cuprous chloride complex. The reduction reaction, as judged by loss in blue color, requires about 24 hrs and is therefore usually carried out overnight at 4°. Such cuprous chloride solutions remain colorless and can be used over a period of several weeks if maintained under strict anaerobic conditions (10^{-4} to 10^{-5} mm Hg).

Apoascorbate Oxidase. This oxidase is prepared by dialysis of 1 volume (usually 1.0–1.5 ml) of enzyme solution at 4° against about 200 volumes of 0.1 M NaCN solution prepared in 0.2 M McIlvaine's buffer, pH 5.6. The resulting buffered cyanide solution has a pH of 6.9–7.0, and the dialysis is continued until the enzyme solution in the dialysis bag becomes colorless (usually 20–24 hrs). The dialysis bag is then rinsed with Cu-free water several times and subjected to a short dialysis against frequent changes of 0.2 M McIlvaine's buffer of pH 5.6 to remove the cyanide–copper complex ions and the excess cyanide ions (usually 5–6 changes of the buffer during a 3-hr period). The resulting cyanide-free apoenzyme solution contains essentially no enzyme activity or copper.

Reconstituted Ascorbate Oxidase. In an open test tube at room temperature, a 1.0–1.5 ml sample of a solution containing 4–5 mg of the freshly prepared apoenzyme is thoroughly mixed with 0.05 ml (5 μg) of the diluted catalase solution. To the system is then added a 0.1-ml sample of the cuprous chloride solution, taken directly from the Thunberg tube, which is opened immediately prior to use. The mixture is incubated at refrigerator temperature (4°) for 30–45 min with occasional shaking. The system (<2 ml) is then transferred to a small dialysis bag and dialyzed against three 400-ml volumes of 0.2 M McIlvaine's buffer, pH 5.6, during a total period of 18–20 hr to remove the unbonded ionic copper.

Blank and control systems are used to evaluate, respectively, the effectiveness of the dialysis processes in removing copper and cyanide ions, and the effect of the dialysis processes on the activity and copper content of the original ascorbate oxidase. For the blank, a 1.0–1.5 ml volume of 0.2 M McIlvaine's buffer, pH 5.6, is dialyzed against cyanide, treated with the cuprous reagent, and redialyzed, under the identical conditions used for preparation of the apoenzyme and the reconstituted enzyme. For the control, a 1.0–1.5 ml volume of the original ascorbate oxidase is likewise exposed to all the conditions used for preparation of the apoenzyme and the reconstituted enzyme, except that the control enzyme system is initially dialyzed against a 0.2 M McIlvaine's buffer, pH 5.6, containing no cyanide.

Properties

Apoascorbate oxidase has the same molecular weight and essentially the same gross quaternary structure as native holoascorbate oxidase.[6] The freshly prepared apoascorbate oxidase exhibits about 10 detectable −SH groups per mole.[20] None can be detected in the native holoenzyme.[23] The removal of the copper from the oxidase protein, and the simultaneous reduction of the disulfide bonds results in an apoenzyme of lower structural stability than the holooxidase.[6] Reconstitution of a fully active oxidase from the inactive apoenzyme involves the oxidative conversion of −SH groups to disulfide bonds.[20,22] The reconstituted ascorbate oxidase has been reported to have a molecular weight of 285,000, suggesting a dimeric form of the holoxidase.[6] The $s^{\circ}_{20,w}$ value of the reconstituted enzyme is reported to be 9.79.[6] The colorless apoascorbate oxidase does not absorb in the visible region. However, the reconstituted oxidase shows a visible spectrum almost identical to that of the original native holoenzyme.[20] The maximum specific activity values (units per milligram of protein or units per microgram of copper) of the reconstituted oxidase are essentially the same as those of the native holoascorbate oxidase.[6]

[23] G. R. Stark and C. R. Dawson, *J. Biol. Chem.* **237**, 712 (1962).

[6] The Preparation of L-Ascorbic Acid [^{35}S]2-Sulfate Having a High Specific Activity

By Richard R. Muccino, Ronald Markezich, Gerald G. Vernice, Clark W. Perry, Joseph Cupano, and Arnold A. Liebman

Early reports describing the preparation of tracer levels of ascorbic acid [^{35}S]2-sulfate[1] generally involved treatment of a 5,6-acetal of L-ascorbic acid[2−6] with an amine-sulfur trioxide complex in a polar apro-

[1] In the early literature, L-ascorbic acid 2-sulfate is incorrectly identified as the 3-sulfate. X-Ray crystallographic data have confirmed the structure to be the 2-sulfuric ester [see reference 2].

[2] A. D. Bond, B. W. McClelland, J. R. Einstein and F. J. Finamore, *Arch. Biochem. Biophys.* **153**, 207 (1972).

[3] L. L. Solomon, *Experientia* **19**, 619 (1963).

[4] E. A. Ford and P. M. Ruoff, *Chem. Commun.* 630 (1965).

[5] T. M. Chu and W. R. Slaunwhite, Jr., *Steroids* **12**, 309 (1968).

[6] R. O. Mumma, A. J. Verlangieri, and W. W. Weber, *Carbohydr. Res.* **19**, 127 (1971).

tic solvent. Although other sulfating agents[6] have been used, these procedures frequently generated poor yields of amorphous products containing considerable amounts of inorganic sulfate. Improved procedures[7,8] for the preparation of L-ascorbic acid 2-sulfate have been reported including the direct sulfation[7] of L-ascorbic acid, but these methods have been directed mainly at the synthesis of unlabeled material.

Outline of Synthesis

The discrete formation of a pyridine (or trialkylamine)-sulfur trioxide complex is unnecessary[9]; the sulfation of 5,6-O-isopropylidene ascorbic

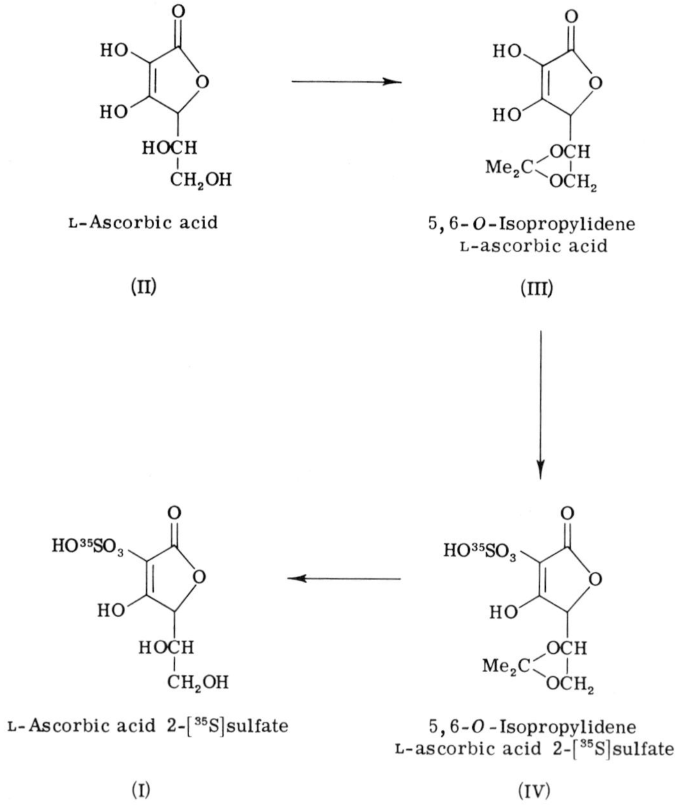

L-Ascorbic acid

(II)

5,6-O-Isopropylidene
L-ascorbic acid

(III)

L-Ascorbic acid 2-[^{35}S]sulfate

(I)

5,6-O-Isopropylidene
L-ascorbic acid 2-[^{35}S]sulfate

(IV)

[7] P. A. Sieb, Y. T. Liang, C. H. Lee, R. C. H. Lee, R. C. Hoseney, and C. W. Deyoe, *J. Chem. Soc. Perkin Trans. 1* 1220 (1974).

[8] S. Farhatulla Quadri, P. A. Sieb, and C. W. Deyoe, *Carbohydr. Res.* **29**, 259 (1973).

[9] R. R. Muccino, R. Markezich, G. G. Vernice, C. W. Perry, and A. A. Liebman, *Carbohydr. Res.* **47**, 172 (1976).

acid (III) can be carried out more efficiently (74% yield after crystallization) with sulfur trioxide in dimethylformamide. Removal of the 5,6-protective isopropylidene group is effected by passage of the labeled substrate (IV) through a column of Dowex 50 (H$^+$) ion-exchange resin,[2] a method that minimizes contamination of the product with inorganic sulfate.[6] The resulting ascorbic acid [^{35}S]2-sulfate can be crystallized as the dipotassium salt or as the barium salt; the insolubility of barium sulfate in water makes purification in this latter case much easier.[10]

Procedure

5,6-O-Isopropylidene L-*Ascorbic Acid.* L-Ascorbic acid (400 mg, 2.27 mmol) in 38 ml of dry acetone is treated with gaseous hydrogen chloride according to the procedure of Solomon.[3] The unrecrystallized product (404 mg, 87% m.p. 225–227°) can be used directly in the next reaction.

Dipotassium L-*Ascorbate* [^{35}S]2-*Sulfate Dihydrate.* A slurry of 5,6-O-isopropylidene ascorbic acid (137 mg, 0.64 mmol) and dry dimethylformamide (4 ml) in a 25-ml round-bottom flask is connected to a glass vacuum line. The mixture is frozen with liquid nitrogen, then the flask is evacuated to 1 μm and sulfur trioxide (66 mg, 0.82 mmol; 630 mCi, 770 mCi/mmol)[11] is introduced by vacuum transfer. The mixture is then warmed to $-15°$ and stirred for 30 min at this temperature; the resulting clear solution (reaction completed) is diluted with water (5 ml). The aqueous solution is adsorbed onto Dowex 50 ion-exchange resin (75 ml wet volume, AG 50W-X8, 100–200 mesh, H$^+$) packed in water in a 28 cm × 1.8 cm column. The 5,6-O-isopropylidene ascorbic acid sulfate is allowed to remain on the column for 45 min. Elution with water (8-ml fractions collected) affords ascorbic acid sulfate contained in fractions 2–5. These fractions are combined and brought to pH 7.7 with 10% aqueous potassium hydroxide. Evaporation of the solvent *in vacuo* at 40° gives an off-white solid ($\sim$300 mg). The solid is redissolved in 1.70 ml of water, and 2.20 ml of dimethylformamide are added to precipitate potassium sulfate and other impurities. Separation of mother liquid and concentration (to $\sim$0.5 ml *in vacuo*) gives a solid ($\sim$210 mg) which is recrystallized from 0.20 ml of water and 0.15 ml of methanol (the solution is allowed to cool overnight at 5°) affording 170 mg (74% yield) of white crystals. A second recrystallization as above from 0.15 ml of water and 0.10 ml of methanol gives 107 mg (213 mCi, ^{35}S was counted using ^{14}C as a secondary stan-

---

[10] The authors have found these procedures to give consistent yields of radioactive material and also to be suitable for large-scale synthesis ($\sim$400 g) of unlabeled ascorbic acid 2-sulfate.

[11] Amersham Corporation, Arlington Heights, Illinois.

dard,[12] 46% chemical yield, 34% radiochemical yield; specific activity 1.99 mCi/mg, 730 mCi/mmol) of dipotassium L-ascorbate [^{35}S]2-sulfate as white prisms with a radiochemical purity of ~98% as determined by thin-layer chromatography on cellulose F (4:3:3, v/v, ethyl acetate/acetic acid/water, R_f 0.3).

Barium L-Ascorbate [^{35}S]2-Sulfate Dihydrate. By the procedure cited above, 5,6-*O*-isopropylidene ascorbic acid (187.4 mg, 0.87 mmol) is treated with sulfur [^{35}S]trioxide (60 mg, 0.75 mmol, 600 mCi; specific activity 800 mCi/mmol).[11] After hydrolysis of the blocking group on the Dowex 50 (H$^+$) ion-exchange column, the collected fractions are combined and stirred for 1 hr with barium carbonate (300 mg, 1.52 mmol). The insoluble materials are removed by filtration, and the mother liquid is concentrated *in vacuo.* The resulting oil is dissolved in water (3.5 ml) (the insolubles being removed by centrifugation) and precipitated with an equal volume of methanol affording 107 mg (33%) of barium L-ascorbate [^{35}S]2-sulfate dihydrate. Recrystallization from water–methanol (1:1) yields ~94 mg of the product (146 mCi, ^{35}S is counted using ^{14}C as a secondary standard[12]; specific activity 1.55 mCi/mg, 658 mCi/mmol). The radiochemical purity is ~98% as determined by thin-layer chromatography on cellulose F [E. Merck, 4:3:3, v/v, ethyl acetate/acetic acid/water, R_f 0.3].

[12] J. P. Buckley, *Int. J. Appl. Radiat. Isot.* **22**, 41 (1971).

[7] The Hydrolysis of Ascorbate 2-Sulfate by Sulfatase A

By A. B. ROY

$$\text{Ascorbate 2-sulfate} + H_2O \rightarrow \text{ascorbate} + SO_4^{2-} + H^+$$

The acid-lability of the sulfate ester bond in ascorbate 2-sulfate resembles that of an aryl sulfate, such as phenyl sulfate, rather than that of a simple carbohydrate sulfate. It is therefore not surprising that ascorbate 2-sulfate is hydrolyzed by certain arylsulfatases. In particular, the arylsulfatases A of mammalian tissues hydrolyze it at a rate comparable to that for nitrocatechol sulfate, the usual synthetic substrate for those enzymes. Ascorbate 2-sulfate bears a closer relationship to such artificial substrates than it does to the physiological substrate of sulfatase A, the galactosyl 3-sulfate residues of certain glycolipids. The ester is also hydrolyzed by sulfatase B, but at only about 1% of the rate of nitrocatechol sulfate.

METHODS IN ENZYMOLOGY, VOL. 62

Substrate

Barium ascorbate 2-sulfate is easily prepared by the sulfation of 5,6-isopropylidene-L-ascorbic acid with pyridine–sulfur trioxide in anhydrous N,N-dimethylformamide[1] or with trimethylamine–sulfur trioxide in aqueous alkali at 70°.[2] The barium salt crystallizes from 50% methanol (about 250 ml/g) as a dihydrate (MW 427.5) which has been kept at 5° over silica gel for more than 3 years with no detectable decomposition. For use as a substrate, the barium salt is converted to the dipotassium salt by passing a 50 mM solution of the former through a column of Dowex 50-X8 (K$^+$ form). The concentration of the resulting solution is determined spectrophotometrically ($\epsilon_{254\ nm}$, 18,500 at pH 7.5 : $\epsilon_{233\ nm}$, 12,000 at pH 1) and adjusted to give 33.3 mM dipotassium ascorbate 2-sulfate in 0.25 mM EDTA, pH about 7. It is preferable to keep the substrate in neutral solution until it is required because there is a slight, but significant, hydrolysis at pH values less than about 4.5.

It should be noted that, although unlabeled barium ascorbate 2-sulfate is quite stable in the solid state, ascorbate 2-[^{35}S]sulfate is not and decomposes rapidly by self-irradiation to give unidentified products,[3] but apparently not $^{35}SO_4^{2-}$.[4]

Method

Assay. The enzymic hydrolysis of ascorbate 2-sulfate is readily followed in the pH-stat by titration of the liberated proton with 15 mM NaOH (carbonate-free) in an atmosphere of CO_2-free N_2. The alkali is conveniently standardized against potassium biiodate. In this laboratory the method has routinely been used with a slightly modified Radiometer PHM26-TT11-SBR2C-ABU12 assembly (Radiometer Ltd., Copenhagen), but any equipment of a similar type should be satisfactory. The standard conditions for the assay of the sulfatase A of ox liver[5] are as follows:

Volume of reaction mixture	10 ml
Dipotassium ascorbate 2-sulfate	33.3 mM
EDTA	0.25 mM
Sulfatase A	0.2–2 μg/ml
pH	4.8
Temperature	37°

[1] S. F. Quadri, P. A. Seib, and C. W. Deyoe, *Carbohydr. Res.* **29**, 259 (1973).

[2] P. A. Seib, Y.-T. Liang, C.-H. Lee, R. C. Hosenay, and C. W. Deyoe, *J. C. S. Perkin I,* 1220 (1974).

[3] S. S. Shapiro and J. P. Poon, *Biochim. Biophys. Acta* **385**, 221 (1975).

[4] G. M. Powell, T. Parry, and C. G. Curtis, *Biochem. Soc. Trans.* **6**, 141 (1978).

[5] L. W. Nichol and A. B. Roy, *J. Biochem. (Tokyo)* **55**, 643 (1964).

The ionic strength of this solution is 0.1. The presence of EDTA is necessary to give reproducible results,[6] presumably because sulfatase A is powerfully inhibited by low concentrations (< 1 mM) of ascorbate in the presence of traces of Cu^{2+}, or perhaps other metals, from the reagents.[7]

After the substrate solution has been brought to 37° in the reaction vessel (5 min equilibration) the pH is adjusted to the appropriate value, 4.8 in the routine assay, and the reaction is started by adding the enzyme, in 10 or 20 μl of 5 mM Tris·HCl buffer, pH 7.5. Recordings are made at a chart speed of 7.5 cm/min for about 3.5 min from the start of the reaction.

Calculation. The reactions catalyzed by sulfatase A are, in general, not of zero order, and this is certainly true for the hydrolysis of ascorbate 2-sulfate. The reaction velocity falls rapidly through a substrate-induced inactivation of the enzyme which has a $t_{1/2}$ of about 4 min under the usual conditions. To determine the initial velocity of the hydrolysis is therefore a matter of some difficulty, especially as the initial stages of a pH-stat recording are, for instrumental reasons,[8] sigmoidal and extrapolation through the origin is not possible.

The simplest way of handling the experimental data is by drawing tangents to the progress curves at some fixed time. This cannot be used to give v_0 because of the irregularities in the initial stages of the curves, but it can give quite precise values of v_1, the velocity at 1 min. The latter can also be determined by the graphical method of Stinshoff,[9] as modified for use with the pH-stat.[10] The data are best handled by fitting to Eq. (1),[10] where u is the amount of alkali consumed at time t, v_0 is the initial velocity, k^* an apparent velocity constant for the inactivation of the enzyme, and C an empirical constant arising from the irregularities in the early stages of the recording. Alone among the methods mentioned, the last gives values

$$u = (v_0/k^*)(1 - e^{-k^*t}) + C \qquad (1)$$

not only for v_0 but also for k^*. In routine assays, 17 readings at 8-sec intervals, starting 1 min after the addition of the enzyme, are taken and fitted to Eq. (1) by a least-squares procedure. Because of the irregularities in the earlier stages of the recording, it is unwise to use data for reaction times of much less than about 1 min.

Whatever method is used, the value of the velocity so obtained is only

[6] A. B. Roy, *Biochim. Biophys. Acta* **377,** 356 (1975).
[7] A. B. Roy, *Biochim. Biophys. Acta* **198,** 76 (1970).
[8] C. F. Jacobsen, J. Léonis, K. Linderstrøm-Lang, and M. Ottesen, *Meth. Biochem. Anal.* **4,** 171 (1957).
[9] K. Stinshoff, *Biochim. Biophys. Acta* **276,** 475 (1972).
[10] R. G. Nicholls, A. Jerfy, and A. B. Roy, *Anal. Biochem.* **61,** 93 (1974).

VALUES OF THE FACTOR α REQUIRED TO CORRECT APPARENT RATES OF
HYDROLYSIS OF ASCORBATE 2-SULFATE FOR THE IONIZATION OF ASCORBATE[a]

pH	α	$1/\alpha$
4.0	0.443	2.26
4.1	0.500	2.00
4.2	0.557	1.79
4.3	0.613	1.63
4.4	0.666	1.50
4.5	0.715	1.40
4.6	0.760	1.32
4.7	0.799	1.25
4.8	0.834	1.20
4.9	0.863	1.16
5.0	0.888	1.13
5.1	0.909	1.10
5.2	0.926	1.08
5.3	0.940	1.06
5.4	0.952	1.05
5.5	0.962	1.04
5.6	0.969	1.03
5.7	0.976	1.03
5.8	0.980	1.02
5.9	0.984	1.02
6.0	0.988	1.01
7.0	0.999	1.00

[a] The experimentally determined value of the velocity must be divided by the appropriate value of α to give the true velocity.

an apparent one because of the different pK values of ascorbic acid and ascorbic acid 2-sulfate, 4.1 and 11.8 and < 1 and 2.75, respectively.[11] The first dissociating group of ascorbic acid is therefore not fully ionized at pH 4.8, and some of the protons released by the hydrolysis of the ester are utilized to maintain the appropriate ratio of the concentrations of the non-ionized and singly ionized forms of ascorbic acid. A correction factor, α, is calculated from Eq. (2),[8] where the pH is that of the reaction mixture and pA is the pK, 4.1, of the appropriate group in ascorbic acid.

$$\alpha = 10^{pH-pA}/(1 + 10^{pH-pA}) \qquad (2)$$

Values of α over the pH range 4.0 to 7.0 are given in the table, and the true reaction velocity is given by dividing that obtained directly from the

[11] A. D. Bond, B. W. McClelland, J. R. Einstein, and J. F. Finamore, *Arch. Biochem. Biophys.* **153**, 207 (1972).

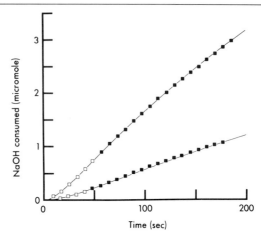

FIG. 1. The hydrolysis of ascorbate 2-sulfate by sulfatase A in the pH-stat. The conditions were: 33.3 mM ascorbate 2-sulfate; 0.25 mM EDTA; pH 4.8; 37°; enzyme concentration 1.7 μg/ml (upper curve) and 0.68 μg/ml (lower curve). The filled squares were used to compute v_0 and k^* by Eq. (1), the lines are computed pH-stat progress curves using these values of v_0 and k^*, and the open squares show the earlier experimental points not used in the computations.

The lines represent the equation [A. B. Roy, *Biochim. Biophys. Acta* **526**, 489 (1978).]

$$u = (v_0/k^*)(1 - e^{-k^*t}) + [v_0/(k^* - \omega)] \cdot (e^{-k^*t} - e^{-\omega t})$$

where u, v_0, and k^* have the same significance as in Eq. (1) and

$$\omega = \kappa(e^{-k^*t}/b)$$

in which κ is a constant incorporating various instrumental factors and b is the buffer capacity of the reaction mixture.

progress curve by the appropriate value of α. The correction is particularly important at low pH values.

Figure 1 shows typical experimental data and the theoretical pH-stat curves computed from the values of v_0 and k^* obtained from the data by using Eq. (1). Ten replicate assays under the above conditions gave a mean (± 1 standard deviation) v_0 of 1.47 $\pm$ 0.08 μmol/min. Values of v_1 obtained from the same data were 1.17 $\pm$ 0.05 μmol/min by the Stinshoff method and 1.15 $\pm$ 0.03 μmol/min by visually drawing tangents to the progress curves.

General Comments

The pH optimum for the hydrolysis of ascorbate 2-sulfate by sulfatase A is about 4.8, and the optimum substrate concentration is approximately 30 mM, above which the velocity falls, at least partly because of the increasing ionic strength due to the substrate. At pH 4.8 the kinetics are not strictly Michaelis in type although they are at pH 5.6. The reason for this difference probably lies in the existence of sulfatase A as a tetramer at pH 4.8 and a concentration of 1 μg/ml, but as a monomer at the same concentration at pH 5.6.[12] At pH 5.6, K_m is 23 mM ascorbate 2-sulfate and V_0 about 80 μmol min^{-1} mg^{-1} (compare with about 230 μmol min^{-1} mg^{-1} for nitrocatechol sulfate): at pH 4.8 precise values cannot be given because of the non-Michaelis kinetics but K_m is about 5 mM and V_0 about 150 μmol min^{-1} mg^{-1}. Sulfate is a competitive inhibitor with a K_i of about 0.5 mM.

Ascorbate 2-sulfate is a much less useful substrate for sulfatase B. At pH 4.8 in 0.15 M KCl, K_m is about 8 mM ascorbate 2-sulfate and V is about 1.2 μmol min^{-1} mg^{-1} (compare with 150 μmol min^{-1} mg^{-1} for nitrocatechol sulfate at pH 5.6). In the absence of KCl, the rate of hydrolysis is much less.

The above method has not been used in this laboratory with other sulfatases, but there is no obvious reason why it should not be applicable to the assay of sulfatases A from other mammalian sources or of any other enzyme hydrolyzing ascorbate 2-sulfate. As mentioned above, it is less likely to be useful for the assay of sulfatases B. Nor has the method been used with unpurified enzymes: again there is no theoretical reason why quite crude enzyme preparations could not be used in the pH-stat, but care would have to be taken that the large amounts of extraneous protein did not increase the buffer capacity of the reaction mixture to unacceptable levels.[8] Also, erroneously high activities would be obtained if any acid metabolite of ascorbic acid, or of ascorbate 2-sulfate, were produced.

In the form described above, the method is not particularly sensitive, requiring about 10 μg of enzyme per assay (about 2 U based on assays with nitrocatechol sulfate), but the sensitivity could be increased, probably by a factor of 10, by using more dilute NaOH as titrant. When working with such dilute solutions of NaOH, the usual precautions would have to be taken. The use of Ba(OH)$_2$ as titrant is not to be recommended because the BaSO$_4$ precipitated during the reaction can inhibit by adsorbing the enzyme unless bovine serum albumin (0.1 mg/ml) is added to the reaction mixture.

Finally, a practical point worthy of note is the inhibition of sulfatase A by at least some calomel electrodes through the diffusion of HgCl into the reaction mixture.[13] Although not usually important in assays lasting only a few minutes, the effect could be important if long times of incubation were used, and it should be eliminated, most simply by frequently changing the saturated KCl solution in the reference electrode.[13]

[12] A. Jerfy, A. B. Roy, and H. J. Tomkins, *Biochim. Biophys. Acta* **422**, 335 (1976).
[13] A. Jerfy and A. B. Roy, *Anal. Biochem.* **49**, 610 (1972).

Section II

Thiamine: Phosphates and Analogs

[8] An Improved Procedure for the Determination of Thiamine

By E. E. EDWIN

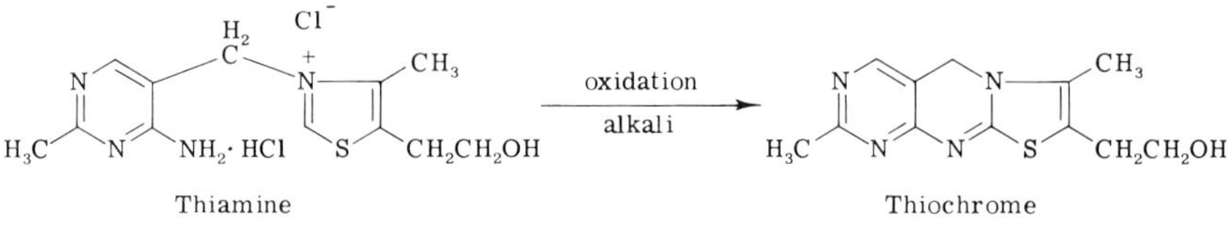

Thiamine can be readily and quantitatively oxidized to its highly fluorescent derivative thiochrome, and this has formed the basis of several procedures for its determination.[1-3] The procedure described here incorporates several modifications to established methods, making it easier and less time-consuming in application. It consists of the following steps: (1) extraction of thiamine and its phosphate esters, (2) dephosphorylation, (3) adsorption of total free thiamine on an ion-exchange material followed by removal of interfering impurities by washing, (4) oxidation under alkaline conditions to thiochrome, (5) extraction of thiochrome into isobutanol, and (6) measurement of fluorescence.

Reagents

Dilute hydrochloric acid, 2 ml of concentrated hydrochloric acid in 1 liter of deionized water

Hydrochloric acid, 0.1 N

Dilute acetic acid, 1 ml of glacial acetic acid in 1 liter of deionized water

Sodium acetate·3H_2O

Sodium chloride

Sodium hydroxide, 7.5 N

Mercuric chloride, 1% aqueous solution

Isobutanol, fluorescence-free grade

Repelcote water repellent (Hopkin and Williams), a 2% solution of dimethyldichlorosilane in carbon tetrachloride

Fluorometer. Any standard photofluorometer with primary filters having a transmission peak at 365–370 nm and secondary filters with a peak at 435–445 nm. Alternatively a spectrofluorometer may be used.

[1] B. C. P. Jansen, *Recl. Trav. Chim. Pays-Bas* **55**, 1046 (1936).
[2] M. Fujiwara and K. Matsui, *Anal. Chem.* **25**, 810 (1953).
[3] E. E. Edwin and R. Jackman, *Analyst* **100**, 689 (1975).

Phosphatase.[4] An acid phosphatase must be used for dephosphorylation, since the stability of thiamine decreases rapidly as the pH of its aqueous solution rises above 7. Several commercially available enzyme products have been found to possess phosphatase activity, e.g., Clarase (Miles Laboratories Inc., Elkhart, Indiana), or Taka-Diastase (Parke-Davis). These are fungal α-amylase preparations, which in addition possess phosphatase activity.

Decalso F. This resin (also called Zepolit S/F) was purchased from BDH Chemicals Ltd. It is a synthetic sodium aluminosilicate cation-exchange resin of mesh size 60–85. For use it must be freed from contaminating particles of iron or rust by running a strong magnet over it. It is then mixed thoroughly with a 25% solution of sodium chloride in dilute hydrochloric acid. After settling, the supernatant liquid is poured off together with much "fines." This treatment is repeated three times; the resin is then washed repeatedly with dilute acetic acid until the washings are free from chloride. At the end of this treatment only the coarser particles will be left, and these should settle to the bottom rapidly. Finally, the resin is spread on a tray and dried at room temperature.

Preparation of Standards. Thiamine chloride hydrochloride is deliquescent and must be stored in a desiccator over phosphorus pentoxide for several days, preferably in the dark. The standards are prepared by dissolving an accurately weighed amount of dry thiamine (say 100 mg) in 100 ml of 0.1 N HCl and are further diluted to give standard solutions of 0.1–1.0 μg/ml. When not in use the standard solutions must be stored in dark bottles at 0°–4°. Since thiamine tends to be absorbed on glass surfaces, the bottles must be treated with Repelcote water repellent. The solutions must be examined periodically for stability.

Procedure 1. Macromethod

Extraction. Solid samples, such as tissues, plant material, or feedstuffs, are comminuted, accurately weighed (1–2 g), and mixed with 0.1 N HCl (5 ml/g). The suspensions, in loosely capped tubes, are heated in a boiling water bath for at least 15 min. Liquid samples, such as urine or tissue fluids, are mixed with 0.1 N HCl (4–5 ml of acid per milliliter of sample) and heated as are the solid samples. This will ensure that proteins in the sample will be denatured and any bound thiamine or its phosphate esters will be released into the aqueous phase.

---

[4] It is advisable to test each new batch for phosphatase activity using thiamine pyrophosphate as substrate and examining the product by paper chromatography.[5] Occasionally some highly purified batches of Taka-Diastase were found to be devoid of phosphatase activity.

[5] D. Siliprandi and N. Siliprandi, *Biochim. Biophys. Acta* **14,** 52 (1954).

Dephosphorylation. Solid sodium acetate is added to the mixture in small amounts and mixed well until the pH of extracts rises to 4.5. Clarase or crude Taka-Diastase is added (100 mg per gram of sample) to the extract, mixed, and incubated at 45° for 3 hr. Alternatively the incubation can be carried out at room temperature (20°) for 16 hr. The mixture is centrifuged at 3000 g, and the clear supernatant is transferred to a 20-ml measuring flask. The residue is reextracted twice with 4-ml lots of 0.1 N hydrochloric acid and centrifuged; the supernatants are combined and made up to the mark on the flask.

Purification. One milliliter of the dephosphorylated extract is pipetted directly on to 1 g[6] of activated Decalso F in a stoppered tube (17 ml capacity) avoiding the sides of the tube. After mixing (mechanical vibrator), the tube is filled with nearly boiling deionized water, stoppered, shaken briefly, and allowed to stand. When all the Decalso particles have settled the supernatant is decanted off and discarded, taking care not to lose any particles. The washing procedure is repeated three times.

Formation of Thiochrome. To the wet Decalso is added 0.3 ml of mercuric chloride solution; after mixing thoroughly, 2 ml of 7.5 N NaOH are added and the mixture is shaken vigorously (mechanical vibrator) for at least 30 sec. Isobutanol (5.0 ml) is added, and the tube is stoppered and shaken vigorously for 2 min. After centrifugation at 3000 g for 5 min the clear isobutanol extract of thiochrome is transferred to a cuvette for measurement of fluorescence.

Oxidizing agents other than mercuric chloride have been used. Thus the 0.3 ml of mercuric chloride[7] can be replaced by 3 ml of a solution of cyanogen bromide[2] (prepared from saturated bromine water and the minimum amount of 10% KCN solution needed for decolorization). Alkaline potassium ferricyanide has also been used for this oxidation; however, its use is not recommended as it tends to give erratic results.[8,9]

Measurement of Fluorescence. With all determinations, a set of standards (at least 3) and a reagent blank are included.[10] The standards are selected to cover the expected thiamine concentration in the sample extract. A convenient range is 0, 0.1, 0.5, and 1.0 μg/ml. Fluorescence is measured at 436 nm with excitation at 365 nm. The readings are plotted on

---

[6] The amount of Decalso is not very critical and may vary between 0.5 and 1.5 g. Normally a level spatulafull is used, but it must be ensured that all the tubes receive similar amounts. The blank value is likely to rise with higher amounts of Decalso.

[7] M. Morita, T. Kanaya, and T. Minesita, *J. Vitaminol. (Kyoto)* **15**, 116 (1969).

[8] H. G. K. Westerbrink and J. Goudsmit, *Recl. Trav. Chim. Pays-Bas*, **56**, 803 (1937).

[9] T. Myint and H. B. Houser, *Clin. Chem.* **11**, 617 (1965).

[10] When the highest standard (1.0 μg/ml) has been set at 100 divisions, the blank reading must be less than 8. If higher readings are obtained, it would be advisable to change the batch of Decalso.

linear graph paper against the appropriate standard to give a reference curve, and the amount of thiamine in the sample tube is read off this.

Procedure 2. Semimicromethod

When the amount of sample available is limiting, or if regional surveys of thiamine concentration in an organ need to be studied, the semimicromethod can be followed. It is fundamentally the same as procedure 1 except for the following modifications. About 100 mg of sample are weighed into a tube and extracted with 2 ml of 0.1 N hydrochloric acid. After adjusting the pH to 4.5, it is dephosphorylated using 10 mg of Clarase. It is not extracted further, but Decalso (about 200 mg) is added directly to the tube. After mixing, the impurities are washed away with boiling deionized water. The rest of the assay is carried out as in procedure 1.

[9] Differential Determination of Thiamine and Its Phosphates and Hydroxyethylthiamine and Pyrithiamine[1]

By C. J. GUBLER and D. S. MURDOCK

Principle. This is a modification and combination of several reported methods.[2-4] It involves acid extraction of all these components from tissues, separation by column chromatography on Amberlite CG-50 and Thiochrome Decalso followed by differential oxidation by $HgCl_2$ or $K_3Fe(CN)_6$ and fluorometric assay at appropriate wavelengths.

Reagents

Permutit T (Thiochrome Decalso), a silicate cation-exchange resin, obtained from Fisher Scientific Co., Fair Lawn, New Jersey, was used without further processing.

Amberlite CG-50, a polymethacrylic acid cation-exchange resin from Mallinckrodt, Inc., St. Louis, Missouri, is washed repeatedly with H_2O to remove fines, dried, treated with acetone until the wash is colorless, washed again with water to remove acetone, then con-

[1] This work was supported by grants Nos. AM-02448 and AM-16897 from the National Institutes of Health.

[2] M. Morita, T. Kanaya, and T. Minesita, *J. Vitaminol. (Kyoto)* **14**, 67 (1968).
[3] M. Morita, T. Kanaya, and T. Minesita, *J. Vitaminol. (Kyoto)* **15**, 116 (1969).
[4] A. Fujita, Y. Nose, K. Ueda, and E. Hasehawa, *J. Biol. Chem.* **196**, 297 (1952).

verted to the Na^+ form with 40% w/v NaOH with stirring for 3 hr. This is then washed several times with H_2O and converted to the H^+ form with 3 N HCl followed by washing until free of Cl^-.

Taka-Diastase, a powdered preparation of *Aspergillus oryzae* from Parke Davis & Co., Detroit, Michigan, is used for hydrolysis of the thiamine phosphate esters.

Reagent grade isobutyl alcohol (Matheson, Coleman & Bell), redistilled to reduce background fluorescence

$HClO_4$, 0.3 M

KOH, 30%

NaOH, 30%

$HgCl_2$, 1%

$HgCl_2$, 0.01%

$K_3Fe(CN)_6$, 1%, in H_2O

$K_3Fe(CN)_6$, 1%, in 10% NaOH

Procedure

Extraction from Tissue

The tissue is removed, and a weighed portion is placed immediately in 10 ml of cold 0.3 M $HClO_4$ per gram of tissue and homogenized in a suitable tissue homogenizer.

The homogenate is then centrifuged at 5900 g for 15 mins. The supernatant is saved, and the pellet is washed by rehomogenizing in 0.2 volume of the acid and centrifuging as above. The first and second supernatants are then combined and neutralized to pH 5–6 with 30% KOH to precipitate $KClO_4$. After centrifugation, the extract is adjusted to pH 4–5 with 4 N, pH 4.5, sodium acetate buffer. This is called the "neutralized tissue extract."

Isolation of ThDP plus ThTP[5]

Approximately one-sixth of the neutralized tissue extract is placed on an 0.8 × 15 cm column of Amberlite CG-50. The ThDP + ThTP is eluted with 4 ml of distilled H_2O, while thiamine and ThMP remain on the column. (The column can be discarded or recovered by washing with 4 ml of 2 N HCl and H_2O until free of Cl^-.)

[5] Abbreviations used: ThDP, thiamine diphosphate; ThTP, thiamine triphosphate; ThMP, thiamine monophosphate; HET, hydroxyethylthiamine; PTh, pyrithiamine.

Preparation for Total Thiamine, HET, and PTh. About one-sixth of the neutralized tissue extract is not passed through the CG-50, but used for dephosphorylation without pretreatment.

Preparation for Free Thiamine Determination. About two-thirds of the neutralized extract was not subjected to dephosphorylation, but placed on the Decalso column directly.

Dephosphorylation Procedure

To appropriate samples for ThDP + ThTP from the CG-50 column eluate or for total thiamine, HET, and PTh from the untreated neutralized extract are added 1.0 ml of 4 M sodium acetate buffer, pH 4.5, and 50 mg of dry Taka-Diastase powder. The tubes are covered and incubated at 48°–50° for 3 hr.

Decalso Purification

About 1.5 g of Decalso are suspended in H_2O and allowed to settle in a column 0.8-cm in diameter with a 25-ml reservoir on top. The column is then washed with 25 ml 0.5% acetic acid followed by 25 ml of H_2O. The whole sample of dephosphorylated ThDP + ThTP, dephosphorylated total thiamine, HET, and PTh, or the untreated sample for free thiamine is placed on a Decalso column and washed with about 60 ml of boiling H_2O. The washings are discarded. The column is then treated with 25 ml of boiling 20% KCl in 0.1 N HCl; the eluate is collected in a 25-ml volumetric flask, and the volume is made up to 25 ml.

Assays

Thiamine Assay

Appropriate aliquots (3–8 ml) of the Decalso column eluates for total thiamine, free thiamine, or the ThDP + ThTP are placed in 50-ml conical tubes and mixed rapidly with 0.3 ml of 1% $HgCl_2$ using a vortex mixer. Two ml of 30% NaOH are then added to each, again with rapid mixing. A blank is prepared using only 2.0 ml of 30% NaOH. Ten milliliters of the distilled isobutyl alcohol are then added to each tube with thorough mixing for 2 min to extract the thiochrome from the aqueous layer. The tubes are centrifuged at 100 g. The fluorescence is then measured with a suitable fluorometer (Turner 110 or 111 with excitation filter No. 7-60 and emission filters Nos. 2A and 47-B or Farrand with 365-nm excitation and 436-nm emission wavelengths).

HET Assay

A 5–8-ml aliquot of the Decalso eluate for total thiamine is placed in a tube as above, and 2.0 ml of 30% NaOH are added with rapid mixing, followed by 2 drops of 0.01% $HgCl_2$, again with rapid mixing. This is followed by 0.3 ml of 1% $K_3Fe(CN)_6$ in H_2O. The preparation is then extracted with 10 ml of isobutyl alcohol, and the fluorescence is determined as above. This procedure destroys the thiamine, but not HET.

PTH Assay

To another aliquot (3–8 ml) of the total thiamine Decalso eluate is added 2.0 ml of 1% $K_3Fe(CN)_6$ in 10% NaOH with thorough and rapid mixing. The resulting pyrichrome is then extracted into 10 ml of isobutyl alcohol as above. In this case, the fluorescence is measured with a spectrophotofluorometer, such as a Farrand, using 430-nm excitation wavelength and 460-nm emission wavelength.

All the above assays are compared to appropriate standards of thiamine chloride·HCl, PTh·HBr, or HET chloride·HCl (supplied by Dr. M. Morita, Japan) treated in a similar manner as the respective samples. Thiamine, PTh, and HET are calculated as nanomoles of thiamine chloride·HCl (MW 337), PTh bromide·HBr (MW 420), and HET chloride·HCl (MW 381), respectively, per gram wet tissue as follows:

$$\text{nanomoles/gram} = A \times B \times (25/D) \times (E/F \times G) \times 1000$$

where A = fluorescent reading of samples blank; B = μg of pure thiamine, HET, or PTh used in the standard divided by the respective fluorescence reading; D = volume of the aliquot of the Decalso eluate used; E = total volume of the neutralized tissue extract; F = aliquot of the extract used for assay; G = the appropriate molecule weight.

The procedures described allow the determination of thiamine, HET, and PTh in the same sample with percentage errors of only 1.8, 2.3, and 3.0, respectively. Added thiamine, HET, and PTh could be recovered to the extent of 99.6, 98.4, and 100%, respectively. However, a 16% loss of HET occurs during the 3-hr incubation at 48°–50° with Taka-Diastase. Hence, a correction must be allowed for this loss. There is no significant interference of thiamine and HET in the PTh assay until the ratio of thiamine + HET to PTh exceeds 6:1. Likewise, the interference of PTh with the HET assay is insignificant below a PTh/HET ratio of 6:1.

[10] Fluorometric Determination of Thiamine and Its Mono-, Di-, and Triphosphate Esters

By H. K. PENTTINEN

Thiamine, when oxidized in an alkaline medium, is converted to the fluorescent thiochrome derivative. This reaction is used in the quantitative estimation of thiamine and its phosphate esters. However, equimolar amounts of these compounds produce unequal intensities of thiochrome fluorescence. Furthermore, addition of ethanol to the oxidation medium favors the production of thiochrome. These observations[1] have made it necessary to modify the previous method of Lewin and Wei.[2]

Reagents

Ethanol, 50% (v/v)

Alkaline ferricyanide [hexacyanoferrate(III)] (15 ml of 15% NaOH and 1 ml of 2% potassium ferricyanide)

Hydrogen peroxide solution, 30%

Quinine sulfate (0.01% quinine sulfate in 0.1 M H_2SO_4)

Thiamine standard solution, 0.3 mM, U.S.P. Reference Standard

Benzenesulfonyl chloride reagent, diluted with ethanol (1:6, v/v), (Koch-Light Laboratories Ltd., England)

Assay Procedure

A 5-μl sample of thiamine or its phosphate ester is added to 3 ml of 50% ethanol and shaken. After a few minutes, 0.5 ml of alkaline ferricyanide is added, and the mixture is agitated for 2 min. Then 10 μl of 30% H_2O_2 is added to destroy the yellow color of ferricyanide, and the fluorescence can be measured. Fluorometric measurements are conducted in a Farrand fluorometer A 4 with PC Corning Filters numbers 7-37 as primary filter and 3-73 and 5-60 as secondary filters, respectively, or in a Zeiss PMQ II spectrophotometer fitted with a ZFM 4 fluorometer attachment. The excitation wavelength is 365 nm, and the emission maximum is 430 nm.

The background fluorescence is determined by omitting the thiamine compound. It is also estimated by dissolving the sample in 2.8 ml of 50% ethanol, then adding 0.2 ml of benzenesulfonyl chloride reagent and stirring the mixture. Determination is then continued as described before.

[1] H. K. Penttinen, *Acta Chem. Scand. B* **30**, 659 (1976).

[2] L. M. Lewin and R. Wei, *Anal. Biochem.* **16**, 29 (1966).

Correction of the Deficient Thiochrome Formation

Because in this oxidation medium, 74% of thiamine and thiamine monophosphate, 85% of thiamine diphosphate, and 93% of thiamine triphosphate are converted to thiochrome, the fluorescence values have to be corrected. When thiamine is used as standard, the fluorescence reading for thiamine monophosphate needs no correction, but the readings for thiamine di- and triphosphate have to be multiplied by 0.87 and 0.80, respectively.[1] These correction factors have been shown to be constant and independent of the concentration of thiamine compounds.

[11] Separation of Thiamine Phosphoric Esters on Sephadex Cation Exchanger

By JULIA M. PARKHOMENKO, ALLA A. RYBINA,
and ASKAR G. KHALMURADOV

The main problem in estimating thiamine phosphoric esters in biological material is their quantitative separation.[1-4] Separating these pure compounds from their mixture presents no difficulties. Several methods have been suggested, and the more suitable appear to be electrophoresis on cellulose polyacetate strips[5] and chromatography on various ion-exchange resins.[6,7] However, high concentrations of inorganic salts and other impurities complicate greatly the separation of thiamine compounds by the above-mentioned methods. Rindi and de Giuseppe[8] and Koike et al.[9] succeeded in purifying a mixture of thiamine esters on a column packed with active charcoal treated with cholesteryl stearate[6] prior to their separation on ion-exchange resins, such as Dowex 1-X8[8] and Dowex 1-X4.[9] Sephadex appeared to be a suitable material for desalting (Sephadex G-25)[10] and partial separation (Sephadex G-10)[11] of the thiamines. However,

[1] A. Rossi-Fanelli, *Science* **116**, 711 (1952).
[2] W. Bartley, *Biochem. J.* **56**, 379 (1954).
[3] A. A. Rybina, *in* "Vitamins," Vol. 4, p. 10. Akad. Nauk USSR, Kiev, 1959.
[4] Y. V. Khmelevskii, *Vopr. Med. Khim.* **8**, 542 (1962).
[5] Y. Itokawa and J. R. Cooper, this series, Vol. 18A, p. 91.
[6] D. Siliprandi and N. Siliprandi, *Biochim. Biophys. Acta* **14**, 52 (1954).
[7] L. de Giuseppe and G. Rindi, *J. Chromatogr.* **1**, 545 (1958).
[8] G. Rindi and L. de Giuseppe, *Biochem. J.* **78**, 602 (1961).
[9] H. Koike, T. Wada, and H. Minakami, *J. Biochem. (Tokyo)* **62**, 492 (1967).
[10] L. Wildemann, *Z. Klin. Chem. Klin. Biochem.* **7**, 509 (1969); cited in *Nutr. Abstr. Rev.* **40**, 832 (1970).
[11] T. Nishimune, M. Abe, and R. Hayashi, *Biochim. Biophys. Acta* **279**, 527 (1972).

better separation was achieved by using ion-exchange Sephadexes.[12,13] The method of separation of thiamine esters from biological materials on Sephadex cation exchanger is described here.

Principle. Using a Sephadex ion-exchange column, it is possible to achieve simultaneous purification of substances and their separation by molecular dimensions and charges of their ionic forms. The major purpose of this method was to obtain sharp separation of thiamine triphosphate (TTP) and thiamine pyrophosphate (TPP); thiamine monophosphate (TMP) isolation as known presents no difficulties. In preliminary investigations it was found that 5–50 mM acetic buffer, pH 3.8, was most suitable. Electrophoretic studies conducted in our laboratory[14] revealed that under these conditions TTP was negatively charged, TPP was electroneutral or weakly positively charged, and TMP and thiamine (T) had some positive charges. Thus TTP was not affected by the negatively charged ionogen groups of Sephadex and emerged first from the column after the void volume, while TPP was affected only slightly and appeared next in the sharp peak. TMP and T combined strongly with the Sephadex and could be eluted at high ionic strength of the buffer. As the eluted TTP fractions were contaminated with other substances that interfered with its quantitative determination by the thiochrome method, additional purification on ion-exchange resin was necessary.

Reagents

Trichloroacetic acid (TCA), 6% and 5%

Diethyl ether, distilled

SE-Sephadex C-25 (a product of Pharmacia Fine Chemicals), 100–270 mesh, sodium form

Sodium acetate buffer, 5 mM, pH 3.8

Sodium acetate buffer, 50 mM, pH 3.8

Sodium acetate buffer, 1 M, pH 3.8, with 1 M NaCl

Extract of Orysin (phosphatase-containing preparation, which may be replaced by Taka-Diastase): 10 mg of enzyme preparation were dissolved in 1 ml of 1 M acetic buffer, pH 4.5 and filtered after standing for 1 hr.

SDV-3, anion exchange resin in acetate form (may be replaced by Amberlite IRC-50[8])

Preparation of Sephadex Ion Exchanger and Column. SE-Sephadex C-25, medium (100–270 mesh) was treated according to the manufacturer's recommendations.[15,16] Sephadex was allowed to swell in distilled

[12] A. Nakamura, K. Sanada, and E. Katsura, *Vitamins* **37**, 1 (1968).

[13] J. M. Parkhomenko, A. A. Rybina, and R. B. Pol'schak, *Ukr. Biochem. J.* **48**, 384 (1976).

[14] J. M. Parkhomenko, unpublished data, 1974.

[15] "Sephadex Filtration in Theory and Practice," Pharmacia Co.

[16] "Sephadex Ion Exchangers, an Outstanding Aid in Biochemistry," Pharmacia Co.

water for 24 hr; during this time any fines were removed by repeated decantation. Then the material was placed to a Büchner funnel and washed repeatedly with 0.5 N HCl (total volume about 0.5 liter). The excess of HCl was removed by rinsing with distilled water. The ion exchanger was treated with 0.5 N NaOH (0.5 liter) and washed with distilled water to remove excess base. Finally the gel was mixed with distilled water and poured carefully into a chromatographic glass tube (1.5 × 35 cm). The slurry was allowed to settle for about 5 min; then the outlet was opened to allow a gentle flow. The gel suspension was gently stirred with a glass rod to remove air bubbles. The final height of the Sephadex bed was 25 cm. It was equilibrated with the starting buffer (5 mM acetic buffer, pH 3.8), 50 ml of buffer being added. The same Sephadex was used for a large number of experiments. For Sephadex regeneration, the column was percolated with 2 M sodium acetate solution after each experiment was completed. For the next regeneration step, the gel was transferred onto a Büchner funnel and washed with distilled water to remove excess salt. Possible lipid impurities were removed by washing with 95° ethanol followed by distilled water. Although the Sephadex ion exchanger in all our experiments was used in the sodium form, its subsequent treatment with 0.5 N HCl and 0.5 N NaOH, as indicated above, was repeated to remove traces of strongly adsorbed impurities.

Procedure. About 5 g of rapidly frozen (liquid nitrogen) liver was powdered in a porcelain mortar. The frozen powder was quickly transferred to a precooled and weighed test tube containing 6% TCA, and the mixture was weighed. TCA solution was added to adjust the final tissue:TCA ratio to 1:3. The thiamine compounds were extracted for 30 min in an ice bath with occasional stirring, and the mixture was centrifuged for 10 min at 2500 g. The precipitate was washed twice with 5% TCA in a ratio of 1:1. The supernatants were combined, and the total volume was measured. TCA was removed from the combined extract by several extractions with ether. The traces of ether were removed by bubbling nitrogen through the solution. A measured volume was concentrated to a thick syrup on a rotatory evaporator at 25°. It was resolved in 5 mM acetic buffer, pH 3.8, taking 1 ml of buffer per 1.5–2.5 g of the initial tissue. About 1.5 ml of such a sample was carefully pipetted onto the top of the Sephadex bed. For elution, buffer solutions of stepwise increasing ionic strength were added as follows: 5 mM acetic buffer, pH 3.8, 25 ml; 50 mM acetic buffer, pH 3.8, 100 ml; 1 M acetic buffer, pH 3.8, with 1 M NaCl, 100 ml. The flow rate was approximately 15 ml/cm^2 per hour, all operations with the column being carried out at 10°.

In these conditions thiamine compounds emerged from the column in the order shown in Fig. 1. Fractions corresponding to individual phosphoric esters of thiamine were combined, the phosphatase solution was

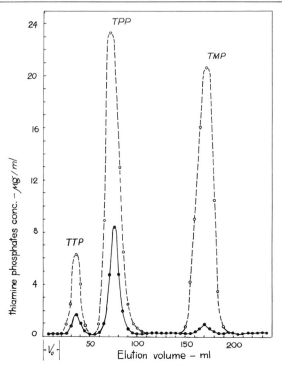

FIG. 1. Separation of the thiamine phosphates on SE-Sephadex C-25. Sample: rat liver, 3.75 g: Sephadex bed size: 1.5 × 25 cm; flow rate: 15 ml/cm² per hour; fraction volume: 5 ml; temperature: 10°. ——, Elution peaks of thiamine esters, when only tissue extract was applied on a column; ___, elution peaks of thiamine esters when pure compounds were added in tissue extract: TTP, 10 μg; TPP, 35 μg; TMP, 65 μg. The ester content in each fraction was determined in the aqueous phase against an appropriate ester standard after oxidation by alkaline potassium ferricyanide.

added, taking 1 ml per 10 ml of eluate, and the mixture was incubated at 37° overnight. The eluate portion, containing TMP and T, was divided into two equal parts, one of which was exposed to phosphatase action. The portion containing TTP, after phosphatase action, was additionally purified on the resin SDV-3 in the acetate form[17] (or on Amberlite IRC-50 in the H⁺ form[8]). The thiamine content in all portions was determined by the thiochrome method,[18] and the TMP by the difference between the thiamine content in the portion exposed to phosphatase action and that without it.

[17] A. A. Dmitrovsky, in "Vitamin Sources and Their Utilisation," p. 100. Akad. Nauk SSSR, Moscow, 1955.
[18] "Method of Vitamin Assay," 3rd ed. (Association of Vitamin Chemists, ed.), p. 123. Wiley (Interscience), New York, 1966.

The thiamine esters of tissue were calculated on the basis of the molecular weight of each compound and expressed in microgram per gram of tissue.

Comments. It was not our aim to separate TMP and T by the described method. They were eluted together with buffer solution of high ionic strength. If needed, however, one may choose the appropriate conditions for differential elution of these compounds as well.

The recovery of each of the esters applied to the Sephadex column in the tissue sample in amounts from 10 to 100 μg was about 98–100%.

[12] High-Pressure Liquid Chromatography of Thiamine, Thiamine Analogs, and Their Phosphate Esters[1]

By C. J. GUBLER and B. C. HEMMING

Of the three thiamine phosphate esters known to occur in biological material, only thiamine diphosphate (ThDP[2]) and thiamine triphosphate (ThTP) appear to have biological activity, free thiamine and thiamine monophosphate (ThMP) being considered only as the hydrolysis products of the more phosphorylated forms of the vitamin. Various separation methods[3] have been used to resolve and to estimate thiamine and its phosphates in biological materials. Simultaneous determination of thiamine, oxythiamine (OTh), and their phosphate esters in rat tissue[4] has been accomplished by taking advantage of the fact that thiamine compounds give the thiochrome reaction, whereas Oth and its phosphate esters do not form the fluorescent product. Separation of thiamine and its esters after conversion to thiochrome has been achieved with conventional column chromatography.[5] High-pressure liquid chromatography (HPLC) has been applied for thiamine determination, generally as a separation method for pharmaceutical application.[6] This report describes HPLC methods for the separation of the phosphate esters of thiamine, Oth, or pyrithiamine and the separation of thiamine and Oth. These methods are combined

---

[1] This work was supported by NIH Grants Nos. AM-02448 and AM-16897.
[2] Abbreviations used: Th-HCl, thiamine hydrochloride; ThMP, thiamine monophosphate; ThDP, thiamine diphosphate; ThTP, thiamine triphosphate; OTh, oxythiamine; HPLC, high-pressure liquid chromotography.
[3] H. Koike and T. Yusa, this series, Vol. 18A [18], p. 105.
[4] L. De Giuseppe and G. Rindi, *Int. Z. Vitamin, forsch.* **34**, 21 (1964).
[5] T. Nishimuna, M. Abe, and R. Horpshi, *Biochim, Biophys. Acta* **279**, 257 (1972).
[6] R. B. H. Wills, C. G. Shaw, and W. R. Day, *J. Chromatogr.* **15**, 262 (1977).

with a selective fluorometric detection system similar to that of Van De Weerdhof *et al.*[7]

Principle. Two methods are presented based on different HPLC chromatographic modes, namely anion exchange (Vydac) and reverse-phase paired-ion chromatography (μBondapak C_{18}). The advantages are the versatility, selectivity, and speed of analysis made possible for numerous samples of biological origin or of complex mixtures.

Reagents

Vydac anion-exchange column, 500 mm × 20 mm (i.d.), a strong anion exchanger supplied in the chloride form, particle size 30–40 μm

Solvents employed in the anion exchange mode:

Solvent A: 1mM KH_2PO_4, pH 6.0

Solvent B: 0.5 M KH_2PO_4, pH 6.0 (UV-absorbing contaminants in the buffers are removed or decreased by the method of Shmukler.[8]

μBondapak C_{18} column, 300 mm × 4.0 mm (i.d.), particle size 10 μm

Solvents employed in the reverse-phase mode:

Solvent A, aqueous 50 mM tetrabutylammonium hydroxide and 1% acetic acid (final pH 4.3)

Solvent B, 80:20 mixture of spectrograde methanol and solvent A

$K_3Fe(CN)_6$, 0.6 mM, in 15% NaOH. This solution serves as the oxidant for the fluorometric detection system.

Deionized distilled water is used in all preparations; all solvents are degassed under vacuum for 5.0 min.

Instrumentation. The conditions of phosphate ester separation require a liquid chromatograph capable of linear gradient formation. The high-pressure liquid chromatograph used is a Varian Model 4200 series equipped with a Model 635 series variable wavelength spectrophotometer (cell volume, 8.0 μl). The effluent oxidizing system is connected as the second detector and consists of a mixing module (Varian Instruments, Part No. 02-0015-75-00) and an oxidant reservoir (Varian Instruments, Part No. 02-001673-00). The oxidant flow rate is a function of the hydrostatic pressure developed from a nitrogen line, which in turn is controlled by a gas regulator gauge (Hoke, Inc., Model No. 5124F4B) and a fine-metering valve (Nupro Co., Part No. SS-25G). The effluent line of the mixing module is connected to a Turner Model 111 fluorometer. The fluorometer flow cell of 600-μl volume is reduced to 100 μl to provide

[7] T. Van De Weerdhof, M. L. Wiersum, and H. Reissenweber, *J. Chromatogr.* **83,** 455 (1973).

[8] H. W. Shmukler, *J. Chromatogr. Sci.* **8,** 581 (1970).

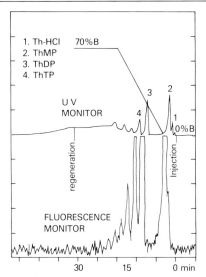

FIG. 1. Ultraviolet (UV) and fluorescence detection of thiamine and its phosphate esters from an anion-exchange column. Conditions are given in the text. The small late-eluting peaks in the UV tracing are not the same material that produces the smaller last two fluorescent peaks, as determined by lapse times between detectors.

greater resolution in the reverse-phase method. Chromatograms are produced by a Rikadenki 2-pen strip-chart recorder.

Procedure. The phosphate ester of thiamine and its analogs, Oth and pyrithiamine, are synthesized by the method of Matsukawa *et al.*[9] The product of the thiamine phosphate ester synthesis reaction is a crude mixture of unreacted thiamine, ThMP, ThDP, ThTP, and higher phosphorylated forms. This mixture is separated by anion exchange directly after redissolving in water (Fig. 1). A linear gradient is initiated 4 min after injection of a 20-μl sample going from 0 to 70% B at a rate of 5% B per minute. A flow rate of 1.0 ml/min produces a pressure drop of 600 psi. Thiamine elutes in the void volume. The time lapse between UV and fluorometric detectors is 1.0 min. Fluorometric detection requires one primary filter (peaks at 360 mm) for excitation and two secondary filters, a sharp-cut filter (passes longer than 415 nm) and a narrow-pass filter (peaks at 436 nm), for emitted light. The fine-metering valve is used to maintain a constant pressure of 20 psi on the oxidation line. The fluorometer slit width is held at maximum sensitivity (30×) when the 600-μl flow cell is used. Chart speed is set at 20 cm/hr. Recorder channels are set to twice the full signal output voltages of both spectrophotometer and fluorometer. This

[9] T. Matsukawa, H. Hirano, and S. Yurugi, this series, Vol. 18A [25], p. 141.

allows both tracings to be placed on the same chart without overlap. UV detection is at 254 nm and 0.5 absorption unit full scale. The responses of the two detectors are linear between 0.250 μg and 1.000 μg of thiamine. Peaks are identified by comparison with retention times of purified commercial standards. Separation of a synthetic mixture of Oth and its phosphate esters under these given conditions provides a chromatogram similar to that of Fig. 1. No fluorescent peaks are obtained for oxythiamine compounds, as noted earlier. Pyrithiamine phosphate esters have not been examined in this chromatographic mode.

The reverse-phase column can be used in an ion-pair chromatographic method for thiamine.[6] The conditions of this method separate mixtures of thiamine and Oth (Fig. 2). Monitoring at 280 nm increases the sensitivity for oxythiamine detection. Under these conditions phosphorylated species elute in the void volume, except for the monophosphates, which are only slightly retained. Pyrithiamine has the same retention time as thiamine under these conditions.

Figure 3 gives the conditions for a separation of thiamine, ThMP, and ThDP standards on the reverse-phase column with tetrabutylammonium ion as the pairing agent. ThTP is eluted (27 min) by use of a linear gradient that commences at a constant rate of 5% B per minute solvent change

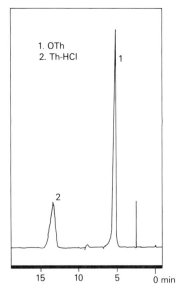

Fig. 2. Isocratic reverse-phase chromatogram of oxythiamine (Oth) and thiamine (Th-HCl). Operating conditions: column, μBondapak; mobile phase, 3:1 water/methanol plus 1% acetic acid and 1 mM heptanesulfonic acid; flow rate, 2.0 ml/min; pressure, 2500 psi; detection wavelength, 280 nm.

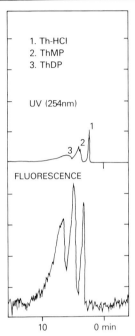

FIG. 3. Reverse-phase chromatography of a thiamine (Th-HCl), thiamine monophosphate (ThMP) and diphosphate (ThDP) ester mixture. Operating conditions: column, μBondapak C_{18}; mobile phase, 50 mM tetrabutylammonium hydroxide and 1% acetic acid (pH 4.3); flow rate, 1.0 ml/min, pressure 1500 psi; detection by UV (254 nm) and fluorescence system with 100-μl flow cell at 10× slit width.

after an initial 10-min time delay (not shown). Separations of synthetic mixtures of pyrithiamine and Oth phosphate esters produce chromatograms and retention times similar to those shown in Figure 3. (Thiamine, pyrithiamine, and Oth elute at void volume (3 min) under these conditions.) The mechanism of retention, therefore, appears to be predominantly ion-pair formation involving the phosphate moieties of the parent compounds.

Comments

Wider use of HPLC fluorescence methods for detection of both naturally fluorescent and selectively derivatized compounds has led to the commercial availability of fluorescent detectors specifically designed for HPLC compatibility. Such detectors, with their inherent lower dead volumes, offer greater resolution and sensitivity, thereby greatly enhancing the possible qualitative and quantitative aspects of these methods.

Together with the selective nature of the thiochrome formation for detection, analysis of these compounds in biological material should be greatly improved by avoidance of problems associated with UV monitoring (e.g., nucleotide interferences).

[13] Electrophoretic Separation of Thiamine and Its Mono-, Di-, and Triphosphate Esters

By H. K. PENTTINEN

Paper electrophoresis is a rapid and convenient method for separating the various thiamine phosphate esters. But because of interference by many substances, e.g., salts, buffers, and organic material, the results tend to be unsatisfactory. As this was the greatest disadvantage with the acetate buffer used before,[1] a new modification has been developed.[2] In this method better separation is achieved by using citrate buffer containing methanol, ethanol, and propanol, which were found to improve the elution of thiamine compounds from electrophoretic paper.

Reagents

Sodium acetate buffer, 50 mM, pH 3.8

Sodium citrate buffer, 50 mM, pH 5.6, containing 0.025 part (v/v) of a mixture of methanol, ethanol, and propanol (1 : 1 : 1, v/v/v)

Alkaline ferricyanide [hexacyanoferrate(III)] reagent [76 ml of 50% ethanol (v/v), 15 ml of 15% NaOH, and 1 ml of 2% potassium ferricyanide]

Electrophoretic paper, Munktells S311 (Grycksbo, Sweden)

Electrophoresis

A sample of 5 μl is applied in the middle of a paper strip (2.5 × 48 cm) that has been soaked in the citrate buffer and blotted. A similar sample containing 5 nmol each of thiamine and its phosphate esters is applied to another strip, which is used later to locate the fluorescent compounds. The strips are subjected to electrophoresis for 45–75 mins at a constant current of 3 mA per strip in a high-voltage apparatus (Analysteknik AB, Vallentuna, Sweden). The voltage ranges from 2 to 4 kV. The electrode vessels contain sodium acetate buffer. The reference strip containing added thiamine compounds is sprayed with alkaline ferricyanide reagent.

[1] Y. Itokawa and J. R. Cooper, in this series, Vol. 18A, p. 91.

[2] H. K. Penttinen, *Acta Chem. Scand. B* **32**, 609 (1978).

The fluorescent bands are visualized with UV light, and according to this strip the other strips are cut into pieces corresponding to the band to be measured. The thiamine compounds are eluted from the pieces of paper with 3 ml of 50% ethanol (v/v) for 45 min. After removal of the paper, fluorometric determination is carried out as described in this volume [10].

Comments

With this method thiamine phosphate esters can be separated from mixtures containing 0.37 M perchloric acid and 70 mM glycylglycine, sodium acetate, potassium phosphate, or sucrose. When thiamine compounds are determined from biological material, the proteins of the sample are first precipitated with perchloric acid. The sample is neutralized with K_2CO_3 and then lyophilized in order to concentrate the contents of thiamine. Blanks are determined using benzenesulfonylchloride to prevent the oxidation of thiamine to thiochrome as described in this volume [10].

[14] Thiamine Phosphate Pyrophosphorylase

By TAKASHI KAWASAKI

$$HMP\text{-}PP^1 + Th\text{-}P \xrightleftharpoons{Mg^{2+}} TMP + PP_i$$

The systematic name for thiamine pyrophosphorylase is 2-methyl-4-amino-5-hydroxymethylpyrimidinepyrophosphate:4-methyl-5-(2'-phosphoethyl)-thiazole   2-methyl-4-aminopyrimidine-5-methenyltransferase (EC 2.5.1.3).

Assay Method

Principle. TMP formed is hydrolyzed to thiamine by Taka-Diastase, oxidized to thiochrome with alkaline cyanogen bromide, and then determined fluorometrically.[2]

Reagents. The reagents for enzyme activity assay are:

[1] Abbreviations used: HMP (thiamine pyrimidine), 2-methyl-4-amino-5-hydroxymethylpyrimidine; Th (thiamine thiazole), 4-methyl-5-hydroxyethylthiazole; HMP-P, HMP phosphate; HMP-PP, HMP pyrophosphate; Th-P, Th phosphate; TMP, thiamine monophosphate.

[2] A. Fujita, Y. Nose, S. Kozuka, T. Tashiro, K. Ueda, and S. Sakamoto, *J. Biol. Chem.* **196,** 289 (1952).

Tris·HCl buffer, 0.5 M, pH 7.5
HMP-PP, 05 mM
Th-P, 0.5 mM
MgCl$_2$, 0.1 M
Enzyme
Those for TMP determination are:
Sodium acetate buffer, 1 M, pH 4.5
Taka-Diastase solution, 0.02%, in 0.1 M sodium acetate buffer, pH 4.5
Cyanogen bromide, 0.3 M
NaOH, 1 M

Procedure. The reaction mixture contains 0.5 ml each of Tris·HCl buffer, HMP-PP, and Th-P, 0.3 ml of MgCl$_2$, and enzyme in a total volume of 4 ml. The enzyme is diluted by 50 mM Tris·HCl buffer, pH 7.5, containing 5 mM 2-mercaptoethanol when necessary. The mixture is preincubated for 5 min at 37°; the reaction is started by the addition of enzyme and allowed to incubate for 5 min. The reaction is stopped by the addition of 0.5 ml of acetate buffer and heated for 15 min at 85°. The mixture is then incubated for 30 min at 45° after adding 0.5 ml of Taka-Diastase and centrifuged for 10 min at 3000 rpm to remove denatured proteins. An aliquot of the supernatant is brought to 3.5 ml with water, and 0.5 ml of cyanogen bromide freshly prepared before use is added and mixed, followed by the addition of 1.0 ml of NaOH. Fluorescence intensity of the solution is determined (excitation, 365 mn; emission, 430 nm) with thiochrome solution as the standard at an appropriate concentration. A blank mixture composed of Tris·HCl buffer and enzyme in the same total volume is treated in the same manner. The blank value is subtracted from that of the sample determined.

Definition of Specific Activity. Specific activity is defined as nanomoles of TMP formed per milligram of protein per minute.

Purification Procedure

A partial purification of this enzyme has been described.[3]

Source and Cultivation of Bacterium. A thiamine regulatory mutant of *Escherichia coli* K12, strain PT-R1, containing a derepressed level of TMP pyrophosphorylase[4] is used. Strain PT-R1 is aerobically grown at 37° in the minimal medium of Davis and Mingioli[5] containing 0.2% glucose as a carbon source; at 15–17 hr after incubation, the cells are harvested by

[3] Y. Kayama and T. Kawasaki, *Arch. Biochem. Biophys.* **158**, 242 (1973).
[4] T. Kawasaki and Y. Nose, *J. Biochem. (Tokyo)* **65**, 417 (1969).
[5] B. D. Davis and E. S. Mingioli, *J. Bacteriol.* **60**, 17 (1950).

centrifugation (12,000 g, 4°) and washed once in 50 mM Tris·HCl buffer, pH 7.5, 5 mM 2-mercaptoethanol, 1 mM EDTA (medium A). The washed cells obtained are approximately 10 g wet weight from 60 liters of the culture medium.

Step 1. Preparation of Crude Extract. The washed cell pastes are resuspended in 100 ml of medium A, and all further steps are carried out below 4°. The cell suspension is subjected to sonic disruption for 20 min and then centrifuged for 20 min at 15,000 g. The sediment is discarded.

Step 2. Removal of Nucleic Acid. The supernatant fluid is adjusted to pH 6.0, and 2% protamine sulfate is added with vigorous stirring at a ratio of 0.1 mg to milligram of protein and then centrifuged. This treatment results in an improved absorption ratio of 280:260 nm of the supernatant to 0.9–1.0.

Step 3. First Ammonium Sulfate Fractionation. Solid ammonium sulfate (24.3 g/100 ml) is added to bring the solution to 40% saturation with constant stirring. The pH of the solution is maintained at 7.5 by dropwise addition of 5 M ammonium hydroxide during the addition of ammonium sulfate. The suspension is allowed to stand for 30 min before centrifugation and the precipitate is discarded. The supernatant is brought to 60% saturation by adding ammonium sulfate (13.2g/100 ml) with constant stirring and standing for another 30 min. The precipitate obtained after centrifugation is dissolved in 10 ml of medium A (ammonium sulfate I fraction).

Step 4. First Sephadex Column Chromatography. The ammonium sulfate I fraction is applied to a Sephadex G-100 column (2.5 × 50 cm) previously equilibrated with medium A and elution is carried out with the same buffer system to collect 10-ml fractions at approximately 2 ml/min. The fractions containing high enzyme activity are combined (Sephadex I fraction).

Step 5. Second Ammonium Sulfate Fraction. Solid ammonium sulfate (35.1 g/100 ml) is added to the Sephadex I fraction to give 55% saturation, while the solution is maintained at pH 7.5 as described above. After standing for 30 min, the precipitate is collected by centrifugation and dissolved in 20 ml of medium A (ammonium sulfate II fraction).

Step 6. DEAE-Cellulose Column Chromatography. The ammonium sulfate II fraction is dialyzed for 3 hr against 5 liters of medium A and applied to a DEAE-cellulose column (2.5 × 50 cm) previously equilibrated with medium A. Elution is carried out with 300 ml of medium A and then with a linear gradient of NaCl consisting of 200 ml of medium A in the mixing flask and an equal volume of medium A containing 0.4 M NaCl in the reservoir flask. Fractions of 5 ml are collected. High enzyme activity is obtained by elution at 0.2–0.25 M NaCl, and these fractions are pooled

PURIFICATION OF *Escherichia coli* TMP PYROPHOSPHORYLASE[a]

Fraction	Total protein (mg)	Total activity (nmol/min)	Specific activity (nmol/mg/min)	Purification (fold)	Yield (%)
Crude extract[b]	16,900	6250	0.37	1.0	100
Ammonium sulfate I[b]	7740	4650	0.60	1.6	74.5
Sephadex I[b]	1920	2550	1.3	3.6	40.8
Ammonium sulfate II[c]	475	1920	4.0	11	30.7
DEAE-cellulose[c]	90	966	11	29	15.4
Sephadex II[c]	5.2	338	65	176	5.4

[a] From Y. Kayama and T. Kawasaki, *Arch. Biochem. Biophys.* **158**, 242 (1973).
[b] Protein was determined by the biuret method [A. G. Gornall, C. S. Bardawill, and M. M. David, *J. Biol. Chem.* **177**, 751 (1941)].
[c] Protein was determined by the method of O. Warburg and W. Christian, *Biochem. Z.* **310**, 384 (1941).

and concentrated to approximately 3 ml in a collodion bag (DEAE fraction).

Step 7. Second Sephadex Column Chromatography. The DEAE fraction is applied to a Sephadex G-100 column (2.5 × 50 cm) previously equilibrated with medium A and eluted with the same medium. The fractions of 5 ml containing high enzyme activity are pooled and concentrated to approximately 1 ml in a collodion bag (Sephadex II fraction).

The purification procedure is summarized in the table.

Properties

Substrate Specificity. The enzyme is specific for both of the substrates, HMP-PP and Th-P. HMP, HMP-P, and Th are not replaced as substrates.

Kinetic Properties. When assayed at pH 8.5 and 40° in the presence of 6 mM MgCl$_2$, the apparent K_m values for HMP-PP and Th-P are calculated to be 8.5×10^{-7} M and 4.0×10^{-7} M, respectively.

Metal Requirement. The purified enzyme requires the presence of divalent cation for its maximal activity. Mg^{2+} is most effective at 6 mM and Mn^{2+} at 6 mM is equally effective. Co^{2+}, Ca^{2+}, and Zn^{2+} are only 20% as effective as Mg^{2+}. The apparent K_m value for Mg^{2+} is calculated to be 6.3×10^{-5} M under the standard assay conditions.

Effect of High-Energy Phosphate Compounds. High-energy phosphate compounds at 4 mM are all inhibitory in the presence of 6 mM MgCl$_2$. Percentage inhibitions by these compounds of the activity are: ATP, 64.7; ADP, 42.4; GTP, 76.7; UTP, 71.6; CTP, 58.7; acetyl phosphate, 82.8; phosphoenol pyruvate, 31.1; and phosphocreatine, 63.8. Acetyl phosphate is the most effective inhibitor and causes 50% inhibition at 0.5 mM. Practically no inhibition of the activity is found with AMP, pyridine nucleotides, and sugar phosphates. Kinetics indicate that the inhibition caused by ATP and acetyl phosphate is uncompetitive for either HMP-PP or Th-P. A partially purified TMP pyrophosphorylase of yeast is also inhibited by high-energy phosphate compounds.[6]

Molecular Weight. The molecular weight of the enzyme measured by the method of Andrews[7] is approximately 17,000.

Other Properties. The purified preparation retains full catalytic and regulatory properties for at least several months when stored at $-20°$. Optimal pH for activity is 8.5 with 50 mM Tris·HCl buffer, and optimal temperature is 40°. The enzyme is completely inactivated when incubated for 5 min at 45° without addition of both substrates. The enzyme is inhibited by inorganic pyrophosphate and 50% inhibition results at 50 μM concentration.

[6] T. Kawasaki and K. Esaki, *Biochem. Biophys. Res. Commun.* **40**, 1468 (1970).
[7] P. Andrews, *Biochem. J.* **91**, 222 (1964).

[15] Synthesis of Pyrimidine ^{14}C-labeled Thiamine

By E. E. EDWIN

Thiamine labeled in its thiazole moiety either with ^{14}C or ^{35}S is commercially available.[1] However, it is difficult to obtain the pyrimidine-labeled compound, and no detailed accounts of its synthesis have been published. Since it is a useful compound in studies in many areas of thiamine metabolism, a procedure that has been successfully used in our laboratory is described. It has been adapted from a number of published methods to suit semimicro and radiochemical techniques. The method yields compounds labeled either on the 2-methyl substituent or the 2'-position of the pyrimidine ring, depending on whether [1-^{14}C]acetonitrile or [2-^{14}C]acetonitrile is used as a starting material.

Principle. [1-^{14}C]Acetonitrile or [2-^{14}C]acetonitrile is converted into acetamidine[2] and condensed[3] with β-ethoxymethylene malononitrile. The resulting cyanomethylpyrimidine is reduced to give the aminomethyl

[1] Radiochemical Centre, Amersham, England.
[2] A. W. Dox, *in* "Organic Syntheses" (A. H. Blatt, ed), Collective Vol. I. p. 5. Wiley, New York, 1941.
[3] R. Greive, *Hoppe-Seyler's Z. Physiol. Chem.* **242**, 5 (1941).

METHODS IN ENZYMOLOGY, VOL. 62

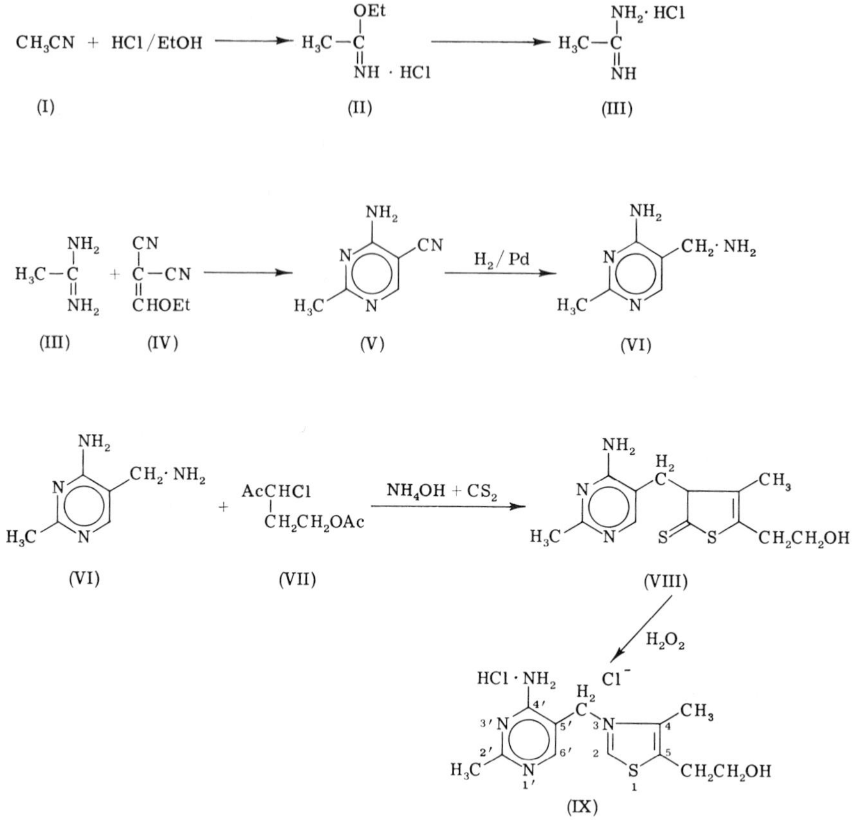

compound and condensed[4] with 3-chloro-4-oxopentyl acetate and CS_2. Oxidation[4] of the resulting thiazolethione gives thiamine chloride hydrochloride, which is purified by paper chromatography. As far as possible, intermediates are not isolated, and the number of reaction vessels must be kept to a minimum.

Reagents

Acetonitrile, 1- or 2-^{14}C labeled (Radiochemical Centre, Amersham, England)

β-Ethoxymethylene malononitrile (Aldrich Chemical Co)

3-Chloro-4-oxopentyl acetate (used as an intermediate in the manufacture of thiamine)

CS_2

Anhydrous ethanol

[4] H. M. Wuest, *in* "The Vitamins," 2nd ed. (W. H. Sebrell, Jr, and R. S. Harris, eds.), Vol. 5, p. 108. Academic Press, New York, 1972.

Saturated solution of HCl in anhydrous ethanol
Saturated solution of NH$_3$ in anhydrous ethanol
Palladium on charcoal, 5%
Hydrogen peroxide, 30%
BaCl$_2$, saturated solution
Alkaline potassium ferricyanide: 1.5 ml of a 1% aqueous solution of
    potassium ferricyanide mixed with 30 ml of 10% solution hy-
    droxide solution.
Repelcote: a 2% solution of dimethyldichlorosilane in carbon tetra-
    chloride (marketed by Hopkin & Williams)

Synthetic Procedure

Acetamidine[2] (III). Labeled acetonitrile (I) (250 μCi) is diluted with
carrier acetonitrile (41 mg, 1 mmol) and treated with an excess of anhydrous ethanol (3 ml) saturated with dry HCl in a stoppered tube. After the
preparation has stood for 3 days, the excess HCl and alcohol are evaporated off under reduced pressure to give the imino ether (II). A saturated solution (3 ml) of gaseous ammonia in dry ethanol is added and allowed to stand for 3 hr with occasional shaking. After this the solvent and
excess ammonia are removed under reduced pressure, to give labeled
acetamidine hydrochloride as a colorless crystalline desposit.

4-Amino-5-cyano-2-methylpyrimidine[3] *(V)*. The acetamidine from the
last step is dissolved in 2 ml of dry ethanol and treated with sodium ethoxide (prepared by dissolving 30 mg (slight excess) of sodium in 1 ml of
dry ethanol). The solution of free acetamidine is filtered directly into a solution of β-ethoxymethylene malononitrile (IV) (122 mg, 1 mmol) in 1 ml
of dry ethanol. The condensation should take place immediately, the
product separating as a crystalline mass, which is filtered and washed
with a little cold dry ethanol (0.1 ml).

4-Amino-5-aminomethyl-2-methylpyrimidine[3] *(VI)*. The cyanomethyl
pyrimidine is dissolved in 20 ml of dry ethanol containing a little dry HCl.
After addition of the catalyst (50 mg of 5% palladium on charcoal) it is hydrogenated at ambient temperature and pressure until no more hydrogen
is absorbed. The catalyst is filtered off, then the solution is evaporated to
about 5 ml.

*(4 - Amino - 2 - methyl - 5 - pyrimidinylmethyl) - 5 - (2- hydroxyethyl -4-
methyl)thiazolethione*[4] *(VIII)*. The solution from the last step is neutralized using ammonia (pH paper), and a further 0.1 ml of 30% aqueous
ammonia is added. This is followed by 3-chloro-4-oxopentyl acetate (180
mg, 153 μl, 1 mmol) and CS$_2$ (70 μl). The mixture is allowed to stand overnight and then dissolved in 1 ml of 20% HCl. It is boiled for 15 min, and

the ethanol is removed under reduced pressure. The residue is made alkaline with 30% NaOH to give a crystalline mass. This is centrifuged, and the supernatant, which contains the excess NaOH, is discarded.

Thiamine Chloride Hydrochloride (IX). The residue from the last step is mixed with 1 ml of 5% HCl and 0.4 ml of 30% H_2O_2. The mixture is heated at 50° for 20 min, and aqueous $BaCl_2$ is added until no more precipitate of $BaSO_4$ appears. The precipitate is filtered, and the filtrate is evaporated to a very small volume. This is mixed with 5 ml of dry ethanol and allowed to stand at 4°. Thiamine chloride hydrochloride precipitates gradually as a fine crystalline powder. This is collected by centrifugation.

The labeled thiamine so formed must be further purified by paper chromatography. The crude thiamine is dissolved in a small amount of water, streaked on a Whatman No. 3 paper and chromatographed for 16 hr (descending: isobutanol:acetic acid:water, 4:1:1). The thiamine can be located under ultraviolet light, being seen as a quenching band; It can be located also by converting it to thiochrome by spraying a narrow strip of the paper with alkaline potassium ferricyanide, in this case it will be seen as a bright blue fluorescent spot under UV light (365 nm). The relevant area on the paper is cut out and eluted with dilute HCl (ca. 1 mM). Determination of the radioactivity of an aliquot, and assay for thiamine content (thiochrome method),[5] will give the specific activity of the solution. A further amount of thiamine can be obtained from the mother liquor from the crystallization of the thiamine. The labeled thiamine solution can be frozen and stored at −20° without undue loss. Before storage it is advisable to treat the containers with Repelcote to prevent absorption of thiamine on the glass surface. Using the above procedure, the overall radioactive yield of labeled thiamine is about 20%.

[5] E. E. Edwin, this volume [8].

[16] Thiamine Transport in *Escherichia coli* Crookes

By FRANKLIN R. LEACH and CORALIE A. C. CARRAWAY

The transport reaction can be written:

$$\text{Thiamine}_{\text{external}} \rightleftharpoons \text{thiamine}_{\text{internal}}$$

and is followed using radioactive thiamine; it can be studied at three levels—whole cells, membrane vesicles, and binding proteins. Table I summarizes results of studies on thiamine transport in bacteria. The apparent K_m's are in the range of $10^{-7}\ M$, and both Mg^{2+} and glucose are

TABLE I

THIAMINE TRANSPORT IN VARIOUS MICROORGANISMS

Organism	Fold concentrated	K_m (M)	pH optimum	Activators	Inhibitors	References
Bacillus cereus	—	1.98×10^{-8}	6.5	K^+, Ca^{2+}, Mg^{2+}	Pyrithiamine, EDTA	a
Saccharomyces cerevisiae	10,000	1.8×10^{-7}	4.5	Glucose	Dinitrophenol, iodoacetamide, dicyclohexyl cardodiimide, pyrithiamine, short-chain fatty acids	b, c d
Escherichia coli KG 33	175	8.3×10^{-7}	6.5	Glucose	Pyrithiamine, oxythiamide, thiamine pyrophosphate	e, f
Lactobacillus fermenti	510	4.8×10^{-7}	6.5	Glucose, Mg^{2+} K^+	Iodoacetamide, pyrithiamine	g, h i
Escherichia coli Crookes	1,200	8.3×10^{-7}	8	Mg^{2+}, glucose	Dinitrophenol, azide, cyanide, fluoride, thiamine pyrophosphate, pyrithiamine, oxythiamine	j

[a] I. Tobüren-Bots and H. Hagedorn, Arch. Microbiol. 113, 23 (1977).

[b] A. Iwashima and Y. Nose, Biochim. Biophys. Acta 399, 375 (1975).

[c] A. Iwashima, Y. Wakabayashi, and Y. Nose, Biochim. Biophys. Acta 413, 243 (1975).

[d] A. Iwashima, H. Nishino, and Y. Nose, Biochim. Biophys. Acta 330, 222 (1973).

[e] T. Kawasaki, I. Miyata, K. Esaki, and Y. Nose, Arch. Biochem. Biophys. 131, 223 (1969).

[f] A. Matsuura, A. Iwashima, and Y. Nose, J. Vitaminol. (Kyoto) 18, 29 (1972).

[g] H. Y. Neujahr, Acta Chem. Scand. 17, 1902 (1963).

[h] H. Y. Neujahr, Acta Chem. Scand. 20, 771 (1966).

[i] H. Y. Neujahr, Acta Chem. Scand. 20, 786 (1966).

[j] C. A. C. Carraway, Ph.D. Thesis, Oklahoma State University, 1974.

required; energy poisons, pyrithiamine, and thiamine pyrophosphate (TPP) are effective inhibitors.

Kaback[1] has described the preparation of membrane vesicles and their use in the study of transport.[2] Their use as a model system for studying active transport has also been reviewed by Kaback.[3,4] The transported substrates studied include amino acids, carbohydrates, peptides, sugar phosphates, bases, cations, and metabolic intermediates (e.g. citrate, pyruvate, lactate); there is no reference to studies using vitamins as the transported compounds.[3,4] The study of thiamine transport in membrane vesicles is presented below.

Neu and Heppel[5] developed an osmotic shock procedure for releasing proteins from the periplasmic space of gram-negative bacteria. Rosen and Heppel[6] review and consider the present status of binding proteins. These proteins exist for sulfate, phosphate, glucose 1-phosphate, arabinose, ribose, maltose, riboflavin, thiamine, cyanocobalamin, and various amino acids.

Binding of other vitamins to osmotic shock fluids has been demonstrated for biotin, pyridoxine, and lipoic acid by Griffith and Leach.[7] Mulligan and Snell[8] did not find a periplasmic binding protein for vitamin B_6 in *Salmonella typhimurium*. Snell and colleagues also do not find a pantothenate-binding protein in osmotic-shock fluids (Snell preliminary observations, personal communication).

Henderson *et al.*[9] have solubilized a folate-binding protein from *Lactobacillus casei* membranes. This protein has been purified 100-fold, has a molecular weight of 25,000, and binds 0.85 mol of folate per mole of protein.[10] A thiamine-binding protein of molecular weight 29,000 has been purified 74-fold from crude membranes of *L. casei*.[11] The thiamine-binding proteins derived from *Escherichia coli* are discussed in detail below.

[1] H. R. Kaback, this series, Vol. 22, p. 99.
[2] H. R. Kaback, this series, Vol. 31, p. 698.
[3] H. R. Kaback, *Science* **186**, 882 (1974).
[4] H. R. Kaback, *Biochim. Biophys. Acta* **265**, 367 (1972).
[5] H. C. Neu and L. A. Heppel, *J. Biol. Chem.* **240**, 3685 (1965).
[6] B. P. Rosen and L. A. Heppel, *in* "Bacterial Membranes and Walls" (L. Leive, ed.), Vol. 1, p. 209. Dekker, New York, 1973.
[7] T. W. Griffith and F. R. Leach, *Arch. Biochem. Biophys.* **159**, 658 (1973).
[8] J. H. Mulligan and E. E. Snell, *J. Biol. Chem.* **251**, 1052 (1976).
[9] G. B. Henderson, E. M. Zevely, and F. M. Huennekens, *Biochem. Biophys. Res. Commun.* **68**, 712 (1976).
[10] G. B. Henderson, E. M. Zevely, and F. M. Huennekens, *J. Biol. Chem.* **252**, 3760 (1977).
[11] G. B. Henderson, E. M. Zevely, and F. M. Huennekens, *J. Supramolec. Struct.* **6**, 239 (1977).

Thiamine Transport in Whole Cells of *E. coli* Crookes

Measurement Technique

E. coli Crookes ATCC 8739 were grown in M-9 minimal medium[12] supplemented with 0.2% glucose on a New Brunswick R-10 reciprocating shaker at 37°. Appropriate samples of the cell suspension, in log growth unless otherwise specified, were filtered onto Gelman Metrical GA-6 filters (0.45 μm) and washed with 5 ml of 50 mM Tris, pH 8.0.

[35S]Thiamine was obtained from Amersham-Searle with the specific activity varying from 55 to 158 mCi/mmol. For measurement of uptake, cells suspended in buffer at a final $A_{620 nm}$ of 0.1 were incubated at 37° with 10 mM glucose, 1 mM MgSO$_4$, and 0.1 mg of chloramphenicol per milliliter for 30 min at 37° on a New Brunswick Metabolyte Water Bath Shaker at 100 rpm. [35S]Thiamine was added to give a final concentration of 1.0 μM. Samples were removed at appropriate intervals with a Cornwall syringe set at 0.5 ml and filtered rapidly on Gelman Metrical GA-6 filters on a Bradley multisample filtration apparatus connected to a water aspirator. The filters were washed with 1 ml of buffer, air-dried, and counted in 10 ml of Bray's[13] scintillation cocktail in a Packard Tri-Carb liquid scintillation spectrometer to a 1% standard counting error. Amount of accumulation was quantitated using an appropriately quenched standard of [35S]thiamine. A standard curve of $A_{620 nm}$ versus dry weight of *E. coli* Crookes was used to express the results per milligram of cells.

Characterization of the Reaction

Figure 1 shows the course of thiamine uptake with thiamine concentrations of 0.1, 0.2, and 1.0 μM. The inset shows that at 0.1 μM the uptake is linear for 10 min, whereas the higher concentrations show irregular curves. Lineweaver–Burk plots show two distinct transport components having apparent K_m's of $5 \times 10^{-8} M$ and $8 \times 10^{-7} M$. This behavior can be interpreted as a high-affinity thiamine transport system observed at low thiamine concentrations and a lower-affinity system predominating at higher thiamine concentrations.

Little uptake of thiamine occurs at 4°, and both the initial rate and the capacity (amount taken up when uptake no longer increases with additional incubation) are approximately doubled by increasing the temperature from 25° to 37°. Glucose is the most effective energy source; the rela-

[12] E. H. Anderson, *Proc. Natl. Acad. Sci. U.S.A.* **32**, 120 (1946).
[13] G. A. Bray, *Anal. Biochem.* **1**, 279 (1960).

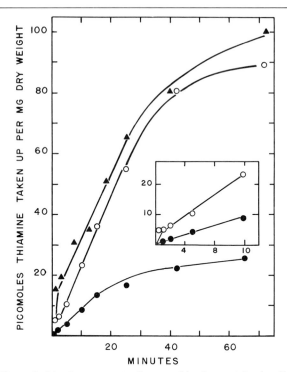

FIG. 1. Effect of thiamine concentration on thiamine uptake by *Escherichia coli* Crookes cells. The thiamine concentrations used are 0.1 μM (●), 0.2 μM (○), and 1.0 μM (△). The inset shows shorter time intervals in greater detail. Thiamine accumulation was determined as described in the measurement section.

tive rates of uptake with 10 mM glucose, 20 mM DL-lactate, and 10 mM pyruvate are $1 : 0.78 : 0.58$, respectively.

Table II summarizes the effects of various inhibitors on the rate of thiamine uptake. Both —SH compounds and —SH-specific reagents reduce uptake. Compounds that prevent energy production also inhibit; thiamine analogs are likewise effective inhibitors.

Effects of Proteolytic Enzymes on Thiamine Transport[14]

A useful technique for studying structure and function is perturbation by exogenous agents and assessment of the effects of the perturbant on the parameter in question. Trypsin treatment is a useful probe for determining protein involvement in thiamine transport.

The cells were grown and prepared as described above. Trypsin solu-

[14] C. A. C. Carraway and F. R. Leach, *Biochem. Biophys. Res. Commun.* **67**, 728 (1975).

TABLE II
EFFECT OF INHIBITORS ON THIAMINE TRANSPORT BY WHOLE CELLS

Inhibition concentration	Relative rate of transport
None	1.00
Azide, 10 mM	0.55
Fluoride, 10 mM	0.75
Cyanide, 10 mM	0.70
2-Mercaptoethanol, 10 mM	0.59
N-Ethylmaleimide, 10 mM	0.78
Thiamine pyrophosphate, 10 pM	0.37
Oxythiamine, 10 pM	0.57
Pyrithiamine, 50 pM	0.10

tion (Worthington Biochemical, 3X crystallized and salt free) was added at final concentrations of 0.1 μg–1 mg/ml, and the mixture is incubated for 15–60 min at 30° or 37° with shaking. The trypsin action was stopped by adding either 2 mM phenylmethanesulfonyl fluoride or soybean trypsin inhibitor (1.5 × the weight of trypsin). Uptake was measured as described above. Cells with different initial levels of thiamine-transport activity responded differently to trypsin treatment. Cells having very high activities (about 4000 pmol per milligram dry weight) showed decreased transport rate and final accumulation under all conditions tested. Mild treatment (30°, 30 min) of cells with normal thiamine-transport activity (about 600 pmol per milligram dry weight) enhanced the velocity of uptake but decreased the amount accumulated at equilibrium. More strenuous treatment (1 mg/ml trypsin for 1 hr at 37°) reduced both parameters. However, cells displaying low initial thiamine transport activity (about 20 pmol per milligram dry weight) showed markedly increased (5- to 8-fold) uptake and accumulation at equilibrium under all but the mildest treatment conditions. These results are consistent with the hypothesis that cells of low and intermediate activity have cryptic transport sites or systems.

Thiamine Transport in Membrane Vesicles

Measurement Technique

The *E. coli* Crookes were grown as described above and harvested by centrifugation. They were treated according to Kaback[1] with lysozyme-EDTA. The final preparation was suspended in 0.5 M potassium phosphate, pH 6.6 (the optimal pH for membrane vesicle thiamine accumulation), and stored in small aliquots at $-15°$. That the vesicle preparations were sealed and retained the ability to concentrate ligands against a gradient is evidenced by their concentrating thiamine 500- to 1200-fold over that in the medium.

The filtration method described above suffered from two drawbacks when applied to membrane vesicles: (1) the small-pore-size filters necessary for retaining the vesicles made filtration slower than desirable for performing uptake studies; (2) the filtrate also contained thiamine-binding activity (which could be thiamine-binding protein). For these reasons the amount of thiamine accumulated by membrane vesicles was determined by equilibrium dialysis. A five-chambered equilibrium dialysis cell from the Chemical Rubber Company with a chamber volume of 1 ml was used with a single thickness of 48 Å-pore-size dialysis tubing (A. H. Thomas N. 3787-D42), which had been boiled in 1 mM EDTA and washed. The suspensions contained 200–300 μg of membrane vesicles protein, 10 mM glucose, 1 mM MgSO$_4$, 114 μg of penicillin G and 0.1 μg of dihydrostreptomycin per milliliter. Stirring was accomplished with a plastic bead inside each chamber while the cell was rocked with a Chemical Rubber Company rocker motor at room temperature (22°). After 12–17 hr (10–12 hr are required to achieve equilibrium under these conditions), samples were taken from each side of the chamber and radioactivity was determined. The net counts per minute on the side containing the protein were converted into picomoles of thiamine accumulated.

Properties of Thiamine Accumulation by Membrane Vesicles

The time course of thiamine accumulation is shown in Fig. 2. There is a rapid accumulation, the maximum level being achieved in 10–15 min. The amount accumulated varied between 600 and 800 pmol/mg with different vesicle preparations. Thiamine accumulation is proportional to membrane vesicle protein concentration from 0.1 to 1.0 mg of protein.

Thiamine accumulation at 22° and at 4° as influenced by varying thiamine concentrations is shown in Fig. 3. At 4° there is a marked (82%) reduction in the amount accumulated and the appearance of a biphasic curve. The curve of thiamine accumulation at 22° is not hyperbolic; a Lineweaver–Burk plot is biphasic. Data obtained at low thiamine concentrations can be extrapolated to give a K_m of 2.5×10^{-8} M. Above 0.1 μM a more complex curve is obtained from which a valid K_m cannot be calculated. This behavior is very similar to that observed for whole cells, in which two transport sites (or systems) are apparent. A high-affinity system operates at thiamine concentrations ≤ 0.1 μM and a lower-affinity site (or system) at higher concentrations.

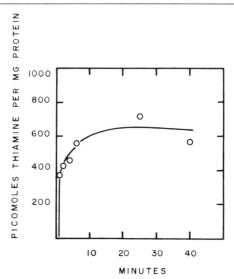

FIG. 2. Time course of thiamine accumulation by membrane vesicles. Thiamine accumulation was determined as described in the measurement section using 0.1 μM thiamine.

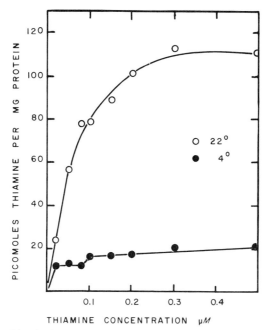

FIG. 3. Effect of incubation temperature on thiamine. The accumulation of thiamine was determined at 22° (○); and 4° (●).

TABLE III
EFFECTS OF METAL IONS, VESICLE TREATMENT, ENERGY SOURCE, AND
INHIBITORS ON THIAMINE ACCUMULATION BY MEMBRANE VESICLES

Source of effect		Relative accumulation
A. Metal ion addition		
10 mM Mg^{2+}		1.0
None		0.19
10 mM EDTA		0.11
10 mM EDTA then Mg^{2+}		0.84
10 mM Ca^{2+}		0.17
B. Energy source, 10 mM		
Glucose		1.0
None		0.41
DL-Lactate		0.46
ATP		0.35
Glycerol		0.33
Acetate		0.33
Succinate		0.29
C. Inhibitors	Concentration, mM	
None		1.0
Azide	10	0.20
Fluoride	10	0.16
Oxamate	10	1.3
TPP	0.001	0.2
D. Vesicle treatment		
Incubate 48 hr at 22°		0.43
Sonic oscillation		0.16
Phenethyl alcohol, 0.25%		0.17
Trypsin, 30 min at 30°		1.3
Trypsin, 60 min at 37°		0.65

Mg^{2+} is required for accumulation; EDTA causes an 89% inhibition, largely reversed by subsequent addition of Mg^{2+}. Ca^{2+} does not stimulate uptake (Table III).

Figure 4 shows the effect of glucose on uptake at varying thiamine concentrations. Without glucose there is 60% reduction in the amount of thiamine accumulated. Table III shows the effects of various energy sources relative to glucose. As found for whole cells, glucose is the most effective energy source tested. Azide and fluoride inhibited accumulation; TPP at a 10-fold greater concentration completely (98%) inhibited accumulation. Oxamate, which is a specific inhibitor of D-lactate dehydrogenase, did not reduce accumulation. Phenethyl alcohol, a chaotropic compound, inhibits accumulation, as does sonic oscillation and incubation for 2 days at 22° (presumably owing to proteolysis by endogenous

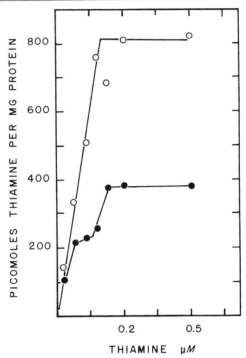

FIG. 4. Effect of glucose on thiamine accumulation by membrane vesicles. Samples of membrane vesicles were incubated as described in the measurement section with (○) and without (●) 10 mM glucose.

proteases). Mild trypsin treatment (30 min, 30°) increases accumulation, while more rigorous treatment (60 min, 37°) decreases it. These observations are consistent with a trypsin unmasking of thiamine-transport system(s) and destruction of the system(s) with more strenuous treatment.

Thiamine-Binding Proteins

Measurement Techniques

The apparatus described by Paulus[15] was used to measure the binding of thiamine by ultrafiltration. The ultrafiltration cells were constructed by Heinz Hall in the Oklahoma State University Chemistry–Physics shop according to the plans of Paulus shown in Fig. 5. Small filter disks were

[15] H. Paulus, *Anal. Biochem.* **32**, 91 (1969).

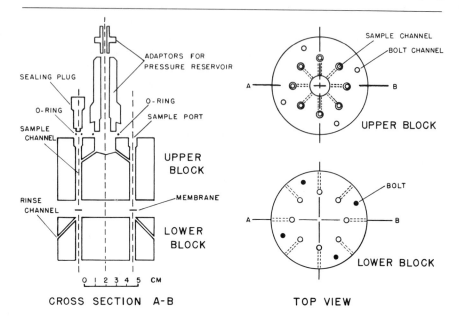

Fig. 5. Plans for the Paulus ultrafiltration cell [H. Paulus, *Anal. Biochem.* **32,** 91 (1969)]. This apparatus is commercially available from MRA Corporation, 1058 Cephas Rd., Clearwater, Florida.

prepared from 48 Å pore size dialysis tubing. Samples of 0.1 ml total volume containing 20 μg of protein, 0.1 M potassium phosphate buffer, pH 6.8, 1 mM MgCl$_2$, and 1 μM [^{35}S]thiamine were filtered using 40 psi nitrogen.

After filtration was completed, 15 ml of ethylene glycol were injected through the rinse channels to wash the bottoms of the disks. The pressure was then released, and the disks were removed with forceps and placed into glass counting vials. Bray's liquid scintillation cocktail (10 ml) was added, and the radioactivity was determined on duplicate samples to a 1% counting error.

Alternatively, equilibrium dialysis was performed as described in the section on measurement of thiamine transport in membrane vesicles.

Purification[7]

E. coli Crookes were grown aerobically in three 12-liter batches (New Brunswick fermentor, stirred at 500 rpm, and aerated at 12 liter/min) on medium M-9 with 0.2% glucose in early stationary phase. The cells were harvested in a continuous-flow, Sharples refrigerated centrifuge.

Osmotic Shock Procedure

About 550 g of *E. coli* cell paste were suspended in 800 ml of 20% sucrose, 30 mM Tris, and 1 mM EDTA at pH 8.0; the mixture was incubated with stirring at room temperature for 10 min, and then the cells were harvested by centrifugation (Sorvall RC-2; 13,000 rpm; 15 min). The cell paste was smeared on the inside of a 2-liter Erlenmeyer flask, 1 liter of cold (4°) 0.5 mM MgCl$_2$ was added, and the cells were suspended by vigorous shaking. After standing for 10 min, the suspension was centrifuged (RC-2; 16,000 rpm; 30 min) and the protein of the supernatant fluid was concentrated by ultrafiltration using an Amicon UM-10 filter.

Purification Procedure

When the concentrated shock fluid was fractionated with (NH$_4$)$_2$SO$_4$, the thiamine-binding activity precipitated between 430 and 600 g of the salt per liter. Approximately 400 mg of protein were applied to a DEAE-cellulose 1.3 × 15 cm chromatographic column, washed with 20 ml of 10 mM Tris, 10 mM MgCl$_2$, and 0.5 mM mercaptoethanol, pH 8.0, and was eluted with a 200-ml linear 0 to 0.15 M NaCl gradient in the above buffer. The fractions containing thiamine-binding activity were pooled, dialyzed against 10 mM potassium phosphate buffer, pH 7.5, concentrated by Diaflo filtration, and chromatographed on a hydroxyapatite column (1.5 × 20 cm BioGel Cellex-HPT and eluted with a linear 0.05 to 0.2 M potassium phosphate buffer gradient, pH 7.5). The fractions with thiamine-binding activity were combined, then dialyzed and concentrated as above, and chromatographed on a BioGel P-150 column (1.5 × 60 cm). Table IV shows that application of this purification scheme resulted in a 230-fold purification of the thiamine-binding protein.

TABLE IV

PURIFICATION OF THIAMINE BINDING PROTEIN[a, b]

Fraction	Volume (ml)	Protein (mg/ml)	Total activity (pmol × 10^{-3})	Specific activity (pmol/mg)	Purification
Crude shock fluid	250	4.0	80	80	1
(NH$_4$)$_2$SO$_4$ (0.7–0.9 saturation)	29	12.7	48	130	1.6
DEAE	16	1.4	27	1,200	15.0
Hydroxyapatite	20	0.26	25	4,800	60
Gel filtration	7.7	0.15	21	18,200	230

[a] Binding assays were performed at 1 μM substrate by ultrafiltration.
[b] T. W. Griffith and F. R. Leach, *Arch. Biochem. Biophys.* **159**, 658 (1973).

Assay of Purity

Polyacrylamide gel electrophoresis in 7% polyacrylamide yielded a major band of protein containing 92% of the protein and two smaller bands containing 6 and 2%. The major band binds thiamine; when the gel was immersed in 1 μM labeled thiamine solution after electrophoresis, washed, sliced into 25 equal parts, and counted, the single peak of radioactivity was coincident with the major protein staining band.

Properties

pH Optimum. The optimum binding pH was determined using a 0.15 M citrate–phosphate–Tris buffer (50 mM in each) over the pH range of 2–12. The thiamine binding protein exhibited a definite and rather sharp pH optimum at 9.2.

Affinity. A hyperbolic saturation curve for thiamine binding was obtained. A double-reciprocal plot yielded a K_D of 5×10^{-8} M.

Exchange. If the thiamine binding protein is involved in transport, the binding must be reversible. Thiamine-binding protein was equilibrated with 0.5 μM labeled thiamine for 11 hr at 4°. A 1000-fold excess of unlabeled thiamine was then added to both sides of the equilibrium dialysis cell; approximately 80% of the labeled thiamine was displaced within 2 hr. At 10 hr the labeled thiamine was distributed equally on both sides of the equilibrium dialysis cell.

Inhibition of Binding by TPP. TPP competitively inhibits thiamine binding to the binding protein. The K_i was calculated from the binding data by assuming all reactions to be in equilibrium. The K_i was 0.5 ± 0.1 μM.

Effect of Other Inhibitors. The effect of various compounds on thiamine binding was determined (Table V). Such inorganic salts as potassium phosphate, sodium pyrophosphate, and magnesium chloride had little effect on thiamine binding. Similarly 2-mercaptoethanol, *N*-ethylmaleimide, and iodoacetic acid had no effect. Some inhibition by the pyrimidines, *viz.* thymine and cytosine, occurred at 100 μM.

Molecular Weight. The molecular weight of the thiamine-binding protein was estimated to be 35,000 from its mobility in 6% polyacrylamide gel electrophoresis in the presence of 0.1% sodium dodecyl sulfate. When the molecular weight was determined by the method of Davis,[16] a value of 43,000 was obtained. Chromatography on a BioGel P-150 column gave a molecular weight of 41,000. The average of these three molecular weights was 40,000.

[16] B. J. Davis, *Ann. N. Y. Acad. Sci.* **121**, 404 (1964).

TABLE V

EFFECT OF VARIOUS COMPOUNDS ON THIAMINE BINDING BY PROTEIN[a]

Additions	Percent of control
None (control)	100
MgCl$_2$, 1 mM	96
KPO$_3^{2-}$, 0.5 M, pH 7.5	94
Na pyrophosphate, 1 mM	101
2-Mercaptoethanol, 0.1 M	94
N-Ethylmaleimide, 10 mM	96
Iodoacetic acid, 10 mM	85
Thymine, 10 μM	104
Thymine, 100 μM	81
Cytosine, 10 μM	104
Cytosine, 100 μM	87

[a] Thiamine-binding protein was exposed to the compounds listed before assaying via the Paulus binding assay at 2 μM thiamine. All values shown are the average of duplicate experiments and are listed as percent of the control.

TABLE VI

AMINO ACID COMPOSITION OF THIAMINE-BINDING PROTEINS

Amino acid	Escherichia coli[a]	Lactobacillus casei[b]
Asp	24	10
Thr	12	12
Ser	11	16
Glu	25	13
Pro	11	10
Gly	13	28
Ala	23	37
Val	16	28
Met	2	—
Ile	3	25
Leu	20	34
Tyr	8	9
Phe	9	19
Lys	15	11
His	1	4
Arg	5	6
Cys	1	—
Trp	ND[c]	7
NH$_3$	31	—

[a] T. W. Griffith, Ph.D. Thesis, Oklahoma State University, 1971.

[b] G. B. Henderson, E. M. Zevely, and F. M. Huennekens, J. Supramolec. Struct. **6**, 239 (1977).

[c] ND, not determined.

TABLE VII

COMPARISON OF *Escherichia coli* THIAMINE BINDING PROTEINS

Group	E. coli	K_D (M)	pH optimum	Specific activity (pmol/mg)	Fold purified	Electrophoretic purity	Molecular weight	Mole thiamine bound per mole protein
Hayashi[a-c]	70–23	2×10^{-8}	7.2	25,900	262	60%[i]	34,000	0.83
Leach[d,e]	Crookes	5×10^{-8}	9	18,200	230	92%[i]	40,000	0.73
Nose[f-h]	KG 33	0.2×10^{-9}	8–9	4,925	90	Single band	39,000	0.19

[a] T. Nishimune and R. Hayashi, Biochim. Biophys. Acta **328**, 124 (1973).
[b] T. Nishimune and R. Hayashi, Biochim. Biophys. Acta **244**, 573 (1971).
[c] T. Nishimune and R. Hayashi, Anal. Biochem. **53**, 282 (1973).
[d] T. W. Griffith, C. A. Carraway, and F. R. Leach, Fed. Proc., Fed. Am. Soc. Exp. Biol. **30**, II, 1115 (1971).
[e] T. W. Griffith and F. R. Leach, Arch. Biochem. Biophys. **159**, 658 (1973).
[f] A. Iwashima, A. Matsuura, and Y. Nose, J. Bacteriol. **108**, 1419 (1971).
[g] A. Matsuura, A. Iwashima, and Y. Nose, Biochem. Biophys. Res. Commun. **51**, 241 (1973).
[h] A. Matsuura, A. Iwashima, and Y. Nose, this series, vol. 34, p. 303.
[i] Calculated as the amount of protein in band that binds thiamine.

Amino Acid Composition. Table VI compares the amino acid composition of the *E. coli* thiamine-binding protein released by osmotic shock and the *L. casei* thiamine-binding protein, which is detergent solubilized from the membrane. As would be expected, there is a greater proportion of hydrophobic amino acid residues in the *L. casei* thiamine-binding protein.

Comparison of Results with Those of Others

Three groups have purified the thiamine-binding protein from *E. coli.* Table VII summarizes some of the major parameters. The specific activities of the preparations of Hayashi's and Leach's groups are comparable; when the difference in molecular weight is considered, the proteins bind 0.83 and 0.73 mol of thiamine per mole of protein. Although the thiamine-binding protein obtained by Nose's group gives a single band on polyacrylamide gel electrophoresis, it binds only 0.19 mol of thiamine per mole of protein. This suggests either considerable inactivation of the binding site or the presence of an inactive protein that is not removed by the purification. The electrophoretic analysis of the Hayashi preparation shows two bands with a 60:40 distribution, the first one binding thiamine. This is inconsistent with the measured binding of thiamine on the assumption that there is one binding site per molecule of protein. Our data show greater purity on electrophoresis, but lower binding; this can be accounted for by the presence of inactive protein.

This is manuscript 3449 of the Oklahoma Agricultural Experiment Station.

[17] Assay of Thiamine-Binding Protein

By TAKAHIRO NISHIMUNE and RYOJI HAYASHI

The assay procedure described below is a modification of the equilibrium dialysis method usually used for the detection of ligand-binding activities. The advantage of the modified method lies in the speed (3 hr) in which the dialysis equilibrium can be attained without decreasing the linearity of the protein–activity relationship. In the usual equilibrium dialysis methods,[1,2] radioactive ligands are placed outside the dialysis tube, with the binding protein inside. When samples of thiamine-binding protein

[1] J. R. Piperno and D. L. Oxender, *J. Biol. Chem.* **241**, 5732 (1966).

[2] Y. Anraku, *J. Biol. Chem.* **243**, 3116 (1968).

are from *Escherichia coli* strains, there is substantially no specific modification activity for the ligand. Therefore the protein sample and the thiamine can be put together in the dialysis tubing so that the concentration difference of the ligand between the inner and outer solutions of the dialysis tube becomes greater with the shorter period of time necessary to attain equilibrium.[3] In the resin methods,[4,5] the ligand is also placed together with the protein and the remaining ligand is taken out of the reaction system by centrifugation. However, the scope of application of this method is so far limited to inorganic ions, such as sulfate and phosphate. In the ultrafiltration method[6] the ligand is likewise mixed with the protein and the excess ligand is removed from the reaction mixture by ultrafiltration. This method is applied to the assay of thiamine-binding protein.[7] However, the method described below can be employed with the shorter time necessary for the assay of samples of large numbers (e.g., 30 or more) at one time without any special apparatus. The most important advantage of the ultrafiltration method is presumably that ligand concentrations exceeding the concentration of protein by several orders of magnitude can be attained easily. This can be accomplished because the absolute amount of ligand bound to the protein is measured by this method instead of the difference of ligand concentrations between the inner and the outer solution of the equilibrium dialysis method. Neither a serological method[8] nor a Millipore-filter adsorption method[8] has been developed for thiamine-binding protein.

Materials

Radioactive thiamine (10 mCi/mmol), 1 mM in 20 mM acetate buffer, pH 4.5

Cellophane tube [diameter 6.4 mm ($\frac{1}{4}$ in) × 12 cm] boiled in 10 mM EDTA, pH 8, and then boiled twice in H_2O for a few minutes each time

Potassium phosphate buffer, 0.1 M, pH 7.2, containing 0.5 mM β-mercaptoethanol

Procedure. To 1.0 ml of sample solution containing binding activity of 20 pmol to 2 nmol (in 0.1 M potassium phosphate buffer of pH 7.2 containing 0.5 mM β-mercaptoethanol) is added 10 μl of 1 mM radioactive thiamine. The mixture is incubated at 37° for approximately 15 min with gentle shaking. At the end of the incubation period, 0.8 ml of the reaction

[3] T. Nishimune and R. Hayashi, *Anal. Biochem.* **53**, 282 (1973).
[4] A. B. Pardee, *J. Biol. Chem.* **241**, 5886 (1966).
[5] N. Medveczky and H. Rosenberg, *Biochim. Biophys. Acta* **211**, 158 (1970).
[6] H. Paulus, *Anal. Biochem.* **32**, 91 (1969).
[7] T. W. Griffith and F. R. Leach, *Arch. Biochem. Biophys.* **159**, 658 (1973).
[8] A. D. Riggs and S. Bourgeois, *J. Mol. Biol.* **34**, 361 (1968).

mixture is transfered into dialysis tubing, and dialysis is continued for 3 hr at 37° against 31 ml of 0.1 M potassium phosphate buffer, pH 7.2 with 0.5 mM β-mercaptoethanol on a shaker. Dialysis can be conveniently performed in a 100-ml polyethylene bottle with a tight lid by shaking at 120–140 strokes/min. The cellophane tubing is hung from the top of the bottle by cotton threads or by the remaining tubing itself. After equilibrium is attained under the above conditions, 0.5 ml of the inner solution and the same amount of the outer solution are pipetted into liquid scintillation vials, and radioactivities are counted after adding 0.5 ml of Hyamine hydroxide and 10 ml of dioxane-based cocktail. The amount of substrate bound to protein is calculated by subtracting the radioactivity outside the tubing from that inside.

Comment. Under the above conditions, the total amount of thiamine added to a sample is ca. 8.0 nmol and the concentration in the outer solution is 0.25 μM in the case when the amount of thiamine bound to protein is negligible. There is a linear relationship between binding activity and protein concentration from 0 to approximately 2 nmol with only a slight deviation from linearity between 2 and 3 nmol.[3] When it is necessary to use a large excess (>50:1) of substrate concentration against the protein, the binding activity of less than 250 pmol can be measured under the above conditions. When dialysis at 4° is preferable, the dialysis equilibrium can be attained after 5 hr.[3] Binding activity measured in a crude sample sometimes has a lower value than expected because the sample contains thiamine phosphates that compete with the radioactive thiamine. This can be avoided by dialyzing the crude sample against an appropriate buffer prior to the activity assay. In addition to the structural analogs of thiamine, acidic pH strongly reduces the binding activity. Alkaline pH is harmful to the stability of thiamine. It is also necessary to keep a buffer concentration when dilute thiamine solutions (<3 μM) are handled to avoid the nonspecific adsorption of the ligand to glass surfaces. However, the temperature difference between 37° and 4° has little effect on the binding activity. In the case when the ultrafiltration method is applied, care has to be taken to exclude artifacts from the results obtained.[9]

[9] E. Heyde, *Anal. Biochem.* **51**, 61 (1973).

[18] Isolation and Characterization of *Escherichia coli* Mutants Auxotrophic for Thiamine Phosphates

By Hideo Nakayama and Ryoji Hayashi

Principle

Mutants of *Escherichia coli* that can grow on thiamine phosphates but not on the free form of thiamine include two types: one requires thiamine pyrophosphate (TPP) and the other thiamine monophosphate (thiamine-P) or TPP. Mutants showing such growth responses have never been obtained from a parental strain prototrophic for thiamine, whereas they can readily be produced by further mutation of a mutant strain auxotrophic for thiamine. This indicates that the requirements for either thiamine-P or TPP are manifested as the result of additional mutations arising in a cell where *de novo* synthesis of thiamine-P has been blocked by the first mutation.[1,2]

It has been shown that in the wild type of *E. coli* exogenous thiamine is accumulated intracellularly in the form of TPP; this is one of the reasons why thiamine has been thought to serve as a true precursor of TPP. Incubation of thiamine with the cell suspensions results in the accumulation of TPP in both the wild type and the thiamine auxotroph, thiamine-P in TPP auxotroph, and the free form of thiamine in thiamine-P auxotroph. The stepwise phosphorylation of thiamine to TPP can be confirmed by enzymic experiments with cell-free extracts of the parental and mutant organisms. In the presence of adenosine triphosphate (ATP), crude extract of the thiamine auxotroph is capable of catalyzing both reactions, from thiamine to thiamine-P and from thiamine-P to TPP. On the other hand, crude extracts of mutant cells are able to catalyze only one of these reactions, i.e., the enzyme from thiamine-P auxotroph converts thiamine-P to TPP but not thiamine to thiamine-P, whereas the enzyme from TPP auxotroph converts thiamine to thiamine-P but not thiamine-P to TPP. Combination of extracts of both types of mutant results in the stepwise formation of TPP from thiamine. Experimental results with the newly isolated mutants indicate that in *E. coli* the free form of thiamine is not involved in the *de novo* synthesis of TPP, but thiamine-P, an exclusive product formed by the reaction between 2-methyl-4-amino-5-hydroxymethylpyrimidine

[1] H. Nakayama and R. Hayashi, *J. Bacteriol.* **109**, 396 (1972).

[2] H. Nakayama and R. Hayashi, *J. Bacteriol.* **112**, 1118 (1972).

METHODS IN ENZYMOLOGY, VOL. 62

pyrophosphate (hydroxymethylpyrimidine-PP) and 4-methyl-5-β-hydroxyethylthiazole monophosphate (hydroxyethylthiazole-P), is directly phosphorylated to form TPP. Exogenous thiamine, on the other hand, is converted to TPP via the intermediate formation of thiamine-P.

The growth of the TPP auxotroph is inhibited by either thiamine or thiamine-P, and the growth of the thiamine-P auxotroph is similarly inhibited by thiamine. In addition, TPP uptake of cells of TPP auxotroph is inhibited by either thiamine or thiamine-P, and the thiamine-P auxotroph by thiamine. Although the TPP uptake of a reverse mutant, selected for prototrophy from the thiamine-P auxotroph, is inhibited by thiamine, its growth is neither inhibited by thiamine nor accelerated by TPP. This indicates that the presence of both the original and either of the two mutations in thiamine phosphate auxotrophs might be required for the growth inhibition by thiamine or thiamine-P.

These phenomena can be explained if one assumes that the growth inhibition is primarily due to the inhibition of the entry of TPP into the cells.[3] In the case of the TPP auxotroph, for example, the phosphorylation steps of hydroxymethylpyrimidine-P and thiamine-P are both blocked. This mutant requires exogenous TPP for growth but cannot transport TPP into the cells when excess thiamine (or thiamine-P) is present in the medium. These results support the assumption that thiamine, thiamine-P, and TPP share a common transport system in *E. coli*.

Procedures

Reagents

Thiamine hydrochloride, thiamine monophosphate (thiamine-P), and thiamine pyrophosphate (TPP), purchased from Sigma Chemical Co.

2-Methyl-4-amino-5-hydroxymethylpyrimidine (hydroxymethylpyrimidine), supplied by Takeda Chemical Industries, Osaka, Japan

2-Methyl-4-amino-5-hydroxymethylpyrimidine monophosphate (hydroxymethylpyrimidine-P) and 2-methyl-4-amino-5-hydroxymethylpyrimidine pyrophosphate (hydroxymethylpyrimidine-PP), prepared according to the procedure of Lewin and Brown[4]

4-Methyl-5-β-hydroxyethylthiazole (hydroxyethylthiazole), supplied by Takeda Chemical Industries, Osaka, Japan

4-Methyl-5-β-hydroxyethylthiazole monophosphate (hydroxyethylthiazole-P), prepared according to the procedure of Miyagawa[5]

[3] H. Nakayama and R. Hayashi, *J. Bacteriol.* **118,** 32 (1974).
[4] L. M. Lewin and G. M. Brown, *Arch. Biochem. Biophys.* **101,** 197 (1963).
[5] K. Miyagawa, *Bitamin* **20,** 255 (1960).

Media. The minimal medium of Davis and Mingioli[6] is used as the basal medium. Solid medium contains 1.5% washed agar. Thiamine-P and TPP are sterilized by filtration with a type HA filter (Millipore Co.). Other supplements, which include thiamine, hydroxymethylpyrimidine, and hydroxyethylthiazole, are sterilized by autoclaving.

Bacterial Strains. *E. coli* W (ATCC 9637) and its derivatives are used. Mutant 70-23 is used as a parental strain. The biochemical lesion of this organism is at the stage in the phosphorylating step of hydroxymethylpyrimidine-P; hydroxymethylpyrimidine-PP is not formed from the monophosphate. Thiamine is required by the organism as the intact molecule because hydroxymethylpyrimidine-PP is not permeable through the cell membrane.[7,8]

Isolation of Mutants Auxotrophic for Thiamine Phosphates. Thiamine-P and TPP auxotrophs are isolated by treatment of mutant 70-23 with N-methyl-N'-nitro-N-nitrosoguanidine by the method of Adelberg, Mandel, and Chen.[9] The treated cells are cultured at 37° for 24 hr in a minimal medium supplemented with 0.1 μM TPP. Prototrophic and parental types of cells are eliminated in a minimal medium containing 0.1 μM thiamine and 200 U of penicillin per milliliter. Mutants that can grow on TPP, but not on thiamine, are selected by the replica plating method of Lederberg and Lederberg.[10] Two types of mutants showing such growth responses can be isolated, but they show different growth response to thiamine-P, one (70-23-102, referred to as thiamine-P auxotroph) responds to both thiamine-P and TPP, whereas the other (70-23-107, referred to as TPP auxotroph) responds to TPP but not to thiamine-P.

Thiamine-P and TPP auxotrophs of *E. coli* may be produced not only from a phosphohydroxymethylpyrimidine kinase-deficient mutant but also from mutants that have lost the ability to perform another indispensable enzymic reaction for thiamine-P biosynthesis.

Isolation of Reverse Mutant Strains from the Thiamine-P Auxotroph. If double mutations, one pertaining to the phosphorylation of hydroxymethylpyrimidine-P (original block) and the other to the biosynthesis of thiamine-P from thiamine, had occurred in the thiamine-P auxotroph, and if the original block were removed by a reverse mutation, the reverse mutant should be a thiamine prototroph although it would still lack thiamine monophosphokinase activity. Such a strain has not been ob-

[6] B. D. Davis and E. S. Mingioli, *J. Bacteriol.* **60,** 17 (1950).
[7] H. Nakayama, *Bitamin* **10,** 359 (1956).
[8] H. Nakayama and R. Hayashi, *J. Vitaminol.* (*Kyoto*) **17,** 64 (1971).
[9] E. A. Adelberg, M. Mandel, and G. C. C. Chen, *Biochem. Biophys. Res. Commun.* **18,** 788 (1965).
[10] J. Lederberg and E. M. Lederberg, *J. Bacteriol.* **64,** 399 (1952).

tained directly from a wild-type organism, whereas it can be readily produced as a reverse mutant from the thiamine-P auxotroph. For this purpose, the thiamine-P auxotroph grown in the minimal medium supplemented with 20 nM TPP is diluted 100 times in sterile water. A 10-ml amount of the bacterial suspension is exposed to a 20-W Germicidal lamp (Toshiba Electric Co.) in an open petri dish in a dark room until surviving cells decrease to 10^{-4} per milliliter. A 1.0-ml amount of the irradiated suspension is transferred to the minimal medium containing thiamine-P and incubated at 37°C for 16 hr. A portion of this culture is streaked on a minimal agar plate, and colonies that grow rapidly on the plate are isolated at random. Pure cultures are obtained by reisolating the colonies on the plate that contained thiamine-P. The growth of these revertants on the minimal medium is comparable to that of the W strain. Although both the original (W strain) and parental (mutant 70-23) organisms are able to take thiamine into the cells and then convert it into TPP, free thiamine is accumulated in the revertant cells; they cannot convert exogenous thiamine into thiamine-P. In the reaction mixture containing hydroxymethylpyrimidine and hydroxyethylthiazole, neither thiamine-P nor TPP is formed by mutant 70-23 cells, whereas TPP is synthesized in significant amount by the cells of W strain and the revertants.

Paper Chromatographic and Bioautographic Assay Method for Thiamine, Thiamine-P, and TPP. A 10-μl portion of samples is spotted on Toyo filter paper No. 50. Paper chromatograms are developed by the ascending technique. The solvents used are (A) isopropyl alcohol–0.3 M (pH 7.0) potassium phosphate buffer (2:1, v/v), and (B) isopropyl alcohol–0.1 M (pH 5.0) acetate buffer–water (7:1:2, v/v/v). The R_f values of thiamine, thiamine-P, and TPP are 0.60, 0.24, 0.08, respectively, with solvent A; they are 0.55, 0.30, 0.18, respectively, with solvent B. Bioautography is used to estimate the extent and the rate of formation of thiamine-P from thiamine and that of TPP from thiamine-P. A 300-ml portion of the minimal medium containing 1.5% agar is melted and cooled in a water bath to 50°C. A 3.0-ml amount of bacterial suspension (ca. 10^9 cells/ml) is added to the agar, which is then poured into a sterile glass dish (300 by 200 by 5 mm). The developed chromatogram is air-dried and placed in contact with the surface of the solid medium. The chromatogram is removed after 10 min, and the plate is covered and incubated at 37° for 20 hr. The inoculum used to seed the plate in these experiments consisted of the suspension of washed cells taken from a 24-hr culture of the test organism grown on the minimal agar slant containing 0.05 μM TPP. Mutant 70-23 is used as a test organism to detect thiamine, thiamine-P, and TPP. The size (average diameter) of the growth zones of this organism, given each thiamine compound, is proportional to the amount of the compounds spotted on the paper. Therefore, it is possible to estimate the amount of thiamine, thiamine-P, or TPP by comparing the size of the growth zones with the size of those derived from a standard amount of the corresponding thiamine compound spotted on the chromatogram. The formation of thiamine-P and TPP can also be confirmed by bioautography with thiamine-P auxotroph and TPP by that with TPP auxotroph.

Assay of TPP Synthesis by Whole Cells. In *E. coli* both hydroxymethylpyrimidine and hydroxyethylthiazole, supplied from the outside, are taken up into the cells and converted to TPP through the *de novo* synthetic pathway. The assay system contains: cell suspension (10 mg/ml, dry weight), 0.3 ml; 1mM hydroxymethylpyrimidine, 0.1 ml; 1 mM hydroxyethylthiazole, 0.1 ml; 10% glucose, 0.1 ml; 1 M potassium phosphate buffer (pH 7.0), 0.2 ml; 0.1 M MgCl$_2$, 0.1 ml; and water, 0.1 ml. After the reaction has proceeded at 37° for 1 hr in a shaking water bath, each tube is adjusted to pH 4.5 with 1 N HCl and heated at 80° for 20 min. The supernatant fraction obtained with centrifugation is spotted on paper. Thiamine and its phosphate esters are assayed by paper bioautography with solvent B. Approximately, 0.2 nmol per milligram dry weight of TPP is formed in cells of W and revertants from thiamine-P auxotroph, but neither thiamine, nor the phosphate esters is formed in mutant 70-23 cells. When thiamine is used in place of the pyrimidine and thiazole moieties, both thiamine-P and TPP are formed in W and mutant 70-23, but not in revertants. In cells of the thiamine-P or TPP auxotroph, neither thiamine-P nor TPP is formed with the former organism and thiamine-P but TPP is formed with the latter one.

TPP Synthesis by Enzyme Preparations. In the system for TPP biosynthesis from the pyrimidine and thiazole moieties, the following enzymic steps are involved: hydroxymethylpyrimidine → hydroxymethylpyrimidine-P (hydroxymethylpyrimidine kinase, EC 2.7.1.49), hydroxymethylpyrimidine-P → hydroxymethylpyrimidine-PP (phosphohydroxymethylpyrimidine kinase, EC 2.7.4.7), hydroxyethylthiazole → hydroxyethylthiazole-P (hydroxyethylthiazole kinase, EC 2.7.1.50), hydroxymethylpyrimidine-PP + hydroxyethylthiazole-P → thiamine-P (thiamine-phosphate pyrophosphorylase, EC 2.5.1.3), thiamine-P → TPP (thiamine monophosphatekinase EC 2.7.4.16). ATP serves as a phosphate donor for each of the enzymes except for thiamine-phosphate pyrophosphorylase. However, if ATP is supplied at the required concentrations for these enzymes, the activity of thiamine-phosphate pyrophosphorylase is strongly inhibited. Therefore, it is almost impossible to estimate TPP formation from the pyrimidine and thiazole moieties employing enzyme preparations that contain five different enzymes. Moreover, the four enzymes involved in the *de novo* synthesis of thiamine-P from the pyrimidine and thiazole moieties are repressible and derepressible in *E. coli,* depending upon the intracellular thiamine level.[11] The formation of thiamine monophosphokinase (EC 2.7.1.89) might be subjected to the same control mechanism. Therefore, cells grown on a limiting amount of thiamine or its phosphate esters should be used as an enzyme source.

Assay of Thiamine Monophosphokinase. A 5.0-g amount of the cell paste, harvested from the culture grown on the minimal medium with 10 nM TPP for mutants auxotrophic for thiamine phosphates under aerobic conditions, is suspended in 15.0 ml of 20 mM potassium phosphate buffer (pH 7.0) that contains 1 mM MgCl$_2$ and 1mM β-mercaptoethanol. The cells in the suspension are disrupted by sonic oscillation (20 kc) for 3 min, and the resulting suspension is centrifuged at 10,000 $\times$ g for 10 min. The supernatant fraction is dialyzed against 2 liters of the same buffer containing 1 mM MgCl$_2$ and 0.1 mM β-mercaptoethanol with two changes. The dialyzed fluid, referred to as a crude extract, containing approximately 40 mg of protein per milliliter, is used to measure the phosphorylating activity of thiamine.

The complete assay system contains: crude extracts, 0.25 ml (10 mg of protein); 0.1 mM thiamine, 0.1 ml; 0.1 M ATP, 0.1 ml; 1 M potassium phosphate buffer (pH 7.0), 0.1 ml; 0.1 M MgCl$_2$, 0.1 ml; and water, 0.35 ml. The reaction is carried out at 37° for 30 min. After the reaction is over, the reaction mixture is adjusted to pH 4.5 by 0.2 N HCl, heated at 90° for 5 min, and centrifuged. A 10-μl amount of the resulting supernatant fraction is spotted on the paper and developed by solvent B. Thiamine and thiamine-P are located by bioautography with mutant 70-23. The activity of this enzyme is lacking in the thiamine-P auxotroph and revertants derived from the same organism.

Assay of Thiamine Monophosphatekinase. This enzyme is assayed by the procedure reported by Nishino *et al.,*[12] who found that either potassium or ammonium ion markedly stimulated the activity. However, enzyme activity is much less apparent than that of thiamine monophosphokinase in an *in vitro* assay, although it is fully active in an *in vivo* assay. A small amount of TPP is contained in the preparation of crude extracts obtained from cells grown with a limiting amount of TPP. Such TPP might be enzyme bound and is difficult to remove by dialysis even after ammonium sulfate fractionation under alkaline conditions. Moreover, when reaction mixtures containing crude extracts, but not substrate, are incubated at

[11] T. Kawasaki, A. Iwashima, and Y. Nose, *J. Biochem. (Tokyo)* **65**, 407 (1969).
[12] H. Nishino, A. Iwashima, and Y. Nose, *Biochem. Biophys. Res. Commun.* **45**, 363 (1971).

37°, the TPP content gradually decreased, with concomitant formation of thiamine-P. Such an apparent phosphatase activity may obscure TPP formation from thiamine-P in an *in vitro* system. These difficulties are overcome to some extent by the following method. The cell paste, harvested from the culture grown on the minimal medium with 10 nM TPP for mutants auxotrophic for thiamine phosphates under aerobic conditions, is washed twice with 10 mM Tris·HCl buffer (pH 7.4) containing 30 mM NaCl and subjected to osmotic shock by the procedure of Nossal and Heppel[13] with an exception that the sucrose-treated cells are dispersed in distilled water instead of $MgCl_2$ solution. Shocked cells are harvested by centrifugation. A 5.0-g amount of the cell paste is suspended in 15.0 ml of 10 mM Tris·HCl buffer containing 1 mM $MgCl_2$ and 1 mM β-mercaptoethanol. The cells in the suspension are disrupted by a sonic oscillator (20 kc) for 2 min and centrifuged at 10,000 × g for 10 min. The extracts, containing 40 mg of protein per milliliter, is used as an enzyme source. The extracts prepared from shocked cells contain less TPP and a decreased level of the phosphatase activity than those prepared from intact cells. The activity of this enzyme cannot be detected in the TPP auxotroph.

A Criterion for Analyzing Metabolic Pathways

The following diagram shows the cases where the biosynthesis of an essential end product E is blocked at different steps in the series of reactions. When compound Z is supplied from outside the cell, E is shown to be alternatively formed via intermediate formation of Y.

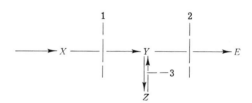

A compound found to accumulate under the condition where step 2 was blocked by a mutation could occupy the position of compound Y, or both Y and Z if the latter was interconvertible with compound Y. If a block existed either at step 1 or at steps prior to the biosynthesis of compound X, the organism should grow on Z as well as Y. Therefore, there is really no way to choose an alternative interpretation as to whether the active compound serves as a true precursor or is a substance capable of con-

[13] N. G. Nossal and L. A. Heppel, *J. Biol. Chem.* **241**, 3055 (1966).

version to a true precursor. Mutants having a mutational block at step 3, responding to Y but not Z, cannot be obtained from an organism capable of forming Y through the *de novo* synthetic pathway, whereas they could be obtained from a mutant organism whose metabolic sequence has been blocked prior to the formation of Y. If the original block has been deleted from the doubly blocked mutant by a reverse mutation or by transduction from the corresponding wild-type allele, the mutant should be prototrophic for compound Z, but it should be still lacking in the ability to go from Z to Y. Consequently, the successful isolation of such mutant can lead to the conclusion that Y, a compound serving as a true precursor of E, could occupy a position where compound Z joins the *de novo* synthetic pathway.

[19] A Radiometric Assay for Thiamine Pyrophosphokinase Activity[1]

By C. J. Gubler

Thiamine pyrophosphokinase (TPK)[2] (ATP: thiamine pyrophosphotransferase, EC 2.7.6.2) plays a very essential role in the utilization of thiamine by converting the inactive free thiamine into the active coenzyme, thiamine diphosphate (ThDP)[2] as follows:

$$[2\text{-}^{14}\text{C}]\text{Thiamine} + \text{ATP} \xrightarrow{\text{Mg}^{2+}} [2\text{-}^{14}\text{C}]\text{ThDP} + \text{AMP}$$

The assay described is used to facilitate the isolation and purification of TPK from animal tissues. It is simpler and far more sensitive than the coupled enzyme assay described earlier[3,4] and is based on the use of [2-^{14}C]thiazole-labeled thiamine as substrate and separation of the [2-^{14}C]ThDP formed from the unchanged [2-^{14}C]thiamine, followed by radioactive counting. This is a slight modification of the method reported earlier.[5]

Materials

Thiamine-[2-^{14}C]thiazole HCl is from Amersham/Searle, Arlington Heights, Illinois. In order to reduce the blank counts, it is treated as follows: A 50-μCi sample is taken up in 10 ml of H_2O and placed

[1] Supported by Grants Nos. AM-02448 and AM-16897 from the National Institutes of Health.

[2] Abbreviations used: ThDP, thiamine diphosphate; TPK, thiamine pyrophosphokinase.

[3] L. R. Johnson, C. J. Gubler, and S. A. Kuby. *Biochim. Biophys. Acta* **156**, 85 (1968).

[4] C. J. Gubler, this series, Vol. 18A [37], p. 219.

[5] J. W. Peterson, C. J. Gubler, and S. A. Kuby, *Biochim. Biophys. Acta* **397**, 377 (1975).

on an Amberlite CG-50 column (2 × 5 cm) in the H$^+$ form. This is then washed with 200–300 ml of H$_2$O, after which the [^{14}C]thiamine is eluted with six 5-ml portions of 0.2 M HCl. The fractions containing significant radioactivity are combined, lyophilized and made to 10 ml with 2 mM cold thiamine which had been neutralized to pH 7.0.

Amberlite CG-50 (200–300 mesh) (Mallinckrodt, Inc., St. Louis, Missouri). This is washed with H$_2$O first to remove fines, dried, treated with acetone until the wash is colorless, washed again with H$_2$O to remove acetone, then converted to the Na$^+$ form with 40% w/v NaOH with stirring for 3 hr. This is washed several times with H$_2$O, converted to the H$^+$ form with 3 N HCl followed by washing with H$_2$O until free of Cl$^-$. It is stored as a slurry under H$_2$O.

Glycylglycine buffer, 0.75 M, pH 7.8

ATP, 0.2 M neutralized to pH 7.4

MgCl$_2$, 0.12 M

Dithioerythritol, 40 mM

Bovine serum albumin, 20 mg/ml

(NH$_4$)$_2$SO$_4$, 0.3 M

Trichloroacetic acid, 50% w/v

Amberlite CG-50 columns (0.5 × 4 cm) prepared in Pasteur pipettes

Assay

The following are placed in a small glass or plastic culture tube: 0.05 ml of buffer, 0.025 ml of 0.12 M MgCl$_2$, 0.05 ml of 0.2 M ATP, 0.05 ml of 4 M (NH$_4$)$_2$SO$_4$, 0.025 ml of 2% bovine serum albumin, 0.025 ml of 10 mM dithioerythritol, and 0 to 0.25 ml of TPK solution plus H$_2$O to a final volume of 0.475 ml. The reaction is started by the final addition of 0.05 ml of the substrate (2 mM thiamine containing ca. 300,000 dpm). The tubes are covered with parafilm and incubated at 30° or 37° for 60 min. The reaction is stopped by the addition of 0.05 ml of 50% trichloroacetic acid with shaking. If sufficient protein is present to give a precipitate, centrifuge at low speed. The supernatant is neutralized with 0.025 ml of 15% NaOH, and 0.2-ml aliquots are placed on the prepared columns. The columns are washed with 4.0 ml of H$_2$O to elute the [^{14}C]ThDP, and a 1.0-ml aliquot of the eluate is placed in a scintillation vial[6] with 10 ml of a suitable scintillation cocktail (Aquasol from New England Nuclear Corporation), and the

---

[6] A significant saving of expensive cocktail can be achieved by using a new system developed by Nalge using Nalge filmware tubes. The 1.0 ml of eluate and only 1.0 ml of aquasol are placed in a special plastic bag, which is then sealed by heat, shaken, and placed in a polyethylene vial for counting. These proportions of aqueous solution plus aquasol form a clear gel on shaking which still gives 50–60% efficiency for ^{14}C.

radioactivity is counted. A blank contains everything but the enzyme and is carried through this procedure. A suitable aliquot of the [2-^{14}C]thiamine standard (0.1 ml diluted to 1.0 ml, then 10 μl) + 0.9 ml of H_2O is placed in a scintillation vial and counted. From an efficiency curve with known ^{14}C standards under the conditions used, the counts per minute (cpm) of the samples and standard are converted to disintegrations per minute (dpm). The 0.05 ml of [^{14}C]thiamine used for the substrate contains 200 nmol of thiamine. Therefore, the nanomoles of ThDP formed per minute per milligram of enzyme protein are calculated as follows:

$$\text{Nanomoles ThDP/min/mg} = \text{dpm of sample} \times \frac{1}{\text{vol. enzyme}}$$
$$\times \frac{0.60}{0.20} \times \frac{4.0}{1.0} \times \frac{1}{\text{mg protein/ml}}$$

The columns can be regenerated and reused many times by washing with 4.0 ml of 0.2 M followed by washing with 4.0-ml aliquots of H_2O (about 5 aliquots) until the washings are free of Cl$^-$.

[20] Assay of Thiamine Pyrophosphokinase (ATP:Thiamine Pyrophosphotransferase, EC 2.7.6.2) Using Anion-Exchange Paper Disks

By BRUNO DEUS and HUBERT E. BLUM

Since the separation of thiamine pyrophosphate (TPP) from an assay mixture by chromatography or electrophoresis on paper (this volume [37]) is rather time consuming, a simple procedure for the rapid radiometric determination of thiamine pyrophosphokinase activity has been designed.[1] Analogous procedures for the assay of thymidine kinase,[2] glycerol kinase, and hexokinase[3] have been described earlier.

Principle. Based on its anionic property, labeled thiamine pyrophosphate, which is formed during the reaction is bound to DEAE-cellulose anion-exchange paper disks, while [^{14}C]thiamine is completely eluted.

Reagents
Glycylglycine buffer, 0.5 M, pH 7.3
ATP, disodium salt, 0.1 M, pH 7.0
MgSO$_4$, 0.1 M

[1] B. Deus and H. E. Blum, *Anal. Biochem.* **57**, 614 (1974).
[2] T. R. Breitman, *Biochim. Biophys. Acta* **67**, 153 (1963).
[3] E. A. Newsholme, J. Robinson, and K. Taylor, *Biochim. Biophys. Acta* **132**, 338 (1967).

Thiamine hydrochloride,[4] 10 mM, labeled; specific activity 1 to 3 × 10[7] dpm per micromole

All reagents are stored at − 15°.

Procedure. The assay for thiamine pyrophosphokinase activity is carried out at 37°. The reaction mixture contains 50 μl of glycylglycine buffer, 20 μl of MgSO$_4$, 30 μl of ATP, and the enzyme. The volume is brought up to 240 μl with water. After equilibration at 37°, the reaction is started by the addition of 10 μl of [14C]thiamine. The reaction is stopped by heating at 100° for 1 min. Precipitated protein is removed by centrifugation. A 50-μl aliquot of the supernatant is applied to a DEAE-cellulose paper disk (Whatman DE-81, 23 mm diameter). [14C]Thiamine is removed by washing the disk with five 20-ml volumes of 0.1 mM sodium acetate on a filter holder (Millipore Filter Corp.). The paper disk is dried and placed in a scintillation vial containing 5 ml of scintillation medium.[5] The amount of [14C]TPP formed is calculated from the disintegrations per minute on the basis of the specific radioactivity of thiamine.

Comments. Under the conditions outlined above, more than 99% of [14C]thiamine is removed from the paper disks. Correction is made for this background radioactivity.

Since the binding capacity of the anion-exchange paper is limited, care must be taken not to overload the disk. To achieve complete retention, the total amount of TPP applied must not exceed 20 pmol in an application volume of up to 50 μl. The range of linearity may be extended by using a stack of two or more paper disks.

Thiamine monophosphate that might be formed by cleavage of TPP in crude enzyme preparations is also retained by the paper disks, even though to a much lesser extent. Therefore the enzyme source must be checked for TPP cleaving activities (this volume [37]).

Anions other than TPP present in the reaction medium compete with TPP for the binding sites on the disk. However, ATP and MgSO$_4$ up to a concentration of 10 mM displace at most 5% of the TPP applied.

The method described gives results which are in close agreement with those obtained by the chromatographic procedure (this volume [37]).

This simple procedure is very useful for the detection and measurement of thiamine pyrophosphokinase activity in biological material (e.g., tissue extracts, subcellular fractions[6]), in eluates of column chromatography, for the elaboration of optimal assay conditions, and for related problems.

---

[4] The Radiochemical Centre, Amersham, England, provides thiaminethiazole-2-14C hydrochloride and 35S-labeled thiamine hydrochloride.

[5] Distilled toluene containing 0.3% 2,5-diphenyloxazole (PPO) and 0.03% *p*-bis-2-(5-phenyloxazolyl) benzene (POPOP).

[6] B. Deus, and H. E. Blum, *Biochim. Biophys. Acta* **219,** 489 (1970).

[21] Affinity Chromatography of Thiamine Pyrophosphokinase from Rat Brain on Thiamine Monophosphate-Agarose

By YASUO WAKABAYASHI, AKIO IWASHIMA, and YOSHITSUGU NOSE

The partial purification of thiamine pyrophosphokinase from several microbial and animal sources has been achieved by multistep sequences of conventional procedures.[1-3] The first homogeneous preparation of thiamine pyrophosphokinase from pig heart muscle was reported by Hamada,[4] and recently Mitsuda et al.[5] have also reported the complete purification of this enzyme from parsley leaf. However, the purification of thiamine pyrophosphokinase from a relatively small amount of animal tissue, such as rat brain, has been further hampered by low enzyme levels in the tissue, and the usual procedures employed in purification become quite disadvantageous if reasonable quantities of enzyme are required for study. It is apparent, therefore, that a single-step procedure based on affinity chromatography should greatly facilitate the purification process as well as enhance the recovery of enzyme activity.

We have previously reported a successful application of thiamine pyrophosphate-agarose to the purification of a thiamine-binding protein of *Escherichia coli*.[6] It has been known that thiamine monophosphate markedly inhibits thiamine pyrophosphokinase purified from parsley leaf[5] as well as the crude enzyme of rat brain.[7]

This chapter demonstrates that thiamine monophosphate-agarose can be effectively applied to purify thiamine pyrophosphokinase from a crude extract of rat brain.

Preparation of Thiamine Monophosphate-Agarose

Twenty-five milliliters of packed Sepharose 6B are suspended in distilled water to achieve a total volume of 50 ml. Crushed CNBr, 1.25 g, is added at once to the suspension, which is vigorously mixed by magnetic stirring. The pH of the suspension is adjusted to and maintained at

[1] Y. Kajiro, *J. Biochem.* **46**, 1523 (1959).
[2] Y. Mano, *J. Biochem.* **47**, 283 (1960).
[3] J. W. Peterson, C. J. Gubler, and S. A. Kuby, *Biochim. Biophys. Acta* **397**, 377 (1975).
[4] M. Hamada, *Seikagaku* **41**, 837 (1969).
[5] H. Mitsuda, Y. Takii, K. Iwami, and K. Yasumoto, *J. Nutr. Sci. Vitaminol.* **21**, 103 (1975).
[6] A. Matsuura, A. Iwashima, and Y. Nose, *Biochem. Biophys. Res. Commun.* **51**, 241 (1973); this series, Vol. 34, p. 303.
[7] S. K. Sharma, and J. H. Quastel, *Biochem. J.* **94**, 790 (1965).

10.5–11.5 by the addition of 7.5 N NaOH. The temperature is maintained at about 20° by the addition of pieces of ice. Upon completion of the reaction (about 10 min, as indicated by the cessation of base uptake), sufficient ice is added to the gel, which is then filtered and washed rapidly on a funnel with 300 ml of ice-cold water and thereafter with 500 ml of ice-cold 0.1 M NaHCO$_3$. The activated, moist agarose cake is suspended in 25 ml of 0.1 M NaHCO$_3$ containing 3 g of ethylenediamine, and the pH of the slurry is immediately adjusted to 10 with concentrated HCl. The slurry is gently stirred at 4° for 16 hr. The resulting aminoethyl derivative of Sepharose is filtered, washed with 500 ml of 0.1 M NaHCO$_3$, and then with 500 ml of water. The filter cake is suspended in 24 ml of water to which is added 400 mg of 1-ethyl-3-(3-dimethylaminopropyl)carbodiimide hydrochloride and 240 mg of thiamine monophosphate. After the pH of the suspension is adjusted to 6.4 with 0.5 N NaOH, the suspension is stirred at 4° for 16 hr. The gel is washed sequentially with 500 ml of 0.1 M NaHCO$_3$ containing 0.5 M NaCl, 300 ml of water, 500 ml of 0.1 M sodium acetate buffer (pH 4.0) containing 0.5 M NaCl, and finally with 300 ml of water until the filtrate is negative with the thiochrome fluorescence reaction.[8]

Purification of Thiamine Pyrophosphokinase from Rat Brain with Thiamine Monophosphate-Agarose

Eight male rats of the Wistar strain, weighing 200–250 g, were killed by decapitation; the brains were homogenized for 2 min in 10 volumes of 20 mM Tris·HCl (pH 7.4) containing 2 mM 2-mercaptoethanol and 1 mM EDTA. The homogenate was then centrifuged at 100,000 g for 60 min. The supernatant was used as the crude extract after dialysis against 5 liters of 20 mM Tris·HCl (pH 7.4) containing 2 mM 2-mercaptoethanol and 1 mM EDTA. Eighteen milliliters of the crude extract were applied to a column of thiamine monophosphate-agarose (0.5 × 6 cm) previously equilibrated with the buffer described above. Virtually all the thiamine pyrophosphokinase activity in the crude extract was adsorbed to the column. After washing with 20 ml of the same buffer, the enzyme was eluted with the buffer containing 10 μM thiamine. Fractions of 2.5 ml were collected at a flow rate of about 10 ml per hour at 4°. The specific activity of the enzyme in the eluate showed a purification of approximately 700-fold with a recovery of 71.6% (table). The enzyme was also recovered with 0.1 mM thiamine monophosphate, but not with 10 μM thiamine monophosphate or with 4 mM Mg^{2+}-ribonucleoside triphosphate, such as ATP, GTP, UTP, and CTP.

---

[8] A. Fujita, this series, Vol. 2, p. 622.

PURIFICATION OF THIAMINE PYROPHOSPHOKINASE FROM RAT BRAIN BY
THIAMINE MONOPHOSPHATE-AGAROSE CHROMATOGRAPHY[a]

Step	Volume (ml)	Total protein (mg)	Total activity (nmol)	Specific activity (nmol/hr/mg)
Crude extract	18.0	55.4	22.5	0.405
Affinity chromatography	10.0	0.057	16.1	282.1

[a] Thiamine pyrophosphokinase activity was assayed as previously described [Y. Waka-bayashi, A. Iwashima, and Y. Nose, *Biochim. Biophys. Acta* **429**, 1085 (1976)].

The substituted agarose is regenerated by washing with 20 volumes of 0.6 M NaCl and equilibrating with the buffer described above, and it remains stable at least for 3 months at 4°.

Preliminary results from this laboratory indicate that thiamine pyrophosphokinase partially purified from pig brain by a conventional method is also adsorbed to the column, and the enzyme eluted with 0.1 mM thiamine exhibits a single band on polyacrylamide gel electrophoresis.[9]

[9] Y. Wakabayashi, in preparation.

[22] Enzymatic Formation of Thiamine Pyrophosphate in Plants

By HISATERU MITSUDA, YUKIO TAKII, KIMIKAZU IWAMI,
KYODEN YASUMOTO, and KENJI NAKAJIMA

Thiamine pyrophosphokinase (EC 2.7.6.2) converts thiamine to thiamine pyrophosphate (TPP) through direct transfer of pyrophosphate from ATP. The activity of the enzyme has been found in brewers' yeast,[1] *Escherichia coli*,[2] animals,[3-5] and a number of plants.[6] The enzyme, with a 8000-fold purification, was recently isolated from parsley leaf.[7] The purification procedures are described below.

Principle. The content of TPP formed during the enzyme reaction is manometrically determined by measuring the amount of the carbon

[1] Y. Kajiro, *J. Biochem.* (*Tokyo*) **46**, 1523 (1959).
[2] J. Miyata, T. Kawasaki, and Y. Nose, *Biochem. Biophys. Res. Commun.* **27**, 601 (1967).
[3] Y. Mano, *J. Biochem.* (*Tokyo*) **47**, 283 (1960).
[4] J. W. Peterson, C. J. Gubler, and S. A. Kuby, *Biochim. Biophys. Acta* **397**, 377 (1975).
[5] M. Hamada, *Seikagaku* **41**, 837 (1969) (in Japanese).
[6] H. Mitsuda, Y. Takii, K. Iwami, and K. Yasumoto, *J. Nutr. Sci. Vitaminol.* **21**, 19 (1975).
[7] H. Mitsuda, Y. Takii, K. Iwami, and K. Yasumoto, *J. Nutr. Sci. Vitaminol.* **21**, 103 (1975).

dioxide evolved in the decarboxylation reaction, in which the substrate, pyruvate, is transferred to acetaldehyde and carbon dioxide by pyruvate decarboxylase as a holoenzyme formed from the coenzyme produced, TPP, and added apoenzyme. The comparable relation between increasing concentrations of TPP and increasing activity of the decarboxylation, i.e., increasing amounts of evolved carbon dioxide, is accomplished under experimental conditions described below.

Assay

The reaction mixture, in a final volume of 2 ml, contains 10 nmol of thiamine, 4 μmol of ATP, 20 μmol of $MgCl_2$, and an aliquot of the enzyme preparation in 50 mM Tris·HCl buffer (pH 8.0). Incubation is performed at 37° for 1 hr. The reaction is terminated by adding 2 ml of 1.0 M citrate buffer (pH 6.0), then heating the mixture at 90° for 5 min. The denatured proteins are removed by centrifugation. The TPP content in the clear solution is determined with a Warburg manometer by measuring the amount of carbon dioxide evolved from pyruvate during incubation at 30° for 30 min. The assay medium for the determination of TPP content, in a final volume of 3 ml, contains 200 μmol of sodium pyruvate, 30 μmol of $MnCl_2$, an aliquot of tested solutions, and 0.1 ml of a freshly prepared apocarboxylase solution in 0.5 M citrate buffer (pH 6.0). Apoprotein of pyruvate decarboxylase (EC 4.1.1.1) is prepared from bakers' yeast according to the procedure of Aoshima.[8] One unit of the enzyme is defined as 1 nmol of TPP formed per hour at 37°. Specific activity is expressed as units per milligram of protein. Protein is estimated by the procedure of Lowry et al.[9] using Lab-Trol as the standard.

Preparation of Thiamine Pyrophosphokinase

All operations are done in the cold (0°–4°) unless otherwise stated.

Step 1. Extraction. Fresh leaves of parsley (800 g) are ground in a Waring blender with 1.5 liters of 50 mM Tris·HCl buffer (pH 7.5) containing 10 mM 2-mercaptoethanol and 1 mM EDTA. The homogenate is squeezed through three layers of gauze.

Step 2. Supernatant. The filtrate obtained (1920 ml) is centrifuged at 5000 g for 15 min to remove cell debris.

Step 3. Ammonium Sulfate Fractionation. Solid ammonium sulfate is added to the solution (1740 ml) to give first 30% and then 45% saturation.

---

[8] Y. Aoshima, Seikagaku 29, 861 (1958) (in Japanese).

[9] O. H. Lowry, N. J. Rosebrough, A. L. Farr, and R. J. Randall, J. Biol. Chem. 193, 265 (1951).

The protein fraction precipitating between 30% and 45% saturation is collected by centrifugation, dissolved in a small volume of the buffer, then dialyzed at intervals of 5 hr against two changes of 2.5 liters of the Tris buffer without EDTA.

Step 4. First DEAE-Cellulose Column Chromatography. The precipitate formed during dialysis is removed by centrifugation, and the supernatant solution is placed on a DEAE-cellulose column (2.5 × 90 cm) previously equilibrated with Tris buffer. The column is washed with 1 liter of the Tris buffer and eluted by a linear gradient of 0 to 0.5 M KCl in the buffer. Activity of the pyrophosphokinase is observed over a concentration range of around 0.2–0.28 M KCl. Active fractions are combined and brought to 60% saturation with ammonium sulfate. The resultant precipitate is obtained by centrifugation. The purification step separates the main fractions of thiamine and TMP kinases from an active fraction of acid phosphatase.

Step 5. Second DEAE-Cellulose Column Chromatography. The obtained precipitate is dissolved in the Tris buffer and dialyzed as described above. The dialyzate is again chromatographed on a column (2.5 × 40 cm) of DEAE-cellulose equilibrated with the buffer, then the protein is eluted by linearly increasing concentrations of KCl to 0.5 M. Active portions are collected and concentrated to a minimal volume by ultrafiltration through reverse osmosis in a collodion bag.

Step 6. Sephadex G-150 Column Chromatography. The condensate is applied to a Sephadex G-150 column (2.2 × 95 cm), which had been previously equilibrated with 50 mM Tris·HCl buffer (pH 7.5). Eluates are collected in 5-ml aliquots. The enzyme activity is detected in fractions 50–69. The protein elution is separated into three major peak fractions, in which the kinase activity is found in the third peak. Thiamine pyrophosphokinase activity appears to be recovered with almost no contamination with related enzymes in this step.

Step 7. Hydroxyapatite Column Chromatography. The combined active fraction at the previous step is concentrated by ultrafiltration, and an aliquot of the solution is passed through a hydroxyapatite column (1.1 × 26 cm) previously equilibrated with 10 mM potassium phosphate buffer (pH 7.2) containing 10^{-2} M 2-mercaptoethanol. Eluates are collected in batches of 5 ml each. Active fractions are encountered in fractions 1 to 11. Most of the enzyme activity is found in the first protein peak, in which a majority of the contaminants remain on the column.

Step 8. Polyacrylamide Gel Electrophoresis. The enzyme preparation after chromatography on hydroxyapatite is subjected to electrophoresis (30 mA for 6 hr) on a preparative polyacrylamide gel slab (1.5 × 0.5 × 17 cm). When completed, the gel slab is sliced into sections about 5-mm

PURIFICATION OF THIAMINE PYROPHOSPHOKINASE FROM PARSLEY LEAF[a, b]

Step	Description	Volume (ml)	Protein (mg)	Activity (units)	Specific activity	Recovery (%)
1	Crude extract	1920	27,200	287	0.01	100
2	Supernatant (5000 g, 15 min)	1740	10,700	243	0.02	84.5
3	Ammonium sulfate (30–45% satd.)	63.5	5850	254	0.04	88.3
4	DEAE-cellulose, 1st	50.6	1950	267	0.14	92.9
5	DEAE-cellulose, 2nd	3.9	484	284	0.59	98.6
6	Sephadex G-150	2.1	21.6	223	10.5	77.6
7	Hydroxyapatite	3.0	7.8	195	25.0	67.9
8	Electrophoresis	2.1	0.37	30.7	83.0	10.7

[a] The assay conditions were the same as described in the text except for a constant level of ATP (5 mM) and pH 7.5.
[b] Fresh parsley leaves, 0.8 kg.

thick. Each is homogenized with 1.0 ml of the Tris buffer and allowed to extract proteins with the buffer at 4° for 15 hr. A single peak having the enzyme activity is formed with the relative mobility $K_B = 0.3$.

The resulting purification is summarized in the table. Thiamine pyrophosphokinase is approximately 8000-fold purified from parsley leaf. The enzyme in the final preparation proved to be homogeneous on analytical polyacrylamide gel electrophoresis.

Properties of the Enzyme

General Properties. The molecular weight of the enzyme, estimated by gel filtration with Sephadex G-150, is approximately 30,000. In 50 mM Tris·HCl, the enzyme activity shows a pH optimum over a range of 8 to 9. A least-squares analysis of Lineweaver–Burk and Hofstee plots gave K_m values of 0.8 mM for ATP and 0.15 μM for thiamine: the latter was much less than 75 μM for the yeast kinase and relatively close to 0.3 μM for the *E. coli* enzyme. Isoelectric focusing in polyacrylamide gel containing 1.0% carrier ampholine (pH 3–10) showed that the isoelectric point of the enzyme is 4.5. An Arrehenius plot of the relative velocity against the temperature gave an activation energy of 4500 for its reaction.

Effect of Various Inhibitors. The enzyme activity is completely suppressed by 0.1 mM p-chloromercuribenzoic acid (PCMB) and 10 mM pyrophosphate. Sulfhydryl reagents, such as N-ethylmaleimide (NEM) and iodoacetamide (10 mM each) inhibited the activity, while KCN, NaF, and molybdate have little effect. The activity suppressed by PCMB is almost completely restored by adding excess dithiothreitol (10 mM). Inorganic monophosphate at 10 mM has no effect on the activity, in contrast to the inhibition by pyrophosphate. EDTA, a predominant chelating agent, exhibits strong inhibition. Thiamine monophosphate (TMP) not only is unable to replace thiamine, but also inhibits strikingly the TPP formation from thiamine.

These findings are consistent with the view that TMP is not directly converted to TPP but after being dephosphorylated by the action of a monoesterase, resulting thiamine is pyrophosphorylated with ATP by thiamine pyrophosphokinase (EC 2.7.6.2) to form TPP; thus giving clear evidence regarding the mechanism of TPP formation in plant tissue.

[23] Preparation of Thiamine Triphosphate

By H. K. PENTTINEN

Thiamine triphosphate can be synthesized by heating a mixture of thiamine hydrochloride and phosphorus pentoxide in orthophosphoric acid. It is purified by chromatography on a strong acid cation ion-exchange resin, Dowex 50-W,[1] instead of the weak acid resin Amberlite IRC-50 used before.[2]

Reagents

Thiamine hydrochloride (Sigma Chemical Company, St. Louis, Missouri)

Phosphorus pentoxide

Orthophosphoric acid

Ethanol

Acetone

Cation-exchange resin, Dowex 50-W X-8 (200–400 mesh)

Procedure. Orthophosphoric acid (77 ml) is heated to 320° and allowed to cool to 100°; thereafter a mixture of 50 g of thiamine hydrochloride and 50 g of phosphorus pentoxide is added in small amounts, and the temperature of the mixture is kept at about 100°–105° for a further 20 min. After cooling, the mixture is dissolved in 100 ml of water and poured slowly into 2 liters of a cold (4°) ethanol–acetone mixture (1:1, v/v). The precipitate is collected and dissolved in 100 ml of water. Precipitation is repeated three times and the final volume of the aqueous mixture is adjusted to 150 ml after removal of the acetone and ethanol under reduced pressure. The mixture is divided into 15-ml portions, which are stored at −18°. One portion is applied to a Dowex column (4 × 16 cm, H form) and eluted at room temperature with water at a rate of 10 ml/min. The effluent between 300–800 ml is collected and lyophilized. The white powder obtained is dissolved in 1 ml of water. Ethanol is added until the solution shows some turbidity. The mixture is kept at 4° for 3 hr and thereafter at −18° overnight. The precipitate is collected and redissolved in water, and precipitation is repeated twice. The crystals are washed with cold ethanol and ether, dried at room temperature and stored at −70°. The column is regenerated with 5 liters of 4 *M* HCl and washed with water. The crystals from 10 chromatographies are pooled and then recrystallized as before. The final yield of pure thiamine triphosphate is 2 g.

[1] H. K. Penttinen, *Finn. Chem. Lett.* 1 (1976).
[2] T. Yusa, *J. Biochem.* (*Tokyo*) **46**, 391 (1959).

[24] Determination of Thiaminase Activity Using Thiazole-Labeled Thiamine

By E. E. EDWIN

This method is based on the rapid radioactive procedure for the determination of thiaminase activity reported by Edwin and Jackman.[1]

Assay of Thiaminase Activity

Principle. The two known types of thiaminases catalyze the cleavage of thiamine at the methylene bridge into its constituent thiazole and pyrimidine moieties, although their reaction mechanisms are different. Thiaminase I [thiamine: base 2-methyl-4-aminopyrimidine-5-methenyltransferase, EC 2.5.1.2] is a transferase and requires a cosubstrate for the reaction to proceed. Aromatic primary amines, heterocyclic bases, or sulfhydryl compounds are known to function as cosubstrates or activators. Thiaminase II [thiamine hydrolase, EC 3.5.99.2) mediates in the hydrolysis of thiamine, the elements of water being added at the meth-

[1] E. E. Edwin and R. Jackman, *J. Sci. Food. Agric.* **25**, 357 (1974).

ylene bridge. In either reaction the thiazole moiety of thiamine is released. By using thiamine labeled in the thiazole moiety with ^{14}C or ^{35}S, and measuring the amount of this compound formed in the reaction, the activity of the enzyme can be assayed. Further, the method can be used to differentiate between the two types of thiaminases or adapted to locate thiaminase producing bacterial colonies grown on semisolid media.

Reagents

Citrate-phosphate buffer, 0.1 M, pH 6.4

Pyridine, analytical grade

Labeled thiamine. [^{14}C] or [^{35}S]Thiazole-labeled thiamine chloride hydrochloride (Radiochemical Centre, Amersham, England) is dissolved in 0.1 M HCl to give a 100 μCi/ml solution.

Ethyl acetate, analytical grade

Liquid scintillator prepared by dissolving 4 g of 2,5-bis[5'-*tert*-butylbenzoxazolyl(2')]thiophen (BBOT) and 80 g of naphthalene in a mixture of 600 ml of toluene (AR) and 400 ml of 2-methoxyethanol (scintillator grade)

Solution of Substrates.[2] A nonradioactive stock solution of substrates containing thiamine (5 μg, 14.8 nmol, per milliliter) and pyridine (3.9 mg, 49.4 μmol per milliliter) is made up in citrate-phosphate buffer. Radioactive solution of substrates is made from this by adding labeled thiamine solution to it at the rate of 0.2 μl/ml.

Preparation of Samples. Solid or semisolid samples are weighed and extracted with citrate–phosphate buffer and centrifuged lightly before use. Liquid samples, such as bacterial culture fluids, are lightly centrifuged, and the supernatants are used as such.

Procedure. Pipette 0.5 ml of the sample and 0.5 ml of the radioactive solution of substrates into a stoppered test tube and incubate at 37°. Include a blank in which the thiaminase-containing sample is replaced by 0.5 ml of buffer solution. At the end of 10 min, pipette 2 ml of ethyl acetate into the tube and shake vigorously, using a mechanical vibrator. Centrifuge the mixture at 2500 g for 5 min to separate phases and ensure that the ethyl acetate layer does not carry over any droplets of the aqueous phase. Pipette 0.5 ml of the ethyl acetate extract of the radioactive thiazole moiety into 5 ml of liquid scintillator in a counting vial; after allowing for a short period of equilibration, measure the radioactivity in a liquid scintillation counter. Thiaminase activity is given by the amount of radioactivity extracted by the ethyl acetate and can be expressed as number of enzyme units per milliliter of sample.

Unit of Enzyme. The unit of enzyme is defined as the amount of en-

[2] B. V. McCleary and B. F. Chick, *Phytochemistry* **16**, 207 (1977).

zyme that will destroy 1 μmol of thiamine (or release 1 μmol of the thiazole moiety) in 1 min under the specified conditions of the assay.

Comment. This procedure can be made more sensitive by increasing the specific activity of the radioactive solution of substrates and by reducing the amount of ethyl acetate used for the extraction of the labeled thiazole set free in the reaction. Since the method depends on the relative solubilities of thiamine and its thiazole moiety in water and ethyl acetate, respectively, great care must be taken that the tip of the pipette used to withdraw the ethyl acetate extract is well clear of the aqueous phase so that minute droplets of water containing unreacted thiamine are not transferred to the scintillator vial.

The radioactive procedure described here can be used for determination of thiaminase activity in crude preparations, such as bacterial cultures, extracts of feces, plants, fish, or gut and rumen contents. It is easy in performance, sensitive, and reproducible and is suitable where large numbers of samples are to be assayed. It has been successfully used in our laboratory for a number of years.

Alternative Methods

Thiaminases have been determined by measuring fluorometrically the amount of thiamine destroyed on incubation with the enzyme.[3] Methods for the assay of thiaminase I have been proposed by Kenten[4] and Wittliff and Airth[5] which depend on the spectrophotometric measurement of the increase in optical density of the pyrimidine-based product when pyridine[4] or aniline[5] is used as cosubstrate in the reaction.

Differentiation between Thiaminase Type I and Type II

Thiaminase I[6] occurs in many species of fish including shellfish, several species of bacteria, and certain plants. Thiaminase II[6] seems to be restricted to various bacteria, such as *Bacillus aneurinolyticus* and related organisms only.

The assay is carried out as described above, but the enzyme preparation must be freed of any naturally occurring cosubstrates that may have been carried over during preparation. This can be done either by dialysis at 0°–4° against water or by ammonium sulfate precipitation of the en-

---

[3] J. L. Wittliff and R. L. Airth, this series, Vol. 18A, [42], p. 234.

[4] R. H. Kenten, *Biochem. J.* **67**, 25 (1957).

[5] J. L. Wittliff and R. L. Airth, this series, Vol. 18A, [41], p. 229.

[6] K. Murata, *in* "Review of Japanese Literature on Beriberi and Thiamine," p. 220. Vitamin B Research Committee of Japan, Tokyo, 1965.

zyme, followed by passage through a Sephadex G-25 column. Enzymes prepared in this way are assayed for thiaminase activity as described under procedure, with and without added cosubstrate. The K_m for cosubstrates may vary over a wide range for different thiaminases, and it is

EXAMPLES OF THIAMINASE ACTIVITY AND EFFECT OF INCREASING
COSUBSTRATE CONCENTRATION

Source[a]	Molar ratio, cosubstrate : substrate	Counts per minute[b]	Micromoles of thiazole formed/min/ml sample $\times 10^{-3c}$
Equisetum arvense	0	653	0.28
(horsetail)	1	884	0.38
	10	1673	0.72
	100	2603	1.12
Pteridium aquilinum	0	773	0.33
(bracken fern)	1	871	0.37
	10	1216	0.52
	100	1212	0.52
Clostridium sporogenes	0	350	0.15
	1	364	0.16
	10	585	0.25
	100	1846	0.79
Bacillus thiaminolyticus	0	427	0.18
	1	418	0.18
	10	584	0.25
	100	986	0.42
Bacillus aneurinolyticus	0	297	0.13
	1	277	0.12
	10	299	0.13
	100	305	0.13

[a] Equisetum leaves or bracken fronds were macerated with citrate–phosphate buffer and centrifuged; the supernatant was used for assay. Bacterial culture fluids were centrifuged, and the clear supernatants were used without further purification.

[b] Counts per minute after subtracting blank. Blank varied between 40 and 50 cpm.

[c] These values were calculated from counts per minute, after correction for efficiency of the scintillation counter for ^{14}C (85%) and amount of [^{14}C]thiazole extracted into ethyl acetate (80%). See E. E. Edwin and R. Jackman, J. Sci. Food Agric. **25**, 363 (1974). The specific activity of the [^{14}C]thiamine in the assay was calculated from the supplier's batch analysis data and adjusted for dilution with nonradioactive thiamine. In the above determinations the specific activity of the [^{14}C]thiamine purchased from Radiochemical Centre was 14 mCi/mmol and that used for assay 1.23 μCi (2.73 $\times$ 10^6 dpm) per micromole. Thus:

micromoles of thiamine destroyed (thiazole formed) per minute per milliliter of sample

$$= \text{cpm} \times \frac{100}{85} \times \frac{100}{80} \times \frac{8 \text{ (dilution factor)}}{10 \text{ (assay time)}} \times \frac{1}{2.73 \times 10^6 \text{ (specific activity)}}$$

$$= \text{cpm} \times 4.3 \times 10^{-7}$$

advisable to try a number at different concentrations. For general purposes pyridine is used, as it is a potent cosubstrate for most thiaminase I enzymes studied. Thiaminase II neither loses activity on dialysis, nor is activated by the addition of cosubstrate, as shown in the table.

Test for Thiaminase Activity in Colonies[7]

The method can be modified for determining thiaminase activity in individual bacterial or fungal colonies and can be applied in screening for thiaminase-producing microorganisms in mixed populations.

Solution of Substrates. This solution contains labeled thiamine and pyridine in a molar ratio of 1 : 1000 in citrate–phosphate buffer. Thus, 1 ml contained 1 μl of the radioactive stock solution of thiamine (0.1 μCi, 7.1 nmol) and 0.56 mg (7.1 μmol) of pyridine.

Procedure. Bacterial or fungal colonies grown on solid media, such as nutrient agar or Sabouraud plates, under the particular conditions necessary for growth are suitable for this. Using a shortened Pasteur pipette, remove a plug (approximately 2 mm in diameter × 3 mm deep) of the supporting agar gel adjacent to colonies and transfer to a stoppered test tube containing 0.2 ml of the solution of substrates. Incubate the tubes for 30 min at 45°. Pipette 1 ml of ethyl acetate and mix vigorously (mechanical vibrator) for 5 sec. Separate the phases by centrifugation (2500 g for 5 min). Transfer 0.5 ml of the organic layer to a scintillation vial containing 5 ml of liquid scintillator and determine the amount of radioactivity. With all determinations, include a blank by taking a plug well removed from bacterial colonies. Thiaminase-producing colonies will show high levels of radioactivity.

Comment. The amount of agar in the medium must be such as to discourage swarming, and the inoculum must be diluted suitably to give discrete well defined colonies without undue crowding. Although plugs can be removed from the middle of colonies, it is advisable to take them from the clear medium adjacent to them, as this will ensure that one is dealing with diffused enzyme only and that the sample is devoid of live cells.

[7] E. E. Edwin, J. Shreeve, and R. Jackman, *J. Appl. Bacteriol.* **44**, 305 (1978).

[25] Membrane-Bound Thiamine Triphosphatase

By ROBERT L. BARCHI

$$\text{ThTP} \xrightarrow{\text{Mg}^{2+}} \text{ThDP} + \text{P}_i$$

Thiamine triphosphate (ThTP) is present in significant quantities in most organisms although its specific role in the cellular economy is unknown. Involvement of this compound in the function of excitable membranes has been postulated, but this hypothesis as yet remains unproved.[1] The membrane thiamine triphosphatase described here was first characterized in rat brain.[2] It catalyzes the hydrolysis of the γ phosphate from ThTP, producing as products ThDP and P_i. A distinct soluble ThTPase has also been described.[3]

Assay

Principle. Enzyme activity is determined by monitoring the quantity of inorganic phosphate released from ThTP during a test interval under conditions in which the subsequent hydrolysis of ThDP can be shown to be negligible. Inorganic phosphate may be determined by any of several colorimetric assays (cf. Vol. 3 of this series).

Reagents

MES [2-(N-morpholine)ethane sulfonate] buffer, 0.5 M, pH 6.5
$MgCl_2$, 50 mM
ThTP, 50 mM (adjusted to pH 6.5)
Trichloroacetic acid, 5% (cold)
Membrane suspension to be assayed (0.5–2.0 mg/ml)

Procedure. A total reaction volume of 1 ml is used. The following quantities of reagents are combined and preincubated at 37° for 5 min: MES, 0.1 ml; $MgCl_2$, 0.1 ml; protein solution (final protein concentration of 50–200 μg/assay tube); and deionized water to a final volume of 0.9 ml. After preincubation at 37°, the reaction is initiated by addition of 0.1 ml of ThTP with rapid mixing and continued incubation in a shaking water bath at 37°. After an appropriate interval (usually 20 min) the reaction is terminated by the addition of 0.5 ml of cold TCA. When multiple samples are being run, care must be taken to evenly space initiation and termination of reactions so that the total incubation time for each tube is the same.

---

[1] A. Von Muratt, *Ann. N. Y. Acad. Sci.* **98,** 499 (1962).
[2] R. L. Barchi and P. E. Braun, *J. Biol. Chem.* **247,** 7668 (1972).
[3] Y. Hashitani and J. R. Cooper, *J. Biol. Chem.* **247,** 2117 (1972).

Reaction tubes are centrifuged briefly, and aliquots of the deproteinized supernatants are removed for colorimetric determination of inorganic phosphate by the method of Baginski *et al.*[4] Separate control tubes must be carried through the assay sequence for ThTP and enzyme. All readings are determined relative to a buffer blank and corrected for P_i in substrate and tissue.

Under the conditions described, the concentration of ThDP formed during the incubation period is sufficiently below the K_m of the diphosphatase (2mM), and the V_{max} of the diphosphatase is sufficiently low, that the further breakdown of ThDP to ThMP and P_i does not contribute significantly to the measured P_i.[5]

Source of the Enzyme

The membrane-associated thiamine triphosphatase has not been isolated in pure form. Based on subcellular fractionation studies in rat brain,[6] the enzyme is present in highest concentrations in the crude nuclear pellet and in the synaptosomal and microsomal fractions.[2] Lowest specific activities are seen in the mitochondrial and soluble fractions. Detectable levels of the enzyme have been demonstrated in crude membrane preparations of most soft tissues of the rat, highest specific activities being found in the intestine and kidney. Standard procedures (cf. Vol. 31 of this series) can be used to prepare membranes enriched in this ThTPase relative to the crude homogenate, and suitable for assay by the procedure described above. Attempts to solubilize this enzyme using a variety of detergents have to date been unsuccessful.

Comments

The membrane-associated ThTPase demonstrates a pH optimum between 6.2 and 7.2.[2] At pH 6.5 the apparent K_m for ThTP in the presence of excess Mg^{2+} is about 1.5 mM. As might be anticipated by analogy with known ATPases, the true substrate for this ThTPase can be shown to be the $Mg^{2+} \cdot$ThTP complex.[7] We have found a dissociation constant for this complex of 65 μM at pH 7.5 and 22°. Excess free Mg^{2+} or free Ca^{2+} in concentrations above 3 mM act as competitive inhibitors of the membrane ThTPase. The apparent inhibitory constant for free Mg^{2+} under standard

[4] E. Baginski, P. Foa, and B. Zak, *Clin. Chim. Acta.* **15,** 155 (1967).
[5] R. L. Barchi and P. E. Braun, *J. Neurochem.* **19,** 1039 (1972).
[6] K. Kataoka and E. De Robertis, *J. Pharmacol. Exp. Ther.* **156,** 114 (1967).
[7] R. L. Barchi and R. O. Viale, *J. Biol. Chem.* **251,** 193 (1976).

assay conditions is 7 mM. Excess free ThTP does not appear to act as a competitive inhibitor.

The membrane ThTPase is inhibited by nucleoside diphosphates and triphosphates. ATP and ADP are most effective, reducing hydrolysis by 50% at concentrations of 20 μM and 75 μM, respectively.[8] Nucleoside monophosphates, free nucleosides, and P$_i$ have no effect on enzyme activity. ThDP and ThMP also fail to inhibit the enzyme in concentrations up to 5 mM. Nonhydrolyzable analogs of ATP and ADP incorporating a methylene phosphonate linkage between terminal phosphates are also potent inhibitors (K_i = 40 μM for AMP-PCP; K_i = 90 μM for A-PCP).[8] Use of these compounds allows regulation of the reaction to be studied without interference from parallel hydrolysis of ATP by its specific membrane-bound enzyme. Under these conditions, inhibition for both compounds is of a mixed noncompetitive type.

A similar nonhydrolyzable methylene phosphonate analog of ThTP (ThMP-PCP) has been synthesized.[2] This compound acts as a competitive inhibitor of the membrane ThTPase with a calculated K_i of 3 mM, close to the apparent K_m for the native substrate. The analog has no effect on the hydrolysis of ATP by its specific membrane enzymes.

---

[8] R. L. Barchi, *J. Neurochem.* **26**, 715 (1976).

[26] Transition-State Analogs of Thiamine Pyrophosphate

By JAN A. GUTOWSKI

The major structural transformations that occur during catalysis of most thiamine pyrophosphate-dependent enzymic reactions have been well documented.[1] Common to all these mechanisms are high-energy intermediates in which the thiazole ring, unlike that of thiamine pyrophosphate itself, is uncharged. Both thiamine thiazolone pyrophosphate (I) (TTPP) and thiamine thiothiazolone pyrophosphate (II) (TTTPP), which have been shown to behave as transition-state analogs for *Escherichia coli* pyruvate dehydrogenase,[2,3] have structures that resemble these postulated intermediates, but specifically the metastable enamine (III) which is the immediate product of the decarboxylation of pyruvate. The binding of (II) and (III) to the *E. coli* pyruvate dehydrogenase complex is so tight

---

[1] L. O. Krampitz, *Annu. Rev. Biochem.* **38**, 213 (1969).

[2] J. A. Gutowski and G. E. Lienhard, *J. Biol. Chem.* **251**, 2863 (1976).

[3] J. R. Butler, F. H. Pettit, P. F. Davis, and L. J. Reed, *Biochem. Biophys. Res. Commun.* **74**, 1667 (1977).

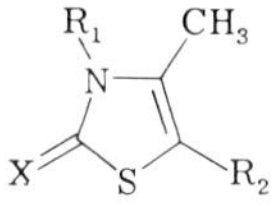

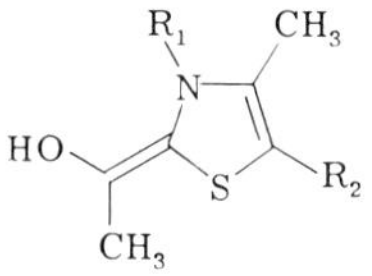

(I): X = O   (III)

(II): X = S

$R_1 =$

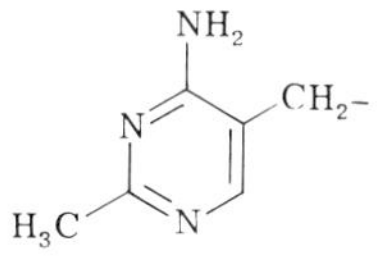

$R_2 = -C_2H_1P_2O_7^{3-}$

that these compounds inhibit stoichiometrically and almost irreversibly at very low concentrations.

Principle

Thiamine thiazolone pyrophosphate can be obtained by allowing thiamine thiazolone to react with pyrophosphoyl tetrachloride[4] and subsequently phosphorylating the resulting monophosphate with crystalline phosphoric acid.[5] The thiazolone is conveniently prepared in high yield by heating thiamine disulfide in isobutanol.[6] Thiamine thiothiazolone can be phosphorylated by essentially the same procedure.

A much more convenient method of pyrophosphorylating thiamine thiazolone has recently been outlined.[3,7] The one-step synthesis based on a procedure of Viscontini *et al.*[8] takes advantage of the relative stability of the thiazolone under strong acidic conditions.

Preparation of Thiamine Thiazolone

   Reagents
    Thiamine disulfide (from ICIN)
    Isobutanol
    Ethanol

[4] S. M. Hecht and S. D. Hawrelak, *Biochemistry* **14**, 974 (1975).
[5] J. W. Kozarich, A. C. Chinault, and S. M. Hecht, *Biochemistry* **14**, 981 (1975).
[6] P. Sykes and A. R. Todd, *J. Chem. Soc. London* p. 534 (1951).
[7] J. R. Butler, unpublished experiments.
[8] von M. Viscontini, G. Bonetti, and P. Karrer, *Helv. Chim. Acta* **32**, 1478 (1949).

Procedure. Thiamine disulfide (5 g) is suspended in isobutanol (100 ml) and heated with stirring under reflux for 2 hr. The disulfide dissolves slowly to give a clear, pale-yellow solution, which is then cooled and left overnight at 4° to yield colorless needles. The crystals are filtered, recrystallized from hot EtOH:H$_2$O (2:7 v/v) at 0.1% concentration, and air dried to give thiamine thiazolone (1.2 g), m.p. 234–236°. The molecular extinction coefficient in phosphate buffer (0.01 M, pH 7.0) at 233 nm is 12,850 mol^{-1} cm^{-1}.

Method I

Preparation of Pyrophosphoryl Tetrachloride.[9]

> *Reagents*
> Carbon tetrachloride
> Phosphorus pentoxide
> Phosphorus trichloride

Procedure. To a 500-ml, three-neck, round-bottom flask containing 100 ml of CCl$_4$ is added P$_2$O$_5$ (71 g) and PCl$_3$ (87.5 ml). The flask is flushed out with nitrogen, and chlorine gas is passed through the stirred solution heated under reflux until an excess is present. The P$_2$O$_5$ slowly dissolves after about 20 min to give a white suspension. The chlorine is stopped, and refluxing is continued for a further 30 min. The contents of the flask are then cooled, filtered, and distilled under reduced pressure to give a colorless liquid of pyrophosphoryl tetrachloride (36 g), b.p. 122–128°/40 mm of Hg, d^{20} 1–832.

Preparation of Thiamine Thiazolone Monophosphate

> *Reagents*
> Thiamine thiazolone
> Pyrophosphoryl tetrachloride
> m-Cresol, freshly distilled and stored under nitrogen
> Ether
> NaOH, 2 M
> Bio-Rad AG 1-X8, (100–200 mesh) formate form suspended in H$_2$O, pH 4
> Formic acid, 0.1 M

Procedure. To a stirred solution under nitrogen of thiamine thiazolone (1.4 g), dissolved with warming (ca. 60°) in m-cresol (20 ml) and cooled to

---

[9] P. C. Croft, I. M. Downie, and R. B. Heslop, *J. Chem. Soc. London,* p. 3673 (1960).

4°, is added dropwise pyrophosphoryl tetrachloride (3.3 ml) in *m*-cresol (5 ml). The resulting yellow solution is stirred at 4° for a further 3 hr, diluted with ground ice (ca. 30 g), and extracted with ether. The aqueous phase is then adjusted to pH 7.0 with NaOH, concentrated by rotary evaporation, and treated with AG 1-X8 resin (ca. 150 ml) until the supernatant has negligible absorbance at 260 nm. After washing with water to remove sodium formate, the resin is adjusted to pH 2.5 in 0.1 M formic acid. The released TTP is washed off with 0.1 M formic acid, the solution is concentrated by rotary evaporation while maintaining the pH near 3 by periodically adding water, and lyophilized to yield 1.4 g (80%) of a 70% pure product.

For synthesizing TTPP, further purification of the monophosphate ester is not required.

Preparation of Thiamine Thiazolone Pyrophosphate

Reagents
TTP
Tri-*n*-butylamine
1,1'-Carboxyldiimidazole
Methanol
Pyridine, freshly distilled over KOH
N,*N*-Dimethylformamide (*N*,*N*-DMF)
H_3PO_4, crystalline
Bio-Rad AG 2-X8 (100–200 mesh), acetate form, suspended in H_2O, pH 6
Bio-Rad AG 50W-X8 (100–200 mesh), ammonium form, suspended in H_2O, pH 8.
Ammonium acetate, 1 M, pH 6
Mono-(tri-*n*-butylammonium) phosphate. This is prepared by dissolving crystalline H_3PO_4 (0.7 g) in H_2O (4 ml) and adding successively pyridine (12 ml) and tri-*n*-butylamine (1.7 ml). The mixture is stirred for 5 min and rotary evaporated to dryness. The gummy, colorless residue is made anhydrous by repeated evaporations with dry pyridine.
All organic solutions are freshly distilled and stored over molecular sieves of 3Å.

Procedure. TTP (0.5 g) is dissolved in water (2 ml) and pyridine (12 ml), and treated with tri-*n*-butylamine (0.7 ml). The solution is rotary evaporated to dryness and rendered anhydrous by repeated evaporations with dry pyridine. Under anhydrous conditions the salt is dissolved in *N*,*N*-DMF (10 ml) containing 1,1'-carboxyldiimidazole (1.1 g) and stirred for 12 hr. The mixture is then treated with dry methanol (0.7 ml) over 30 min and subsequently with the mono-(tri-n-butylammonium) phosphate in N,N-DMF (5 ml). After stirring for 24 hr, dry methanol (10 ml) is added over 30 min. A precipitate that forms is filtered off, and the remaining solution is evaporated to dryness. The remaining residue is dissolved in water, adjusted to pH 6 with HCl, and applied to a column (1.2 × 9 cm) of AG 2-X8. The column is washed with water until the eluent has negligible absorbance at 260 nm. TTPP is eluted between 0.8 and 1.6 liters of a linear gradient of 1 liter of water against 1 M ammonium acetate, pH 6. Ammonium acetate is removed by rotary evaporation of the pooled fraction, which is then lyophilized to yield 0.8 g of a 50% pure product.

The crude material is purified by column chromatography and recrystallization. An aqueous solution of the pyrophosphate ester is adjusted to pH 8 with ammonia and passed down a short column of AG 50W-X8. The eluate is lyophilized, redissolved in several milliliters of water, and adjusted to pH 8 with ammonia. A mixture of acetone: EtOH (9:1 v/v) is added at 60° with vigorous mixing until a slight precipitate remains. The TTPP precipitates on cooling the mixture to 4°. It is filtered, washed briefly with acetone, and dried over P_2O_5 under vacuum to yield 350 mg (52%) of a pure product. This procedure can also be used to purify TTP.

Method 2

Preparation of Thiamine Thiazolone Pyrophosphate

Reagents
Thiamine thiazolone
Orthophosphoric acid, 85%
NaOH, 6 M
Amberlite CG-50 (100–200 mesh), H$^+$ form, suspended in H_2O

Procedure. Orthophosphoric acid (85%, 2 ml) is heated over an open flame until it becomes faintly cloudy. To the syrupy liquid, cooled to room temperature, is added thiamine thiazolone (0.12 g), and the mixture is heated in an oil bath at 100° with occasional stirring with a glass rod. After 15 min the heat is removed and H_2O (5 ml) containing 1.5 drops of orthophosphoric acid is added. The solution is stirred at room temperature for a further 2 hr. The pH is then adjusted to 1–2 with NaOH (6 M), and the solution is applied to a column (2.6 × 59 cm) of Amberlite CG-50. The column is eluted with H_2O, and the absorbance is monitored at 234 nm. The peak fractions are analyzed by thin-layer chromatography at pH 4, as in Method 1, and the samples containing a single ultraviolet-absorbing component at R_f 0.48–0.50 are combined and rotary evaporated to dryness to yield 97 mg (55%) of the pyrophosphorylated thiamine thiazolone.

Analysis. When 4 mM TTPP is treated with *E. coli* alkaline phosphatase (0.5 mg/ml) in 8 mM MgCl$_2$/210 mM ammonium sulfate/17 mM Tris·HCl, pH 8.5, the compound is found (by thin-layer chromatography of aliquots taken at various time intervals) to hydrolyze completely within 2 hr to thiamine thiazolone (R_f 0.89; Eastman cellulose plates with fluorescent indicator; EtOH/n-butanol/0.15 M sodium citrate, pH 4 (10:1:6, v/vv) as solvent) via the monophosphate ester. The monophosphate and pyrophosphate esters have R_f values of 0.67 and 0.50, respectively, in the same system. Paper electrophoreses at pH 4 and 7 exhibit single ultraviolet-absorbing components with mobilities similar to those of AMP and ADP, respectively. The ultraviolet spectrum of TTP and TTPP in 10 mM potassium phosphate, pH 7, exhibit λ_{max} at 234 nm with molar extinction coefficients of 11,900 and 10,600 mol^{-1} cm^{-1} calculated on a weight basis. The proton magnetic resonance spectrum of the TTPP in deuterium oxide at pH 3, taken at 60 MHz, shows the following signals expressed as parts per million downfield from tetramethylsilane as external standard: 2.60 (singlet), 3.02 (singlet), 3.36 (triplet, coupling constant of 6 Hz), 4.51 (multiplet), 5.32 (singlet), 8.26 (singlet). The relative integrated intensities of these signals is 3:3:2:2:2:1, respectively. By comparison with the spectrum of thiamine,[10] they can be assigned to the two CH$_3$ groups, the CH$_2$ on position 5 of the thiazolone nucleus, $-$CH$_2$OP, the CH$_2$ linking the pyrimidine and thiazolone rings, and the aromatic proton at position 6 of the pyrimidine ring, respectively.

Both TTTP and TTTPP can be characterized similarly. The ultraviolet spectrum of thiamine thiothiazolone, TTTP, and TTTPP in 10 mM potassium phosphate, pH 7.0, reveals that each have maxima at 319 and 231 nm. Their molar extinction coefficients at 319 nm based on weight are 12,400, a 12,800, and 10,900 mol^{-1} cm^{-1}, respectively.

[10] J. J. Mieyal, G. Bantle, R. G. Votaw, I. A. Rosner, and H. Z. Sable, *J. Biol. Chem.* **246,** 5213 (1971).

Section III

Lipoic Acid and Derivatives

[27] Chromatographic and Spectral Properties of Lipoic Acid and Its Metabolites

By Harold C. Furr, Jason C. H. Shih, Earl H. Harrison, Hsien-Hsin Chang, Joseph T. Spence, Lemuel D. Wright, and Donald B. McCormick

Although a dietary requirement for lipoic acid has not been established for any vertebrate species, the metabolism of this cofactor is of interest. Several studies on its metabolism in humans[1,2] and in rats[3-6] have shown that very little administered lipoate is excreted in unaltered form. More extensive research has been carried out on the catabolism of lipoate by a bacterium, *Pseudomonas putida* LP, capable of growing on lipoate as its sole source of carbon, energy, and sulfur.[7-9]

Relative mobilities of metabolites of lipoic acid on paper and thin-layer chromatographies are given in Table I[3-7,10-15] for several of the solvent systems that have proved most useful in separating lipoate derivatives. Lipoamide and its disulfhydryl and thiolsulfinate forms are included, even though they have not been isolated from the excreta of mammals or culture filtrates of bacteria. Lipoic acid and its derivatives in which the disulfide bond is intact (including the thiolsulfinates, or monoxides) are readily

[1] E. L. Patterson, H. P. Broquist, M. H. VonSalza, A. Albrecht, E. L. R. Stokstad, and T. H.·Jukes, *Am. J. Clin. Nutr.* **4,** 269 (1956).

[2] T. Nakamura, T. Kusunoki, S. Konishi, H. Kato, and A. Mibu, *J. Vitaminol. (Kyoto)* **11,** 37, 54, 60 (1965).

[3] E. M. Gal and D. E. Razevska, *Arch. Biochem. Biophys.* **89,** 253 (1960).

[4] M. Wada, Y. Shigeta, and K. Inamori, *J. Vitaminol. (Kyoto)* **7,** 237 (1960).

[5] E. H. Harrison and D. B. McCormick, *Arch. Biochem. Biophys.* **160,** 514 (1974).

[6] J. T. Spence and D. B. McCormick, *Arch. Biochem. Biophys.* **174,** 13 (1976).

[7] J. C. H. Shih, L. D. Wright, and D. B. McCormick, *J. Bacteriol.* **112,** 1043 (1972).

[8] J. C. H. Shih, M. L. Rozo, L. D. Wright, and D. B. McCormick, *J. Gen. Microbiol.* **86,** 217 (1975).

[9] H.-H. Chang, M. L. Rozo, and D. B. McCormick, *Arch. Biochem. Biophys.* **169,** 244 (1975).

[10] M. Silver and D. P. Kelly, *J. Chromatogr.* **123,** 479 (1976).

[11] E. L. Patterson, J. V. Pierce, E. L. R. Stokstad, C. E. Hoffmann, J. A. Brockman, F. P. Day, M. E. Macchi, and T. H. Jukes, *J. Am. Chem. Soc.* **76,** 1823 (1954).

[12] L. J. Reed, B. G. DeBusk, C. S. Hornberger, and I. C. Gunsalus, *J. Am. Chem. Soc.* **75,** 1271 (1953).

[13] J. C. H. Shih, P. B. Williams, L. D. Wright, and D. B. McCormick, *J. Heterocycl. Chem.* **11,** 119 (1974).

[14] H. C. Furr and D. B. McCormick, *Int. J. Vit. Nutr. Res.* **48,** 68 (1978).

[15] H. C. Furr, H.-H. Chang, and D. B. McCormick, *Arch. Biochem. Biophys.* **185,** 576 (1978).

TABLE I

CHROMATOGRAPHIC MOBILITIES OF LIPOIC ACID AND ITS METABOLITES[a]

R_f values from ascending chromatography[b]

Compound	Whatman No. 1 paper					Toyo Roshi No. 51 paper[c]		Polygram Sil-N-HR TLC plate[d]		Silica Gel 60 F_{254} TLC plate[e]			
	A	B	C	D	E	F	G	H	I	J	K	L	M
Lipoic acid	0.65(11)	0.85(12)	0.46(3)	0.50(7)	0.91(5)	0.57	0.82	0.73(7)	0.50(15)				
Bisnorlipoic acid			0.38(5)	0.40(7)	0.90(5)			0.65(7)	0.50(15)				
Tetranorlipoic acid			0.23(5)	0.30(7)	0.92(7)			0.63(13)					
					0.66(5)								
Dihydrolipoic acid			0.73[f]		0.96[f]			0.77[f]	0.56[f]				
Lipoic thiolsulfinate	0.30(11)	0.65(12)	0.13(3)		0.66(5)	0.23	0.72	0.68(15)	0.13(15)				
			0.23(5)		0.85(6)								
Bisnorlipoic thiolsulfinate			0.16(6)		0.82(6)			0.66(15)	0.11(15)				
Tetranorlipoic thiolsulfinate			0.09(6)		0.69(6)								

Compound	A	B	C	D	E	F	G	H	I	J
Lipoamide	0.90[f]	0.88[f]	0.85	0.95	0.74[f]	0[f]	0.90	0.54	0.88	0.81
Dihydrolipoamide	0.88[f]	0.88[f]			0.74[f]	0[f]	0.95	0.62	0.86	0.81
Lipoamide thiolsulfinate	0.69[f]	0.73[f]	0.66	0.87	0.63[f]	0.36[f]				
β-Hydroxybisnorlipoic acid	0.31(6)	0.70(6)			0.46(9)					
Bisnorlipoate methyl ketone	0.90(14)	0.91(14)			0.87(14)	0.70(14)				
Lipoate							0.77	0.45	0.90	0.93
Dihydrolipoate							0.77	0.48	0.92	0.93
Dihydrolipoate persulfide							0.80	0.45	0.60	0.43
Dihydrolipoamide persulfide							0.95	0.52	0.60	0.43

[a] The numbers given in parentheses are references.

[b] Specified solvents (ratios by volume) are: A, n-butanol saturated with 0.5 N NH_4OH; B, 2,6-lutidine:water (65:35); C, n-butanol saturated with 2 N NH_4OH; D, n-butanol:ethanol:NH_4OH (4:1:1); E, n-butanol:acetic acid:water (4:1:5, upper phase); F, n-butanol:30% NH_4OH (4:1); G, n-butanol saturated with water; H, chloroform:methanol:formic acid (8:1:1); I, benzene:ether (1:1); J, n-propanol; K, n-butanol; L, n-propanol:acetic acid:35% NH_4OH (7:3:2); M, n-butanol:acetic acid:35% NH_4OH (7:3:2).

[c] Values taken from Wada et al. (4).

[d] Precoated TLC plates, Brinkmann Instrument.

[e] Aluminum, Merck AG; values taken from Silver and Kelly (10).

[f] H. C. Furr, unpublished observations.

TABLE II

DISTINCTIVE SPECTRAL CHARACTERISTICS OF LIPOIC ACID AND ITS METABOLITES[a]

Compound	UV (nm)	IR (cm^{-1})	NMR (ppm)	Other
Lipoic acid	333 (ϵ = 150)(19)	2970, 3300–2500, 1710, 1410, 1280, 1250, 945, 665(20)	1.6(m) 1.8–2.0(m) 2.2–2.6(m) 2.9–3.3(t) 3.4–3.7(q) 10.5 (broad)(13)	pK_a' 4.76 (20) Mass spectrum: P = 206 B = 81(13)
Bisnorlipoic acid	330 (ϵ = 160)(13)	2900, 1710, 1440, 1215, 945, 675(13)	1.8–2.2(m) 2.4–2.7(m) 3.1–3.3(t) 3.4–3.9(q) 10.7 (broad)(13)	Mass spectrum: P = 178 B = 71(13)
Tetranorlipoic acid	280(21) (ϵ = 226)(13) 333 (ϵ = 161)	2900, 1690, 1405, 1295, 1215, 900(22)	2.57(m) 3.29(m) 4.31(q) 9.7 (broad)(22)	pK_a' 3.42(21) Mass spectrum: P = 150 B = 150

Compound	λ_{max}	IR	NMR	Mass spectrum
β-Hydroxybisnorlipoic acid	335(9)	3300, 2570–2700(9)	2.2–2.4(m) 2.5–2.6(m) 3.1–3.3(t) 3.5–3.9(m)(9)	Mass spectrum: P = 194 B = 194 pK$_a'$ 5.0(23)
Lipoic thiolsulfinate	244	(ε = 1020)(19) (ε = 875)(15)	3300–3050, 1740, 1040(23)	1.4–2.0(m) 2.3–2.6(m) 2.6–2.9(m) 3.2–3.4(m) 3.4–3.8(m)(15)	Mass spectrum: P = 222 B = 81
Bisnorlipoic thiolsulfinate	245	(ε = 875)(15)	1000, 1025 (s-1 thiolsulfinate)(15) 1035, 1075 (s-2 thiolsulfinate)(15)	1.7–2.0(m) 2.2–2.7(m) 2.7–2.9(m) 3.1–3.3(m) 3.4–3.9(q)(15)	Mass spectrum: P = 194 B = 71
Bisnorlipoate methyl ketone	320(s) 270(s) 235	(ε = 150) (ε = 395) (ε = 1650)(14)	2930, 1710, 450(14)	1.7–2.0(m) 2.15(s) 2.4–2.9(m) 3.0–3.2(t) 3.5–3.8(m)(14)	Mass spectrum: P = 176 B = 43

[a] The numbers given in parentheses are references.

visualized (at 5–10 μg) as dark spots on developed paper or thin-layer chromatograms under ultraviolet light; the disulfhydryl compounds do not absorb in the near ultraviolet and so are not detectable by this means. The disulfhydryl compounds react readily with spray reagent of nitroferricyanide (1.5 g of sodium nitroferricyanide dissolved in 5 ml of 1 M sulfuric acid, then diluted with 95 ml of methanol and 10 ml of 28% ammonium hydroxide and filtered), giving a transient red color; disulfide compounds react with nitroferricyanide only after also spraying with cyanide (2 g of sodium cyanide in 5 ml of water, diluted to 100 ml with methanol), as do the thiolsulfinates. Thiolsulfinate compounds can be detected by exposing chromatograms to hydrochloric acid fumes, followed quickly by spraying with a solution of 1% potassium iodide and 1% soluble starch.[16] The platinum chloride spray reagent (a spray of equal parts of 3.3 mM chloroplatinic acid and 66 mM potassium iodide plus six parts of water, followed promptly by a spray of concentrated hydrochloric acid)[17] is useful for sulfur compounds in general. Compounds on thin-layer chromatograms may be visualized by exposure to iodine vapor in an enclosed chamber; the disulfide and thiolsulfinate compounds absorb iodine readily and darken more rapidly than does the surrounding chromatographic support, whereas the disulfhydryl spots remain bleached for a time, presumably because of reduction of molecular iodine to iodide.[18] Radioautography[3] and bioautography[12] have also been used to determine relative mobilities on paper chromatograms.

Both anion-exchange and gel filtration (on Sephadex LH-20, Pharmacia Fine Chemicals) have proved useful as column chromatographic techniques for separating catabolites of lipoic acid.[9,13]

Spectral characteristics of lipoic acid and its known metabolites are given in Table II.[9,13–15,19–24] It is apparent that weak ultraviolet absorption at 330 nm is characteristic of the unmodified dithiolane ring, as is a peak at m/e 105 in the electron-impact ionization mass spectra. Ultraviolet absorption at 245 nm and infrared absorption at 1000 to 1075 cm^{-1} distin-

[16] D. Barnard and E. R. Cole, *Anal. Chim. Acta* **20**, 540 (1959).

[17] H. M. Winegard, G. Toennies, and R. J. Block, *Science* **108**, 506 (1948).

[18] P. R. Brown and J. O. Edwards, *J. Chromatogr.* **43**, 515 (1969).

[19] J. A. Barltrop, P. M. Hayes, and M. Calvin, *J. Am. Chem. Soc.* **76**, 4348 (1954).

[20] L. J. Reed, I. C. Gunsalus, G. H. F. Schnakenberg, Q. F. Soper, H. E. Boaz, S. F. Kern, and T. V. Parke, *J. Am. Chem. Soc.* **75**, 1267 (1953).

[21] G. Claeson, *Acta Chem. Scand.* **9**, 178 (1955).

[22] G. Claeson, *Ark. Kemi* **30**, 277 (1969).

[23] J. A. Brockman, E. L. R. Stokstad, E. L. Patterson, J. V. Pierce, and M. E. Macchi, *J. Am. Chem. Soc.* **76**, 1827 (1954).

[24] I. Saito and S. Fukui, *J. Vitaminol. (Kyoto)* **13**, 115 (1967).

guish the thiolsulfinates. The optical rotation of (+)-lipoic acid has been given as $[\alpha]^{23}D + 104°$ (c 1.88, benzene), that of (−)-lipoic acid, $[\alpha]^{23}D -113°$ (c 1.88, benzene).[25]

[25] E. Walton, A. F. Wagner, F. W. Bachelor, L. H. Peterson, F. W. Holley, and K. Folkers, *J. Am. Chem. Soc.* **77**, 5144 (1955).

[28] Thin-Layer Chromatography of Lipoic Acid, Lipoamide, and Their Persulfides

By MARVIN SILVER

The persulfides of lipoic acid and lipoamide have been reported to be the initial intermediates formed in the scission of thiosulfate by the rhodanese-catalyzed reaction[1] in the presence of dihydrolipoate and dihydrolipoamide, respectively. As the maximum absorbance of α-lipoate, lipoamide, and their persulfides is at 330–340 nm, it is not possible to distinguish them by spectrophotometry.[2] A system to separate these compounds on silica gel thin-layer sheets was developed[3] in which the solvent systems used were neither very acidic, which would cause the hydrolysis of the sulfide moiety, nor very alkaline, which would cause the oxidation of the persulfides of dihydrolipoate and dihydrolipoamide to α-lipoate and lipoamide, respectively, with the evolution of sulfide.

Materials and Reagents

Silica gel, 60 F_{254}, 20 × 20 cm, 0.2 mm thick, precoated aluminum sheets (Merck A. G., Darmstadt, Germany)

n-Propanol–acetic acid–NH$_3$ (35%) (7:3:2)

n-Butanol–acetic acid–NH$_3$ (35%) (7:3:2)

n-Propanol

n-Butanol

DL-α-Lipoic acid (Sigma), 0.1 M in 80% (v/v) ethanol

DL-Lipoamide (Sigma), 40 mM in 80% (v/v) ethanol

DL-Dihydrolipoate (Sigma), 0.1 M in 0.2 N NaOH

Dihydrolipoamide, 0.1 M, is prepared by a modification of the method of Reed *et al.*[4] by dissolving 205.3 mg of DL-lipoamide (Sigma) in 5 ml 80% (v/v) ethanol at 0° before adding 300 mg of NaBH$_4$ in 1 ml of H$_2$O. The reaction mixture is stirred under

[1] M. Villarejo and J. Westley, *Fed. Am. Soc. Exp. Biol.* **22**, 412 (1963).
[2] M. Villarejo and J. Westley, *J. Biol. Chem.* **238**, 4016 (1963).
[3] M. Silver and D. P. Kelly, *J. Chromatogr.* **123**, 479 (1976).
[4] L. J. Reed, M. Koike, M. E. Levitch, and F. R. Leach, *J. Biol. Chem.* **232**, 143 (1958).

oxygen-free nitrogen for about 1 hr until clear, adjusted to a pH lower than 3 with 1 N HCl, extracted in a separatory funnel with reagent grade CHCl$_3$, dried *in vacuo,* and dissolved in 10 ml of 0.1 N NaOH.

Persulfides of dihydrolipoate and dihydrolipoamide are prepared by a modification of the method of Villarejo and Westley.[2] A 10mM solution of DL-dihydrolipoate (Sigma) or dihydrolipoamide is allowed to react with 10 mM Na$_2$S in 0.1 N NaOH for 3 hr at room temperature under oxygen-free nitrogen.

Procedure

Five to fifty microliters of each of the reference compounds is applied 1.0 cm from the lower edge of a chromatographic sheet by means of a micropipette and an air dryer. The chromatographic sheet is then subjected to development in either of the solvents for a distance of 10–15 cm and air dried. α-Lipoate and lipoamide can be detected by ultraviolet light, as can the persulfides if the chromatograms are developed in the absence of oxygen. In the presence of oxygen, the persulfides are destroyed with the production of α-lipoate or lipoamide. Residual dihydrolipoate persulfate and dihydrolipoamide persulfate can be detected with iodine vapor. These spots remain lighter than the background, as do the spots of dihydrolipoate and dihydrolipoamide up to 5 min after the initial exposure to iodine vapor, after which they become darker. The spots of α-lipoate and lipoamide are always darker than the background.

Both solvent systems can resolve dihydrolipoate persulfide from dihydrolipoate and α-lipoate, and dihydrolipoamide persulfide from dihydrolipoamide and lipoamide (table). These solvent systems can also separate dihydrolipoate and α-lipoate from dihydrolipoamide and lipoamide, but cannot separate the oxidized and reduced forms of these compounds

R_f VALUES OF LIPOATE, LIPOAMIDE, AND THEIR PERSULFIDES

Compound	Solvent A[a]	Solvent B[b]
DL-α-Lipoate	0.90	0.93
DL-Dihydrolipoate	0.92	0.93
Dihydrolipoate persulfide	0.60	0.43
DL-Lipoamide	0.88	0.81
Dihydrolipoamide	0.86	0.81
Dihydrolipoamide persulfide	0.60	0.43

[a] Solvent A: n-propanol–acetic acid–NH$_3$ (35%) (7:3:2).
[b] Solvent B: n-butanol–acetic acid–NH$_3$ (35%) (7:3:2).

from each other. Nor are the persulfides successfully separated from each other using these solvent systems. These can be separated by chromatography using n-propanol as a solvent system, with which dihydrolipoate persulfide and dihydrolipoamide persulfide have R_f values of 0.80 and 0.95, respectively, or by using n-butanol as a solvent system, with which dihydrolipoate persulfide and dihydrolipoamide persulfide have R_f values of 0.45 and 0.52, respectively.

Both the persulfides of dihydrolipoate and dihydrolipoamide are labile in the presence of molecular oxygen. The stability of these compounds can be greatly increased by chromatography in an atmosphere of oxygen-free nitrogen in a glove box. Using this method, it was possible to detect the formation of the persulfides of dihydrolipoate and dihydrolipoamide when these compounds were used as the acceptors of sulfane sulfur from thiosulfate in the reaction catalyzed by rhodanese from *Thiobacillus* A2.[5]

[5] M. Silver and D. P. Kelly, *J. Gen. Microbiol.* **97**, 277 (1976).

[29] Preparation and Separation of S-Oxides of α-Lipoic Acid[1,2]

By ROBERT W. MURRAY, FRANK E. STARY, and SOHAN L. JINDAL

When lipoic acid (I) is isolated from beef liver, it is accompanied by a monooxidation product.[3,4] This product, originally called β-lipoic acid, (II) or (III), has two structural possibilities depending upon which S atom is oxidized. In addition, (II) and (III) each consists of a stereoisomeric pair (IIa), (IIb) and (IIIa), (IIIb). In this early work, it was not possible to determine which monooxidation product had been formed or whether the material was naturally occurring or was produced in the work-up.

Oxidation of lipoic acid has been accomplished by several chemical reagents. These studies have variously reported one or two monooxida-

[1] This study was supported by grants from the National Science Foundation (Nos. GP 29373 X and CHE 75 23074 A01).
[2] Original source: F. E. Stary, S. L. Jindal, and Robert W. Murray, *J. Org. Chem.* **40**, 58 (1975). Portions of the original source including Fig. 1 and Table I are reprinted with permission of the American Chemical Society. Copyright by The American Chemical Society.
[3] E. L. Patterson, J. A. Brockman, Jr., and F. P. Day, *J. Am. Chem. Soc.* **73**, 5919 (1951).
[4] L. J. Reed, I. C. Gunsalus, G. H. F. Schnakenberg, Q. F. Soper, H. E. Boaz, S. F. Kernan, and T. V. Parke, *J. Am. Chem. Soc.* **75**, 1267 (1953).

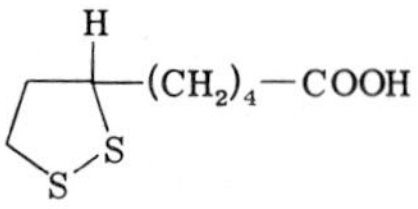

H
├─(CH₂)₄─COOH

S S

(I)

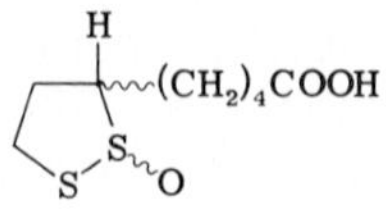

H
├∿(CH₂)₄COOH

S S O

(IIa, IIb)

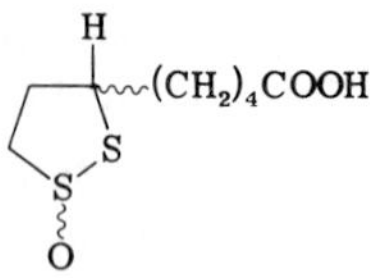

H
├∿(CH₂)₄COOH

S S

O

(IIIa, IIIb)

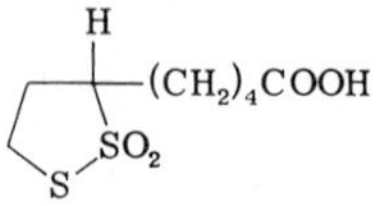

H
├─(CH₂)₄COOH

SO₂
S

(IV)

H
├─(CH₂)₄COOH

S S

O₂

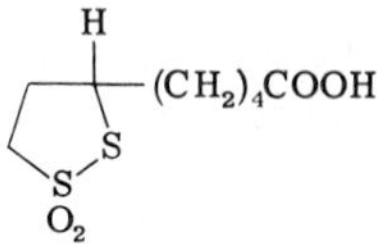

(V)

tion products with no definitive structural assignments. In addition to the four monooxidation products, thiolsulfinates (IIa), (IIb), (IIIa), and (IIIb), lipoic acid oxidation can also give two thiolsulfonate products, (IV) and (V).

We have developed a technique for determining the true number of oxidation products formed under various oxidation conditions. The technique combines the use of thin-layer chromatography (TLC) and the use of nuclear magnetic resonance (NMR) chemical shift reagents. The technique permits a determination of the relative amounts of isomers formed, but specific structural assignments within thiolsulfinate and thiolsulfonate groups must await the results of ongoing ^{13}C NMR studies. To facilitate work-up, all procedures were carried out with the methyl ester of lipoic acid.

Oxidation of Methyl α-Lipoate

Reagents
Benzene
Chloroform
Diethyl ether
Ethanol
Ethyl acetate
Methanol
Methylene blue
Methylene chloride
Chloroform-d
Methyl α-lipoate
Tris(1,1,1,2,2,3,3-heptafluoro-7,7-dimethyl-d_6-4,6-octanedione-d_3)–

Europium(III) [Eu(fod)$_3$-d_{27}]
Triphenyl phosphite
Ammonium persulfate
tert-Butyl hydroperoxide
Aqueous peracetic acid, 40%

Instrumentation

The NMR spectra were measured on a Varian T-60 high-resolution NMR spectrometer. The spectra were measured in CDCl$_3$ solution. Chemical shift values are δ values relative to internal TMS. Infrared spectra were measured on a Perkin–Elmer Model 137 infrared spectrophotometer. The mass spectral analyses were carried out on an AEI MS-12 mass spectrometer and were run at 70 eV. The gas-liquid phase chromatographic (GLPC) analyses were performed on a Varian-Aerograph Model 705 gas chromatograph, using a 0.25 inch × 6 foot column of 10% Carbowax on 60–80 mesh Chromosorb, operated at 140° with an He flow rate of 40 ml/min. The photolysis apparatus was similar to one described in the literature,[5] and used a General Electric DWY 650-W lamp without filter. Ozone was produced in a Welsbach Model T-408 Laboratory Ozonizer.

Procedure

Sensitized Photooxidation. A solution of 1.10 g (4.59 mmol) of methyl α-lipoate and 0.0522 g of Methylene Blue in 200 ml of CHCl$_3$ is photooxidized for ca. 10 min, by which time oxygen absorption will have essentially ceased. A total of 49 ml (2.19 mmol) of O$_2$ is absorbed. The temperature of the reaction solution is maintained at 3°–5°. The chloroform solvent is then removed on the rotary evaporator to give ca. 1.17 g of residue. The residue is analyzed by TLC as described below. A similar oxidation can be carried out using methanol as solvent.

Ammonium Persulfate. To a solution of 0.102 g (0.463 mmol) of methyl α-lipoate in 2.5 ml of diethyl ether and 5 ml of 90% ethanol is added 470 μl (0.47 mmol) of 1 *M* aqueous ammonium persulfate. The reaction solution is allowed to stand at room temperature for 16 hr and then analyzed by preparative TLC.

tert-Butyl Hydroperoxide.[6,7] A solution of 0.102 g (0.463 mmol) of

[5] C. S. Foote, S. Wexler, W. Ando, and R. Higgins, *J. Am. Chem. Soc.* **90**, 975 (1968).

[6] I. Saito and S. Fukui, *J. Vitaminol.* (*Kyoto*) **13**, 115 (1967).

[7] M. W. Bullock, J. A. Brockman, Jr., E. L. Patterson, J. V. Pierce, M. H. von Saltza, F. Sanders, and E. L. R. Stokstad, *J. Am. Chem. Soc.* **76**, 1828 (1954).

methyl α-lipoate in 2.5 ml of diethyl ether is added to 8 ml of methanol. To this solution is added a solution of 0.0416 g (0.463 mmol) of *tert*-butyl hydroperoxide in 2 ml of methanol. The resulting solution is allowed to stand overnight at room temperature. The volume of solution is reduced to ca. 2 ml on the rotary evaporator and then analyzed by preparative TLC.

Peracetic Acid. To a solution of 0.102 g (0.463 mmol) of methyl α-lipoate in 2.5 ml of diethyl ether, cooled to 0°, are added slowly 74 μl (0.46 mmol) of 40% aqueous peracetic acid. After addition of the oxidant is complete, the solution is allowed to warm to room temperature and stand for 17 hr. A similar oxidation can be carried out using a combination of 2.5 ml of diethyl ether and 10 ml of methanol as solvent for the methyl α-lipoate. In both cases, the final reaction solutions are analyzed by preparative TLC.

Triphenyl Phosphite Ozonide. To 50 ml of methylene chloride at −78° saturated with ozone is added a solution of 0.310 g (1 mmol) of triphenyl phosphite in 10 ml of methylene chloride over a period of ca. 30 min. Ozone is always maintained in excess as indicated by the blue color of the solution. After addition of the triphenyl phosphite is complete, the reaction mixture is flushed with nitrogen to remove excess ozone. A solution of 0.22 g (1 mmol) of methyl α-lipoate in 5.4 ml of diethyl ether is added to the phosphite ozonide solution over a period of 2 min and in the dark. The resulting solution is allowed to warm to room temperature and is then evaporated to give a residue, which is analyzed by TLC.

Chromatography

The TLC analyses are carried out on Merck precoated silica gel, F-254, 5 × 10 cm plates of 0.25 mm thickness. Preparative TLC plates are made from Merck silica gel, PF-254, on 20 × 20 cm glass plates ca. 1 mm thick. All plates are activated by heating in an oven at 125° for a minimum of 3 hr. Products are visualized by ultraviolet radiation.

The product residues from the oxidation reactions are then analyzed using the TLC plates to determine optimum conditions, including developing solvent. The chromatographic separations are then carried out on preparative TLC plates. A sample procedure, which can be used for all the reaction products described above, is given here. The product residue used in this sample procedure is that from the photosensitized oxidation of methyl α-lipoate in chloroform.

The residue is dissolved in chloroform and then applied to the preparative TLC plates using a commercial applicator. The plates are developed using ethyl acetate–benzene (1:1). The separated bands are extracted from the TLC plates using chloroform. The extracted materials are then weighed, and spectroscopic information is obtained. A sample analysis using 40.2 mg of the crude reaction product from photosensitized oxidation of methyl α-lipoate in chloroform leads to the following results:

Band No.	R_f	Weight (mg)	% of total	m/e	Infrared
1	0.27	13.4	36	236	1070 cm^{-1} (thiolsulfinate)
2	0.39	10.4	28	236	1080 cm^{-1} (thiolsulfinate)
3	0.68	10.2	26	252	1310, 1130 cm^{-1} (thiolsulfonate)
4	0.77	2.8	6	279	Unidentified

The results shown are optimal TLC results. Despite repeated attempts it was not possible to further separate products into more than two thiolsulfinate and one thiolsulfonate band. Column and dry column chromatography do not disclose the presence of more products.

mass spectra of the separated components were also determined with

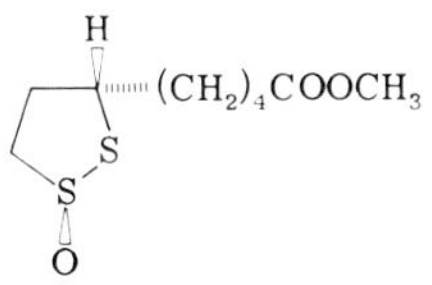

(VI)

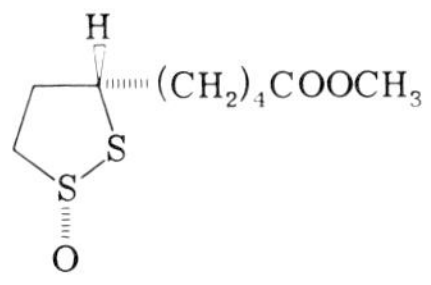

(VII)

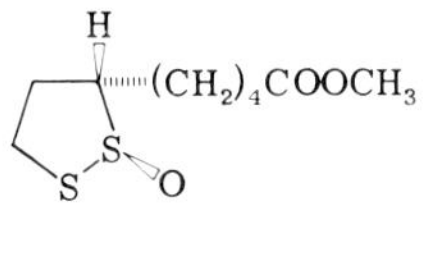

(VIII)

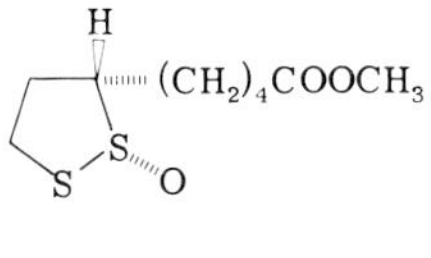

(IX)

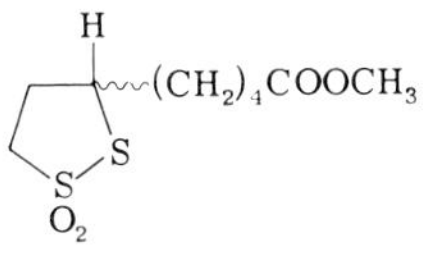

(X)

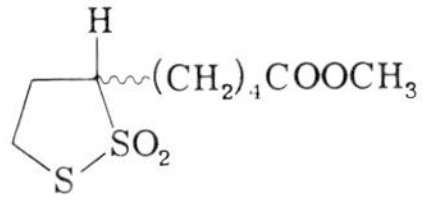

(XI)

the following results. Band 1 material had absorptions at 3.65(S, 3H, CH_3), 3.2(t), 2.4(m), 1.6(broad S), 1.3(S), and 0.9(m). Band 2 material had absorptions at 3.65(S, 3H, CH_3), 3.6–2.6(m), 2.4(t), and 1.6(broad S). Band 3 material had absorptions at 4.3(broad m), 3.65(S, 3H, CH_3), 3.6–3.1(m) 2.9(d), 2.8(d), 2.4(t), and 1.6(broad S).

The separated fractions are then analyzed further using an NMR chemical shift reagent as described below. This step is necessary since there are four possible thiolsulfinate oxidation products, (VI)–(IX), and two possible thiolsulfonate products, (X) and (XI). Despite the fact that the chromatographic analysis used here and those used by earlier workers always suggest less than six oxidation products, such analyses should not be regarded as conclusive.

Use of an NMR Chemical Shift Reagent

The materials in the TLC bands with R_f values of 0.27, 0.39, and 0.68 are then examined further using the NMR shift reagent tris(1,1,1,2,2,3,3-heptafluoro-7,7-dimethyl-d_6-4,6-octanedione-d_3) europium(III), Eu(fod)$_3$-d_{27}. The material with an R_f of 0.39 is dissolved in

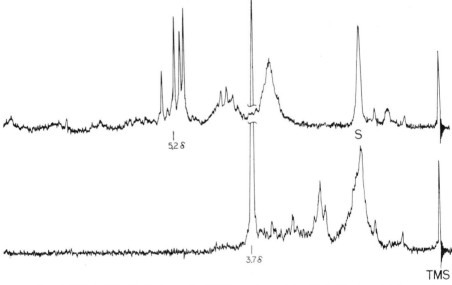

FIG. 1. Effect of Eu-(fod)$_3$-d_{27} on the NMR spectrum of the thiolsulfinate oxidation products of methyl α-lipoate. The bottom spectrum is the unshifted spectrum. The top spectrum was obtained with a ratio of shift reagent to substrate of 0.67. The peak marked S is due to the shift reagent. From F. E. Stary, S. L. Jindal, and Robert W. Murray, *J. Org. Chem.* **40**, 58 (1975). Reprinted with permission.

CDCl$_3$ and treated with varying ratios of shift reagent to substrate. The shift reagent causes all NMR absorptions to shift to lower field. At all ratios of shift reagent to substrate up to 1.0, the methyl group NMR absorption remains a single peak indicating the presence of a single thiolsulfinate of methyl α-lipoate. When the material with R_f value of 0.27 is similarly treated with shift reagent, the methyl group absorption is shifted downfield and splits into three separate peaks. The optimal ratio of shift reagent to substrate for disclosing the presence of three peaks is 0.67. This TLC band thus contains three thiolsulfinates. It is not presently possible to assign structures [(VI)–(IX)] to the thiolsulfinates. The thiolsulfinate bands (R_f values of 0.27 and 0.39) in each of the oxidations is combined and treated with shift regaent at a ratio of 0.67. This allows a determination of the relative distribution of thiolsulfinates, which are arbitrarily designated A–D, with A at lowest field and D at the highest field in

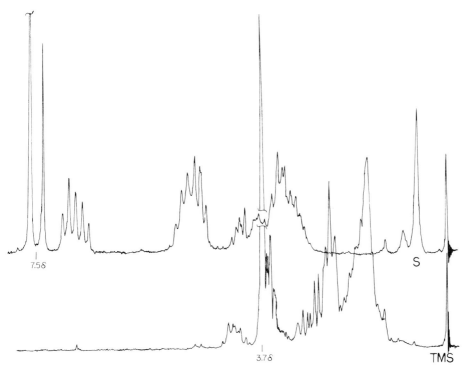

FIG. 2. Effect of Eu-(fod)$_3$-d_{27} on the NMR spectrum of the thiolsulfonate oxidation products of methyl α-lipoate. The bottom spectrum is the unshifted spectrum. The top spectrum was obtained with a ratio of shift reagent to substrate of 0.27. The peak marked S is due to the shift reagent.

SUMMARY OF RESULTS OF OXIDATION OF METHYL α-LIPOATE[a]

Oxidant	Solvent	Percentage distribution of thiolsulfinates				Total yield of thiolsulfinates (%)	Percentage distribution of thiol-sulfonates[b]		Total yield thiolsulfonates (%)
		A	B	C	D		E	F	
O_2, hv, S	$CHCl_3$	9	48	28	16	64	59	41	25.7
O_2, hv, S	MeOH	14	29	25	32	75.4	14	86	15.4
$(NH_4)_2S_2O_8$	90% EtOH	15	30	25	30	21	—	—	Trace
t-BuOOH	MeOH	10	35	20	34	69	—	—	Trace
$CH_3C(=O)OOH$	Et_2O	4	37	26	33	42	10	90	ND[c]
$CH_3C(=O)OOH$	MeOH	11	29	26	34	52	—	—	Trace
$(\phi O)_3PO_3$	CH_2Cl_2	12	33	23	32	26	—	—	Trace

[a] From F. E. Stary, S. L. Jindal, and Robert W. Murray, J. Org. Chem. 40, 58 (1975). Reprinted with permission.
[b] Where no data are given, there was insufficient material for an accurate analysis.
[c] ND, not determined.

the shifted spectrum. Figure 1 illustrates the effect of shift reagent on the thiolsulfinate spectrum.

The TLC band with an R_f value of 0.68 was also treated with shift reagent. In this case the optimal ratio of shift reagent to substrate is 0.27. At this ratio the methyl absorption is seen to shift downfield and clearly split into two absorptions indicating the presence of both thiolsulfonates, (X) and (XI) (Fig. 2). Again it is not presently possible to assign structures to the two thiolsulfinate NMR absorptions, and they are arbitrarily designated as thiolsulfonates E and F, with E at lowest field in the shifted spectrum.

This procedure using shift reagent is repeated for the products of all the oxidations described above. The results are contained in the table. The results indicate that, in general, the photooxidation conditions produce all six possible oxidation products, whereas chemical oxidation conditions usually produce only the four thiolsulfinate oxidation products.

[30] Photochemical Preparation of Acyl Lipoic Acids

By Makoto Takagi

Acetyl dihydrolipoic acids (6,8-dimercaptooctanoic acid, 6,8-dithioloctanoic acid; 6- and 8-S-acetyl isomers) have been prepared by chemical or enzymic acetylation.[1,2] The 8-S-acetyl isomer is obtained from chemical procedures, while the 6-S-acetyl isomer results from enzymic action. The lipoamide counterparts have been prepared enzymically.[3,4]

8-S-Acetyl-6,8-dimercaptooctanoic acid is conveniently prepared in high yield from lipoic acid by irradiating the solution in acetaldehyde.[5] The reaction can be extended to other simple aliphatic aldehydes, producing the corresponding 8-S-acylated compounds. The reaction proceeds by a radical chain mechanism, and the induction periods sometimes appear, depending on the purity of the aldehydes.

Procedure. An irradiating light ($\lambda \geq 330$ nm) is obtained from a 100 W immersion-type high-pressure mercury lamp doubly jacketed with Pyrex

[1] I. C. Gunsalus and W. R. Razzell, this series, Vol. 3 [138], p. 941.

[2] I. C. Gunsalus, L. S. Barton, and W. Gruber, *J. Am. Chem. Soc.* **78,** 1763 (1956).

[3] J. V. Pauksteilis, E. F. Byrne, T. P. O'Connor, and T. E. Roche, *J. Org. Chem.* **42,** 3941 (1977).

[4] T. C. Linn, J. W. Pelley, F. H. Pettit, F. Hucho, D. D. Randall, and L. S. Reed, *Arch. Biochem. Biophys.* **148,** 327 (1972).

[5] M. Takagi, S. Goto, and T. Matsuda, *J. Chem. Soc., Chem. Commun.* 1976, 92.

METHODS IN ENZYMOLOGY, VOL. 62

$$\underset{S-S}{\overset{\text{COOH}}{\diagup\diagdown\diagup\diagdown}} + \text{RCHO} \xrightarrow{h\nu} \underset{\text{RCOS} \quad \text{SH}}{\overset{\text{COOH}}{\diagup\diagdown\diagup\diagdown}}$$

tubes. The internal jacket contains circulating cold water, and the external jacket is filled with an optical filter solution prepared from 500 g of sodium bromide, 3 g of lead nitrate, and 1 liter of distilled water. The irradiating assembly is combined with a reaction vessel, to which 100 ml of freshly distilled (under nitrogen or argon) aldehyde and 2.0 g of commercial DL-lipoic acid are placed and stirred. The amount of lipoic acid is reduced if it exceeds the solubility in aldehyde. The atmosphere is then replaced with nitrogen or argon by refluxing under reduced pressure at room temperature. The solution is cooled externally by ice water ($0°-10°$) and irradiated until the yellowish color of the solution disappears (40–90 min). The reaction may be followed by thin-layer chromatography on silica gel (chloroform–ethyl acetate). The residual oil after concentration of the reaction mixture under reduced pressure was further evacuated (≤ 0.01 mm Hg) for several hours and then chromatographed on silica gel (28 mm × 20 cm). After a small amount of impurity has been washed out with chloroform, the product is eluted with chloroform–ethyl acetate (20:1–2, v/v). The main band is separated from the succeeding minor band, which contains by-products as well as the unreacted lipoic acid. Removal of the solvent under reduced pressure gives an acylated dihydrolipoic acid as colorless oil (50–70% yield for C_2-C_5 aldehydes). The product is stored under an inert atmosphere in the dark to prevent air-oxidation. The reaction is clearer (yield $\geq 70\%$) when a reduced amount of lipoic acid (0.2–0.5 g in 100 ml of aldehyde) is used. A single-jacketed irradiation assembly ($\lambda \geq 300$ nm) without a liquid filter may be used with a slight reduction in yield. The identification of the product is achieved by means of elemental analysis, IR, and ^{1}H NMR. 8-S-Acetyl-dihydrolipoic acid: IR, cm^{-1}: ν_{OH} (3300–2500), ν_{SH} (2580, very weak), ν_{COOH} and $\nu_{RCOSR'}$ (1695, 1710). ^{1}H NMR, δ(ppm from TMS in CDCl$_3$): 1.35 (d, $J = 7.0$ Hz,

$\diagup\!\!\!\diagdown$CH—SH, The signal disappears on exchange with D$_2$O), 2.31 (s,

CH$_3$CO—), 3.01 (t, $J = 7.0$ Hz, small splittings with $J \leq 2$ Hz, —COSCH$_2$—). The other signals are similar to those of dihydrolipoic acid. The identification through derivatization by carbodiimide[2] is not very successful. The purity based on —SH determination by means of 5,5'-dithiobis(2-nitrobenzoic acid)[6] is 85–90%. 8-S-n-Butyryldihydrolipoic acid: IR: similar to the acetyl derivative; ^{1}H NMR: 0.95 (t, $J = 7.2$ Hz,

—CH$_2$CH$_3$), 1.39 (d, $J = 7.5$ Hz, $\diagup\!\!\!\diagdown$CH—SH), 3.02 (m, —COSCH$_2$—),

2.55 (t, $J = 7.5$ Hz, —CH$_2$CH$_2$COS—).

[6] G. L. Ellman, *Arch. Biochem. Biophys.* **82,** 70 (1959).

[31] Lipoyl Disulfide Reducing Polymers

By MARIAN GORECKI and ABRAHAM PATCHORNIK

Reduction of disulfide bonds in small organic molecules such as cystine-containing peptide hormones or cofactors may be facilitated by use of the mild polymeric reducing agent that can be easily removed, leaving the reduced thiol in solution.

Different thiol-containing polymers have been recently described in which the monothiol, such as *N*-acetylhomocysteine,[1-3] glutathione-2-pyridyl disulfide,[4] or cysteine,[4] were attached covalently to different polymeric matrices. Some of these conjugates were subsequently used as antioxidants for thiol-containing proteins[1] and for affinity chromatography of SH-containing peptides[5] and proteins.[6,7]

In this chapter, the preparation of thiolated polymers capable of reducing S-S bonds is described. They are easily synthesized by covalent attachment of lipoic acid to different insoluble matrices such as Sephadex, Sepharose, cellulose, polyacrylamide, and polystyrene.[8] The disulfide moiety on the polymer is then reduced to dihydrolipoamide. Owing to the low redox potential of dihydrolipoamide, these conjugates are superior reducing agents for disulfide bonds. The reducing action of dihydrolipoyl polymers is shown below. At the end of reduction, the polymer is removed simply by centrifugation or filtration.

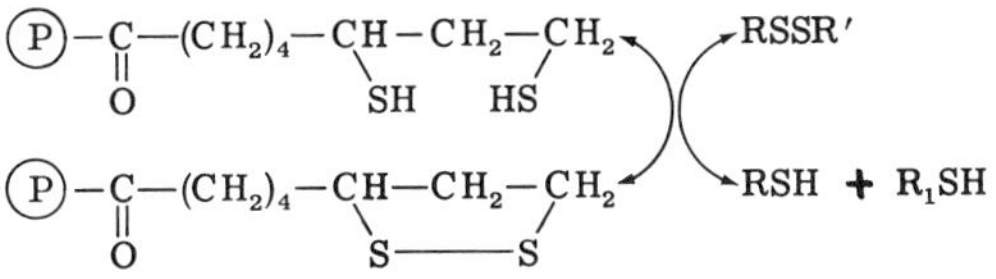

Preparation of Lipoyl Polymers

Lipoic acid is attached to various insoluble polymers via its carboxylic group, utilizing an active ester, or through the amine group, utilizing 2-(lipoamido)ethylamine.

[1] L. Eldjarn and E. Jellum, *Acta Chem. Scand.* **17**, 2610 (1963).

[2] P. Cuatrecasas, *Nature (London)* **228**, 1327 (1970).

[3] P. Cuatrecasas, *J. Biol. Chem.* **245**, 3059 (1970)

[4] K. Brocklehurst, J. Carlsson, M. P. J. Kierstan, and E. M. Crook, *Biochem. J.* **133**, 573 (1973).

[5] T. A. Egorov, A. Svenson, L. Ryden, and J. Carlsson, *Proc. Natl. Acad. Sci. U.S.A.* **72**, 3029 (1975)

[6] E. Jellum, *Acta Chem. Scand.* **18**, 1887 (1964).

[7] J. Carlsson, I. Olsson, R. Axen, and H. Drevin, *Acta Chem. Scand.* **330**, 180 (1976).

[8] M. Gorecki and A. Patchornik, *Biophys. Biochim. Acta* **303**, 36 (1973).

METHODS IN ENZYMOLOGY, VOL. 62

Procedure for Lipoyl Succinimide Ester. One-tenth mole of lipoic acid, *N*-hydroxysuccinimide, and dicyclohexyl carbodimide are dissolved in dioxane–ethyl acetate (3:1, v/v). The mixture is stirred overnight in the dark and the resulting dicyclohexylurea is filtered off. The solvent is then evaporated at low pressure, and the residue is dissolved in ethyl acetate. The small amount of residual dicyclohexylurea is filtered off, and the solvent is evaporated. The compound is recrystallized from isopropanol. The yield is 80%; m.p. 98°–99°).

Procedure for 2-(Lipoamide)ethylamine. One-tenth mole each of lipoic acid, benzyloxycarbonyl ethylenediamine·HCl,[9] dicyclohexylcarbodiimide, and triethylamine are dissolved in ethyl acetate–chloroform and stirred overnight in the dark. Dicyclohexylurea is filtered off, and the solvent is evaporated. The residue is taken up in ethyl acetate, washed successively with 1 M NaHCO$_3$, 1 M HCl, and water, and dried over Na$_2$SO$_4$. Crystallization from ethyl acetate gives 2-(lipoamido)benzyloxycarbonylethylamine in 60% yield (m.p. 79–80°). The benzyloxycarbonyl blocking group is removed with HBr in acetic acid for 1 hr. The hydrobromide is precipitated with ether, washed with ether, and dried in a desiccator. The compound is hygroscopic. The neutral equivalent was found to be 318 (calculated 329) by titration in ethanol with 0.1 M sodium methoxide using thymol blue as indicator.

Lipoyl Aminoethyl Sephadex. A swollen slurry of aminoethyl Sephadex,[1] coarse grade (25 g), is thoroughly sucked on a Büchner funnel and suspended in 15 ml of dimethylformamide. Lipoyl succinimide ester (4.5 g) in 20 ml of dimethylformamide and 10 ml of 1 M NaHCO$_3$ is slowly added. The reaction mixture is shaken for 10 hr. The substituted Sephadex is washed with dimethylformamide, 50% aqueous dimethylformamide, water, and ethanol, and finally dried in a desiccator. The sulfur content (2.3%), as well as the ninhydrin test, indicates that all the amino groups of the starting materials are lipoylated. An alternative route for the synthesis of lipoyl aminoethyl Sephadex is based on coupling of 2-(lipoamide) ethylamine with CNBr-activated Sephadex.[10] CNBr-activated with washed Sephadex (4 g) is suspended in 10 ml of 0.1 M NaHCO$_3$, and stirred with 10 mmol of 2-(lipoamide) ethylamine·HBr in 10 ml of dimethylformamide. After reaction for 16 hr at 4°, the gel is washed with large volumes of dimethylformamide, 50% dimethylformamide, water, ethanol–water, and absolute ethanol, and finally dried. About 0.6 mmol of lipoic acid is bound per gram of polymer.

A similar procedure is applied for the preparation of lipoyl aminoethyl

[9] W. B. Lawson, M. D. Leafer, Jr., A. Tewes, and G. J. S. Rao, *Hoppe-Seyler's Z. Physiol. Chem.* **349,** 251 (1968).
[10] J. Porath, R. Axen, and S. Ernbäck, *Nature (London)* **215,** 1491 (1967).

Sepharose, but the gel is not washed with ethanol or dried. The substituted Sepharose is stored swollen at 4°. This treatment results in a derivative having about 25 μmol of lipoic acid per milliliter of Sepharose.

Lipoyl Aminoethyl Cellulose. This is prepared from aminoethyl cellulose (Serva, 0.28 mequivalent of amine per gram of cellulose) as described for lipoyl aminoethyl Sephadex. The degree of substitution is 85–95%.

Lipoyl Aminoethyl Polyacrylamide. This is also prepared by substitution of the aminoethyl derivative of polyacrylamide[11] with *N*-lipoyl succinimide. The starting polymer could be either swollen in water or suspended in dimethylformamide, and the reaction with *N*-lipoyl succinimide proceeds identically to that described for the preparation of lipoylated Sephadex. When a 5-fold excess of active ester is used, all the amino groups are substituted. With one equivalent, only 50% of the aminoethyl chains are lipoylated and the remaining free amino groups could be carbamylated or acetylated.

Lipoyl Aminoethyl Polystyrene. Aminoethyl polystyrene is prepared by heating chloromethyl polystyrene (2% cross-linked, 2.24 mmol of Cl per gram) with ammonia in an autoclave for 48 hr and results in a polymer containing 2 mmol of amine per gram of polymer. This is lipoylated following the procedure for lipoyl aminoethyl Sephadex and gives a polymer with 0.25 mmol of lipoic acid per gram of dry polymer.

Reduction of Lipoyl Polymers to Dihydrolipoyl Polymers. The polymer-bound dithiolane is reduced with $NaBH_4$. The procedure described below is used for all the lipoyl polymers: The lipoyl polymer is swollen in water and 50 mg of $NABH_4$ dissolved in water is added per milliliter of gel. The reaction is allowed to proceed for 20–40 min. The reduced polymer is then filtered on a Büchner funnel and washed under suction with large volumes of cold water, 0.1 M acetic acid, and again with water. If the polymers are to be used immediately, final washing is performed with the appropriate buffer. Otherwise the reduced polymer (except for dihydrolipoylated Sepharose) is washed with ethanol and dried in a vacuum desiccator.

The substituted polymers are stable between pH 2 and 10 and can be used repeatedly without loss of lipoic acid and without destruction of the grain form. Prolonged exposure to alkali causes destruction of the disulfide group. The polymers in their reduced, dihydrolipoyl form are stable for several weeks if kept dry. For reductions, pH 8-buffered suspensions of the polymers should be used within a few hours; otherwise they undergo relatively fast oxidation with a decrease in the number of active thiol groups.

[11] J. K. Inman and H. M. Dintzis, *Biochemistry* **8**, 4074 (1969).

Blocking of Unreacted Amine Groups. The lipoamide polymers contain free amino groups, which, if left unreacted, can cause the undesired absorption of compounds destined for reduction. To derivatize these unreacted amine groups, the lipoyl polymers are treated with a 10-fold excess of acetic anhydride in 0.1 M NaHCO₃ for 5 hr or with a 5-fold excess of KCNO in 1 M phosphate buffer pH 7.5 for 15 hr.

Sulfhydryl content of polymers is determined with 5,5'-dithiobis-(2-nitrobenzoate)[12] and free amines with ninhydrin.[13]

Reduction of Disulfides in Batch. Cystamine, cystine, oxidized mercaptoethanol, and oxidized glutathione are easily reduced by simply mixing the disulfide with the polymer. The procedure is described for a batchwise reduction of oxidized glutathione. GSSF (3 μmol) is dissolved in 0.5 ml of 50 mM phosphate buffer (pH 8.0) and stirred or shaken gently at room temperature with an amount of reducing polymer, containing 30 μmol of SH groups, swollen in the reaction buffer before use. A stream of nitrogen is passed over the reaction mixture. After a desired time, the polymer is filtered off and washed twice with 0.2 ml of buffer; the combined filtrate is acidified to pH 3.0 by the addition of 1 M HCl. Under these conditions, dihydrolipoyl aminoethyl polyacrylamide gives quantitative reduction of the glutathione in 5 min. The reducing polymers derived from aminoethyl Sephadex and aminoethyl cellulose are less reactive, and the full reduction of glutathione is achieved after 7 and 15 min, respectively. The maximum rate of reaction is obtained in the pH range of 7.5–8.5. At pH values higher than 9.0, the rate of reduction is slower, probably owing to fast reoxidation of the resulting thiols. The reduction is carried out in 0.05–1 M buffers to avoid nonspecific absorption of the reduced polymers. An exception is dihydrolipoyl ethylaminopolyacrylamide which does not adsorb even at 0.1 M buffer concentration.

A similar immobilized lipoyl derivative attached to propyl glass beads has been recently synthesized[14] and used for affinity chromatography of lipoamide dehydrogenase. After reduction of the N-propyllipoyl glass, the dihydrolipoyl derivative was reported to reduce effectively oxidized glutathione. However, the degree of substitution of glass beads is severalfold less than that bound to agarose or polyacrylamide, and the amount of reducing polymer that has to be used should be considerably greater.

Reduction on the Column. The reduction is carried out by incubation of the peptide in the column containing reducing polymer.[15] Oxytocin (15 μmol, 5.04 mg) is dissolved in 50 μl of 0.1 M hydrochloric acid and

[12] G. L. Ellman, *Arch. Biochem. Biophys.* **82,** 70 (1959).
[13] A. M. Crestfield, S. Moore, and W. H. Stein, *J. Biol. Chem.* **238,** 622 (1963).
[14] W. H. Scouten, this series, Vol. 34, p. 288.
[15] R. Sperling and M. Gorecki, *Biochemistry* **13,** 2347 (1974).

brought to pH 8.0 with 1 M Tris solution. The solution is applied to a 0.5 × 5 cm column packed with the dihydrolipoyl polyacrylamide P-6. The flow is halted, and the peptide is allowed to stand in the closed column at 23° for 3 hr. Oxytocin is eluted from the column with 50 mM Tris·Cl, pH 8.0, into a volumetric flask, to which 5 μl of 4 M HCl per milliliter of eluted effluent is added to ensure a lowering of the pH to pH 2.0. The elution products could be kept for a few hours without detectable oxidation. The peptide emerging just after the void volume of the column is quantitatively eluted with 12 ml of eluent (80% can be collected in the first 4 ml). Even when oxytocin is not completely dissolved before application, the recovery of oxytoceine is quantitative. Under these conditions, complete reduction of oxytocin is achieved, as determined by the comparison of amino acid content of the eluted protein with its sulfhydryl content.

Activation of Papain and Mercuripapain and Reduction of Disulfide Bonds in Proteins. Papain, a thiol enzyme, must be activated by thiol reagents before use. Dihydrolipoyl polymers can serve as good activators for papain. When 5–10 mg of a reducing polymer is added to a reaction mixture containing N^α-benzoylarginine ethyl ester and papain, full esteratic activity of the enzyme is recovered within a few seconds. Dihydrolipoyl aminoethyl Sephadex is effective also in the activation of mercuripapain. Titration of the SH group obtained by demercuration of mercuripapain with dihydrolipoyl ethylamino Sephadex gives a thiol content corresponding to a molecular weight of 23,000.[16] The reducing polymers can also be used for reduction of exposed bridges in proteins. Thus, dihydrolipoyl Sepharose was found to be the most suitable reducing polymer for proteins, probably because Sepharose has a pore size large enough to admit most of proteins. Incubation of ribonuclease in 0.5 M Tris, pH 8.0, with dihydrolipoyl-Sepharose for 16 hr, followed by removal of reagents, results in preferential reduction of exposed S-S bridges. This was shown by alkylation of the reduced protein with radioactive iodoacetic acid, followed by tryptic peptide analysis by two-dimensional paper chromatography and electrophoresis. Quantitative reduction of S-S bonds in ribonuclease is achieved by incubation with dihydrolipoyl-Sepharose in 8 M urea.

[16] S. Blumberg, I. Schechter, and A. Berger, *Eur. J. Biochem.* **15,** 97 (1970).

[32] Biosynthesis of Lipoic Acid via Unsaturated Fatty Acids

By Jean-Paul Carreau

From *in vivo* experiments performed on the rat,[1-4] it has been suggested that a metabolic relationship exists between unsaturated fatty acids and lipoic acid. According to this hypothesis, essential fatty acids may be the preferential precursors for lipoic acid biosynthesis through a yet unknown metabolic pathway.

Principle. [U-^{14}C]linoleic acid and [U-^{14}C]oleic acids are injected into 30 6-day-old rats. The analysis of the aqueous extracts of rat livers 3 hr after the injection reveals that lipoic acid and its degradative products are 7–10 times more radioactive with [U-^{14}C]linoleic acid than with [U-^{14}C]oleic acid.

Materials and Methods

Purity of the Labeled Fatty Acids

The fatty acids are obtained from the International Chemical and Nuclear Corporation.

The specific radioactivity of [U-^{14}C]linoleic acid and [U-^{14}C]oleic acid is 250–500 mCi/mmol.

The purity of the fatty acids can be checked by radio-gas chromatography, after esterification with anhydrous MeOH–HCl (1.1 N).[5]

Analytical Conditions. Gas chromatograph: Microteck 220. Column: 200 cm × 4.5 cm. Liquid phase: DEGS (20%) on Chromosorb 60 × 80. Temperature of the column: 200°. Nitrogen pressure: 2 bars. Radioactive detector: Packard; sensitivity, 5 K, time constant, 5%.

Preparation of the Animals and Injection of Labeled Fatty Acids

Experiments on Pregnant Females. Pregnant rats (Wistar) are fed after day 10 of gestation with a fat-free diet containing (per kg): fat-free

[1] J. P. Carreau, D. Lapous, and J. Raulin, *J. Physiol. (Paris)* **71,** 327A (1975).
[2] J. P. Carreau, D. Lapous, and J. Raulin, *C. R. Acad. Sci. (Ser. D),* **281,** 941 (1975).
[3] J. P. Carreau, Doctoral thesis, University of Paris VI, 1976.
[4] J. P. Carreau, D. Lapous, and J. Raulin, *Biochimie* **59,** 487 (1977).
[5] M. Loury, Rev. *Fr. Corps Gras* **14,** 383 (1967).

VITAMIN MIX

Vitamin[a]	
A	5000 IU
D_3	400 IU
K	0.5 mg
E	2.0 mg
B_1	2.0 mg
B_2	4.0 mg
B_6	2.0 mg
Inositol	20.0 mg
PP	2.0 mg
B_{12}	1.1 μg
Biotin	200 μg
Folic acid	100 μg
Casein, fat-free, Q.S.	1.0 g

[a] Vitamin A: 1 IU = 0.3 μg; vitamin D_3: 1 IU = 0.025 μg.

casein, 250 g; sucrose, 630 g; cellulose, 20 g; salt mix, 150 g (according to the formula of Osborne and Mendel), and vitamin mix, 10 g, choline Cl, 500 mg.

Experiments on Animals during the Nursing Period. Labeled fatty acids (1 μCi per 3 g of rat) are dissolved in 0.3 ml (times the number of rats) of an emulsion (500 mg of glucose, 200 mg of Tween 20, 0.1 ml of sunflower oil, and 10 ml of water). The radioactive mixture is emulsified with a homogenizer. The radioactive mixture (0.3 ml) is injected intraperitoneally.

Extraction of Rat Livers

Protocol. The animals are decapitated, and the livers are excised and treated by the solvent mixture proposed by Bligh and Dyer.[6]

The homogenization is performed at 4° with a Potter–Elvehjem apparatus.

For 250 mg of tissue (containing 200 mg of water), 1.8 ml of water is added. After homogenization the following steps are carried out:

1. Addition: 4.4 ml of methanol; homogenization.
2. Addition: 2.4 ml of chloroform; homogenization. At this step, the mixture is homogeneous.
3. Addition: 2.0 ml of chloroform. The mixture is now heterogeneous.
4. Addition: 2.0 ml of water; homogenization.
5. Centrifugation at 2000 rpm for 20 min.

[6] E. G. Bligh and W. J. Dyer, *Can. J. Biochem. Physiol.* **37,** 911 (1959).

Three different phases are observed: (H) = upper methanolic aqueous phase. (M) = fluffy phase (insoluble). (L) = lower chloroformic phase containing the total lipids.

Purification of the Three Phases

PHASE (M). Phase (M) is purified with 6.5 ml of the heterogeneous mixture of extraction (A) (prepare separately by adding 100 ml of chloroform plus 90 ml of water), i.e., 4.3 ml of upper phase and 2.2 ml of lower phase. This step is repeated until the radioactivity determined with Bray's mixture[7] is negligible in the aliquots of the (H) and (L) phases of the washings. Five extractions have been found to be necessary to obtain this result.

PHASES (H) AND (L). The phases (H) of the first extraction and the phases (H) collected during the purification steps of the phase (M) are pooled and washed with a half-volume of the lower phase of the extraction mixture (A) until the radioactivity is negligible in this lower phase.

The phases (L) of the first extraction and purification of (M) and (H) are pooled and purified with 2 volumes of the upper phase of the extraction mixture (A) until the radioactivity extracted in the upper phase is negligible.

FRACTIONATION OF THE PHASE (H). The pooled phase (H) is lyophilized then treated with anhydrous methanol (2.5 ml of methanol per 250 mg). This treatment is repeated twice. Each decantation is preceded by a centrifugation (2000 rpm for 10 min).

Two partial extracts are obtained: (1) (H_{MeOH}) soluble phase in anhydrous methanol; (2) (H_W) insoluble phase in methanol and soluble water (Scheme 1).

The insoluble residue is treated twice with water (i.e., 0.1 ml/250 mg of tissue). Aliquots are kept for protein content determination[8] and radioactivity counting.[7]

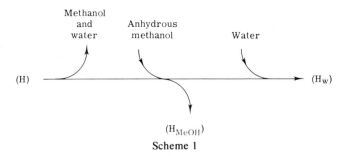

Scheme 1

[7] G. A. Bray, *Anal. Biochem.* **1**, 279 (1960).
[8] O. H. Lowry, N. J. Rosebrough, A. L. Farr, and R. J. Randall, *J. Biol. Chem.* **193**, 265 (1951).

For further steps all the phases (H_W) from the livers of the 30 rats are pooled.

Treatment of the Secondary Extract (H_W)

Total Hydrolysis. One milliliter of 6 N HCl is added to 20 μl (H_W). The tube is left for 1 hr with a stream of nitrogen to eliminate traces of dissolved oxygen. Hydrolysis is then performed in a sealed tube for 24 hr at 110°. After cooling, the acid solution is extracted twice with 1.0 ml of chloroform–hexane (29:71, v/v). The solution is then evaporated under

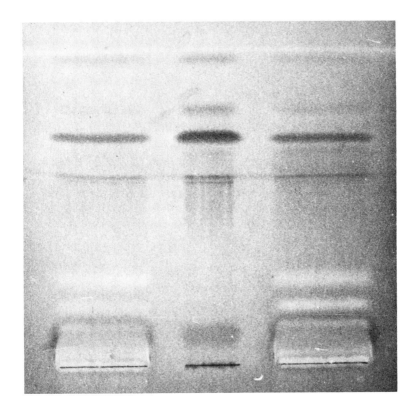

A   **B**   **C**

FIG. 1. Thin-layer chromatography on silica gel of the extracts (M_W) of rat livers treated for 24 hr at 110° with 6 N HCl. Lipoic acid treated in the same way was taken as reference. (A) H_W of [U-^{14}C]linoleic acid-treated rats. (B) Lipoic acid (standard). (C) H_W of [U-^{14}C]oleic acid-treated rats.

vacuum and treated with 1.0 ml of anhydrous methanol–hydrochloric acid for 5 min at 60°, then again evaporated under vacuum.

Partial Hydrolysis. One milliliter of 6 N HCl is added to 100 μl of the extract, and the mixture is left for 16 hr at 110°. After evaporation of the hydrochloric acid, the sample is exracted three times with 1 ml of chloroform. The chloroform is evaporated under vacuum. The lipid residue is treated for 5 min with 1.0 ml of methanol–hydrochloric acid, then evaporated under vacuum.

Treatment of the Lipoic Acid Used as Reference. Lipoic acid (50 mg, Merck) is treated for 24 hr with 3 ml of 3 N hydrochloric acid at 110°. The mixture is then extracted five times with 3 ml of chloroform–hexane (29:71, v/v). The insoluble residue is treated with methanol–hydrochloric acid as indicated above under "Total Hydrolysis."

Analysis of Hydrolysis Products

Analysis of the Water-Soluble Residue Obtained after Total Hydrolysis of (H_w). This analysis is performed by TLC on precoated thin-layer plates of silica gel G-60 (Merck, F 254; 0.25 mm thick). The residues obtained after labeling experiments with the animals and the lipoic acid treated solutions (see above) are chromatographed in the following mixture of solvents: chloroform–methanol–water–formic acid (75:25:4:1). The thin-layer plates are immersed after migration in the following solution: palladium chloride, 0.5 g; water, 100 ml; ethanol, 400 ml; concentrated hydrochloric acid, 10 drops.

The plates are dried and viewed under UV light. The reddish spots, which have the same R_f as the degradative products of lipoic acid, are scraped off, and their radioactivity is determined with Bray's mixture.[7]

Figure 1 reveals in the residues the presence (see above) of com-

TABLE I

INCORPORATION OF [U-^{14}C]LINOLEIC (LI) OR [U-^{14}C]OLEIC (OL) ACIDS IN THE
DEGRADATION PRODUCTS OF LIPOIC ACID AND IN METHYL LIPOATE ISOLATED
FROM WATER-SOLUBLE EXTRACTS (H_w) OF RAT LIVER

	Degraded lipoic acid		
	Total radioactivity (dpm)	Radioactivity per mg of protein of H_w (dpm)	Lipoic acid methyl ester radioactivity (dpm)
LI	67,000	206	480
OL	7,200	22	70
LI/OL	9.6	9.4	6.8

pounds having the same R_f and the same color under UV light as the degradative products of the lipoic acid (Fig. 1B).

The determination of these spots (Table I) shows that the areas isolated from rat livers treated with [U-^{14}C]linoleic acid (Fig. 1A) are almost 10 times more radioactive than those from rat livers injected with [U-^{14}C]oleic acid (Fig. 1C).

Analysis of the Lipid Residue Obtained after Partial Hydrolysis of (H_w)

GAS–LIQUID CHROMATOGRAPHY. The analysis is performed on an SE-30 column.

Analytical conditions: column 120 × 4.5 cm. Liquid phase: SE-30 silicon (10%) on Chromosorb W 60/80. Temperature: column 190°, injection 230°. Nitrogen pressure: inlet 0.5 bar, outlet atmospheric pressure.

Figure 2 (B and C) indicates that the lipid residues of the partial hydrolysis of (H_w) from rat liver contain a compound having the same retention time as the methyl lipoate (Fig. 2A) used as standard.

THIN-LAYER CHROMATOGRAPHY. The mixture of solvents for the development of the chromatography is made as described by Brown.[9] The area with the same R_f as the methyl lipoate is scraped off. The lipoate is

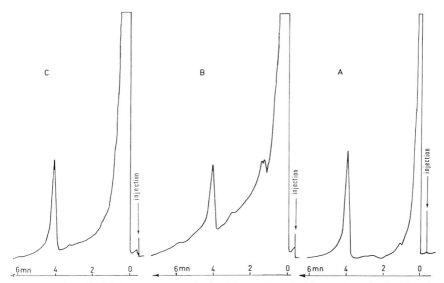

FIG. 2. Gas–liquid chromatograms obtained from (A) methyl lipoate (standard), (B) and (C) methyl lipoate isolated from extracts (H_w). (B) [U-^{14}C]linoleic acid-treated rats; (C) [U-^{14}C]oleic acid-treated rats .

[9] J. L. Brown and J. M. Johnston, *J. Lipid Res.* **3**, 480 (1962).

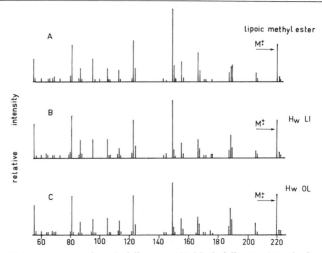

FIG. 3. Mass spectrum of methyl lipoate. (A) Methyl lipoate (standard); (B) and (C) methyl lipoate isolated from extracts (H_W); (B) [U-^{14}C]linoleic acid-treated rats; (C) [U-^{14}C]oleic acid-treated rats.

eluted with chloroform–methanol (1:1). After centrifugation for 5 min at 1000 g, the supernatant is collected and the solvents are evaporated.

The methyl lipoate is suspended in 0.2 ml of chloroform and passed through a microcolumn of silicic acid, 120 mesh (0.5 × 3 cm). The methyl lipoate is eluted with 3 ml of chloroform. The solvent is evaporated, and the samples are analyzed by mass spectrometer TSN 218 by direct injection of the purified methyl esters.

The compounds isolated by TLC and purified on silicic acid columns from the residues (H_W) (Fig. 3B) and (Fig. 3C) have the same mass spectrum as the methyl lipoate (Fig. 3A) used as standard.

Table I shows that the methyl lipoate isolated from the lipid residue (H_W) of [U-^{14}C]linoleic acid-treated rats is 7 times more radioactive than that isolated from (H_W) of [U-^{14}C]oleic acid-treated rats.[10]

Acknowledgments

We are grateful to Professor G. Mouvier and Dr. B. Verneuil (Paris VII University) for their help in the mass spectrometry analysis.

[10] The present procedures were developed in the course of research supported by Equipe de Recherche 91 (CNRS) and Contrat INSERM No. 74.4.083.7.

[33] Turnover of Protein-Bound Lipoic Acid

By FRANKLIN R. LEACH

The turnover of proteins has been extensively studied and reviewed.[1-3] Abnormal proteins turn over faster than normal ones, those of larger size are degraded faster than smaller ones, and those enzymes that are important in control turn over faster than the enzymes that do not regulate. There are proteolytic enzymes that specifically degrade enzymes containing a specific prosthetic group.[4] Many enzymes are regulated by covalent modifications of specific groups.[5] A more economical way of regulation involves the turnover of covalently bound cofactors.

Existence of the Required Enzymes for Attaching and Removing Vitamins

Activating Enzymes

Lipoic Acid. In *Streptococcus faecalis* 10C1 an enzyme system catalyzes the ATP-dependent activation of lipoic acid and its transfer to the apopyruvate dehydrogenase complex to produce the holoenzyme. This system was discovered by Reed and co-workers[6,7] and has been reviewed in this series.[8] The system catalyzes the following reaction sequence:

$$E_1 + ATP + \text{lipoic acid} \rightleftharpoons E_1\text{—lipoyl—AMP} + PP \qquad (1)$$

$$E_1\text{—lipoyl—AMP} + E_2 \rightleftharpoons E_2\text{—lipoyl} + AMP \qquad (2)$$

$$E_2\text{—lipoyl} + A\text{—PCD-NH}_2 \rightarrow PDC\text{—NH—lipoyl} + E_2 \qquad (3)$$

In *Escherichia coli* extracts, however, the activation system has not been resolved into two protein components.

Biotin. Biotin is also covalently linked through the ϵ-amino group of lysine to form the functional holoenzymes.[9-11] Coon's group[12] showed

[1] R. T. Schimke and N. Katunuma, eds., "Intracellular Protein Turnover." Academic Press, New York, 1975.
[2] A. L. Goldberg and A. C. St. John, *Annu. Rev. Biochem.* **45,** 747 (1976).
[3] A. L. Goldberg and J. F. Dice, *Annu. Rev. Biochem.* **43,** 835 (1974).
[4] N. Katunuma, *Curr. Top. Cell Regul.* **7,** 175 (1973).
[5] H. Holzer and W. Duntze, *Annu. Rev. Biochem.* **40** 345 (1971).
[6] F. R. Leach, K. T. Yasunobu, and L. J. Reed, *Biochim. Biophys. Acta* **18,** 297 (1955).
[7] L. J. Reed, F. R. Leach, and M. Koike, *J. Biol. Chem.* **232,** 123 (1958).
[8] F. R. Leach, this series, Vol. 18A, p. 282.
[9] D. P. Kosow and M. D. Lane, *Biochem. Biophys. Res. Commun.* **7,** 439 (1962).
[10] M. D. Lane, D. L. Young, and F. Lynen, *J. Biol. Chem.* **239,** 2858 (1964).

METHODS IN ENZYMOLOGY, VOL. 62

that two soluble enzyme fractions are required to form propionyl-CoA holocarboxylase. One is the apocarboxylase, and the other is an "apocarboxylase-activating enzyme." ATP is required for the activation,[13] which occurs by the following reaction scheme[14–16]:

$$ATP + biotin \xrightleftharpoons{\text{Mg}^{2+}} d\text{-biotinyl } 5'\text{-adenylate} + PP_i \qquad (4)$$

d-Biotinyl $5'$-adenylate + propionyl-CoA apocarboxylase
$$\rightarrow propionyl\text{-CoA holocarboxylase} + AMP \qquad (5)$$

Lane and Lynen[10,11] purified biotin-apotranscarboxylase synthetase [(+)-biotin:methylmalonyl-CoA–pyruvate apocarboxyltransferase ligase (AMP)] and established the following reaction mechanism:

$$ATP + R\text{-}CO_2^- + synthetase \xrightleftharpoons{\text{Mg}^{2+}} RCO\text{-}5'\text{-AMP-synthetase} + PP_i^- \qquad (6)$$

$R\text{-}CO_2^- = (+) \text{ biotin}$

$$RCO\text{-}5'\text{-AMP synthetase} + H_2N\text{-enz} \rightarrow RCO\text{-}NH\text{-enz} + 5'\text{-AMP} + synthetase \qquad (7)$$

$H_2N\text{-enzyme} = apocarboxylase;$ and $RCO\text{-}NH\text{-enz} = holocarboxylase$

4'-Phosphopantetheine. Elovson and Vagelos[17] purified an acyl carrier protein holoprotein synthetase 780-fold from *E. coli,* which catalyzes the following reaction:

$$CoA + apo\text{-}ACP \xrightarrow{\text{Mg}^{2+}} holo\text{-}ACP + 3',5'\text{-adenosine diphosphate} \qquad (8)$$

As would be expected for an enzyme concerned with vitamin metabolism, it is present in small amounts and is also unstable. Not until Elovson and Vagelos found that reduced CoA stabilizes the enzyme in the presence of Mg^{2+} was extensive purification achieved.

Flavin. Walker and Singer[18] have shown that riboflavin of succinic dehydrogenase is covalently linked to the N-3 of histidine. This attachment seems to occur after the apoenzyme is formed (McCormick, personal communication). Nothing is known of the reaction mechanism, but it would be surmised that an enzymic activation is involved.

[11] M. D. Lane, K. L. Rominger, D. L. Young, and F. Lynen, *J. Biol. Chem.* **239,** 2865 (1964).

[12] J. L. Foote, J. E. Christner, and M. J. Coon, *Biochim. Biophys. Acta* **67,** 676 (1963).

[13] D. P. Kosow and M. D. Lane, *Biochem. Biophys. Res. Commun.* **5,** 191 (1961).

[14] H. C. McAllister and M. J. Coon, *J. Biol. Chem.* **241,** 2855 (1966).

[15] L. Siegel, J. L. Foote, and M. J. Coon, *J. Biol. Chem.* **240,** 1025 (1965).

[16] J. E. Christner, M. J. Schlesinger, and M. J. Coon, *J. Biol. Chem.* **239,** 3997 (1964).

[17] J. Elovson and P. R. Vagelos, *J. Biol. Chem.* **243,** 3603 (1968).

[18] W. H. Walker and T. P. Singer, *J. Biol. Chem.* **245,** 4224 (1970).

Removing Enzymes

Lipoic Acid. Lipoamidase (lipoyl-X hydrolase) removes the covalently bound lipoic acid from the ϵ-NH_2 of lysine on the pyruvate dehydrogenase complex by the following reaction:

$$PDC\text{-}NH\text{-}lipoyl + H_2O \rightarrow A\text{-}PDC\text{-}NH_2 + lipoic\ acid \qquad (9)$$

The enzyme has been purified 100-fold from *S. faecalis*[19-21] and has been reviewed in this series.[22] Saito[23,24] has observed that humans, rabbits, and rats all contain enzymes that hydrolyze lipoamide. Seaman[25] has described a similar enzyme in yeast.

Biotin. Thoma and Peterson[26] developed a microbiological assay to measure the release of protein-bound biotin using the fact that *Lactobacillus arabinosus* responds mainly to free biotin, while *Saccharomyces cerevisiae* and *L. casei* use both free and combined forms of biotin. A hog liver enzyme that converts soluble bound biotin into free biotin was named biotinidase. This enzyme was purified 20-fold; it appears to be a peptidase. Koivusalo *et al.*[27] purified biotinidase 700-fold from extracts of *S. faecalis* 10C1. The enzyme is specific for hydrolysis of simple biotin esters and amides.

4'-Phosphopantetheine. Vagelos and Larrabee[28] purified an acyl carrier protein hydrolase 30-fold from extracts of *E. coli*. The Mn^{2+} (or other divalent metal ion)-dependent enzyme, which is stimulated by —SH compounds, catalyzes the following reaction:

$$Holo\text{-}ACP + H_2O \xrightarrow{Mn^{2+}} apo\text{-}ACP + 4'\text{-}phosphopantetheine \qquad (10)$$

Turnover of Covalently Bound Vitamins

Powell *et al.*[29] studied the synthesis and turnover of the prosthetic group of acyl carrier protein *in vivo* in *E. coli*. They postulated the following model[29]:

---

[19] L. J. Reed, M. Koike, M. E. Levitch, and F. R. Leach, *J. Biol. Chem.* **232**, 143 (1958).
[20] M. Koike and L. J. Reed, *J. Biol. Chem.* **235**, 1391 (1960).
[21] K. Suzuki and L. J. Reed, *J. Biol. Chem.* **238**, 4021 (1963).
[22] M. Koike and K. Suzuki, this series, Vol. 18A, p. 292.
[23] J. Saito, *Bitamin* **21**, 359 (1960).
[24] J. Saito, *Bitamin* **39**, 317 (1969).
[25] G. R. Seaman, *J. Biol. Chem.* **234**, 161 (1959).
[26] R. W. Thoma and W. H. Peterson, *J. Biol. Chem.* **210**, 569 (1954).
[27] M. Koivusalo, C. Elorriaga, Y. Kaziro, and S. Ochoa, *J. Biol. Chem.* **238**, 1038 (1963).
[28] P. R. Vagelos and A. R. Larrabee, *J. Biol. Chem.* **242**, 1776 (1967).
[29] G. L. Powell, J. Elovson, and P. R. Vagelos, *J. Biol. Chem.* **244**, 5616 (1969).

$$\text{Medium} \rightleftharpoons 4\text{PP} \underset{\longleftarrow}{\longrightarrow} \text{DCoA} \longrightarrow \text{CoA} \qquad (11)$$
$$\searrow \downarrow$$
$$\text{ACP}$$

where 4PP-pantetheine-P, DCoA-dephospho-CoA, CoA-coenzyme A; ACP-acyl carrier protein, and they obtained kinetic isotope evidence (pulse and chase) consistent with the model.

Powell et al.[29] suggested that since biotin and lipoic acid also covalently bound to their apoproteins, and since enzymes specifically catalyzing the synthesis and hydrolysis of those covalent linkages exist, turnover of those prosthetic groups could occur analogously to that of pantetheine-P of ACP. However, they state: "This may be an imperfect analogy because of the special role of ACP and its precursor, CoA, as coenzymes. The turnover of the pantetheine-P prosthetic group in ACP and CoA may be a special consequence of the particular role of these coenzymes in metabolism."

The above question of Vagelos initiated an examination of the turnover of protein-bound lipoic acid.

Measurement of Turnover

Radioactive Lipoic Acid

[^{35}S]Lipoic acid (35 Ci/mol) is prepared as described previously.[30]

Preparation of Cells

Escherichia coli Crookes strain (ATCC 8739) are grown in M-9 medium[31] at 37° with 0.2% glucose as the carbon source to mid-log phase, harvested by centrifugation, and suspended in fresh M-9 medium. This strain synthesizes lipoic acid, but preferentially incorporates exogenous lipoic acid into the pyruvate and α-ketoglutarate dehydrogenases.[32]

Processing of Samples

In most experiments incorporation is stopped by pipetting 1-ml samples into 2 ml of 95% ethanol. This procedure gives results equivalent to those obtained by boiling. The samples are centrifuged for 10 min at 10,000 g at 4°, and the supernatant solutions are removed and discarded.

[30] F. R. Leach, this series, Vol. 18A, p. 276.
[31] E. H. Anderson, *Proc. Natl. Acad. Sci. U.S.A.* **32**, 120 (1946).
[32] H. Nawa, W. T. Brady, M. Koike, and L. J. Reed, *J. Am. Chem. Soc.* **82**, 896 (1960).

The centrifugation is repeated, and the precipitate is suspended in water, transferred to vials, and counted to a 1% counting error in 10 ml of Bray's scintillation fluid.[33] In turnover experiments, loss of the bound radioactive material is the measured parameter. In experiments in which both the free pool and protein-bound lipoic acid are measured, the uptake is stopped by pipetting 0.5-ml aliquots onto 1 ml of frozen, crushed M-9 and centrifuging as above. The cells are washed once at 4° and then suspended in 0.5 ml of water. Ethanol (1 ml of 95%) is added with vigorous mixing on a Vortex mixer. Centrifugation gives the lipoic acid pool in the supernatant solution and the protein-bound lipoic acid in the precipitate.

Pulse Experiments

E. coli cells, obtained as described above, are divided into three parts. Addition of labeled and unlabeled lipoic acid is made, the samples are incubated for 10 min at 37°, and 1-ml aliquots are taken at short intervals for determination of the amount of protein-bound lipoic acid. Further additions are made at 10 min, and samples are taken at appropriate times.

The experimental protocol for lipoic acid addition is:

Sample	0-Time addition	10-Minute addition
1	2 μM labeled	2 μM labeled
2	2 μM labeled	50 μM unlabeled
3	50 μM unlabeled	2 μM labeled

Chase Experiments

Two types of experiments are done.

1. Long-term experiments with low lipoic acid concentration: E. coli cells are suspended in fresh M-9 medium at an $A_{600\,nm}$ of 0.1. For measurement of protein-bound lipoic acid, the sample is divided into two portions, incubated for 200 min, sedimented by centrifugation, and suspended in fresh M-9 medium. The following is the schedule of lipoic acid additions:

Sample	0-Time addition	200-Minute addition
1	2 μM labeled	500 μM unlabeled
2	None	2 μM labeled

[33] G. A. Bray, Anal. Biochem. **1**, 279 (1960).

At appropriate times the incorporation into the protein-bound form of lipoic acid is measured as described above. Growth is followed by A_{600} measurements.

2. Shorter-term experiments with high lipoic acid concentration: Log-phase cells are harvested by centrifugation and suspended in fresh M-9 medium. The sample is divided into three portions and incubated for 15 min at 37° with the components indicated for the first addition. The cells are harvested by centrifugation and suspended in fresh M-9; the second additions are made, and at the indicated intervals the amount of protein-bound lipoic acid is determined as described above. The following is the lipoic acid addition schedule:

Sample	First addition	Second addition
1	0.5 mM labeled	0.5 mM unlabeled
2	0.5 mM unlabeled	0.5 mM labeled
3	None	0.5 mM labeled

Blocking Incorporation with Chloramphenicol

Chloramphenicol prevents the incorporation of lipoic acid into its protein-bound form but has no effect upon the accumulation into the free pool. During a 20-min incubation, 148 pmol are incorporated in the absence of chloramphenicol, while in the presence of 400 μg of chloramphenicol per milliliter only 8 pmol are incorporated. The respective pool levels are 37 and 35 pmol. If any labeled lipoic acid were released from the bound form, its reincorporation would be prevented by chloramphenicol.

Osmotic Shock Procedures

For osmotic shocking[34] about 1 g of cells is suspended in 40 ml of 20% sucrose, 30 mM Tris·HCl, and 1 mM EDTA at pH 8.0, stirred for 10 min at room temperature (20°–25°), and centrifuged at 13,000 g for 15 min. The well-drained pellet is rapidly suspended in 40 ml of ice cold 0.5 mM $MgCl_2$. After standing 10 min at 4°, cells are harvested by centrifugation and suspended in M-9. Viability is 85–96%.

[34] H. C. Neu and L. A. Heppel, *J. Biol. Chem.* **240**, 3685 (1965).

Results with Lipoic Acid[35]

Pulse Experiment

No evidence was obtained for release of protein-bound lipoic acid during 18 min of incubation under the conditions described above under Pulse Experiments.

Turnover under Growth Conditions

With low lipoic acid concentration (2 and 50 μM) and with high lipoic acid concentration (0.5 mM), treatment as described above under Chase Experiments reveals no lipoic acid turnover during 2–5 hr of growth.

Turnover under Perturbed Conditions

1. When incorporation is blocked with chloramphenicol. Under the conditions described above under Blocking Incorporation with Chloramphenicol, there is no reduction in the amount of protein-bound lipoic acid during a 2-hr incubation.
2. When the pool is reduced by osmotic shock. The amount of lipoic acid in the pool can be markedly reduced (50%) by the osmotic shock procedure described above under Osmotic Shock Procedures. When this treatment precedes measurement of bound lipoic acid, again no turnover of protein-bound lipoic acid is observed.

Comparison with Other Systems

Our observations give validity to the reservations of Powell et al.[29] about comparing turnover of lipoic acid (and biotin) to that of 4'-phosphopantetheine and strengthen their suggestion of the uniqueness of the latter.

Reed et al.[7] found no exchange of the protein-bound lipoic acid and free material when 5000–17,000 times as much free material as bound was used in the pyruvate dehydrogenase, dihydrolipoic transacetylase, and dihydrolipoic acid dehydrogenase reactions in a cell-free system. Green and Oda[36] and Koike et al.[37] have postulated that the dithiolane ring of lipoic acid is located on the end of a 14 Å arm of lipoyllysine extending from the

[35] T. W. Griffith and F. R. Leach, Arch. Biochem. Biophys. 162, 215 (1974).
[36] D. E. Green and T. Oda, J. Biochem. 49, 742 (1961).
[37] M. Koike, L. J. Reed, and W. R. Carroll, J. Biol. Chem. 238, 30 (1963).

backbone of the transacetylase peptide and that the movement of this swinging arm from the reductive acylation site on the pyruvate decarboxylase to the oxidative site on dihydrolipoic dehydrogenase is an essential part of the reaction mechanism. Evidence has been obtained by Ambrose and Perham[38] that the lipoyl moiety has this high mobility relative to the protein and thus has unrestricted freedom of motion for channeling of substrates between the subunits. This would not require release of the protein-bound lipoic acid.

In mammals the pyruvate dehydrogenase complex contains two regulating enzymes, a kinase and a phosphatase.[39] While in bacterial systems the regulation is by product inhibition and feedback regulation by nucleotides.

Dietrich and Henning[40] found that the pyruvate dehydrogenase in *E. coli* K12 is inducible by pyruvate and other α-keto acids. Using thiamine-requiring mutants, they distinguished enzymic activity from inducibility because an apoenzyme complex was formed. The pyruvate dehydrogenase components of the complex participate in the regulation of the synthesis of the complex. There was no decrease in the enzymic activity of the complex within stationary cells incubated with aeration for 8 hr; however, this observation does not eliminate the possibility of degradation and resynthesis or of turnover of the individual components. Our results show that the lipoic acid does not turn over.

Lipoamidase activity was not detectable in homogenates of bovine heart, liver, or kidney or in *E. coli* Crookes strain extracts.[21,22] Whether the lack of lipoamidase accounts for finding no turnover of lipoic acid or whether other factors are involved is not clear at present.

This is manuscript No. 3448 of the Oklahoma Agricultural Experiment Station.

[38] M. C. Ambrose and R. N. Perham, *Biochem. J.* **155**, 429 (1976).
[39] L. J. Reed, *Acc. Chem. Res.* **7**, 40 (1974).
[40] J. Dietrich and U. Henning, *Eur. J. Biochem.* **14**, 258 (1970).

[34] The Radial Diffusion Assay of Lipoamide Dehydrogenase

By JASON C. H. SHIH

Lipoate + NADH $\rightleftharpoons$ reduced lipoate + NAD

Principle. The principle of this technique is to allow the enzyme to diffuse in an agarose gel containing lipoate. The enzyme, after 24 hr of diffusion and an additional 2 hr of reaction with NADH, can be determined by the size of a dark or fluorescence-quenching zone in the gel when illuminated with ultraviolet light. The diameter of the quenching zone, which indicates the enzymic oxidation of NADH, is linearly proportional to the logarithm of the enzyme concentration.

> *Reagents*
> Agarose, low EEO
> Potassium phosphate buffer, 50 mM, pH 6.6, containing 1 mM EDTA
> Lipoate solution, 0.1 M in 0.2 M NaHCO$_3$, freshly made
> NADH, 20 mM in 0.1 M KPO$_4$, pH 8.0, freshly made
> Enzyme solutions: Purified yeast lipoamide dehydrogenase from the commercial source is used as the enzyme standard, and a series of 2-fold dilutions from 0.2 mg/ml are made with the buffer.
> Tissue homogenate: The freshly excised chick liver is homogenized with 4 × volume of the buffer containing 0.5% Triton X-100. The homogenate goes through one cycle of freezing (− 15%) and thawing before assay.

Procedures

Lipoate Gel Plate. One gram of agarose is dissolved in 95 ml of the buffer in a boiling water bath. Five milliliters of the lipoate solution are then mixed thoroughly into the agarose solution at 50–60°. The mixture is poured rapidly onto a clean glass plate (20 × 20 cm) and is spread evenly. After the gel solidifies, small wells are punched with glass tubing (7.0 mm o.d.). The small wells can then be filled with 50 μl of an enzyme source to be assayed. The preparations of the lipoate solution and the lipoate gel plates are conducted in a semidark room to minimize photolysis and polymerization of lipoate.[1] For a smaller number of enzyme samples, the gel plate can be made smaller in a plastic petri dish, with the reagents reduced proportionately.

Enzyme Assay. Fifty microliters of an enzyme preparation, either a solution of purified enzyme or a tissue homogenate, are pipetted into the well in the gel. The plate is kept in a dark moist chamber at room temperature (20°) for 24 hr to allow the enzyme to diffuse radially. At the end of the diffusion period, the wells in the plate are filled with a molten 1% agarose solution (50 μl each well). When the filling agar solidifies, 2.5 ml of NADH solution are evenly spread over the plate with a glass rod. The gel plate, spread with NADH and incubated for an additional 2 hr in the

---

[1] A. F. Wagner, E. Walton, G. E. Boxer, M. P. Pruss, F. W. Holly, and K. Folkers, *J. Am. Chem. Soc.* **78**, 5079 (1956).

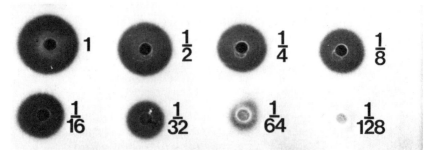

FIG. 1. The radial diffusion zones of a series of 2-fold dilutions of yeast lipoamide dehydrogenase (original concentration: 0.2 mg/ml).

moist chamber, is placed under a long-wavelength UV light[2] in a dark room. The gel exhibits a yellowish-green fluorescence owing to the presence of NADH, but around the wells where enzyme is added there are dark or quenched circles. The diameters of the circular zone are measured with calipers. The plate may be photographed for a permanent record.

Standard Curves. Purified yeast lipoamide dehydrogenase is used to establish the relationship between the enzyme concentrations and the sizes of the fluorescence-quenching zones (Fig. 1). A semilogarithmic linear relationship is demonstrated in Fig. 2. A similar linear relationship is obtained with various levels of chick liver homogenate (Fig. 3). Here Triton X-100 permits the detection of higher levels of enzyme, since it solubilizes the mitochondrial membranes and releases lipoamide dehydrogenase. Enzyme activity of these preparations is also measured by the

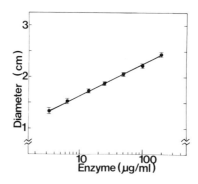

FIG. 2. A semilogarithmic relationship between the diameters of the diffusion zones and the concentrations of yeast lipoamide dehydrogenase. Vertical bars on data points are SEM.

[2] Ultra-Violet Products, Black-ray lamp, Model XX-15.

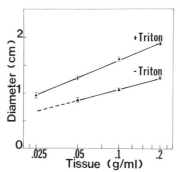

FIG. 3. Determination of lipoamide dehydrogenase in the chick liver homogenate. Triton X-100 permits the detection of higher level of enzyme. Vertical bars on data points are SEM.

conventional spectrophotometric method[3,4] to ensure that the activities are proportional to the concentrations.

Comments

Applications. The radial diffusion assay of enzymes is simple and inexpensive. No sophisticated equipment is needed and a large number of samples can be screened in a short period of work. These advantages have been illustrated by the application of this technique in the studies of several kinds of hydrolytic enzymes.[5–7]

The determination of dehydrogenase activity by the radial diffusion technique has potential importance in clinical and nutritional applications. Under experimental conditions, riboflavin deficiency may be detected by measuring decreased lipoamide dehydrogenase, a flavoprotein, in the livers of experimental chicks.[8] A part of the results is shown in Fig. 4, where the decreased enzyme in the deficient group is easily visualized. For practical application, the enzymic diagnosis of riboflavin deficiency by means of the diffusion assay of two flavoproteins, i.e., lipoamide dehydrogenase and glutathione reductase, in the chicks fed with practical diets has recently been studied.[9]

Specificity. The purified yeast lipoamide dehydrogenase is specific for lipoate or lipoamide but unreactive toward other disulfide compounds

[3] M. Koike and T. Hayakawa, this series, Vol. 18A [50].
[4] J. C. H. Shih, M. Sandholm, and M. L. Scott, *J. Nutr.* **107**, 1583 (1977).
[5] M. Sandholm, R. R. Smith, J. C. H. Shih, and M. L. Scott, *J. Nutr.* **106**, 761 (1976).
[6] M. Ceska, *Clin. Chim. Acta* **33**, 135 (1971).
[7] J. M. Goldberg and P. Pagast, *Clin. Chem.* **22**, 633 (1976).
[8] J. C. H. Shih, *Anal. Biochem.* **89**, 103 (1978).
[9] J. C. H. Shih and P. B. Hamilton, *Poult. Sci.* **57**, 1163 (1978).

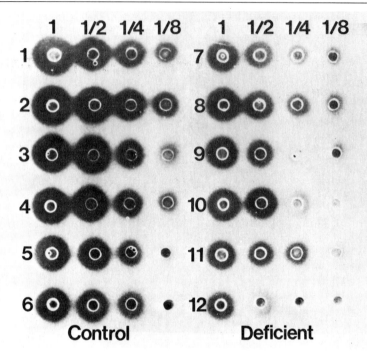

Fig. 4. Radial diffusion assays of lipoamide dehydrogenase of the control and the riboflavin-deficient chick liver homogenates. Integral numbers mark the individual chicks, and fraction numbers on the top indicate the dilution of the tissue from 0.2 g/ml.

tested (table). Since lipoamide is not readily soluble in the gel, it is not the substrate of choice in the standard procedure. NADH, but not NADPH, is reactive with the purified enzyme. The chick liver homogenate is reactive not only toward lipoate or lipoamide with NADH, but also toward oxidized glutathione with NADPH. This is due to the presence of glutathione reductase in the tissue. Therefore, when a different substrate is incorporated in the gel, a different enzyme in the tissue homogenate can be detected accordingly.

Sensitivity. The radial diffusion assay is approximately fifty times less sensitive than the spectrophotometric method.[3,4] Hence this technique is a supplemental method serving a different purpose, rather than a replacement for the spectrophotometric method, which is highly sensitive and more accurate. On the other hand, the diffusion assay is extremely useful in large-scale screening, and lower sensitivity can also be an advantage because no exhaustive dilution is necessary in preparing the tissue samples.

SUBSTRATE AND COENZYME SPECIFICITIES FOR THE RADIAL DIFFUSION
ASSAY OF LIPOAMIDE DEHYDROGENASE

Substrate	Yeast LipDH		Liver homogenate[a]	
	NADH	NADPH	NADH	NADPH
Lipoate	+[b]	−	+	−
Lipoamide	+	−	+	−
Glutathione (ox.)	−	−	−	+[c]
Cystine	−	−	−	−
Pantothine	−	−	−	−

[a] Liver from a 2-week-old chick.
[b] +, Detection of the zone of quenched fluorescence; −, no quenching of fluorescence.
[c] Detection of glutathione reductase.

Reverse Reaction. Lipoamide dehydrogenase may be analyzed by either a forward or reverse reaction on the gel plate. When reduced lipoate, prepared from the sodium borohydride reduction,[10] is included in the gel and NAD is spread after diffusion, a fluorescent zone indicating enzymic reduction of NAD to NADH against a dark background can be detected under a UV lamp. However, to avoid the additional step of reduction of lipoate, the lipoate–NADH-quenching (substrate–coenzyme detection) procedure is recommended and described here as the standard method for the assay of lipoamide dehydrogenase.

Comparison of Enzymes. It should be remembered that a level of enzyme in chick liver should not be compared for its absolute concentration with the standard curve of a yeast enzyme. The enzymes of two species are different in physical properties, especially their molecular weights, and therefore possess different diffusion rates in the gel. When the two zones of enzymes from different sources are of the same size, they may not be at the same concentration.

Other Oxidoreductases. The same technique has recently been applied to six other NAD- and NADP-dependent oxidoreductases, namely, alcohol, glucose 6-phosphate, lactate, malate, and hydroxysterol dehydrogenases, and glutathione reductase.[11] The results are satisfactory and demonstrate that this simple method may have broad application to all NAD- and NADP-dependent enzymes.

[10] I. C. Gunsalus, L. S. Barton, and W. Gruber, *J. Am. Chem. Soc.* **78**, 1763 (1956).
[11] J. C. H. Shih and E. P. Teulings, *Clin. Chem.* **24**, 2176 (1978).

[35] Asparagusate Dehydrogenase and Lipoyl Dehydrogenase from Asparagus

By Hiroshi Yanagawa

Asparagusate dehydrogenase:

Asparagusate + NADH + H$^+$ $\rightleftharpoons$ dihydroasparagusate + NAD$^+$

Lipoyl dehydrogenase:

Lipoate + NADH + H$^+$ $\rightleftharpoons$ dihydrolipoate + NAD$^+$

Asparagusate dehydrogenase and lipoyl dehydrogenase are essential components of asparagus pyruvate dehydrogenase complexes,[1] which catalyze a CoA- and NAD-linked oxidative decarboxylation of pyruvate.[2] These two enzymes are flavoproteins and catalyze the oxidoreduction of asparagusate[3] and lipoate.[4]

Assay Method

Principle. Asparagusate dehydrogenase and lipoyl dehydrogenase activities are measured spectrophotometrically by the decrease in absorption at 340 nm due to NADH oxidation in the presence of asparagusate or lipoate.

Reagents

Sodium phosphate buffer, 0.6 M, pH 5.9

EDTA, 7.5 mM

Asparagusic acid[3,5] in 10% ethanol, 24 mM, freshly prepared every day

DL-lipoic acid (Tokyo Kasei Co., Ltd.) in ethanol, 48 mM, freshly prepared every day

NADH, 3 mM

NAD, 3 mM

Procedure. Assays are carried out in cuvettes of 10-mm light path maintained at 25° and containing 0.05 ml each of sodium phosphate buffer and EDTA and 0.02 ml each of NADH, NAD, and asparagusic acid or DL-lipoic acid. The reaction is started by adding enzyme. Absorbancy at

[1] H. Yanagawa, Ph.D. dissertation, University of Tohoku, 1973.

[2] H. Yanagawa, T. Kato, and Y. Kitahara, *Plant Cell Physiol.* **14,** 1213 (1973).

[3] H. Yanagawa, T. Kato, Y. Kitahara, N. Takahashi, and Y. Kato, *Tetrahedron Lett.,* 2549 (1972).

[4] U. Schmidt, P. Grafen, K. Atland, and W. Goedde, *Adv. Enzymol.* **32,** 423 (1969).

[5] H. Yanagawa, T. Kato, H. Sagami, and Y. Kitahara, *Synthesis* **9,** 607 (1973).

340 nm is measured with a Gilford spectrophotometer, Model 2400-2 with a recorder. A linear relationship generally is maintained for the first 10 min.

Units. A unit of enzyme activity is defined as that amount of enzyme which oxidizes 1 nmol of NADH per minute under the conditions of assay specified above. Specific activity is defined as units per milligram of protein; protein is determined by the method of Lowry *et al.*,[6] using crystalline bovine serum albumin as a standard.

Purification Procedure

A procedure for the preparation of asparagusate dehydrogenases and lipoyl dehydrogenase from asparagus mitochondria has been reported.[7] All operations are carried out at 4°. The centrifugations and concentrations are performed in a Kubota KR-200A centrifuge and an Amicon PM-10 filter. Phosphate buffer is used throughout purification because the enzymes are more stable in phosphate buffer.

Preparation of Mitochondrial Fraction.[2] Fresh etiolated or green asparagus (*Asparagus officinalis* L.) shoots are washed with distilled water, and moisture is removed by pressing them gently between sheets of absorbent paper. In a typical preparation, 2 kg of the tissues are chopped, then homogenized in a Waring blender in a medium containing cold 0.30 *M* sucrose and 70 m*M* phosphate buffer, pH 7.4. The homogenate is squeezed through gauze and centrifuged at 2700 *g* for 15 min to remove whole cells, cell debris, nuclei, and chloroplasts. The supernatant liquid is centrifuged at 15,000 *g* for 15 min to sediment mitochondrial particles. The particles are resuspended in the homogenizing medium and again sedimented at 15,000 *g* for 15 mins. The mitochondrial pellet thus obtained is suspended in 67 m*M* phosphate buffer, pH 7.0, containing 10 m*M* EDTA to give a protein concentration of about 30 mg/ml.

Solubilization of Lipoyl and Asparagusate Dehydrogenases. Mitochondrial suspension is treated by freezing and thawing. Freezing of the mitochondrial suspension at −50° and thawing at 37° is repeated at least twice over a period of 20 min. The solubilized enzyme is then separated by centrifugation at 36,000 *g* for 30 min. The supernatant is dialyzed overnight against 67 m*M* phosphate buffer, pH 7.0, containing 10 m*M* EDTA; the small amount of precipitate formed during dialysis is removed by centrifugation at 36,000 *g* for 60 min. The precipitates are subsequently treated with sodium dodecyl sulfate (SDS) to solubilize asparagusate de-

[6] O. H. Lowry, N. J. Rosebrough, A. L. Farr, and R. J. Randall, *J. Biol. Chem.* **193,** 265 (1951).

[7] H. Yanagawa and F. Egami, *Biochim. Biophys. Acta* **384,** 342 (1975).

hydrogenase. The SDS concentration used is 0.5% in the ratio of 0.5 mg of SDS per milligram of mitochondrial protein. The solubilization is performed at 37° in a shaking water bath for 20 min. The solubilized enzyme is then separated by centrifugation at 36,000 g for 30 min.

Lipoyl Dehydrogenase

Step 1. Gel Filtration on Sephadex G-200. A fraction of the enzyme from the freeze–thaw treatment is concentrated to 8 ml and filtered through a Sephadex G-200 column (3 × 80 cm). The column is eluted with 67 mM phosphate buffer, pH 7.0. Some impurities are eluted in front of the enzyme; the active fractions are combined and then concentrated to 5 ml. The enzyme is purified 3-fold by this step.

Step 2. Second Gel Filtration on Sephadex G-200. The concentrated enzyme fraction is refiltered through a Sephadex G-200 column (3 × 80 cm) which is eluted with 67 mM phosphate buffer, pH 7.0. The enzyme is eluted in a single peak. This step affords a further 8-fold purification of the solubilized enzyme.

Step 3. Chromatography on DEAE-Cellulose. The enzyme from the previous step is concentrated and adsorbed onto a DEAE-cellulose (ED-32) column (1.2 × 50 cm) equilibrated with 10 mM phosphate buffer, pH 7.0. The column is washed with 10 mM phosphate buffer, pH 7.0, to remove unadsorbed proteins; a linear phosphate gradient from 10 to 200 mM is then applied to the column. The fractions containing lipoyl dehydrogenase activity are eluted with 120 mM buffer in a single peak. These fractions are pooled and concentrated.

Step 4. Chromatography on Calcium Phosphate Gel Cellulose. The concentrated lipoyl dehydrogenase-containing fraction is applied to a calcium phosphate gel-cellulose column (1.8 × 12 cm) previously equilibrated with 10 mM phosphate buffer, pH 7.0, containing 10 mM EDTA. The column is washed with 100 ml of the same buffer, and elution is begun with 400 ml of a linear phosphate gradient in a concentration range from 10 mM to 300 mM, pH 7.0. The fractions containing lipoyl dehydrogenase activity are eluted at phosphate concentrations of 0.12 to 0.16 M as a single peak. These fractions are pooled and concentrated. The purification scheme employed is summarized in Table I.

Asparagusate Dehydrogenases

Step 1. Gel Filtration on Sephadex G-200. A fraction of the enzymes from the treatment of sodium dodecyl sulfate is concentrated to 4 ml and then applied to a Sephadex G-200 column (3 × 80 cm). The column is eluted

TABLE I
SUMMARY OF THE PURIFICATION OF LIPOYL DEHYDROGENASE OF
ASPARAGUS MITOCHONDRIA

Purification step	Volume (ml)	Total protein (mg)	Total activity (units)	Specific activity (units/mg)	Yields (%)	Fold purification
1. Extract by freezing and thawing	50	244	96.40	0.40	100	1
2. Sephadex G-200	33	72	75.10	1.05	78	3
3. Sephadex G-200	14	7.4	68.50	9.25	71	23
4. DEAE-cellulose	65	0.32	64.50	201.00	67	502
5. Calcium phosphate gel cellulose	5	0.04	55.20	1317.70	57	3294

with 67 mM phosphate buffer, pH 7.0. The fractions are analyzed for protein and two dehydrogenase activities. Asparagusate and lipoyl dehydrogenase activities are eluted within the same fraction. This step results in 4- and 3-fold purification for asparagusate and lipoyl dehydrogenase activities.

Step 2. Second Gel Filtration on Sephadex G-200. Fractions from the preceding step are combined and concentrated to 3 ml. The resulting solution is filtered through a Sephadex G-200 column (3 × 80 cm). The column is eluted with 67 mM phosphate buffer, pH 7.0. Asparagusate and lipoyl dehydrogenase activities are eluted in the same fraction.

Step 3. Chromatography on DEAE-Cellulose. Fractions possessing asparagusate and lipoyl dehydrogenase activities from the preceding step are combined and concentrated to 8 ml. The concentrate is then applied to a DEAE-cellulose column (1.2 × 37 cm) previously equilibrated with 10 mM phosphate buffer, pH 7.2. After the sample has been adsorbed on the top of the column, a few milliliters of 10 mM phosphate buffer, pH 7.20, are added, and elution is begun with 600 ml of a linear phosphate gradient in a concentration range from 10 to 500 mM phosphate buffer, pH 6.95. The fractions are analyzed for protein and the two dehydrogenase activities. Asparagusate dehydrogenase activities are eluted at phosphate concentrations of 150 to 160 mM (fraction I) and 165 to 190 mM (fraction II), respectively; furthermore, two lipoyl dehydrogenase activities are also eluted at the same phosphate concentration. Thus, each protein peak contains both asparagusate and lipoyl dehydrogenase activities.

Step 4. Second Chromatography on DEAE-Cellulose. Two fractions, fractions I and II from the preceding step, are combined, dialyzed, and concentrated to 10 ml. The resulting solution is adsorbed on a DEAE-cellulose column (1.4 × 37 cm), and the enzymes are further purified by a gradient of phosphate buffer. The phosphate concentration varies linearly from 10 to 500 mM. The first (fraction I) and second (fraction II) protein peaks both coincide with the asparagusate dehydrogenase activity; furthermore, each protein peak also possesses lipoyl dehydrogenase. Results obtained after each step of the preparation leading to the two fractions are summarized in Table II. In the final purification step, fraction I reveals specific activities of 425 and 1081.6 units/mg toward asparagusic acid and DL-lipoic acid and fraction II shows specific activities of 214.9 and 546.4 units/mg toward both the acids. In both fractions I and II, the ratio of specific activity toward asparagusic acid to that toward DL-lipoic acid is 2.55:1. Fractions I and II thus isolated are tentatively named asparagusate dehydrogenases I and II, respectively, based on the assumption that asparagusic acid may be the natural substrate of the enzymes. In view of specificity, these two enzymes are novel enzymes.

Properties[8]

Criteria of Purity. Ultracentrifugation studies show that asparagusate dehydrogenases I and II and lipoyl dehydrogenase move as a single boundary during the entire run at 60,000 rpm. Purity of each enzyme preparation is also routinely examined by polyacrylamide disc gel electrophoresis in Tris-glycine buffer, pH 8.3. Approximately 100–150 μg of protein are applied to a 7.5% gel (0.5 × 5 cm). After the run, gels are stained with nitroblue tetrazolium and NADH as well as with Amido Black. Each purified preparation exhibits a single protein band by Amido Black-staining which corresponds to the activity-staining band. These preparations may be regarded, therefore, to be homogeneous.

Sedimentation Coefficient. The sedimentation coefficients ($s_{20,w}$) of asparagusate dehydrogenase I and II and lipoyl dehydrogenase are calculated to be 6.22 S, 6.39 S, and 5.91 S.

Estimations of Molecular Weight and Partial Specific Volume. In the meniscus depletion sedimentation equilibrium runs, the plots of the logarithm of the vertical displacement of a single fringe versus the square of the radial distance are virtually linear, indicating homogeneity of the proteins. The molecular weights of asparagusate dehydrogenases I and II and lipoyl dehydrogenase are estimated to be 111,000, 110,000, and 95,000. Further, the molecular weight of asparagusate dehydrogenases I and II and lipoyl dehydrogenase are estimated to be 112,000, 112,000, and 92,000 by gel filtration on Sephadex G-200. The partial specific volumes of asparagusate dehydrogenases I and II and lipoyl dehydrogenase are calculated to be 0.719, 0.715, and 0.733 ml/g from the amino acid composition.

[8] H. Yanagawa and F. Egami, *J. Biol. Chem.* **251**, 3637 (1976).

TABLE II

SUMMARY OF THE PURIFICATION OF ASPARAGUSATE DEHYDROGENASES OF ASPARAGUS MITOCHONDRIA

Purification step	Volume (ml)	Total protein (mg)	Asparagusate dehydrogenase				Lipoyl dehydrogenase			
			Total activity (units)	Specific activity (units/mg)	Yield (%)	Fold purification	Total activity (units)	Specific activity (units/mg)	Yield (%)	Fold purification
1. 0.5% Sodium dodecyl sulfate extract	50	315	58.80	0.19	100	1	85.40	0.27	100	1
2. Sephadex G-200	135	129	98.40	0.76	167[a]	4	117.15	0.91	137[a]	3
3. Sephadex G-200	60	50	70.30	1.95	119[a]	10	339.52	4.73	397[a]	18
4. DEAE-cellulose I	24	0.282	12.45	44.12	21	236	31.98	113.40	37	418
II	30	0.308	18.54	60.23	32	322	47.16	153.22	55	565
5. DEAE-cellulose I	5	0.024	10.05	425.00	17	2237	25.96	1081.6	30	4006
II	5	0.071	15.32	214.90	26	1131	38.79	546.4	45	2024

[a] Anomalous excess yield of the activity in these steps is probably due to removal of sodium dodecyl sulfate.

Isoelectric Point. Asparagusate dehydrogenases I and II and lipoyl dehydrogenase are slightly acidic proteins with an isoelectric point of 6.75, 5.75, and 6.80, respectively.

Flavin Component. The yellow color and fluorescence of the three enzyme preparations suggest the presence of flavin. The enzyme preparations in the oxidized form display a characteristic flavoprotein absorption spectrum, with maxima at 272, 356, 453 nm for both asparagusate dehydrogenases I and II; 278, 355, 454 nm for lipoyl dehydrogenase. The absorption at 450 nm is significantly decreased by the addition of sodium dithionite. The flavin released from asparagusate dehydrogenases and lipoyl dehydrogenase by immersion in boiling water has the same retention volume as standard FAD by high-speed liquid chromatography. Moreover, the spectrum of this peak material is identical to that of FAD. From the measurements by the method of Beinert and Page,[9] the flavin contents of asparagusate dehydrogenases I and II and lipoyl dehydrogenase are calculated to be 10.22, 12.13, and 10.94 nmol per milligram of protein, respectively. These values correspond to minimal molecular weights of 98,000, 82,000, and 91,000. Taking into consideration the molecular weights obtained by physical methods, it may be considered that each enzyme preparation possesses 1 mol of FAD per mole of protein.

Amino Acid Composition. The numbers of total residues for asparagusate dehydrogenases I and II and lipoyl dehydrogenase are 935, 947, and 754, and furthermore each minimum molecular weight calculated from the amino acid content is 112,000, 112,000, and 94,000. Although the amino acid compositions of the asparagusate dehydrogenases I and II and lipoyl dehydrogenase appear to be similar, there are several amino acid residues that vary in amount, which would indicate some differences between the enzymes. Glutamic acid, aspartic acid, glycine, and alanine are contained in high percentages in the three proteins. Serine is found to be present in a higher amount only in asparagusate dehydrogenase II.

Reversibility of Asparagusate Dehydrogenase and Lipoyl Dehydrogenase Reactions. In asparagusate dehydrogenase II, the rates of the oxidation of dihydroasparagusic and dihydrolipoic acids are very slow. The reduction rates of asparagusic and lipoic acids by asparagusate dehydrogenase and lipoyl dehydrogenase are very rapid. Lipoyl dehydrogenase also gives a similar result. It is more convenient to follow the reaction in the direction of NADH oxidation for two reasons. First, at neutral and acidic pH, the equilibrium is in favor of reduction of asparagusic and lipoic acids and substrate is stable. Second, it avoids the necessity for fresh preparation of both unstable dihydro forms in aqueous solution.

[9] H. Beinert and E. Page, *J. Biol. Chem.* **225,** 479 (1957).

Substrate and Cofactor Specificities. Lipoyl dehydrogenase is highly specific for lipoic acid and does not show activity for asparagusic acid and other disulfides, cystine, and oxidized glutathione. Asparagusate dehydrogenases I and II are specific for both asparagusic and lipoic acids, which cannot be replaced by other disulfides; the relative rates of NADH oxidation are in the ratio 1:2.5 for asparagusic acid and lipoic acid. For reduction of asparagusic and lipoic acids by asparagusate dehydrogenases I and II and lipoyl dehydrogenase, NADH is an active hydrogen donor, but NADPH is inactive. The K_m values (about 3 mM) of asparagusate dehydrogenase for lipoic acid are lower than those (about 20 mM) for asparagusic acid. In the case of lipoyl dehydrogenase, the K_m value for lipoic acid is 5.0 mM. The optimum concentration of NADH on NADH-asparagusic acid and NADH-lipoic acid reductase activities of asparagusate dehydrogenase II is 0.1 mM. A high concentration of NADH considerably inhibits.

Effect of NAD^+ on Rate of NADH Oxidation by Asparagusic and Lipoic Acids. With asparagusate dehydrogenase and lipoyl dehydrogenase from asparagus mitochondria, however, a less pronounced lag period is observed when the asparagusate and lipoyl dehydrogenase activities are measured in a reaction mixture containing no NAD^+. Duration of the lag period varies with different concentration of NADH. However, when NAD^+ is added to a reaction mixture, the lag period is eliminated and the oxidation of NADH by asparagusic and lipoic acids takes place immediately. Maximum lipoyl dehydrogenase activity is achieved at a concentration above 0.3 mM. In asparagusate dehydrogenase, NAD^+ exerts a biphasic effect on the reductase activity. At 0.2 mM NADH, the reductase activity is stimulated by concentrations of NAD^+ up to 0.2 mM. However, at higher NAD^+ concentrations, a decrease in stimulation and inhibition is observed. The stimulatory effects of NAD^+ on two reductase activities cannot be replaced by the addition of $NADP^+$ or FAD. On the contrary, FAD strongly inhibits the two reductase activities. Thus, both NADH-lipoic acid and NADH-asparagusic acid reductase activities are dependent on NAD^+, and the integrity of the NAD^+ molecule is essential for the stimulatory effect.

Effect of pH on Asparagusate Dehydrogenase and Lipoyl Dehydrogenase Activities. The pH optima of the asparagusate dehydrogenase and lipoyl dehydrogenase activities in asparagusate dehydrogenases I and II are approximately 5.9. The pH optimum of the lipoyl dehydrogenase activity in lipoyl dehydrogenase is almost the same as that of asparagusate dehydrogenase.

Diaphorase Activities of Asparagusate Dehydrogenases and Lipoyl Dehydrogenase. The diaphorase activities of the asparagusate dehydrogenases I and II and lipoyl dehdrogenase are tested with $K_3Fe(CN)_6$ as an electron acceptor for NADH oxidation at an optimum pH of 5.25. Under this condition, asparagusate dehydrogenases I and II and lipoyl dehydrogenase reduce $K_3Fe(CN)_6$ with NADH. K_m values for acceptor in the $K_3Fe(CN)_6$-NADH activity of asparagusate dehydrogenases I and II and lipoyl dehydrogenase are 0.9, 0.8, and 0.9 mM.

Effect of Temperature on Reaction Velocity. Around 40° to 45°, both activities are maximal. From Arrhenius plots for reductase activities of both asparagusic and lipoic acids, the apparent energies of activation of asparagusate dehydrogenase II are calculated to be approximately 10.0 kcal for asparagusic acid and 3.3 kcal for lipoic acid.

Effects of Thiol Inhibitors on Dehydrogenase Activities. Treatment of the enzyme with pCMB, Hg^{2+}, and N-ethylmaleimide results in a decrease of both activities. The inhibitory effects of IAA and arsenite on both activities are much smaller than those of pCMB, Hg^{2+}, and N-ethylmaleimide. IAA only slightly inhibits the lipoyl dehydrogenase activity. However, treatment of this enzyme with arsenite in the presence of NADH causes a considerable decrease of both activities. The diaphorase activity is strongly inhibited by pCMB and Hg^{2+}, but is not significantly inhibited by N-ethylmaleimide, IAA, and arsenite. Addition of Arsenite in the presence of NADH does not cause marked decrease of diaphorase activity. The difference of these inhibitory effects is due to the properties of the inhibitor and catalytic sites; however, these findings indicate the involvement of protein disulfide linkage or thiol group in the catalytic site.

Effects of Phospholipids and Surfactants on Asparagusate Dehydrogenase. Natural and synthetic lecithins and Tween-80 cause a drastic activation of lipoyl dehydrogenase activity, but they have no effect on asparagusate dehydrogenase activity. Triton X-100, SDS, and deoxycholate do not appreciably affect either dehydrogenase activity, but cetyltrimethyl ammonium bromide inhibits only lipoyl dehydrogenase activity.

Inhibition of Lipoyl Dehydrogenase Activity of Asparagusate Dehydrogenase by Asparagusic Acid. Lipoyl dehydrogenase activity is strongly inhibited by asparagusic acid.[10] Kinetic experiments indicate that the inhibition by asparagusic acid is a mixture of competitive and noncompetitive reactions with lipoic acid.

Stability. Preparations of asparagusate dehydrogenases I and II and lipoyl dehydrogenase in 67 mM phosphate buffer, pH 7.0, are very unstable at 4° but stable for at least one month at $-80°$. EDTA and phosphate ion stablize the enzymes. After heating at 50° for 5 min, both asparagusate and lipoyl dehydrogenase activities are completely retained. Moreover, the asparagusate dehydrogenase activity is considerably activated by heat

[10] H. Yanagawa and F. Egami, *J. Biochem.* **78**, 1153 (1975).

treatment at 60°–70°, whereas lipoyl dehydrogenase activity gradually decreases above 50° and is completely lost at 90°.

Distribution

The *in vivo* activities of asparagusate dehydrogenase and lipoyl dehydrogenase in etiolated and green shoots of asparagus (*Asparagus officinalis* L.) were measured.[11] The region around the apices of the shoots has higher levels of asparagusate and lipoyl dehydrogenases than other regions. The levels of asparagusate and lipoyl dehydrogenases in the etiolated shoots are higher than those in the green shoots. The two enzyme activities are localized in the mitochondrial fraction.

[11] H. Yanagawa, *Plant Cell Physiol.* **17**, 931 (1976).

[36] Preparation and Determination of Asparagusic Acid

By HIROSHI YANAGAWA

Asparagusic acid has recently been found in etiolated young asparagus (*Asparagus officinalis* L.) shoots together with the related compounds, dihydroasparagusic acid, *S*-acetyldihydroasparagusic acid, and asparagusic acid *anti*- and *syn*-*S*-oxides; their chemical structures have been elucidated.[1,2] Recent evidence suggests that asparagusic acid may participate in enzymic reactions in asparagus, including stimulation of pyruvate oxidation,[3] as the substrate of asparagusate dehydrogenase,[4,5] and in the inhibition of lipoyl dehydrogenase.[6] At present, asparagusate dehydrogenase is thought to function in metabolic regulation in the oxidative decarboxylation complexes of α-keto acids of asparagus mitochondria.[7]

Recently, a convenient procedure[8] for the synthesis of asparagusic acid has been found. Asparagusic acid can be synthesized using commercially available ethyl malonate as a starting material: ethyl malonate is al-

[1] H. Yanagawa, T. Kato, Y. Kitahara, N. Takahashi, and Y. Kato, *Tetrahedron Lett.* 2549 (1972).
[2] H. Yanagawa, T. Kato, and Y. Kitahara, *Tetrahedron Lett.* 1073 (1973).
[3] H. Yanagawa, T. Kato, and Y. Kitahara, *Plant Cell Physiol.* **14**, 1213 (1973).
[4] H. Yanagawa and F. Egami, *Biochim. Biophys. Acta* **348**, 342 (1975).
[5] H. Yanagawa and F. Egami, *J. Biol. Chem.* **251**, 3637 (1976).
[6] H. Yanagawa and F. Egami, *J. Biochem.* **78**, 1153 (1975).
[7] H. Yanagawa, Ph.D. dissertation, University of Tohoku, 1973.
[8] H. Yanagawa, T. Kato, H. Sagami, and Y. Kitahara, *Synthesis* **9**, 607 (1973).

lowed to react with formalin to form ethyl bis (hydroxymethyl)malonate,[9] which is heated with hydriodic acid to form β, β'-diiodoisobutyric acid. This acid is treated with sodium trithiocarbonate[10] and aqueous sulfuric acid to give dihydroasparagusic acid. Dihydroasparagusic acid is easily oxidized with dimethyl sulfoxide to afford asparagusic acid.

A method[11] for high-speed liquid chromatographic determination of asparagusic acid in asparagus tissues has been developed. The *in vivo* concentration of asparagusic acid in etiolated and green shoots of asparagus was measured by the method. The region around the apices of the shoots has a higher level of asparagusic acid than other regions.

Preparation

Isolation of Asparagusic Acid from Asparagus Shoots. Fresh etiolated asparagus (*Asparagus officinalis* L.) shoots are washed with distilled water, and moisture is removed by pressing them gently between sheets of absorbent paper. The tissues are chopped and then immersed in methanol at 4° for a week. The methanol solution is separated from the asparagus chips by filtration. After concentration of the methanol solution, the remaining aqueous liquid is adjusted to pH 2.0 with dilute hydrochloric acid, after which it is extracted three times with ether. The ether phase is concentrated and extracted with 5% aqueous sodium bicarbonate. The aqueous phase, after acidification to pH 2.0, is extracted four times with equal volumes of ether, followed by evaporation of the ether under reduced pressure at room temperature. The remaining acidic mixture is separated by preparative thin-layer chromatography (silica gel, 0.1 × 20 × 20 cm, toluene : ethyl formate : formic acid, 5 : 4 : 1). The portion at R_f 0.65 is taken from the thin-layer plates and extracted with methylene chloride. After drying with anhydrous magnesium sulfate, methylene chloride solution is evaporated under reduced pressure at room temperature. The remaining yellowish oily substances are dissolved in a mixed solvent of benzene and cyclohexane and permitted to stand at 4°. The yellow prismatic crystals of asparagusic acid are recrystallized from the mixed solvent of benzene and cyclohexane. Asparagusic acid, m.p. 75.7–76.5°, has the molecular formula $C_4H_6O_2S_2$; ν_{max}^{KBr} at 3500–2500, 1700, 1230, 915, 874, 835, and 785 cm^{-1}; λ_{max}^{EtOH} 325 nm (log ϵ 2.33). Asparagusic acid is methylated with diazomethane to give a methyl ester that shows a molecular peak at m/e 164 with significant fragmentation peaks at 133, 132, 105, 104, 86, 59, and 41. The NMR spectrum of asparagusic acid

[9] K. N. Welch, *J. Chem. Soc.*, p. 257 (1930).
[10] D. J. Martin and C. C. Greco, *J. Org. Chem.* **33**, 1275 (1968).
[11] H. Yanagawa, *Plant Cell Physiol.* **17**, 931 (1976).

$$
\begin{array}{c}
\underset{\text{(I)}}{\text{CH}_2\!\!\left(\begin{array}{l}\text{COOC}_2\text{H}_5\\[-2pt]\text{COOC}_2\text{H}_5\end{array}\right.}
\xrightarrow{\text{HCHO}}
\underset{\text{(II)}}{\text{C}\!\!\left(\begin{array}{l}\text{HOCH}_2\quad\text{COOC}_2\text{H}_5\\[-2pt]\text{HOCH}_2\quad\text{COOC}_2\text{H}_5\end{array}\right.}
\xrightarrow{\text{HI}}
\\[12pt]
\underset{\text{(III)}}{\text{CHCOOH}\!\!\left(\begin{array}{l}\text{ICH}_2\\[-2pt]\text{ICH}_2\end{array}\right.}
\xrightarrow{\text{Na}_2\text{CS}_3,\ \text{H}^+}
\underset{\text{(IV)}}{\text{CHCOOH}\!\!\left(\begin{array}{l}\text{HSCH}_2\\[-2pt]\text{HSCH}_2\end{array}\right.}
\xrightarrow{\text{DMSO}}
\\[12pt]
\underset{\text{(V)}}{\left(\begin{array}{c}\text{S}-\text{CH}_2\\[-2pt]\ \ |\qquad\ \ \text{CHCOOH}\\[-2pt]\text{S}-\text{CH}_2\end{array}\right)}
\end{array}
$$

methyl ester [in C_6D_6 containing tris(dipivalomethanato)europium, $Eu(dpm)_3$] shows 4.75 ppm (2H, double doublet), 6.77 ppm (2H, double doublet), 7.00 ppm (1H, quintet), and 7.25 ppm ($COOCH_3$, singlet). Asparagusic acid in solution is very unstable to light, heating, oxygen, and alkali and easily changes to polymeric substances. Asparagusic acid is soluble in benzene, ethyl acetate, methanol, slightly soluble in water, and insoluble in petroleum ether and hexane.

Synthesis of Asparagusic Acid. Ethylmalonate (I) (1120 g, 6.99 mol) and 40% formalin (2000 g, 26.67 mol) are mixed. Ten percent NaOH is added dropwise to the mixture in an ice bath with stirring and maintained at pH 8–9. After 2 hr the ice bath is removed and the mixture is further stirred at room temperature for 48 hr. The reaction mixture is extracted with petroleum ether and dried with anhydrous magnesium sulfate. Evaporation of the extract gives colorless ethyl bis(hydroxymethyl)malonate (II); yield: 1200 g (78.7%). A mixture of (II) (88 g, 0.4 mol) and 57% hydriodic acid (270 g, 1.2 mol) is refluxed for 5 hr, during which period volatile materials are removed in a hood to make the final volume two-thirds of the starting volume. Upon cooling overnight at room temperature, crude β,β'-diiodoisobutyric acid (III) deposits. It is recrystallized from carbon tetrachloride to afford pure (III) (colorless plate); yield: 68 g (50%); m.p. 127–129.5°. Aqueous sodium trithiocarbonate (4.2 g, 8.9 mmol) is added dropwise to (III) at 25° with stirring under a nitrogen atmosphere; the reaction mixture is heated at 50° for 5 hr. After acidification with 6 N sulfuric acid, the aqueous layer is extracted three times with ethyl acetate. The ethyl acetate extracts are dried with anhydrous magnesium sulfate and concentrated to give crystals of dihydroasparagusic acid (IV), which are recrystallized from benzene–cyclohexane; yield: 0.38 g [88% from (III)]; m.p. 59.5–60.5°. Compound (IV) (1 g, 6.6 mmol) is dissolved in dimethyl sulfoxide (10 ml) and then stirred at 70°–75° under nitrogen. The reaction is monitored by silica gel thin-layer chromatography to prevent further oxidation. The reaction is usually completed within 5 hr. The reaction mixture is poured into cold water, and extracted three times with benzene, the benzene layer being washed with water, and dried on anhydrous magnesium sulfate. Evaporation of the extract gives crude asparagusic acid (V), which is recrystallized from benzene–cyclohexane to afford pure yellowish prismatic crystals; yield: 0.65 g (70%).

Determination

For chromatographic identification and determination of asparagusic acid, high-speed liquid chromatography with a Waters Associate Model ALC/GPC 244 is used. The apparatus is equipped with two Model 6000 A solvent delivery systems and a Model 660 solvent programmer. The asparagusic acid-containing methanol extract is chromatographed using a mobile phase of a linear gradient (H_2O/10 mM $NH_4H_2PO_4$, pH 6.00, 10 min) and a column (4 mm × 30 cm) packed with μBondapak-NH_2 at a flow rate of 2 ml/min. The peak is monitored with two ultraviolet detectors of 245 and 330 nm (absorption maximum of asparagusic acid). The peak material is collected in a flow cell of a Varian spectrophotometer, Variscan, and the spectrum is run to confirm identity. The amount of asparagusic acid is calculated from the standard curve obtained by plotting peak height against concentration of asparagusic acid. The limit of detection is 0.2 nmol of asparagusic acid.

Distribution

The *in vivo* concentration of asparagusic acid in etiolated and green shoots of asparagus (*Asparagus officinalis* L.) were measured by the high-speed liquid chromatographic method.[11] The region around the apices of the shoots had higher levels of asparagusic acid than other regions. The level of asparagusic acid in the green shoots was higher than that in the etiolated shoots. Asparagusic acid was localized in the soluble fraction of the cell. Asparagusic acid was also detected in other asparagus species, i.e., *A. sprengeri* Regel, *A. myriocladus* Hort, *A. falcatus* L., *A. plumosus* Baker *var. nanus* Nichols, and *A. ablight.*

[37] Methodology Employed for Anaerobic Spectrophotometric Titrations and for Computer-Assisted Data Analysis

By CHARLES H. WILLIAMS, JR., L. DAVID ARSCOTT, ROWENA G. MATTHEWS, COLIN THORPE, and KEITH D. WILKINSON

Many proteins that catalyze oxidoreductions have chromophoric prosthetic groups that permit the state of reduction of the protein to be monitored spectrophotometrically. In our laboratory, a simple titration assembly has been developed by which we can monitor absorbance changes during an anaerobic titration of a protein with reducing substrates or chemical reductants such as dithionite. This titration assembly permits large numbers of experimental points to be determined with a single solution of enzyme. It also permits replacement of the titrant solution so that the enzyme may be reduced, and then back titrated, or the reduced enzyme may be titrated with ligands to determine its binding properties.

The basic titration assembly was first described by Burleigh, Foust, and Williams.[1] Lambeth and Palmer[2] suggested elimination of the rubber septum and gas-flow lock. The apparatus we presently use differs only in detail, and a diagram is shown in Fig. 1. The cuvette unit fits directly into the cuvette holder of a Cary Model 118C or Model 14 spectrophotometer. Since the unit is taller than the sample compartment of the spectrophotometer, it is covered by a black Lucite or wooden box constructed so as to exactly cover the sample compartment. Increments of titrant are delivered from a Hamilton gastight syringe, modified as described in the legend to Fig. 1 and fitted with either a screw-thread plunger or a repeating dispenser such that precise increments of titrant may be delivered.

The flask in which the titrant is prepared and from which syringes are filled is shown in Fig. 2. The arrangement by which these units are connected to gas and vacuum lines is shown in Fig. 3. Construction and testing of a variety of anaerobic gas trains have been reviewed by Beinert et al.[3]

The comparison of spectral data generated in chemical manipulations of flavoproteins (or any solution with visible or ultraviolet absorbance) has been facilitated by interfacing a recording spectrophotometer with a minicomputer as shown in Fig. 4. Spectra can be stored on magnetic tape in a standard format and can be recalled for comparisons, mathematical

[1] B. D. Burleigh, Jr., G. P. Foust, and C. H. Williams, Jr. *Anal. Biochem.* **27**, 536 (1969).

[2] D. O. Lambeth and G. Palmer, *J. Biol. Chem.* **248**, 6095 (1973).

[3] H. Beinert, W. H. Orme-Johnson, and G. Palmer, this series, Vol. LIV, p. 111.

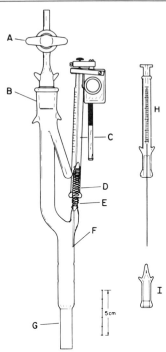

Fig. 1. Cuvette unit. (A) A high-vacuum stopcock, with spring-loaded, hollow plug, 2 mm in diameter, straight bore (Eck and Krebs No. P5004), is attached to a 19/22 standard-taper male ground-glass joint with glass hooks; (B) 19/22 standard-taper female ground-glass joint with glass hooks; (C) Hamilton gastight syringe of the desired volume, fitted with a 4-inch Hamilton SN needle (point style 2, 22 gauge) and with a repeating dispenser (Hamilton PB600-1). The syringe is modified as described below (H); (F) The syringe needle passes through a standard-taper 7/25 male ground-glass joint with glass hooks (E) having a 1 mm i.d. capillary (Kontes K661250), and the needle should be bent so that the tip just rests against the glass wall of the titration assembly. The male joint should be at a 1° or 2° angle to the vertical to allow adequate clearance between the stopcock and the syringe assembly. (G) A quartz cuvette (10 mm light path) is attached via an 18-mm o.d. graded seal (Suprasil) to the remainder of the titration assembly. If a standard absorbance cuvette with two frosted sides is used, the frosted sides should be in the plane of the paper. Alternatively, a quartz fluorescence cuvette with four optical faces may be used. (H) A Kontes 7/25 standard-taper female ground-glass joint with glass hooks having an internal diameter just sufficient to accommodate the outer diameter of the Hamilton syringe barrel is cut to a length of about 8 cm. The ground-glass area of the fitting is lightly greased, and the syringe barrel is coated with an even layer of epoxy cement and inserted into the 7/25 sleeve to form an airtight seal. A slot is cut in the threaded retaining ring of the dispenser to allow attachment and removal of the dispenser from syringes modified in this way. (I) A Kontes 7/25 standard-taper female ground-glass joint with glass hooks is sealed to form an airtight cap, and the dead volume is reduced by filling with epoxy cement.

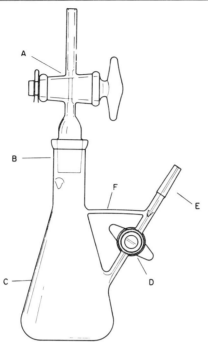

Fig. 2. Flask for anaerobic titrants. (A) As described in legend to Fig. 1 (A). (B) 19/22 standard-taper female ground-glass joint with glass hooks attached to a 125-ml Erlenmeyer flask (C). (D) High-vacuum stopcock (15/35, Kontes, K-841250), joined to a standard-taper 7/25 (Kontes K 661250, 1 mm i.d.) male ground-glass joint (E). (F) Solid glass buttress.

manipulations, or hard-copy output. Spectra are visualized on the display terminal for comparison or manipulation, and hard-copy is made on the X-Y recorder.

The program has been written in BASIC, which is easy to learn yet sophisticated enough to handle a variety of peripheral devices. OS/8 BASIC with LAB8/E functions was obtained from Digital Equipment Corp. (DEC) and modified to accommodate the display terminal and the X-Y recorder and to allow more efficient information transfer within the computer.

The Cary 118C was equipped with a buffer binary-coded decimal accessory on the digital panel meter and a standard scan adapter accessory. In order to transmit sufficiently accurate data to the computer (optical densities to four significant figures), it was necessary to utilize both the digital I/O and the A/D converter. Specification of the optical density requires 18 bits of information; 12 bits can be transmitted to the digital I/O, and the additional 6 bits are transmitted to the A/D converter. The

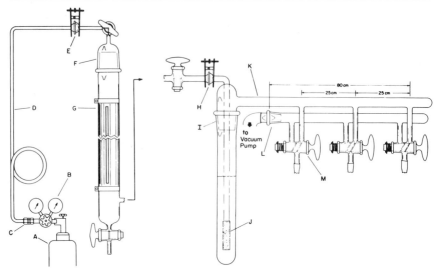

FIG. 3. Anaerobic gas-train. (A) Ultra-high-purity nitrogen cylinder. (B) High-quality regulator with low leak rate (Matheson High Purity Regulator, Model 3500, with CGA connection 590). (C) Flare nut fitting. (D) ¼-inch copper tubing soldered to 18/9 standard brass ball joint. (E) 18/9 glass socket joint fitted to the brass ball with clamp. (F) Adapter, female 45/50 with hooks. (G) Heated catalyst column (Ace Glass 7818-34) equipped with a 45/50 male joint and two 4-mm stopcocks, containing BASF catalyst (Ace 7818-60). (H) 18/9 ball and socket ground-glass joint. (I) 34/45 standard-taper joint with tube attached to female part. (J) Fritted-glass fitting (Ace Glass 7202); the tube is partially filled with water. (K) 18 mm o.d. Pyrex tubing. (L) 19/38 standard-taper joint. (M) High-vacuum two-way stopcock with spring-loaded hollow plug, and 3-mm oblique bore (Eck and Krebs 5020-3).

connection of these to the buffer accessory is shown in Fig. 5. The core memory capacity for our computer is 16,000 (12 bit) words, and the mass memory is supplied by a dual DECtape having a capacity of 190,000 words per tape. The program, CARY18, occupies 14,000 words of which 4000 words are the run-time overhead of OS/8 BASIC. This program has 17 directly addressable commands as outlined in the table. Data for 3 spectra can be resident in core at any one time, file Ø, 1 and 2. Data for approximately 150 spectra can be stored on a DECtape.

The scan (SC) command initiates a scan on the Cary of any 400 nm and stores an absorbance value every nanometer in file 2. An initial dialogue fixes the starting and ending wavelengths, scan speed, and absorbance range. The spectrum is displayed on the Cary recorder and on the display terminal. A final dialogue fixes the nanomoles of enzyme and the volume. The absorbance data are converted to extinction coefficients and stored in file Ø or 1. The absorbance data are erased from the terminal, and the converted spectrum is displayed. The data can then be manipulated, stored

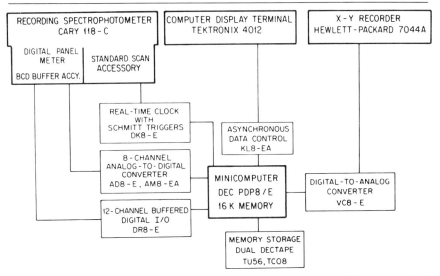

FIG. 4. Block diagram showing interconnections between components of the computerized spectrophotometer system.

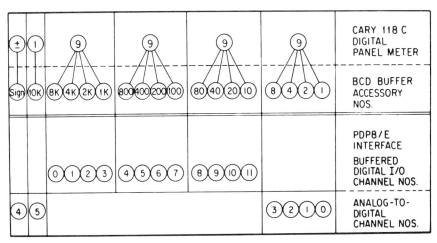

FIG. 5. Connections between the spectrophotometer BCD buffer accessory and the computer via Digital I/O and A/D converter. Optical density readings displayed on the Cary 118C digital panel meter are shown in the upper row. The left-hand columns indicate the sign and integer portion (0-1, 1-2) of the reading. The second row identifies the 18 pins provided by the buffer accessory on the back of the digital panel meter. The third row indicates the 12 pins on the Digital I/O interface. Each pin in the third row is connected to the BCD buffer accessory pin shown directly above it. The fourth row indicates the 6 pins in the A/D converter. Each pin is connected to the BCD buffer accessory pin shown directly above it. Connections are made using a Belden No. 8775 shielded cable.

CARY 18 Subroutines[a]

Command	Description
SC	SCan—Initiates a scan on the Cary and stores the data
SS	Special Scan—Transfer existing Cary spectra into computer by manual curve tracing
RE	REad—Reads a spectrum from DECtape into the computer, file Ø or 1 and displays it on the terminal
RB	Read Baseline—Reads a baseline from DECtape into the computer, file 2 and displays it on the terminal
SA	SAve—Stores a spectrum on DECtape after assignment of a code
CA	CAlibrate—Sets up the coordinates on the X-Y recorder
PL	PLot—Plots a spectrum on the X-Y recorder
AD	ADd—Adds 2 spectra together
SU	SUbtract—Subtracts one spectrum from another
SM	SMooth—Smooths a noisy spectrum by a digital filter routine
MU	MUltiplies—Multiplies a spectrum by a constant
ER	ERase—Erases the display terminal screen and replots files Ø and 1
PR	PRint—Prints relevant information for a spectrum
SL	ScaLe—Changes display terminal parameters
SE	Subtract Extinction—Subtracts a constant extinction from a spectrum
EC	Extinction Coefficient—Displays the extinction coefficient at any requested wavelength
SB	Subtract Baseline—Subtracts a baseline in file 2 from a spectrum in file Ø or 1

[a] A copy of the program is available.

on DECtape or output onto hard-copy as detailed in Table I. Data are saved as a 6 significant-figure floating-point value, thus minimizing rounding errors in subsequent manipulations. Each spectrum stored on DECtape is identified by a 6 alphanumeric code.

The procedures used for preparation of the anaerobic solutions for titration and the subsequent analysis of the data are illustrated by describing in detail the protocol for the measurement of proton release associated with two electron reduction of lipoamide dehydrogenase by dihydrolipoamide. In these experiments, the enzyme is dissolved in an unbuffered solution containing phenol red as a pH indicator and is then reduced with dihydrolipoamide under anaerobic conditions. After each addition of dihydrolipoamide the difference spectrum (reduced enzyme minus oxidized enzyme) is determined. If no proton release occurs on reduction, the shape of these difference spectra will reflect only the conversion of oxidized to reduced enzyme. If proton release does occur on reduction, the difference spectra will be distorted by the conversion of the alkaline form of phenol red (maximum absorbance at 558 nm) to the acid form. The anaerobic solution can then be titrated with standardized NaOH until the reduction difference spectrum is returned to a shape characteristic of reduction at that pH in the absence of phenol red.

The working procedure is as follows:

1. Anaerobic solutions of the titrants, 10 mM dihydrolipoamide and 1 mM NaOH, are prepared in titrant flasks like the one shown in Fig. 2. Air-stable titrant solutions are prepared directly in the flask, and a magnetic stirring bar is inserted. The stopcock in the sidearm is closed, and the upper stopcock is connected with the shortest feasible length of butyl rubber tubing (Sargent No. S-73658) to the anaerobic train. All ground-glass joints are sealed with Apiezon N stopcock grease. The entire assembly is clamped above a magnetic stirring motor. Using the 2-way stopcock on the anaerobic train (M, Fig. 3), the solution is alternately degassed and equilibrated with nitrogen for six cycles while continuous stirring is maintained. Nitrogen is admitted to the flask as the final step in the degassing procedure. [If a titrant solution is to be prepared which is unstable in air, such as dithionite, the solid titrant may be inserted at this point in a small boat. A plastic serum bottle cap is appropriate for this purpose. With the stirring motor turned off and the flask under nitrogen, the upper stopcock is closed and removed, and the boat is floated on the solution using a long pair of tweezers. The stopcock joint is regreased and resealed, and two evacuation/nitrogen cycles are performed without stirring. The boat is then tipped by stirring rapidly.]

2. The titrant solutions are standardized by appropriate methods. In each case an aliquot of anaerobic solution is withdrawn from the titrant flask into a modified Hamilton gas tight syringe. Withdrawal is accomplished by opening the sidearm stopcock while the flask is connected to the nitrogen line so that a steady flow of nitrogen emerges from the open stopcock. The Hamilton syringe is then inserted into the stopcock fitting and secured with small springs or rubber bands. The flask is now tilted so that the solution covers the needle tip, and the syringe barrel is alternately filled and emptied several times before being filled with the solution to be withdrawn. The filled syringe is then removed, the stopcock in sidearm is closed, and the solution is again carried through three cycles of degassing and equilibration with nitrogen. Standardization of dihydrolipoamide solutions is accomplished by titration of aliquots into a solution containing 5,5′-dithiobis(2-nitrobenzoic acid),[4] and standardization of NaOH is accomplished by titration into a solution of potassium hydrogen phthalate.[5] [Dithionite solutions can be conveniently standardized by performing an anaerobic titration of a solution of lumiflavin-N(3)-acetic acid

[4] R. G. Matthews and C. H. Williams, Jr. *J. Biol. Chem.* **251**, 3956 (1976).

[5] R. G. Matthews, D. P. Ballou, C. Thorpe, and C. H. Williams, Jr. *J. Biol. Chem.* **252**, 3199 (1977).

and following the absorbance at 443 nm as successive aliquots of dithionite are added. The concentration of lumiflavin-N(3)-acetate at each stage of the titration is determined using a difference extinction coefficient of 11.3 mM^{-1} cm^{-1} for oxidized minus reduced lumiflavin-N(3)-acetate.[6] (The difference extinction coefficient given previously for this compound[7] is now known to be incorrect.) Details of this procedure have been described by Foust et al.[7]]

3. The enzyme solution to be titrated is prepared in the cuvette. Enzyme that has been extensively dialyzed against 50 mM NaCl is carefully titrated with 10 mM NaOH to bring the pH to the desired value. The spectrum is recorded to determine the enzyme concentration, and then the solution is brought to 16 μM in phenol red by addition of a 1 mM stock solution of phenol red neutralized to the desired pH. The pH of the solution is again checked, and the pH is adjusted if necessary by addition of 1 mM NaOH. A 2.5-ml volume of this solution is transferred to the cuvette, the sidearm is covered by a small glass cap of the type shown by (I) of Fig. 1, and the joint is secured with springs or rubber bands. The spectrum is again recorded. The upper stopcock is attached to a two-way stopcock on the anaerobic train using the minimum length of butyl rubber tubing, and the solution is alternately degassed and flushed with oxygen-free nitrogen for six cycles terminating with nitrogen. Degassing is assisted by agitating the cuvette with a vortex mixer or by gently tapping the cuvette while the solution is allowed to warm to room temperature. Equilibration with nitrogen is assisted by tilting the cuvette so as to increase the surface area of the solution and immersing the cuvette in ice. While the cuvette is under nitrogen, and is attached to the nitrogen line, the sidearm cap is removed and a steady stream of nitrogen is allowed to flow out of the sidearm. A freshly loaded syringe of titrant is inserted and affixed to the cuvette by springs or rubber bands, the upper stopcock is closed, and the cuvette is detached from the train. A spectrum of the anaerobic enzyme solution is recorded at this time to determine any change in concentration due to evaporation and for subsequent use in generating difference spectra. In this case, since absorbance changes of the dye may occur owing to slight shifts in pH on removal of carbon dioxide, changes in concentration are determined at 478 nm where alkaline and acid phenol red are isosbestic.

4. The enzyme is reduced by serial additions of dihydrolipoamide, and a spectrum is recorded and stored on magnetic tape after each addition. [Enzyme and dihydrolipoamide are in equilibrium with reduced enzyme and lipoamide, so reduction is not stoichiometric. Additions are

[6] S. G. Mayhew, and V. Massey, Biochim. Biophys. Acta 315, 181 (1973).
[7] G. P. Foust, B. D. Burleigh, Jr., S. G. Mayhew, C. H. Williams, Jr., and V. Massey, Anal. Biochem. 27, 530 (1969).

stopped when a 1.5-fold excess of reductant over enzyme has been added.] The cuvette is then removed from the spectrophotometer and again connected to the anaerobic train. After evacuating the connecting tubing and establishing positive nitrogen pressure the syringe is removed from the sidearm and replaced by another syringe containing anaerobic 1 mM NaOH. We do not observe reoxidation of the enzyme during syringe transfer. The enzyme solution is then titrated with successive 10-μl aliquots of NaOH, and spectra are recorded after each addition and stored on magnetic tape. Proton release associated with reduction is determined by comparing successive computer-generated difference spectra (reduced enzyme after NaOH addition minus initial oxidized enzyme) with a previously determined difference spectrum obtained in the absence of phenol red (reduced enzyme minus oxidized enzyme) adjusted to the same pH. Proton release during reduction distorts the shape of the difference spectrum and when the shape of the spectrum (reduced enzyme after NaOH addition minus oxidized enzyme) is returned to a shape characteristic of the difference spectrum in the absence of phenol red, the protons released on reduction of lipoamide dehydrogenase by dihydrolipoamide will have been exactly neutralized. Details from difference spectra generated from an experiment performed at pH 7.95 are shown in Fig. 6. Data from such an experiment could conceivably be interpreted without the aid of a computer, but the calculations would be laborious and time consuming. The actual concentration of reduced enzyme produced by addition of dihydrolipoamide may be determined from the neutralized difference spectrum if the extinction difference associated with enzyme reduction at some suitable wavelength is known. In this case the $\Delta \epsilon_{530}$ associated with enzyme reduction at pH 7.95 is known to be 3250 M^{-1} cm^{-1}.[4] Proton release per mole of enzyme reduced may then be calculated.

5. Proton release associated with the addition of substrate or product to an unbuffered solution containing phenol red must also be determined in independent experiments. In this case the ionizations of the thiol groups of dihydrolipoamide are determined spectrophotometrically[5] and proton release expected from the residual dihydrolipoamide in the equilibrium mixture of oxidized and reduced enzyme is calculated. In this manner values can be obtained for proton release associated with the reduction of lipoamide dehydrogenase at each pH value.

6. In attempting to measure proton release associated with reduction of a colored protein by this method, it is important to establish that the pH indicator dye does not complex with the enzyme so as to distort the results. Preliminary experiments should be performed to show that the ionization of the pH indicator shows the same pH dependence in the presence or the absence of the enzyme, and that the spectra of enzyme and

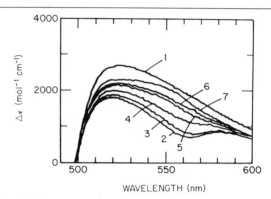

WAVELENGTH (nm)

FIG. 6. Details of difference spectra generated during a proton-release measurement. Curve 1 is the difference spectrum associated with 2-electron reduction of 80% of an enzyme solution at pH 8.0 in the absence of phenol red. Curve 2 is the difference spectrum (reduced enzyme–oxidized enzyme) observed when 260 nmol DL-dihydrolipoamide are added to 164 nmol of oxidized enzyme in the presence of 16.2 μM phenol red, 50 mM NaCl, pH 7.95. Curves 3–7 represent difference spectra (reduced enzyme after NaOH addition minus oxidized enzyme) generated after addition of anaerobic 1.14 mM NaOH: curve 3, 17 nmol of NaOH; curve 4, 40 nmol of NaOH; curve 5, 63 nmol of NaOH; curve 6, 74 nmol of NaOH; and curve 7, 97 nmol of NaOH. Curve 6 shows the shape closest to that of curve 1 and represents the neutralized difference spectrum. The change in extinction at 530 nm is 2210, and formation of 100% two electron reduced enzyme at this pH would have an extinction change of 3250 mol^{-1} cm^{-1}. In this experiment 68% (111 nmol) of the enzyme was reduced on addition of 1.5 equivalents of dihydrolipoamide. Proton release associated with reduction is 0.67 H$^+$ per mole of reduced enzyme and includes protons released by ionization of residual dihydrolipoamide. All curves have been corrected for dilution.

dye are not perturbed when the two are mixed. In addition, interactions between titrants or products and the indicator dye must be excluded. Preliminary experiments in which dithionite was used as a reductant in the presence of phenol red showed that dithionite bleaches phenol red, and is thus an unsuitable reductant for proton-release measurements using this pH indicator.

Two further examples follow of manipulations that are greatly facilitated by this system. Enzymes frequently form spectrophotometrically detectable complexes with ligands. If binding is not tight, large excesses of ligand must be used in the titration and still the spectrum of the complex may be only approximated. The spectrum of the fully formed complex can be deduced using this system. By plotting the reciprocal of the absorbance versus the reciprocal of the concentration of ligand added, the change at infinite ligand concentration can be determined. The wavelength selected must be one at which there is no interference from ligand absorption. In order to generate the spectrum of a fully formed complex, an intermediate spectrum is selected, for example, one in which the complex is 0.25 formed. This spectrum is subtracted from the starting (free enzyme) spectrum using the SU command. This difference spectrum is expanded 4-fold via the MU command generating the difference spectrum between the fully formed complex and the starting enzyme. Addition (AD command) of the expanded difference spectrum to the starting spectrum gives the spectrum of the fully formed complex. This technique has facilitated the identification of a suspected covalent adduct between the C(4a) position of FAD and an active center thiolate in pig heart lipoamide dehydrogenase.[8,9]

Computer analysis of spectrophotometric data also facilitates calculation of the spectra of intermediates seen in the course of titrations. Where an enzyme is capable of successive reductions (as in lipoamide dehydrogenase which contains two reducible active groups), the two phases of reduction may overlap.[4,10] In Fig. 7, data obtained during a dithionite titration of *E. coli* lipoamide dehydrogenase are shown.[11] This reduction takes place in two distinct phases, with an intermediate region where all three enzyme species (oxidized, two electron, and four electron reduced) are present. The spectrum of fully formed two electron reduced enzyme may be determined by essentially the same method as that described above for complex formation: Two actual spectra from the linear portion of the titration, where disproportionation is minimal, are used to generate a difference spectrum. This difference spectrum is expanded to the magnitude that would characterize complete conversion of oxidized enzyme to the two electron reduced state, and then added back to the spectrum of oxidized enzyme. The appropriate expansion factor is determined by extrapolation of the two linear phases to their intersection, as in Fig. 7.

Plots of absorbance change versus titrant added often show a lag. This can be due to residual oxygen and/or to free FAD. In the case of an enzyme such as lipoamide dehydrogenase, these artifacts can be distinguished. Residual oxygen leads to a lag at both 455 nm and at 530 nm, while free FAD leads to a lag only at 530 nm. The free FAD then can be determined from the lag at 455 nm or from the difference between the lags at 455 nm and at 530 nm if both artifacts are present. Subsequent correction for the free FAD is greatly facilitated by the computer system.

The computer system can be used to resolve the spectrum of a mixture of two components into the spectra of the components where the mole frac-

[8] C. Thorpe, and C. H. Williams, Jr. *J. Biol. Chem.* **251**, 7726 (1976).
[9] C. H. Williams, Jr., C. Thorpe, and L. D. Arscott, *in* "Mechanisms of Oxidizing Enzymes" (T. P. Singer and R. Ondarza, eds.), p. 3. Elsevier–North Holland, New York, 1978.
[10] S. G. Mayhew, G. P. Foust, and V. Massey, *J. Biol. Chem.* **244**, 803 (1969).
[11] K. D. Wilkinson and C. H. Williams, Jr. *J. Biol. Chem.* **254** (1979).

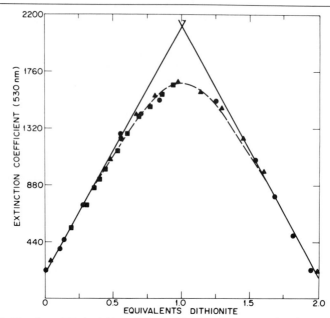

Fig. 7. Titration of *Escherichia coli* lipoamide dehydrogenase with dithionite: 40.5 μM (■) or 18.3 μM (●) and (▲) enzyme-bound FAD in pH 7.6 phosphate buffer, 25°, was titrated with standardized dithionite. The solid lines are extrapolations of the linear phases (where disproportionation is minimal), and the dashed line is calculated from the extrapolated end point and a formation constant,

$$K_f = \frac{[2e^- \text{ reduced}]^2}{[\text{oxidized}][4e^- \text{ reduced}]}$$

of 50.

tions are known, provided that the spectrum of the mixture can be observed under two sets of conditions where the mole fractions are different.[11] The spectra of the components themselves must not change with the change in conditions. This is best illustrated with an example. Lipoamide dehydrogenase from *E. coli* at the two electron reduced level in the presence of 0.2 *M* guanidinium chloride is a mixture of two species. One of the species is fluorescent, and the other has a strong charge-transfer absorption. The ratio of the species varies with pH; low pH favors the fluorescent species, and high pH the charge-transfer species. Thus the mole fractions of the two species can be determined provided the molar fluorescence of the fluorescent species and the molar extinction of the charge-transfer species are known. Since the mole fractions of the two species vary with pH, the properties of the pure species can be estimated by extrapolation of a plot of fluorescence versus extinction coefficient

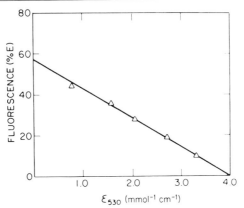

FIG. 8. Relationship of fluorescence to 530 nm extinction of two electron reduced *Escherichia coli* lipoamide dehydrogenase. Each value of 530 nm extinction and of fluorescence was obtained from extrapolation of the results of a dithionite titration at the indicated pH to full formation. The pH's are, from left to right: 5.7, 6.1, 6.5, 7.1, and 7.6.

at various pH values as shown in Fig. 8. At zero fluorescence the extinction coefficient is that of the charge-transfer species, and at zero extinction coefficient the fluorescence is that of the fluorescent species. The spectrum of the two components can be resolved from the spectrum of the mixture observed at two pH values. If we refer to the observed extinction at any given wavelength at pH x as ϵ_x^{obs} and that at pH y as ϵ_y^{obs} then,

$$\epsilon_x^{obs} = F_x^I \epsilon_\lambda^I + F_x^{II} \epsilon_\lambda^{II}$$

and

$$\epsilon_y^{obs} = F_y^I \epsilon_\lambda^I + F_y^{II} \epsilon_\lambda^{II}$$

where F_x^I and F_x^{II} are the mole fractions of the components at pH x and F_y^I and F_y^{II} are the mole fractions at pH y, and where ϵ_λ^I and ϵ_λ^{II} are the extinctions at any given wavelength of the two components. Solving the two equations at each wavelength for ϵ_λ^I and ϵ_λ^I

$$\epsilon_\lambda^I = \frac{F_x^{II}}{F_x^{II}F_y^I - F_x^I F_y^{II}} \epsilon_y^{obs} - \frac{F_y^{II}}{F_x^{II}F_y^I - F_x^I F_y^{II}} \epsilon_x^{obs}$$

$$\epsilon_\lambda^{II} = \frac{F_y^I}{F_x^{II}F_y^I - F_x^I F_y^{II}} \epsilon_x^{obs} - \frac{F_x^I}{F_x^{II}F_y^I - F_x^I F_y^{II}} \epsilon_y^{obs}$$

Evaluations of such equations at all wavelengths yields the spectra of the components. Thus, the observed spectra at the two pH values are multiplied by the respective collection of mole fractions, and the results are subtracted from one another to give the spectrum of one pure species. Without the computer system this would be a formidable task.

Acknowledgments

The authors would like to thank Dr. David Ballou for many helpful discussions, and Dr. Graham Palmer for allowing us to read his manuscript[3] (with Drs. Beinert and Orme-Johnson) prior to publication.

This work has been supported by the Medical Research Service of the Veterans Administration and in part by Grant GM-21444 from the National Institute of General Medical Sciences, Public Health Service.

Section IV

Pantothenic Acid, Coenzyme A, and Derivatives

[38] Microbiological Assay of Pantothenic Acid

By OTTO SOLBERG and IDA K. HEGNA

Turbidimetric methods are the commonly used microbiological assays for vitamin determination; however, the diffusion methods used in the bioassay for penicillin can also be adapted to the assay of vitamins.[1] The diffusion methods are somewhat lacking in sensitivity, but the speed with which they may be carried out compensate for this.[1,2] *Lactobacillus plantarum* is the test organism recommended when testing for pantothenic acid.[3] Another organism, *Pediococcus acidilactici* NCIB 6990 also requires pantothenic acid for growth,[4,5] and has been found to be more sensitive than *L. plantarum* when a diffusion method is used.[6]

Determination of Pantothenic Acid

Maintenance of Stock Culture. The test organism *Pediococcus acidilactici* NCIB 6990, is maintained on blood agar (Oxoid Blood Agar Base No. 2, with 6% citrated horse blood included). When transferred, the culture is incubated for 20 hr at 37° and stored at 4° for no longer than 14 days.

Assay Medium. Solutions of vitamins (Table I) are aseptically mixed with melted basal medium (Table II) together with sterile glucose solution to a final concentration of 4% w/w, adding sterile redistilled water to 100% to make the assay medium. The medium is autoclaved for 5 min at 110°.

Standard Vitamin Solution. The pantothenic acid stock solution has the following composition: calcium D-(+)-pantothenate (vacuum dried), 0.1088 g; redistilled water, 470 ml; and ethyl alcohol (96%), up to 1000 ml. This solution will keep for 4 weeks at 4°.[7] The concentration of pantothenic acid is 100 μg/ml. Nine different dilutions are made from the stock

---

[1] A. L. Bacharach and W. F. J. Cuthbertson, *Analyst* **73**, 334 (1948).

[2] S. Morris and A. Jones, *Analyst* **78**, 15 (1953).

[3] "Methods of Vitamin Assay" (The Association of Vitamin Chemists. Inc., ed.; M. Freed, chairman), p. 201. Wiley (Interscience), New York, 1966.

[4] O. Solberg and O. G. Clausen, *Publications de la faculté des sciences de l'université J. E. Purkyné, Brno, Tchécoslovaquie,* Serie k47, Číslo 509–514, p. 131 (1970).

[5] O. Solberg and O. G. Clausen, *J. Inst. Brew. London* **79**, 227 (1973).

[6] O. Solberg, Ida K. Hegna, and O. G. Clausen, *J. Appl. Bacteriol.* **39**, 119 (1975).

[7] "Pharmacopoea Nordica" Editio Norvegica, Vol. IV, p. 117. Universitetsforlaget, Oslo, 1966.

TABLE I
VITAMINS

Component	Amount present (mg/1000 g assay medium)
D-Biotin	0.2
Thiamine	2.0
Riboflavin	2.0
Pyridoxine hydrochloride	2.0
Nicotinamide	2.0

solution with the following pantothenic acid concentrations (μg/ml): 6.4; 3.2; 1.6; 0.8; 0.4; 0.2; 0.1; 0.05; 0.025.

Inoculum. Prior to the assay, the test organism is cultivated in tomato–malt broth (Table III) for 20 hr at 37°. The culture is centrifuged,

TABLE II
BASAL MEDIUM[a]

Constituent	Weight (g)
KH_2PO_4	1.50
K_2HPO_4	0.50
Sodium acetate·$3H_2O$	10.0
NaCl	0.40
$MgSO_4·7H_2O$[b]	0.40
$CoSO_4·7H_2O$[b]	0.001
$CuSO_4·5H_2O$[b]	0.001
$ZnSO_4·7H_2O$[b]	0.001
$FeSO_4·7H_2O$[b]	0.001
$CaCl_2·2H_2O$[c]	0.004
$MnCl_2·4H_2O$[c]	0.002
$Na_2MoO_4·2H_2O$[c]	0.0001
Adenine	0.01
Guanine[d]	0.01
Cytosine	0.01
Thymine	0.01
Xanthine	0.01
Casamino acids (vitamin free, Difco)	12.0
DL-Tryptophan	0.20
L-Cystine	0.20
Agar (Difco)	18.0
Redistilled water to	800

[a] pH before autoclaving at 120° for 15 min was 6.1.
[b] Salt solution I.
[c] Salt solution II.
[d] Separately dissolved in 1 ml of redistilled water + 0.3 ml of 2.5 M NaOH.

TABLE III
TOMATO–MALT BROTH[a]

Constituent	Amount present
Tomato juice (Oxoid)	200 ml
Malt extract (Difco)	5 g
Yeast extract (Difco)	5 g
Potato extract (Difco)	10 g
Soytone (Difco)	3 g
Tryptone (Difco)	3 g
Peptonized milk (Difco)	3 g
NaCl	1 g
KH_2PO_4	1 g
Distilled water to	1000 g

[a] pH 6.1 prior to autoclaving (at 120° for 15 min). Final pH: 5.7–5.8.

washed three times, and suspended in sterile 0.9% (w/w) NaCl to an optical density of 0.250 at 620 nm measured in a Pyrex tube (18 × 180 mm) in a Coleman Junior spectrophotometer.

Assay Procedure. A portion of 20 ml of the inoculum is added to 1000 g of melted assay medium (48°). The inoculated medium is immediately transferred to sterile petri dishes (30 ml) to an agar depth of 4 mm. The test can be carried out using either the agar-cup method (holes 10 mm in diameter made in the agar) or the cylinder method (Penicylinder, Fisher Scientific Co., inner diameter 6 mm, height 10 mm) putting the cylinders on the agar. The plates are dried at 37° for 1 hr before each hole is filled with 0.2 ml of pantothenic acid solution. The plates rest for 15 min at

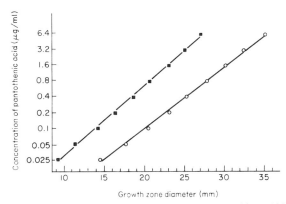

FIG. 1. The average growth zone diameters of *Pediococcus acidilactici* NCIB 6990 corresponding to the different pantothenic acid concentrations employed. ■, Cylinder method; ○, agar-cup method. From O. Solberg, I. K. Hegna, and O. G. Clausen, *J. Appl. Bacteriol.* **39**, 119 (1975).

room temperature (22°) and are then incubated at 37° for 24 hr. The growth zones are recorded as the average of two right-angled diameters using a zone reader. The standard curve should be run at least in duplicate with each assay.

Notes and Interpretation. *P. acidilactici* shows traces of growth down to 0.00625 μg/ml, whereas *L. plantarum* shows no growth below the concentration of 0.05 μg/ml. However, 0.025 μg/ml is the lowest concentration of pantothenic acid that may be assayed with acceptable accuracy using *P. acidilactici* as test organism. Throughout the whole range of pantothenic acid concentration levels tested there is high correspondence between the regression line and observation (Fig. 1). There is also a high correlation (r) between growth zone diameter and concentration both when using the cylinder method ($r = 0.999$) and the agar-cup method ($r = 0.998$) for 8 parallels at each concentration level. Even with 4 parallels, the correlation ($r = 0.996$) is high.

[39] Purification and Properties of Ketopantoate Hydroxymethyltransferase*

By SUE GLENN POWERS and ESMOND E. SNELL

Ketopantoate hydroxymethyltransferase (5,10-methylene tetrahydrofolate:α-ketoisovalerate hydroxymethyltransferase) catalyzes the first committed step in pantothenate biosynthesis, the reversible formation of ketopantoate according to Eq. (1).[1] The functioning of this enzyme in the normal biosynthetic pathway leading

$$\text{Methylenetetrahydrofolate} + \alpha\text{-ketoisovalerate} \rightleftharpoons \text{tetrahydrofolate} + \text{ketopantoate} \quad (1)$$

to pantothenate, rather than a previously described enzyme (designated enzyme I), that catalyzes reaction (2) irreversibly,[2] is supported by the following evidence:

$$\text{HCHO} + \alpha\text{-ketoisovalerate} \rightarrow \text{ketopantoate} \quad (2)$$

the transferase is absent from the ketopantoate auxotroph, *Escherichia coli* 99-4, while enzyme 1 is found in the same amounts as in wild-type *E.*

---

* Ketopantoate is the trivial name used for 2-oxo-3,3-dimethyl-4-hydroxybutyrate, the immediate precursor of D-pantoate (2-hydroxy 3,3-dimethyl-4-hydroxybutyrate).

[1] J. H. Teller, S. G. Powers, and E. E. Snell, *J. Biol. Chem.* **251**, 3780 (1976).

[2] E. N. McIntosh, M. Purko, and W. A. Wood, *J. Biol. Chem.* **228**, 499 (1957).

coli[1]; enzyme 1 is reported[2] to require unphysiologically high concentrations of formaldehyde ($K_m \gtreqqless 10$ mM) and α-ketoisovalerate ($K_m > 100$ mM) whereas the transferase functions at much lower concentrations of substrates[3]; and, like other known enzymes that utilize formaldehyde as substrate, ketopantoate hydroxymethyltransferase requires tetrahydrofolate for activity.[1,3]

Assay Method

Principle. The enzymic breakdown of ketopantoate to formaldehyde via methylene tetrahydrofolate is followed by the Nash reaction,[4] in which the formaldehyde forms a colored adduct with 2,4-pentanedione and ammonium ion; 5,10-methylene tetrahydrofolate shows the same reactivity as free formaldehyde in the Nash reaction.[5] Enzyme activity may also be determined by measuring [^{14}C]ketopantoate formation from H^{14}CHO and α-ketoisovalerate.[1]

Reagents

Formaldehyde is prepared by acid hydrolysis of paraformaldehyde and steam distillation[6]; its solutions are standardized gravimetrically by reaction with dimedon.[7]

d,l-Tetrahydrofolate is prepared by catalytic hydrogenation of folic acid in glacial acetic acid[8] and purified on DEAE-cellulose[9]; its solutions are prepared in 0.1 M potassium ascorbate or in 50 mM 2-mercaptoethanol and standardized by enzymic assay with formyltetrahydrofolate synthetase.[10] Both solid and solutions are stored under nitrogen at $-10°$.

Ketopantolactone is synthesized from pantolactone by the method of Lipton and Strong.[11] Potassium ketopantoate is formed by adjusting solutions of ketopantolactone to pH 7 with KOH.

Procedure. The reaction mixture contains, in micromoles: potassium phosphate (pH 6.8), 75; MgSO$_4$, 1; tetrahydrofolate, 0.5; potassium ketopantoate, 5; and 0.0002–0.002 unit of enzyme in a total volume of 0.7 ml.

[3] S. G. Powers and E. E. Snell, *J. Biol. Chem.* **251**, 3786 (1976).

[4] T. Nash, *Biochem. J.* **55**, 416 (1953).

[5] M. J. Osborn, P. T. Talbot, and F. M. Huennekens, *J. Am. Chem. Soc.* **82**, 4921 (1960).

[6] M. J. Boyd and M. A. Logan, *J. Biol. Chem.* **146**, 279 (1942).

[7] W. R. Frisell and C. G. Mackenzie, *in* "Methods of Biochemical Analysis" (D. Glick, ed.), Vol. 6, p. 67. Wiley (Interscience), New York, 1958.

[8] B. L. O'Dell, J. M. Vandenbelt, E. S. Bloom, and J. J. Pfiffner, *J. Am. Chem. Soc.* **69**, 250 (1947).

[9] N. P. Curthoys and J. C. Rabinowitz, *J. Biol. Chem.* **246**, 6942 (1971).

[10] J. C. Rabinowitz and W. E. Pricer, Jr., *J. Biol. Chem.* **237**, 2898 (1962).

[11] S. H. Lipton and F. M. Strong, *J. Am. Chem. Soc.* **71**, 2364 (1949).

The reaction is run at 37° for 15 min and quenched with 0.3 ml of cold 15% trichloroacetic acid. A 0.5-ml aliquot of the reaction mixture is added to 0.5 ml of Nash reagent (2 M ammonium acetate/50 mM acetic acid/2 mM 2,4-pentanedione[4]), and incubated for 10 min at 58°. The adduct formed is measured at its absorbance maximum, 412 nm, and formaldehyde formation is calculated from the known extinction coefficient of 8000 mol^{-1}.[4] It should be noted that the concentrations of reducing agents normally used to protect tetrahydrofolate solutions from oxidation have to be lowered considerably because these compounds inhibit the Nash assay; noninhibitory levels of reducing agents are 0.5 mM 2-mercaptoethanol, 5 μM dithiothreitol, and 5 mM potassium ascorbate.

Definition of Unit and Specific Activity. One unit of ketopantoate hydroxymethyltransferase activity is defined as the amount of enzyme that catalyzes the transformation of 1 μmol of substrate per minute under the conditions specified. For impure preparations, protein is determined by the method of Lowry *et al.*[12] with bovine serum albumin as standard. The concentration of homogeneous enzyme is calculated from its absorbance; a solution containing 1 mg/ml was found to have an absorbance of 0.85 at 280 nm by the specific refractive index increment method.[13]

Purification Procedure

E. coli K12 is grown at 37° with vigorous shaking in Vogel–Bonner citrate medium[14] containing 0.5% glucose. Large-scale cultures (200 liters) are grown from 2.5% inoculum in a CF250 fermenter with an aeration rate of 8 cubic feet/min, harvested in a Sharples centrifuge, and stored frozen.

The purification procedure is as previously described[1] and is summarized in the table. Unless specified otherwise, all operations are performed at 4°.

Step 1. Preparation of Crude Extract. Cells of *E. coli* K12 (300 g, freshly harvested or thawed after being stored at $-10°$) are suspended in 2 volumes of 50 mM potassium phosphate, pH 6.8, containing 1 mM EDTA and 10 mM 2-mercaptoethanol (buffer A). The suspension is passed twice through a Manton-Gaulin homogenizer. (*Note.* Disruption with a sonic oscillator or protein extraction into buffer A from an acetone powder are equally effective means of cell breakage.) DNase and RNase (approximately 0.1 mg and 10 mg per liter of extract, respectively) are added to the

[12] O. H. Lowry, N. J. Rosebrough, A. L. Farr, and R. J. Randall, *J. Biol. Chem.* **193**, 265 (1951).
[13] E. G. Richards, D. C. Teller, and H. K. Schachman, *Biochemistry* **7**, 1054 (1968).
[14] H. J. Vogel and D. M. Bonner, *J. Biol. Chem.* **218**, 97 (1956).

SUMMARY OF PURIFICATION PROCEDURE FOR
KETOPANTOATE HYDROXYMETHYLTRANSFERASE

Step	Total protein (mg)	Specific activity (μmol min^{-1} mg^{-1})	Overall yield (%)	Purification (fold)
1. Crude extract	27,600	0.03	—	—
2. DEAE-cellulose	288	0.18	62	60
3. 70% $(NH_4)_2SO_4$	105	0.34	43	114
4. BioGel A 0.5 m + 4 M urea	33.9	0.61	25	200
5. Heat treatment	14.8	1.22	22	410
6. BioGel A 1.5 m	2.94	7.30	26	2450

viscous crude extract, which is then incubated overnight. Cell debris is then removed by centrifugation at 13,200 g for 30 min.

Step 2. DEAE-Cellulose Chromatography. The crude extract from step 1 (1150 ml) is dialyzed overnight against a 15-fold excess of buffer A containing 0.2 M KCl and applied to a DEAE-cellulose column (5 × 33 cm) that has been equilibrated with the same buffer–KCl mixture. The column is washed with 5 liters of this equilibration buffer (or until the A_{280} of the eluate is very low), and developed with a linear gradient prepared with 2 liters of 0.2 M KCl in buffer A and 2 liters of 0.6 M KCl in buffer A. Fractions (16.5 ml) are collected at a flow rate of 1 ml/min. The enzyme elutes at 0.4 M KCl (usually fractions 85–130).

Step 3. Ammonium Sulfate Concentration. The hydroxymethyl-transferase-containing pool from step 2 is brought to 70% saturation by the slow addition of solid ammonium sulfate and stirred for 1 hr. The precipitate is recovered by centrifugation (30 min at 13,200 g), resuspended in the minimal amount (about 7 ml) of 100 mM potassium phosphate (pH 6.8), 1 mM EDTA, and 0.5 mM dithiothreitol (buffer B), and dialyzed overnight against 2 liters of this same buffer.

Step 4. Agarose Chromatography in 4 M Urea. An equal volume of 8 M urea in 100 mM potassium phosphate, pH 6.8, is added to the dialyzed protein, and the resulting solution is left at room temperature (approximately 20°) for 2 hr. It is then applied to a BioGel A 0.5-m column (75 × 2 cm), previously equilibrated with 4 M urea in 100 mM potassium phosphate, pH 6.8, and eluted with the same buffer. The chromatography is also carried out at room temperature. Fractions (2.5 ml) are collected at a flow rate of 0.36 ml/min. The enzyme-containing pool (usually fractions 27–34) is dialyzed overnight against 3 liters of buffer B. The dialyzed fraction (30 ml) is concentrated to 3.9 ml in a Sartorius ultrafiltration device.

Step 5. Heat Treatment. The concentrated preparation from step 4 is heated for 5 min in an 80° water bath, and the denatured protein is removed by centrifugation at 27,000 g for 30 min. (*Note*. The enzyme is considerably more heat labile in crude extracts.)

Step 6. BioGel A 1.5-m Chromatography. The protein solution is applied to a BioGel A 1.5-m column (84 × 1.5 cm) previously equilibrated with buffer B and is eluted with the same buffer. Fractions (2.4 ml) are collected at a flow rate of 0.27 ml/min. The enzyme-containing pool (usually fractions 28–44) is concentrated in a Sartorius ultrafiltration apparatus.

The transferase is stored at 4° in buffer B and retains nearly full activity for at least 6 months.

Properties

Homogeneity. Enzyme obtained from step 6 of the purification procedure gives one major band and a fainter broad band on polyacrylamide gel electrophoresis at pH 8.7.[15] Both bands have transferase activity when eluted from the gel, and an eluate of each band gives rise to both when subjected separately to a second electrophoresis under the same conditions. Thus the faint band is apparently a dissociated form of the enzyme. The same pattern of a major and a minor band is obtained at acrylamide concentrations of 4, 5, 7.5, and 10%, all at pH 8.7, and on a 7.5% acrylamide gel at pH 7.5.[16] The final preparation also migrates as a single band at pH 8.8 on an 8% gel containing 0.1% sodium dodecyl sulfate. (SDS)[17] Other indications of homogeneity are a single zone with symmetrical sedimenting boundaries during analytical ultracentrifugation, linearity in plots of ln concentration versus (radial distance)[2] obtained from sedimentation equilibrium ultracentrifugation, and unique NH$_2$-terminal and penultimate amino acid residues.

Molecular Properties.[3] The hydroxymethyltransferase has a molecular weight by sedimentation equilibrium of 255,000, a sedimentation coefficient ($s_{20,w}$) of 11 S, and a partial specific volume of 0.74 ml/g. Polyacrylamide gel electrophoresis in SDS[18] and amino acid analysis give a subunit molecular weight of 27,000 and 25,700, respectively; both procedures indicate the presence of 10 identical subunits. Electron micrographs are consistent with a 10-subunit structure in which the oligomer consists of two layers of pentamers. The NH$_2$-terminal sequence[19] is Met-Tyr. . .

[15] B. J. Davis, *Ann. N. Y. Acad. Sci.* **121**, 404 (1964).

[16] H. R. Mauer, "Disc Gel Electrophoresis," p. 44. Gruyter, New York, 1971.

[17] G. Ames, *J. Biol. Chem.* **249**, 634 (1974).

[18] K. Weber and M. Osborn, *J. Biol. Chem.* **244**, 4406 (1969).

[19] A. M. Weiner, T. Platt, and K. Weber, *J. Biol. Chem.* **247**, 3242 (1972).

The isoelectric point is 4.4, and the enzyme is stable and active over a broad pH range, with an optimum from 7.0 to 7.6.

Catalytic Properties.[3] The forward and reverse reactions catalyzed by this enzyme occur at similar maximum velocities of approximately 8 μmol of ketopantoate formed or decomposed per minute per milligram of enzyme. The forward reaction is partially inhibited by concentrations of formaldehyde (0.8 mM) and tetrahydrofolate (0.38 mM) below or near the K_m values; apparent K_m values are 0.18, 1.1, and 5.9 mM for tetrahydrofolate, α-ketoisovalerate, and formaldehyde, respectively. For the reverse reaction, apparent K_m values are 0.16 and 0.18 mM, respectively, for ketopantoate and tetrahydrofolate, and the saturation curves for both substrates show positive cooperativity.

Only the *l*-isomer of tetrahydrofolate is active with the enzyme, but it is effectively replaced with conjugates containing 1–6 additional glutamate residues; of these, tetrahydropteroylpenta-, tetra-, and triglutamate, the major conjugates of tetrahydrofolate present in *E. coli,* are effective at lower concentrations than tetrahydrofolate itself. α-Ketobutyrate, α-ketovalerate, and α-keto-β-methylvalerate can replace α-ketoisovalerate as substrate with a corresponding change in the structure of the product. The enzyme may be classified as a class II aldolase[20]: it requires Mg^{2+} for activity, with Mn^{2+}, Co^{2+}, and Zn^{2+} being progressively less active, and it is not inactivated by borohydride in the presence of excess substrates.

Ketopantoate hydroxymethyltransferase has the regulatory properties expected of an enzyme catalyzing the first committed step in a biosynthetic pathway. Pantoate (≥ 50 μM), pantothenate (≥ 500 μM), and coenzyme A (above 1 mM) all inhibit; the V_{max} is decreased, K_m is increased, and the cooperativity is enhanced. Since preparative step 4 involves fairly harsh treatment, a modified protocol that did not use this step was devised to partially purify the enzyme; this preparation showed kinetic and regulatory properties identical within experimental error to those described above.

[20] W. J. Rutter, *Fed. Proc., Fed. Am. Soc. Exp. Biol.* **23,** 1248 (1964).

[40] Ketopantoyl Lactone Reductases

By DAVID R. WILKEN, HARLEY L. KING, JR., and
ROBERT E. DYAR

Fungi are reported to metabolize ketopantoyl lactone to a nonracemic mixture of DL-pantoyl lactone.[1] The physiological relevance of this capa-

[1] R. P. Lanzilotta, D. G. Bradley, and K. M. McDonald, *Appl. Microbiol.* **27,** 130 (1974).

METHODS IN ENZYMOLOGY, VOL. 62

bility has not been established. However, formation of D-(−)-pantoyl lactone may relate to coenzyme A metabolism since on hydrolysis it forms D-(+)-pantoic acid, a precursor of coenzyme A. Purification and properties of two forms of ketopantoyl lactone reductase, which convert ketopantoyl lactone to D-(−)-pantoyl lactone [Eq. (1)], are described.

$$
\begin{array}{l}
\underset{\substack{\text{H}_3\text{C}-\text{C}-\text{C}=\text{O}\\ \text{H}_2\text{C}-\text{O}-\text{C}=\text{O}}}{\overset{\text{CH}_3}{}} + \text{NADPH} + \text{H}^+ \longrightarrow \underset{\substack{\text{H}_3\text{C}-\text{C}-\text{C}-\text{OH}\\ \text{H}_2\text{C}-\text{O}-\text{C}=\text{O}}}{\overset{\text{CH}_3\ \ \text{H}}{}} + \text{NADP}^+ \quad (1)
\end{array}
$$

Assay Method[2]

Principle. Ketopantoyl lactone reductase activity is assayed by spectrophotometric measurement of ketopantoyl lactone-dependent oxidation of NADPH.

Reagents

Potassium phosphate buffer, 1 M, pH 7.0

NADPH, 10 mM in 0.2 mM KOH

Ketopantoyl lactone,[3] 0.1 M in 15 mM HCl

Ketopantoic acid, 0.1 M. Ketopantoyl lactone is hydrolyzed at room temperature by adding an equivalent amount of Tris.

Procedure. The assay is performed in a 1-cm light path quartz cuvette at 25°. Enzyme, 0.2 M phosphate buffer, pH 7, 0.1 mM NADPH and 1 mM ketopantoyl lactone are mixed in a final volume of 1 ml. Ketopantoyl lactone is added last to initiate the reaction. The rate of change in absorbance at 340 nm, divided by the millimolar absorption coefficient for NADPH, 6.22×10^3 mM^{-1} cm^{-1}, equals the units of enzyme activity being assayed. The assay for ketopantoyl lactone reductase in crude extracts is complicated by the occurrence of NADPH oxidase and ketopantoic acid reductase in the extracts. Ketopantoic acid reductase is detected in this assay system, since ketopantoic acid is formed during the assay by rapid spontaneous hydrolysis of ketopantoyl lactone.[2] Ketopantoyl lactone reductase activity in crude extracts (step 1) or in partially purified fractions that may contain ketopantoic acid reductase (step 2) is estimated from the rate of NADPH oxidation in the presence of 1 mM ketopantoyl lactone, minus the rate of NADPH oxidation in the presence of 1 mM ketopantoic acid.

[2] H. L. King, Jr., R. E. Dyar, and D. R. Wilken, *J. Biol. Chem.* **249**, 4689 (1974).

[3] Ketopantoyl lactone is prepared by the method of S. H. Lipton and F. M. Strong [*J. Am. Chem. Soc.* **71**, 2364 (1949)], but is recrystallized from ethyl ether and dried over calcium chloride.

Definition of Unit and Specific Activity. A unit of enzyme is that amount of ketopantoyl lactone reductase that reduces 1 μmol of ketopantoyl lactone to pantoyl lactone per minute. Specific activity is expressed as the units per milligram of protein measured by the method of Lowry *et al.*[4]

Growth and Harvest of Yeast[2]

Saccharomyces cerevisiae (NRRL Y-2034, ATCC-24702) is grown at 30° and aerated in a medium composed of 0.3% yeast extract, 0.3% malt extract, 0.5% peptone, and 1% glucose. An inoculum (1.5 liters) of cells grown for 24 hr is added to 150 liters of medium. After 12 hr, cells are harvested by centrifugation in a Sharples centrifuge type AS-16P (15,000 rpm). Cells are resuspended in 25 liters of cold distilled water and recentrifuged. About 1.7 kg of cell paste are obtained.

Purification Procedure[2]

All pH measurements are performed at room temperature, except where indicated. All other operations are performed at 4°. Active fractions from each chromatographic column are concentrated to less than 20 ml, using an Amicon ultrafiltration cell and UM-10 membrane, followed by flow dialysis in a 100-ml chamber at 2 ml/min against 2 liters of 10 mM Tris·HCl buffer, pH 7.4, in 30% glycerol.[5]

Step 1. Cell-Free Extract. A suspension of cell paste (1600 g) and 120-μm glass beads (960 ml) in 0.1 M potassium phosphate buffer, pH 7.0 (1600 ml), is ground 1 hr in an Eppenbach colloid mill at power setting 85 with a 0.03-inch gap. Beads are removed by decantation and washed with water using 1.5-fold the volume of the first decantate. The combined decantates are centrifuged 10 min at 27,000 g. The supernatant fluid is recentrifuged 60 min at 27,000 g. The resulting supernatant fluid (cell-free extract) is removed by aspiration.

Step 2. DEAE-Cellulose Chromatography at pH 7.4. Cell-free extract from two 800-g portions of yeast is treated as follows: Extract from each portion is diluted 5-fold with water. It is then stirred for 30 min with 223 g

[4] O. H. Lowry, N. J. Rosebrough, A. L. Farr, and R. J. Randall, *J. Biol. Chem.* **193**, 265 (1951). The reagent blanks and standards are adjusted to the same concentrations of Tris·HCl buffer and glycerol that are in the sample to be analyzed.

[5] Amicon membranes are preflushed with 1 mM EDTA, pH 7.6, before use. Dialysis tubing is heated twice to 90° for a few minutes in 100 mM EDTA, pH 7.6, washed with distilled water, stored overnight in 0.05% bovine serum albumin, rinsed with distilled water, and stored in 1 mM EDTA, pH 7.6, until used.

(suction-filtered moist weight) of the chloride form of DEAE-cellulose that has been equilibrated with Tris·HCl buffer, pH 7.4. The DEAE-cellulose is collected by suction filtration, washed on the filter with 2 liters of 10 mM Tris·HCl buffer, pH 7.4, and resuspended in 2 liters of the same buffer. The slurry of DEAE-cellulose that remains after decantation is placed on a prepacked column (5 × 68 cm) of DEAE-cellulose (pH 7.4). The column is washed with 2 liters of 10 mM Tris·HCl buffer, pH 7.4, followed by an 8-liter linear gradient of 60 to 125 mM KCl in 10 mM Tris·HCl buffer, pH 7.4. Fractions of 19 ml each are collected at 2 ml/min. Fractions (251–292) from two columns that contain ketopantoyl lactone reductase are combined, concentrated, and dialyzed. This preparation is essentially free of ketopantoic acid reductase, which is eluted mainly in fractions 300 to 380.

Step 3. Hydroxylapatite Chromatography. Dialyzed enzyme from step 2 is applied to a hydroxylapatite column (2.5 × 36 cm) equilibrated with 10 mM potassium phosphate buffer, pH 7.0.[6] The column is washed with 120 ml of the same buffer, followed by a 1600-ml linear gradient of 10 to 125 mM potassium phosphate buffer, pH 7.0. Fractions of 21.5 ml are collected at 1 ml/min. Although enzyme activity occurs in two partially separated areas, all fractions that have activity (43–55) are pooled, concentrated, and dialyzed.

Step 4. Sephadex G-100 Gel Filtration. Dialyzed enzyme from step 3 is applied by upward flow to a Sephadex G-100 column (3.2 × 90 cm) equilibrated with 10 mM Tris·HCl buffer, pH 7.4, in 30% glycerol and eluted with the same solution. Fractions of 10 ml are collected at 0.12 ml/min. Fractions that have activity (51–59) are combined, concentrated, and dialyzed.

Step 5. DEAE-Cellulose Chromatography at pH 8.2. Dialyzed enzyme from step 4 is adjusted to pH 8.4 at 4° with 1 M Tris·HCl buffer, pH 9.0. It is applied to a DEAE-cellulose column (1.9 × 92 cm) equilibrated with 10 mM Tris·HCl buffer, pH 8.2, in 30% glycerol. The column is developed with a 1600-ml linear gradient of 60 to 150 mM KCl in 10 mM Tris·HCl buffer, pH 8.2, in 30% glycerol. Fractions of 10 ml are collected at 0.5 ml/min. Two forms of ketopantoyl lactone reductase are almost completely separated. Fractions containing form A (62–74) and form B (89–104) are combined, concentrated, and dialyzed.

Table I shows results of purifying ketopantoyl lactone reductases from 1600 g of yeast.

---

[6] It is recommended that new hydroxylapatite be used for each enzyme preparation. Buffers used for step 3 are placed under reduced pressure to degas them before use.

TABLE I
PURIFICATION OF 2-KETOPANTOYL LACTONE REDUCTASES
FROM *Saccharomyces cerevisiae* Y-2034

Purification step	Volume (ml)	Protein (mg)	Specific activity (units/mg)	Recovery (%)
1. Cell-free extract	5580	73,600	0.009	100
2. DEAE-cellulose (pH 7.4)	18.5	148	1.78	40
3. Hydroxyapatite	7.2	11.2	13.8	23
4. Sephadex G-100	9.9	2.5	41.2	15
5. DEAE-cellulose (pH 8.2)				
Form A	6.2	0.81	51.0	6.4
Form B	5.6	0.71	69.2	7.2

Properties

Stability. Glycerol (30%) enhances the stability of the reductases. Variable losses of activity approaching 50% may occur during the final concentration of the enzymes. Concentrated reductase form B from step 5 may be stored at $-20°$ for a year with little (15%) loss in activity. Concentrated reductase form A is somewhat less stable.

Substrate Specificity.[7,8] Both forms of the reductase are specific for NADPH. Keto-ω-methylpantoyl lactone is the only analog of ketopantoyl

TABLE II
INHIBITION OF KETOPANTOYL LACTONE REDUCTASES
FROM *Saccharomyces cerevisiae* Y-2034

Compound	Concentration	% Inhibition[a] Form A	Form B
2-Keto-4-hydroxy-3-methylbutyric acid-γ-lactone	$10^{-5} M$	96	54
	$10^{-6} M$	87	39
2-Keto-4-hydroxybutyric acid-γ-lactone	$10^{-4} M$	98	70
	$10^{-5} M$	89	24
1,2-Cyclopentanedione	$10^{-4} M$	85	37
	$10^{-5} M$	52	9
2-Ketopantoic acid	$10^{-3} M$	36	7
2-Ketoisovaleric acid	$10^{-3} M$	55	31

[a] Percent inhibition was measured at pH 7.0 in the standard assay.

[7] H. L. King, Jr., and D. R. Wilken, *J. Biol. Chem.* **247**, 4096 (1972).
[8] D. R. Wilken, H. L. King, Jr., and R. E. Dyar, *J. Biol. Chem.* **250**, 2311 (1975).

lactone tested that is a substrate. Neither ketopantoic acid nor several other 2-keto acids are substrates.

Effects of Inhibitors. Table II shows inhibitory effects of various compounds. The most potent inhibitors closely resemble ketopantoyl lactone and are capable of forming enol tautomers. Structurally similar 2-keto acids are inhibitory at relatively high concentrations. Form A of the enzyme is more susceptible to inhibition than form B.

pH Optimum, Kinetic Properties, and Molecular Weight. The activity of both forms of the reductase is only slightly affected by pH between pH 5.2 and 7.0 when assayed in 0.2 M potassium phosphate buffer. Table III summarizes the apparent pH optimum, K_m values,[2] and molecular weight[2] for each form of the enzyme.

Stereospecificity. Both forms of the reductase transfer the 4-*pro*-R hydrogen[9] of NADPH; therefore, they belong to the A-side stereospecific class of dehydrogenases.[8] D-(−)-pantoyl lactone is produced by both forms of the enzyme.[10]

Ketopantoyl Lactone Reductases from Other Microorganisms.[11] Ketopantoyl lactone reductases have been partially purified from an *Erwinia* species, *S. cerevisiae* A364A, *E. coli* 9637, and *E. coli* AB1899 by chromatography on DEAE-cellulose at pH 7.4.[12] Two forms of the reductase were obtained from *E. coli* 9637. The substrate specificity of the reduc-

TABLE III
pH OPTIMUM, K_m VALUES, AND MOLECULAR WEIGHT OF KETOPANTOYL
LACTONE REDUCTASES FROM *Saccharomyces cerevisiae* Y-2034

Property	Form A	Form B
pH optimum	5.1–5.6	5.6
$K_m{}^a$		
Ketopantoyl lactone	14 μM	31 μM
NADPH	62 μM	39 μM
Molecular weight	27,400	27,000

^{a} Measured in 0.2 M phosphate buffer, pH 7.0.

[9] This terminology conforms to: IUPAC Commission on Nomenclature of Organic Chemistry, Tentative Rules, Section E, of Fundamental Stereochemistry, *Eur. J. Biochem.* **18**, 151 (1971).

[10] D. R. Wilken and R. E. Dyar, *Arch. Biochem. Biophys.* **189**, 251 (1978).

[11] Sources of the microorganisms were: *E. coli* 9637, American Type Culture Collection; *E. coli* AB1899, *E. coli* Genetic Stock Center, Yale University; *S. cerevisiae* A364A, Dr. L. H. Hartwell, University of Washington; *Erwinia* species, Dr. E. A. Grula, Oklahoma State University; and *S. cerevisiae* Y-2034, USDA Agricultural Research Service, Peoria, Illinois.

[12] D. R. Wilken and R. E. Dyar, unpublished results.

tases from each of these sources is similar to the specificity of the reductases from *S. cerevisiae* Y-2034. The reductases from the *Erwinia* species and *S. cerevisiae* A364A resemble those from *S. cerevisiae* Y-2034 in the degree of inhibition by compounds listed in Table II, but reductases from *E. coli* are not inhibited by these compounds. Also, in contrast to the reductases from *S. cerevisiae* Y-2034, the two forms of ketopantoyl lactone reductase from *E. coli* 9637 transfer the 4-*pro*-S hydrogen of NADPH; therefore, they are B-side stereospecific dehydrogenases.[8]

[41] Pantothenate Synthetase from *Escherichia coli* [D-Pantoate: β-Alanine Ligase (AMP-Forming), EC 6.3.2.1][1]

By Kazutaka Miyatake, Yoshihisa Nakano, and Shozaburo Kitaoka

$$\text{D-Pantoate} + \text{ATP} + \beta\text{-alanine} \rightarrow \text{pantothenate} + \text{AMP} + \text{PP}_i$$

Assay Method

Maas and Novelli[2] and Novelli[1] reported a microbioassay technique with a pantothenate auxotrophic mutant of *E. coli* for the determination of the enzyme activity, and Pfleiderer *et al.*[3] used a spectrophotometric method in a system of enzyme reactions, in which AMP formed during the condensation of pantoate and β-alanine was converted into ADP (myokinase) and then into ATP (pyruvate kinase), and the liberted pyruvate was allowed to oxidize NADH in the presence of lactate dehydrogenase. The following is a modification of our isotopic method of enzyme assay.[4,5] β-[1-^{14}C]-Alanine is used for the enzyme reaction, and radioactivity of the formed pantothenate is determined after its separation by ion-exchanger column chromatography.

Procedure. The reaction mixture contains 10 mM ATP (dipotassium salt), 5 mM potassium D-pantoate, 10 mM MgSO$_4$·7H$_2$O, 100 mM Tris·HCl buffer (pH 10.0), 0.25 μCi/5 mM β-[1-^{14}C]alanine and enzyme solution, and distilled water is added to make the total volume 1 ml. The

[1] A previous article on this enzyme in this series: G. D. Novelli, Vol. 2 [102].
[2] W. K. Maas and G. D. Novelli, *Arch. Biochem. Biophys.* **43**, 236 (1953).
[3] G. Pfleiderer, A. Kreiling, and T. Wieland, *Biochem. Z.* **333**, 302 (1960).
[4] K. Miyatake, Y. Nakano, and S. Kitaoka, *Agric. Biol. Chem.* **37**, 1205 (1973).
[5] K. Miyatake, Y. Nakano, and S. Kitaoka, *Bull. Univ. Osaka Pref. Ser. B* **27**, 57 (1975).

mixture is incubated at 30° for 10 min with shaking, and the reaction is stopped by adding 0.1 ml of N HCl and heating at 100° for 5 min. After cooling in water, the reaction mixture is charged onto a Dowex 50 (H^+ form) column (0.6 × 6.0 cm), and the column is eluted with 5 ml of distilled water. Radioactivity of the synthesized pantothenate in the eluate is measured with a gas-flow counter or liquid scintillation counter.

Definition of Unit and Specific Activity. One unit of enzyme activity is 1 nmol of pantothenate formed per minute under the above conditions. Specific activity is expressed as units per milligram protein. Protein is determined by the method of Lowry et al.[6] using bovine serum albumin as a standard.

Purification Procedure

Partial purification of this enzyme from *E. coli* has been reported by Kreiling et al.[7] and Miyatake et al.[5] The following purification[8,9] includes use of preparative polyacrylamide gel electrophoresis and provides a pure enzyme preparation, homogeneous in analytical disc gel electrophoresis. All operations are conducted between 0° and 4°.

Step 1. Bacterial Cultivation and Preparation of Crude Enzyme. E. coli B (IFO 13168) is grown at 30° for 16 hr in a synthetic medium containing 4 g of glucose, 1 g of sodium aspartate, 2 g of polypeptone, 2 g of KH_2PO_4, 0.5 g of $(NH_4)_2SO_4$, 0.2 g of $MgCl_2 \cdot 6H_2O$, a trace of $Fe_2(SO_4)_3 \cdot 7H_2O$, and water to total 1 liter, which is adjusted to pH 7.0 with N KOH, in a 100-liter jar fermentor. Cells (about 1 kg) are harvested by centrifugation, washed twice with deionized water, and suspended in 5 volumes of 10 mM Tris·HCl buffer, pH 7.2. The suspension is sonicated for 30 min at 0° with stirring. After removing cell debris by centrifugation at 6000 g for 30 min, the supernatant is dialyzed against 20 volumes of deionized water and the dialyzate is used as the crude enzyme.

Other strains of *E. coli* may also be employed.

Step 2. Celite Column Chromatography. To the crude enzyme (6.3 liters) are added with stirring, over a period of 10 min, 1260 g of Celite-535 (10 g per gram of protein), and sufficient ammonium sulfate to obtain 60% saturation. After further stirring for 30 min, the suspension is packed into a column (6 × 50 cm). The column is eluted with 50 mM Tris·HCl buffers,

[6] O. H. Lowry, N. J. Rosebrough, A. L. Farr, and R. J. Randall, *J. Biol. Chem.* **193**, 265 (1951).
[7] A. Kreiling, W. Ludwig, and G. Pfleiderer, *Biochem. Z.* **336**, 241 (1962).
[8] H. Wada, H. Kagamiyama, K. Miyatake, Y. Nomata, and H. Takao, *Med. J. Osaka Univ.* **25**, 129 (1975).
[9] K. Miyatake, Y. Nakano, and S. Kitaoka, *J. Biochem.* **79**, 673 (1976).

pH 7.5, containing 50, 40, and 30% of ammonium sulfate, successively. The flow rate is maintained at 45 ml/hr, and 15-ml fractions are collected. The fractions with enzyme activity are combined, and a sufficient amount of ammonium sulfate is added to produce a precipitate that is collected by centrifugation, dissolved in 500 ml of 10 mM Tris·HCl buffer, pH 7.2, and dialyzed against 100 volumes of deionized water.

Step 3. DEAE-Cellulose Column Chromatography. The enzyme solution (700 ml) is applied to a column (5 × 100 cm) of DEAE-cellulose (0.89 meq/g) equilibrated with 50 mM Tris·HCl buffer, pH 7.5, and eluted stepwise with 0.1–0.3 M potassium chloride. The flow rate is maintained at 80 ml/hr, and 15-ml fractions are collected.

Step 4. Ammonium Sulfate Fractionation. Combined active fractions (650 ml) from step 3 are dialyzed against deionized water, and fractionated with ammonium sulfate. The fractions precipitating by 35–55% saturation of the salt are combined, and the precipitate is collected by centrifugation. It is dissolved in 50 ml of 10 mM Tris·HCl buffer, pH 7.2, and dialyzed against deionized water for 48 hr. The enzyme solution (75 ml) is lyophilized and stored at −20°.

Step 5. First Preparative Polyacrylamide Gel Electrophoresis. An apparatus for large-scale preparative gel electrophoresis is employed. The one[8] employed here contains a device for preventing dilution of eluted fractions by programming elution time, but any similar apparatus may also be used. In pilot experiments, use of 5% gel concentration, 2% cross linkage, and 15 mM acetate-Tris buffer, pH 8.0, have been found to give the best separation of this enzyme.

About 150 mg of the lyophilized enzyme dissolved in 5 ml of 40% sucrose solution is layered gently on the top of the gel (height 10.5 cm). The current is fixed at 200 V and electrophoresis is carried out for 24 hr. Fractions are collected at 1-hr intervals, and those showing enzyme activity are combined and lyophilized.

Step 6. Second Preparative Polyacrylamide Gel Electrophoresis. A second preparative gel electrophoresis is performed in a manner similar to the first except that a 7.5-cm gel height is used. The lyophilized enzyme (71 mg) from step 5 is dissolved in 3 ml of 40% sucrose and submitted to electrophoresis for 12 hr. Fractions are collected at 30-min intervals and active fractions combined and lyophilized.

Step 7. Sephadex G-200 Gel Filtration. The lyophilized enzyme (22 mg) is dissolved in 2 ml of 0.1 M phosphate buffer, pH 7.0, and loaded onto a column (2.6 × 100 cm) of Sephadex G-200. The column is eluted with the same buffer and 10-ml fractions are collected. Active fractions are combined, dialyzed against deionized water, and lyophilized.

This purified enzyme protein shows a single band in an analytical disc

SUMMARY OF THE PURIFICATION PROCEDURE FOR PANTOTHENATE
SYNTHETASE OF *Escherichia coli*[a]

Procedure	Protein (mg)	Total activity (unit)	Specific activity (unit/mg protein)	Yield (%)
Crude extract	126,000	519,000	4.2	100
$(NH_4)_2SO_4$ back-gradient fractionation	35,440	420,552	12.1	81.0
DEAE-cellulose chromatography	3,179	328,030	103.2	63.2
$(NH_4)_2SO_4$ fractionation	1,597	206,659	129.4	39.8
1st PPAGE[b]	71	66,336	931.7	12.8
2nd PPAGE[b]	22	45,205	2,018.1	8.7
Sephadex G-200 gel filtration	18	37,310	2,050.5	7.2

[a] K. Miyatake, Y. Nakano, and S. Kitaoka, *J. Biochem.* **79**, 673 (1976).
[b] Preparative polyacrylamide gel electrophoresis.

gel electrophoresis with good correspondence to the peak of enzyme activity.[9] The enzyme has been purified about 500-fold over the crude enzyme. A summary of a typical purification procedure is given in the table.

Properties[9,10]

Stabilities. Lyophilized pantothenate synthetase is very stable; activity is hardly lost on storage at $-20°$ for 6 months. The enzyme in solution is stable up to $37°$, but inactivated by heating at $60°$ for 10 min. The enzyme is stable to incubation for 15 min at $37°$ in the range of pH 5.0 to 10.0 when subsequently assayed at pH 10.0.

Optimum Temperature and pH. The temperature optimum for this enzyme is $30°$ and the optimum pH is 10.0. Bringing the temperature to $47°$ or pH to 8.5 or 11.0 is accompanied by about a 50% loss of activity.

Substrate Specificity. The enzyme requires ATP; other triphosphonucleotides are almost inactive. Many analogs of β-alanine and pantoate are inactive as substrates.

Activators and Inhibitors. Pantothenate synthetase requires a divalent cation, Mg^{2+} or Mn^{2+}, and a monovalent cation, K^+ or NH_4^+, as activators. The optimum concentrations of these are 10 mM for Mg^{2+}, 5 mM for Mn^{2+}, and 15 mM for K^+ or NH_4^+. Above these concentrations, Mg^{2+} and Mn^{2+} are inhibitory while K^+ and NH_4^+ are not. At the optimum concentrations, the combination of Mg^{2+} and K^+ gives a 30% higher enzyme activity than that of Mn^{2+} and NH_4^+. The K_m values toward Mg^{2+} and K^+ are $2.0 \times 10^{-3} M$ and $5.5 \times 10^{-3} M$, respectively.

Such other metal ions as Ag^+, Co^{2+}, Cu^{2+}, Zn^{2+}, Ba^{2+}, Ca^{2+} and Hg^{2+},

[10] K. Miyatake, Y. Nakano, and S. Kitaoka, *J. Nutr. Sci. Vitaminol.* **24**, 243 (1978).

at 5 mM, cause 30–95% inhibitions; Sn^{2+}, Fe^{2+}, Ni^{2+}, Sr^{2+}, and Pb^{2+} are only slightly inhibitive. Chelating agents, such as EDTA, EGTA, and o-phenanthroline show considerable inhibitory action. Sulfhydryl compounds and inhibitors have no effect on the enzyme. At 1 mM, substrate analogs cause the following percent inhibition: taurine, 52; γ-amino-β-hydroxybutyrate, 32; γ-aminobutyrate, 27; aspartate, 24; methylmalonate, 38; gluconate, 32; glycolate, 30.

Kinetics. The apparent K_m values for substrates under optimum reaction conditions as determined from Lineweaver–Burk's plots are as follows: D-pantoate 6.3×10^{-5} M, β-alanine 1.5×10^{-4} M and ATP $1.0 \times 10^{-4} M$. Maas and Novelli[2] and Pfleiderer *et al.*[3] have reported lower affinities of substrates for the enzyme, but their results are apparently due partly to erroneous pH and temperature optima they employed.

Application of Cleland's method[11] has revealed that the enzyme reaction proceeds by a Bi Uni Bi Ping-Pong type mechanism. The order of binding of substrates to and releasing of products from the enzyme is ATP, D-pantoate, pyrophosphate, β-alanine, pantothenate, and AMP. Results support the following reaction mechanism proposed by Maas.[12]

Enzyme (E) + ATP + D-pantoate → E-pantoyladenylate + PP_i
E-Pantoyladenylate + β-alanine → pantothenate + AMP + E

Molecular Weight. Sedimentation equilibrium ultracentrifugation data give a molecular weight of 70,000, and Sephadex G-150 gel filtration a value of 69,000 for this enzyme. The minimum molecular weight calculated from the amino acid composition of the enzyme protein is 19,700, which is in reasonable accord with the value of the enzyme subunit as 18,000 obtained by polyacrylamide gel electrophoresis in the presence of sodium dodecyl sulfate. The enzyme has an amino terminal glycyl residue and a Leu-Ala-SerOH sequence at the carboxyl end.

Other Physical Constants. The enzyme protein has a sedimentation coefficient $s_{20,w}^0$ of 4.9 and a diffusion constant of 5.88×10^{-7} cm^2/sec. The partial specific volume $\bar{v}$ is calculated to be 0.71 cm^3/g, and the isoelectric point is pH 4.6. Extinction of $E_{1\,cm,280}^{1\,\%}$ is 11.02 at pH 7.0 in 10 mM potassium phosphate buffer.

[11] W. W. Cleland, *Biochim. Biophys. Acta* **67**, 104, 173, 188 (1963).
[12] W. K. Maas, *Proc. Int. Congr. Biochem., 4th*, Vol. 11, p. 161 (1960).

[42] Microbial Synthesis of Sugar Derivatives of D-Pantothenic Acid

By FUSAKO KAWAI, HIDEAKI YAMADA, and KOICHI OGATA

D-Pantothenic Acid β-Glucoside[1]

Preparation

Principle. A β-glucoside of D-pantothenic acid (PaA-β-G) is formed by many β-glucosidases from D-pantothenic acid (PaA) and various β-glucosyl donors. Almond β-glucosidase forms large amounts of the derivative (800–900 μg/ml) under optimal conditions. The derivative is separated from PaA by paper chromatography and confirmed by bioautography. The derivative is quantitatively assayed as PaA by microbiological assay after hydrolysis by almond β-glucosidase.

Assay. A β-glucoside of PaA formed in a reaction mixture is identified by a bioautographic technique as follows. A suitable amount of the supernatant fluid is spotted on Toyo filter paper No. 53 (2 cm in width), and chromatographed with a solvent system of *n*-butanol–acetic acid–water (4:1:1, by volume). The paper strip is then cut into four 0.5-cm-wide strips, which are subjected to bioautography with *Saccharomyces carlsbergensis* ATCC 9080[2] or *Lactobacillus plantarum* ATCC 8014.[3] A β-glucoside of PaA is quantitatively assayed as follows. The compound, separated and confirmed on a paper chromatograph, is extracted with water at 30° for 24 hr and hydrolyzed by almond β-glucosidase. The reaction mixture consists of PaA-β-G, 1–5 μg as PaA; acetate buffer (pH 5.0), 50 μmol; and almond β-glucosidase, 10 mg; in a total volume of 1.0 ml. The reaction is carried out at 37° overnight and is stopped by heating the mixture in a boiling water bath for 1 min. Denatured protein is centrifuged off, and PaA in the supernatant fluid is quantitatively bioassayed by the paper disk method, using *L. plantarum.*[3]

Reagents

β-Glucosidases:

Almond β-glucosidase (Sigma Chem. Co.)

Cellulase type II, from *Aspergillus niger* (Sigma Chem. Co.)

[1] F. Kawai, H. Yamada, and K. Ogata, *Agric. Biol. Chem.* **38**, 831 (1974).

[2] L. Atkin, W. L. Williams, A. S. Schultz, and C. N. Frey, *Ind. Eng. Chem., Anal. Ed.* **16**, 67 (1944).

[3] H. R. Skeggs and L. D. Wright, *J. Biol. Chem.* **156**, 21 (1944).

Cellulase type III, from a *Rhizopus* mold (Sigma Chem. Co.)
Naringinase, from *A. niger* (Sigma Chem. Co.)
Hesperiginase, from *A. niger* (Sigma Chem. Co.)
Calcium D-pantothenic acid
β-Glucosyl donors:
    Cellobiose, phenyl-β-D-glucoside, salicin, 4-methylumbelliferyl-β-D-glucoside, and naphthol AS-BI-β-D-glucoside
Buffers:
    Acetate buffer, pH 5.0, or phosphate buffer, pH 7.0

Enzymic Reaction. The reaction mixture is composed of calcium PaA, 23.1 μmol; β-glucosyl donor, 300 μmol; buffer, 50 μmol; and an appropriate amount of enzyme in a final volume of 1 ml. Glucose is used for the blank. The incubation is carried out at 37° overnight (25 hr). The reaction is stopped by immersing a reaction tube in a boiling water bath for 3 min. The supernatant fluid obtained by centrifugation is assayed for PaA-β-G, and results are given in Table I.

Isolation of PaA-β-G

The reaction mixture contains calcium PaA, 50 μmol/ml; cellobiose, 880 μmol/ml; acetate buffer (pH 5.0), 50 μmol/ml; and almond β-glu-

TABLE I

ENZYMIC FORMATION OF D-PANTOTHENIC ACID β-GLUCOSIDE
BY VARIOUS β-GLUCOSIDASES

Enzyme	Reaction pH	β-Glucosyl donor	PaA-β-G formed (μg)
Almond β-glucosidase	5.0	Cellobiose	665
(10 mg)		Phenyl-β-D-glucoside	425
		Salicin	517
		4-Methylumbelliferyl-β-D-glucoside	305
		Naphthol AS-BI-β-D-glucoside	Trace
Cellulase type II[a]	7.0	Cellobiose	313
(100 mg)		Phenyl-β-D-glucoside	1260
		Salicin	997
Cellulase type III	7.0	Cellobiose	132
(150 mg)		Phenyl-β-D-glucoside	179
		Salicin	231
Naringinase (150 mg)	5.0	Cellobiose	1100
		Phenyl-β-D-glucoside	204
		Salicin	553
Hesperiginase (2.5 mg)	5.0	Salicin	405

[a] The powdered cellulase type II preparation was dissolved in water and the supernatant obtained by centrifugation was used for the reaction.

cosidase, 20 mg/ml; in a total volume of 100 ml. The incubation is carried out at 37° overnight and is stopped by immersing the incubation mixture in a boiling water bath for 3 min. Denatured enzyme is removed by centrifugation. The clear supernatant solution contains approximately 800 μg of the derivative per milliliter. The supernatant solution is adjusted to pH 4.0, and active charcoal (15 g) is added to the solution. After stirring at room temperature overnight, the charcoal solution is packed into a column and washed with one volume of 10 mM hydrochloric acid and 4 volumes of water. A mixture of PaA and PaA-β-G is eluted from the charcoal with 10 volumes of acetone and 5 volumes of 50% acetone. The eluate is evaporated *in vacuo* to a small volume. The concentrated solution is subjected to paper chromatography with Toyo filter paper No. 526, with a solvent system of *n*-butanol–acetic acid–water (4:1:1, by volume). PaA-β-G is confirmed by bioautography using *S. carlsbergensis*.[2] The active zone which differs from that of PaA is extracted with 50% ethanol. This process is repeated twice to remove PaA completely. The combined extracts are concentrated *in vacuo* and desalted with Amberlite CG-50 (H$^+$ type) and DEAE-cellulose (OH$^-$ type). The desalted solution is concentrated *in vacuo* and applied on a Sephadex G-10 column (1 × 105 cm, 3 ml/fraction), which is washed with water. The eluate (fractions Nos. 19–29) is evaporated and chromatographed on Toyo filter paper No. 53 with the solvent system described above. The active zone confirmed by bioautography is extracted with 50% ethanol, and the extracts are evaporated. The concentrated extract is applied on a DEAE-cellulose column (2 × 14 cm, 5 ml/fraction) buffered with 30 mM ammonium carbonate and eluted with 30 mM ammonium carbonate. The eluate (fractions Nos. 10–29) is desalted with Amberlite CG-50 (H$^+$ type), evaporated *in vacuo,* and lyophilized. The lyophilized powder is dried *in vacuo* over sodium hydroxide (yield, approximately 65 mg; free from PaA).

Identification and Characterization

The compound is hydrolyzed by almond β-glucosidase, cellulases type II and type III, and naringinase, but not by yeast α-glucosidase (Boehringer Mannheim Japan K. K.). The enzymic hydrolysis of the compound releases PaA and glucose, which are identified by bioautography and paper chromatography. The molar ratio of the PaA and glucose moieties is approximately 1:1 as determined by the glucostat method,[4] the anthrone method,[5] and bioassay. No reducing activity of the isolated

[4] A. Sols and G. de la Fuente, "Methods in Medical Research," Vol. 9. Year Book Publishers, Chicago, 1961.
[5] D. L. Morris, *Science* **107**, 254 (1948).

compound is found by Somogyi–Nelson's method.[6] From these results, the compound is inferred to be β-D-glucopyranosyl-D-pantothenic acid. The specific microbiological activity for strain WNB-75,[7] optical rotation, and NMR spectrum of the compound are used in order to determine at which hydroxyl group of C-2' and C-4' in the PaA moiety the glucosyl residue is linked. The compound shows a similar microbiological activity to 4'-O-(β-D-glucopyranosyl)-D-pantothenic acid[8] for a strain WNB-75, which requires the compound as a specific growth factor. The specific optical rotation of the isolated compound is $[\alpha]_D^{29}$ $-8.4°$ (c = 1.6, H_2O), which agrees with that of 4'-O-(β-D-glucopyranosyl)-D-pantothenic acid. The NMR signals of the compound are slightly different from those of authentic 4'-O-(β-D-glucopyranosyl)-D-pantothenic acid, but NMR spectroscopy with a mixture of equal weights of the isolated and authentic compounds results in the coincidence of signals (in D_2O, δ ppm): 0.92 (3H, s), 0.99 (3H, s), 2.60 (2H, t, J = 6.5 Hz), 3.2–4.0 (m), 4.05 (1H, s), and 4.40 (1H, d, J = 6.5 Hz) with sodium 2,2-dimethyl-2-silapentane 5-sulfonate as an internal standard. From these results, the compound is determined to be 4'-O-(β-D-glucopyranosyl)-D-pantothenic acid.

In microbiological growth studies by the paper disk method, growth of *Saccharomyces carlsbergensis* ATCC 9080[2] and *Lactobacillus plantarum* ATCC 8014[3] on the compound is found to be 10% of that on an equimolar amount of PaA.

D-Pantothenic Acid α-Glucoside[9,10]

Preparation

Principle. An α-glucoside of PaA (PaA-α-G) is formed by several microbial cells or α-glucosidases from PaA and maltose.[10] Washed cells of *Sporobolomyces coralliformis* IFO 1032 forms large amounts of the derivative (maximum 5.74 mg/ml) under the optimal conditions.[10] The derivative is separated from PaA by paper chromatography and confirmed by

[6] M. Somogyi, *J. Biol. Chem.* **195**, 19 (1952).

[7] H. Yoshizumi and T. Amachi, *Agric. Biol. Chem.* **33**, 18 (1969). The microorganism was kindly offered by Drs. Yoshizumi and Amachi of the Central Research Institute of Suntory, Ltd.

[8] This volume [43]; T. Amachi, S. Imamoto, H. Yoshizumi and S. Senoh, *Tetrahedron Lett.* **56**, 4871 (1970). The compound was kindly given by Drs. Yoshizumi and Amachi of the Central Research Institute of Suntory, Ltd., and Dr. Imamoto of the Institute of Food Chemistry.

[9] F. Kawai, K. Maezato, H. Yamada, and K. Ogata, *Biochim. Biophys. Acta* **286**, 91 (1972).

[10] F. Kawai, K. Maezato, H. Yamada, and K. Ogata, *Agric. Biol. Chem.* **42**, 1675 (1978).

bioautography. The derivative is quantitatively assayed as PaA by microbiological assay after hydrolysis by yeast α-glucosidase.

Assay. An α-glucoside of PaA formed in a reaction mixture or culture broth is identified by a bioautographic technique following the procedure described for assay of PaA-β-G. The compound, separated and confirmed on a paper chromatograph, is extracted with 3 ml of water at 37° for 4 hr and hydrolyzed by mold maltase (grade II, from *Aspergillus niger*, Sigma Chem. Co.). The reaction mixture consists of PaA-α-G, under 30 μg as PaA; acetate buffer (pH 5.0), 50 μmol; 2-mercaptoethanol, 10 μmol; the enzyme, 10 mg; in a total volume of 1.0 ml. The reaction is carried out at 37° for 24 hr and is stopped by heating the mixture in a boiling water bath for 3 min. PaA-α-G is quantitatively assayed as PaA by a paper disk method, using *Saccharomyces carlsbergensis* ATCC 9080[2] or *Lactobacillus plantarum* ATCC 8014.[3]

Cultivation. *Sporobolomyces coralliformis* IFO 1032 is grown on a medium (pH 6.0) containing maltose, 5%; peptone, 0.5%; yeast extract, 0.5%; K_2HPO_4, 0.5%; KH_2PO_4, 0.2%; and $MgSO_4 \cdot 7H_2O$, 0.02% in tap water. Cultivation is performed for 48 hr at 28° with 500 ml of the medium in a 2-liter shaking flask on a reciprocal shaker. The cells are collected by centrifugation at 5000 rpm, washed once with 0.9% saline solution, and suspended in the same solution.

Procedure.[9] REACTION. Twenty-one grams (dry weight) of intact cells are incubated aerobically for 40 hr at 28° with 40 g of maltose, 4 g of PaA, 8 mmol of potassium phosphate buffer (pH 7.0), and 0.4 g of cetyl trimethyl ammonium bromide in a final volume of 400 ml. The reaction is stopped by immersing the incubation mixture, adjusted to pH 7.0, in boiling water for 5 min, then the cells are removed by centrifugation at 10,000 rpm for 20 min.

ISOLATION. The supernatant solution is adjusted to pH 4.0 and left at 4° overnight. Any material precipitated is removed by filtration. Active charcoal (60 g) is added to the filtrate. After stirring the filtrate for 1 hr, the charcoal is separated by filtration, washed with 10 mM hydrochloric acid and water, then with 5 volumes of acetone. The acetone eluate is evaporated *in vacuo* to a small volume. The concentrated solution is subjected to paper chromatography with Toyo filter paper No. 526, and the position of the product is confirmed following the methods described in *Assay.* The active zone, which differs from that of PaA, is extracted with 50% ethanol. The extract is concentrated *in vacuo* and desalted with Amberlite CG-50 (H$^+$ type). The concentrated effluent is applied to a Sephadex G-10 column (1 × 105 cm) and eluted with water. Active fractions confirmed by paper disk bioassay are adsorbed on DEAE-cellulose (OH$^-$ type), then eluted with 20 mM lithium chloride. The eluted fraction is desalted with Amberlite CG-50(H^+ type), and adjusted to pH 5.0 with 1 M hydrochloric acid. Saturated $Ba(OH)_2$ solution is added to the fraction to adjust the pH to 7.0–7.2. Excess $Ba(OH)_2$ is removed by repeated concentration and filtration. The completely dried barium salt of the compound is dissolved in methanol. A fluffy precipitate is obtained by adding ether. The precipitate is dissolved in water and subjected to paper chromatography using Toyo filter paper No. 53 in the solvent system described above. The active zone is extracted with 50% ethanol, evaporated, lyophilized, and dried *in vacuo* over phosphorus pentoxide and potassium hydroxide (yield, 21.9 mg as the barium salt; free from PaA).

Identification and Characterization[9,10]

R_f values of the compound on bioautography in various solvent systems differ from those of known metabolites of PaA, as shown in Table II. The acid and enzymic hydrolysis of the compound releases PaA and glucose on the bioautogram and paper chromatogram. The molar ratio of the two components is 1 : 1, as determined by the glucostat method[4] and

TABLE II

R_f Values from Bioautography of Pantothenic Acid α-Glucoside and Other Metabolites of Pantothenic Acid

Compound	R_f values[a] in solvent systems[b]				
	I	II	III	IV	V
D-Pantothenic acid α-glucoside	0.49	0.36	0.63	0.52	0.48
D-Pantothenic acid	0.76	0.50	0.76	0.85	0.69
β-Alanine	0.29	0.13	0.54	0.39	0.62
4-Phosphopantothenic acid[c]	0.45	0.19	0.28	0.40	0.38
Panthenol			No response		
Pantethine			No response		
Pantoyl lactone			No response		

[a] Paper chromatography was carried out by the ascending method on Toyo filter paper No. 53. Location of compounds on a paper chromatogram was determined by bioautography using *Saccharomyces carlsbergensis* ATCC 9080 [L. Atkin, W. L. Williams, A. S. Schultz, and C. N. Frey, *Ind. Eng. Chem., Anal. Ed.* **16**, 67 (1944)].

[b] Solvent I is *n*-butanol–acetic acid–water (4 : 1 : 1, by volume); solvent II, *n*-butanol–pyridine–water (6 : 4 : 3, by volume); solvent III, *n*-propanol–28% ammonia water–water (6 : 3 : 1, by volume); solvent IV, *n*-butanol–ethanol–water (5 : 1 : 4, by volume upper layer); solvent V, isobutyric acid–0.5 M ammonia water–0.1 M EDTA (100 : 80 : 1.6, by volume).

[c] This volume [44]; K. Ogata, S. Shimizu, and Y. Tani, *Agric. Biol. Chem.* **36**, 84 (1971).

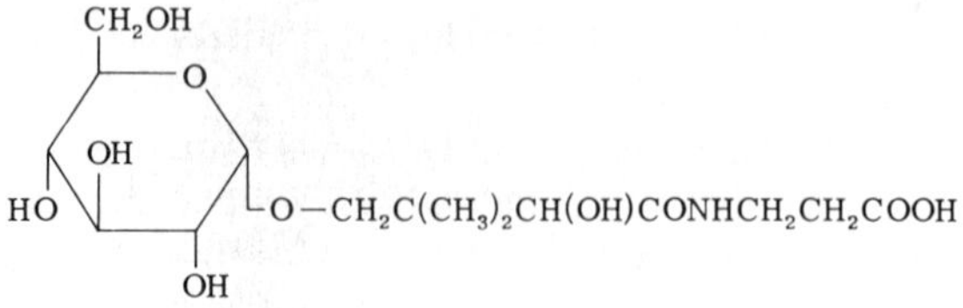

CH₂OH

HO

OH

OH

O—CH₂C(CH₃)₂CH(OH)CONHCH₂CH₂COOH

4'-O-(α-D-Glucopyranosyl)-
D-pantothenic acid

(I)

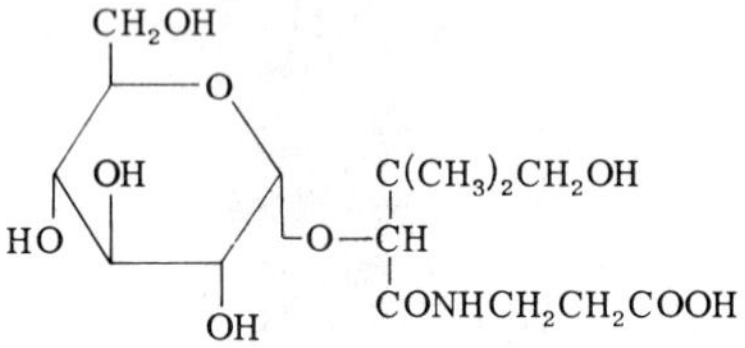

CH₂OH

HO

OH

OH

C(CH₃)₂CH₂OH
|
O—CH
|
CONHCH₂CH₂COOH

2'-O-(α-D-Glucopyranosyl)-
D-pantothenic acid

(II)

Fig. 1. Proposed structures for D-pantothenic acid α-glucoside.

by bioassay. No reducing activity of the isolated compound is found by Somogyi–Nelson's method.[6] The compound is hydrolyzed by yeast α-glucosidase and mold maltase, but not by cellulase type II and almond β-glucosidase. The NMR spectrum (in D_2O, δ ppm) shows two methyl signals (C-3', 0.98 and 0.99, 6H, s), a methylene signal of C-2 (2.42, 2H, t, $J = 6.5$ Hz), a methylene signal of C-3 (3.45, 2H, t, $J = 6.0$ Hz), and a methylene signal of C-4' and ring protons of the sugar moiety (3.55–3.90, m), a methyne signal (C-2', 4.12, 1H, s), and one anomeric proton signal of the α-glucosidic linkage (4.89, 1H, d, $J = 3.5$ Hz). Comparison of the integral values of the two methyl groups of the pantothenic acid residue with that of δ 3.55–3.90 ppm shows that the ratio of the PaA and glucose moieties is approximately one to one. The proposed structures for the compound are shown in Fig. 1. Comparison of the methyne signal (δ 4.12 ppm) of the compound with the signals[11] of the 2'-O-(β-D-glucopyran-osyl)-D-pantothenic acid and 4'-O-(β-D-glucopyranosyl)-D-pantothenic acid suggests that the glucosidic linkage might be present at C-4'.

In microbiological growth studies by the paper disk method, growth of *Saccharomyces carlsbergensis* ATCC 9080[2] on the compound reaches

---

[11] This volume [43]; T. Amachi, S. Imamoto, and H. Yoshizumi, *Agric. Biol. Chem.* **35**, 1222 (1971).

9–10% of that on an equimolar amount of PaA at 20 hr of incubation and increases remarkably to 70–80% by 30 hr of incubation. Growth of *Lactobacillus plantarum* ATCC 8014[3] on the compound is 0.1–0.3% at 20 hr of incubation and increases slightly to 0.2–0.6% by 30 hr of incubation. The authentic 4′-*O*-(α-D-glucopyranosyl)-D-pantothenic acid[12] shows the same activities for *S. carlsbergensis* and *L. plantarum* as the isolated compound.[10]

From these results, D-pantothenic acid α-glucoside is thought to be 4′-*O*-(α-D-glucopyranosyl)-D-pantothenic acid.

Addendum[10]

The optimal reaction conditions of *Sporobolomyces coralliformis* IFO 1032 are as follows. The reaction mixture contains PaA, 10 mg/ml; sucrose, 100 mg/ml; potassium phosphate buffer (pH 7.0), 20 μmol/ml; cetyl trimethyl ammonium bromide, 1 mg/ml; and intact cells (dry weight), 40–50 mg/ml and is incubated for 48 hr at 28° with shaking. Over 5 mg of the derivative is formed per milliliter under these conditions. Yeast α-glucosidase, mold maltase, and cell-free extract of *S. coralliformis* forms small amounts of the derivative.

[12] This volume [43]; S. Imamoto, T. Amachi, and H. Yoshizumi, *Agric. Biol. Chem.* **37**, 545 (1973). The compound was kindly given by Dr. S. Imamoto of the Institute of Food Chemistry and Drs. Yoshizumi and Amachi of the Central Research Institute of Suntory, Ltd.

[43] Syntheses of Glycosyl Derivatives of Pantothenic Acid and Pantetheine

By Shoji Imamoto, Teruo Amachi, and Hajime Yoshizumi

It is well known that wine maturation involves a bacterial conversion of malic acid to lactic acid, which process is designated Malo-lactic fermentation (MLF). MLF occurs following an alcoholic fermentation in wine making, it not only reduces the acidity of wine, but also improves its quality. Many MLF bacteria have been isolated to date, in which WNB-75, isolated by one of the authors[1] and classified as *Leuconostoc oenos* by Garvie, requires some essential factor contained in grape, tomato, and other fruits or vegetables for its growth.[2] The growth factor has

[1] H. Yoshizumi, *Agric. Biol. Chem.* **27**, 590 (1963).

[2] H. Yoshizumi and T. Amachi, *Agric. Biol. Chem.* **33**, 18 (1969).

METHODS IN ENZYMOLOGY, VOL. 62

been isolated from tomato juice[3] and named TJF; it was characterized as
4′-O-(β-D-glucopyranosyl)-D-pantothenic acid (I-D).[4]

(I-D)

The growth factor (TJF) can be substituted for with a large amount
(about 100 times) of D-pantothenic acid for its growth. D-Pantothenic acid
is widely distributed in nature and well known to be a precursor of coenzyme A. The metabolism of pantothenic acid to coenzyme A in microorganisms has been established; the initial reaction is phosphorylation at the

MICROBIAL ACTIVITIES OF GLYCOSYL PANTOTHENIC ACIDS
AND RELATED COMPOUNDS[a]

	WNB-75	Leuconostoc oenos[b]	ML-34[c]	IFO-3345[d]	IFO-3070[e]
4′-O-(β-D-Glucopyranosyl)-D-pantothenic acid(TJF)	0.05	0.05	0.25	25	4.0
D-Pantothenic acid	5.0	5.0	5.0	0.1	0.05
D-Pantethine	5.0	5.0	5.0	2.0	0.2
2′-O-(β-D-Glucopyranosyl)-DL-pantothenic acid	5.0	5.0	5.0	—	25
4′-O-(β-D-Galactopyranosyl)-DL-pantothenic acid	2.0	20	10	25	4.0
4′-O-(β-D-Ribofuranosyl)-DL-pantothenic acid	>20	—	—	—	—
4′-O-(β-Maltosyl)-DL-pantothenic acid	20	20	—	25	25
4′-O-(β-D-Cellobiosyl)-D-pantothenic acid	0.5	5.0	10	—	—
4′-O-(β-D-Glucopyranosyl)-DL-pantethine	0.1	1.0	1.0	—	—

[a] The activity is shown by the minimum amount of compound (μg/ml of medium) for
adequate growth.
[b] Isolated by E. I. Garvie, England.
[c] Leuconostoc citrovorum ML-34, isolated by R. E. Kunkee, United States.
[d] Lactobacillus brevis IFO-3345.
[e] Lactobacillus plantalum IFO-3070.

[3] T. Amachi and H. Yoshizumi, Agric. Biol. Chem. 33, 139 (1969).
[4] T. Amachi, S. Imamoto, and H. Yoshizumi, Agric. Biol. Chem. 35, 1222 (1971).

C-4' hydroxyl group of pantothenic acid.[5] It is reasonable to presume that TJF is also metabolized to coenzyme A, but the role of the glucosyl group in TJF has not been clarified. In order to elucidate the contribution of the glucosyl residue in the metabolism of pantothenic acid, various glycosyl analogs of pantothenic acid and pantetheine have been synthesized, and their biological activities for the growth of some microorganisms have been examined[6] (table). Recently, Kawai *et al.* have reported the enzymic formation of TJF from D-pantothenic acid and some glucosyl donors by various glucosidases.[7]

Syntheses by a General Method

$$
\overset{\text{OBzl}}{\underset{|}{}}
$$

$$R'-X + HO-CH_2-C(CH_3)_2-CH-CONH-CH_2-CH_2-COOBzl \xrightarrow{Hg(CN)_2}$$

R' = 2,3,4,6-Tetra-*O*-acetyl-D-glucopyranosyl; X = Br
R' = 2,3,4,6-Tetra-*O*-acetyl-D-galactopyranosyl; X = Br
R' = Hepta-*O*-acetylmaltosyl; X = Br
R' = Hepta-*O*-acetylcellobiosyl; X = Br
R' = 2,3,5-Tri-*O*-benzoyl-D-ribofuranosyl; X = Cl

$$
\overset{\text{OBzl}}{\underset{|}{}}
$$

$$R' \overset{\beta}{-} O-CH_2-C(CH_3)_2-CH-CONH-CH_2-CH_2-COOBzl \xrightarrow{H_2/Pd\text{-}black}$$

$$
\overset{\text{OH}}{\underset{|}{}}
$$

$$R' \overset{\beta}{-} O-CH_2-C(CH_3)_2-CH-CONH-CH_2-CH_2-COOH \xrightarrow[\text{or NaOCH}_3]{Ba(OCH_3)_2}$$

$$
\overset{\text{OH}}{\underset{|}{}}
$$

$$R \overset{\beta}{-} O-CH_2-C(CH_3)_2-CH-CONH-CH_2-CH_2-COOH$$

    (I):   R = D-Glucopyranosyl
    (V):   R = D-Galactopyranosyl
(VIII):   R = Maltosyl
  (IX):   R = Cellobiosyl
 (XII):   R = D-Ribofuranosyl

4'-O-(β-D-Glucopyranosyl)-DL-, D-, and L-Pantothenic Acid

A mixture of benzyl 2'-*O*-benzyl-DL-pantothenate[8] (7.0 g), pulverized mercuric cyanide (4.7 g), and calcium sulfate (10 g, preheated for 2 hr at

[5] G. M. Brown, *J. Biol. Chem.* **234**, 370 (1959).
[6] S. Imamoto, T. Amachi, and H. Yoshizumi, *Agric. Biol. Chem.* **37**, 545 (1973).
[7] F. Kawai, H. Yamada, and K. Ogata, *Agric. Biol. Chem.* **38**, 831 (1974).
[8] J. Baddiley and E. M. Thain, *J. Chem. Soc.*, p. 1610 (1953).

240°) in dry benzene (25 ml), and nitromethane (25 ml) is stirred at room temperature for 1 hr with exclusion of moisture. Acetobromoglucose[9] (10 g) is added, and the mixture is gently refluxed with vigorous stirring for 20 hr. After cooling, the reaction mixture is filtered and thoroughly washed with benzene. The filtrate combined with washing is concentrated *in vacuo,* and the residue is dissolved in 100 ml of benzene. The benzene solution is washed successively with 1 M potassium bromide, saturated sodium bicarbonate, water, and brine and then dried with anhydrous sodium sulfate. After concentration *in vacuo,* the residue is chromatographed on silicic acid eluting with 2.5% ethanol in chloroform to afford a colorless syrup (4.3 g) of benzyl 2'-*O*-benzyl-4'-*O*-(2,3,4,6-tetra-*O*-acetyl-β-D-glucopyranosyl)-DL-pantothenate (II), R_f[10] (30% ethyl acetate in benzene) = 0.33, $[\alpha]_D$ − 14.1° (c = 1, CHCl₃).

Compound (II) (4.3 g) is dissolved in 40 ml of acetic acid and treated with 800 mg of palladium black. Hydrogenation is permitted to proceed at room temperature under an atmospheric pressure of hydrogen until no more hydrogen is absorbed (6 hr). After removal of the catalyst by filtration, the filtrate is concentrated under reduced pressure and the residue is chromatographed on silicic acid, eluting with 5% ethanol in chloroform to afford 4'-*O*-(2,3,4,6-tetra-*O*-acetyl-β-D-glucopyranosyl)-DL-pantothenic acid (III-DL, 3.15 g).

To a cold solution of 1.5 g of (III-DL) dissolved in 30 ml of anhydrous methanol is added 10 ml of 0.2 M barium methoxide; the mixture is stored at room temperature for 24 hr. Water (30 ml) is added, and the mixture is concentrated *in vacuo* to an aqueous solution, which is applied to a column of Dowex 1-X8 (OH⁻). After a thorough washing with water, the column is eluted with 0.5 M acetic acid and the eluate is lyophilized to give a white, hygroscopic powder (950 mg) of 4'-*O*-(β-D-glucopyranosyl)-DL-pantothenic acid (I-DL), $[\alpha]_D$ − 18.2° (c = 1, H₂O). (I-DL) has half-activity of TJF for the growth of bacterium WNB-75.

Optical resolution can be carried out at the stage of the acetate (III-DL) as follows. Compound (III-DL) (1.9 g) is dissolved in 20 ml of boiling ethyl ether and the solution is stored at room temperature for 2 days to precipitate colorless needles, which are collected and recrystallized twice from ethyl ether to afford 650 mg of 4'-*O*-(2,3,4,6-tetra-*O*-acetyl-β-D-glucopyranosyl)-D-pantothenic acid (III-D), m.p. 135°–136°C, $[\alpha]_D$ − 21.2° (c = 1, CHCl₃). The mother liquor from the first crystallization is concentrated to about 1 ml and stored at 0° for a couple of days. The colorless needles precipitated are recrystallized from ethyl ether–petroleum ether to give 500 mg of 4'-*O*-(2,3,4,6-tetra-*O*-acetyl-β-D-

[9] R. L. Whistler and M. L. Wolfrom, eds., "Methods in Carbohydrate Chemistry," Vol. II, p. 221. Academic Press, New York, 1963.
[10] On silicic acid plates.

glucopyranosyl)-L-pantothenic acid (III-L), m.p. 109°–113°, $[\alpha]_D$ −28.6° (c = 1, CHCl$_3$). Deacetylation of either (III-D) or (III-L) in the way described above gives 4'-O-(β-D-glucopyranosyl)-D-pantothenic acid (I-D) and 4'-O-(β-D-glucopyranosyl)-L-pantothenic acid (I-L), respectively; (I-D): $[\alpha]_D$ −11.2° (c = 1, H$_2$O), (I-L): $[\alpha]_D$ −23.5° (c = 1, H$_2$O). Optical purity of (I-L) is about 90% because its microbial activity for the growth of bacterium WNB-75 is one-tenth of TJF.

4'-O-(β-D-Galactopyranosyl)-DL-pantothenic Acid

Benzyl 2'-O-benzyl-DL-pantothenate (2.0 g) and acetobromogalactose[11] (3.5 g) are allowed to react in nitromethane (20 ml) and benzene (10 ml) in the presence of mercuric cyanide (1.9 g) and calcium sulfate (5.0 g), and the reaction mixture is worked up in the manner described above. Chromatography of the crude product on silicic acid with 2.5% ethanol in chloroform as a solvent affords 2.15 g of benzyl 2'-O-benzyl-4'-O-(2,3,4,6-tetra-O-acetyl-β-D-galactopyranosyl)-DL-pantothenate (IV), R_f (40% ethyl acetate in benzene) = 0.44. Hydrogenation and deacetylation of (IV)(2.15 g) under the conditions described above afford 850 mg of 4'-O-(β-D-galactopyranosyl)-DL-pantothenic acid (V), as a hygroscopic powder, by lyophilization.

4'-O-(β-Maltosyl)-DL-pantothenic Acid

Benzyl 2'-O-benzyl-DL-pantothenate (3.2 g) and hepta-O-acetylmaltosyl bromide[12] (5.6 g) are allowed to react in nitromethane (20 ml) and benzene (10 ml) in the presence of mercuric cyanide (2.1 g) and calcium sulfate (5.0 g) in the same manner. After working up, the reaction mixture is chromatographed on silicic acid eluting with 2.5% ethanol in chloroform to afford 3.1 g of benyl 2'-O-benzyl-4'-O-(hepta-O-acetyl-β-maltosyl)-DL-pantothenic acid (VI) as a syrup, R_f (40% ethyl acetate in chloroform) = 0.45. Compound (VI) (1.6 g) is hydrogenated over palladium black in acetic acid and then hydrolyzed with 0.5 M sodium methoxide (4 ml) in anhydrous methanol (20 ml) at room temperature for 3 hr. Working up of the product in the usual manner affords 700 mg of 4'-O-(β-maltosyl)-DL-pantothenic acid (VII).

4'-O-(β-Cellobiosyl)-D- and L-Pantothenic Acid

Benzyl 2'-O-benzyl-DL-pantothenate (6.0 g) and hepta-O-acetylcellobiosyl bromide[12] (10.5 g) are allowed to react under the conditions

---

[11] "Methods in Carbohydrate Chemistry,"[9] Vol. II, p. 336.

[12] "Methods in Carbohydrate Chemistry,"[9] Vol. II, p. 214.

described above; after working up, the reaction mixture is chromatographed on silicic acid, eluting with 2.5% ethanol in chloroform to afford a syrup (4.1 g) of benzyl 2'-O-benzyl-4'-O-(hepta-O-acetyl-β-cellobiosyl)-DL-pantothenate (VIII-DL). TLC (developed with 40% ethyl acetate in chloroform) of the product (VIII-DL) shows two spots at $R_f = 0.46$ and 0.40. Careful separation of the two by column chromatography on silicic acid with 40% ethyl acetate in chloroform as a solvent yields benzyl 2'-O-benzyl-4'-O-(hepta-O-acetyl-β-cellobiosyl)-D-pantothenate (VIII-D) first and then benzyl 2'-O-benzyl-4'-O-(hepta-O-acetyl-β-cellobiosyl)-L-pantothenate (VIII-L). Each of them is recrystallized from 90% methanol; (VIII-D): needle, m.p. 138°–140°, $[\alpha]_D$ −40.3° ($c = 1$, CHCl$_3$); (VIII-L): needle, m.p. 137.5°–140°, $[\alpha]_D$ −1.8° ($c = 1$, CHCl$_3$). Each of (VIII-D) and (VIII-L) is hydrogenated and then deacetylated in the usual manner described above to afford 4'-O-(β-cellobiosyl-)-D-pantothenic acid (IX-D) and 4'-O-(β-cellobiosyl)-L-pantothenic acid (IX-L), respectively; IX-D: $[\alpha]_D$ −18.6° ($c = 1$, methanol); (IX-L): $[\alpha]_D$ + 0.8° ($c = 1$, methanol).

4'-O-(β-D-Ribofuranosyl)-DL-pantothenic Acid

To a stirred mixture of benzyl 2'-O-benzyl-DL-pantothenate (4.0 g), mercuric cyanide (2.8 g), and calcium sulfate (10 g) in nitromethane (30 ml) and benzene (10 ml) is added a benzene (10 ml) solution of 2,3,5-tri-O-benzoyl-D-ribofuranosyl chloride,[13] freshly prepared from 4.8 g of 2,3,5-tri-O-benzoyl-D-ribose; the mixture is refluxed with stirring for 20 hr. After work-up in the usual manner, the reaction mixture is chromatographed on silicic acid eluting with 1% ethanol in chloroform to afford benzyl 2'-O-benzyl-4'-O-(2,3,5-tri-O-benzoyl-β-D-ribofuranosyl)-DL-pantothenate (X), (6.9 g). Compound (X) (6.9 g) is hydrogenated over palladium black, and the product, 4'-O-(2,3,5-tri-O-benzoyl-β-D-ribofuranosyl)-DL-pantothenic acid (XI), is recrystallized from ethyl ether–petroleum ether (5.2 g), m.p. 128°–135°. Compound (XI) (3.2 g) is dissolved in dry methylene chloride (2 ml) and anhydrous methanol (8 ml); the solution is cooled to 0°, to which 15 ml of 0.5 M sodium methoxide in methanol is added. After standing at room temperature for 30 hr, the mixture is neutralized with dilute HCl and then concentrated *in vacuo*. The residue dissolved in 30 ml of water is extracted with ethyl ether to remove methyl benzoate and the aqueous layer is chromatographed on a Dowex 1-*X*8 (OH⁻) column in the usual manner to yield 1.6 g of 4'-O-(β-D-ribofuranosyl)-DL-pantothenic acid (XII) as a hygroscopic powder after lyophilization.

[13] "Methods in Carbohydrate Chemistry,"[9] Vol. II, p. 109.

2'-O-(β-D-Glucopyranosyl)-DL-pantothenic Acid

$$\text{CH}_2-\text{C(CH}_3)-\overset{\overset{\displaystyle \text{OBzl}}{|}}{\text{CH}}-\text{CO} \quad + \quad \text{H}_2\text{N}-\text{CH}_2-\text{CH}_2-\text{COOC}_2\text{H}_5$$

$$\underset{\text{O}}{\underline{\qquad\qquad}}$$

$$\Big\downarrow \quad 100°$$

$$\text{HO}-\text{CH}_2-\text{C(CH}_3)-\overset{\overset{\displaystyle \text{OBzl}}{|}}{\text{CH}}-\text{CONH}-\text{CH}_2-\text{CH}_2-\text{COOC}_2\text{H}_5$$

$$\Big\downarrow \quad \begin{array}{l}\text{1. Ac}_2\text{O/pyridine}\\\text{2. H}_2/\text{Pd-black}\end{array}$$

$$\text{AcO}-\text{CH}_2-\text{C(CH}_3)-\overset{\overset{\displaystyle \text{OH}}{|}}{\text{CH}}-\text{CONH}-\text{CH}_2-\text{CH}_2-\text{COOC}_2\text{H}_5$$

(XIII)

$$\Big\downarrow \quad \begin{array}{l}\text{1. acetobromoglucose}\\\text{2. Ba(OMe)}_2\end{array}$$

$$\text{Glu}\overset{\beta}{\underline{\qquad\qquad}}\text{O}-\overset{\overset{\displaystyle \text{CONH}-\text{CH}_2-\text{CH}_2-\text{COOH}}{|}}{\underset{\underset{\displaystyle \text{C(CH}_3)-\text{CH}_2-\text{OH}}{|}}{\text{CH}}}$$

(XV)

Ethyl 2'-O-benzyl-DL-pantothenate (10 g), prepared from 2-O-benzyl-DL-pantolactone and β-alanine ethyl ester, is acetylated in acetic anhydride (50 ml) and pyridine (50 ml) at room temperature for 3 days. The acetate is purified by column chromatography on silicic acid with 30% ethyl acetate in chloroform as a solvent and then hydrogenated over palladium black (1 g) in 50 ml of ethanol at room temperature for 5 hr. After removal of the catalyst by filtration, the filtrate is concentrated *in vacuo* and purified on a silica-gel column eluting with chloroform–ethyl acetate (1:1) to afford 7.0 g of ethyl 4'-O-acetylpantothenate (XIII). Acetobromoglucose (7.0 g) and (XIII) (2.9 g) are allowed to react in the presence of mercuric cyanide (5.1 g) and calcium sulfate (10 g) in 50 ml of nitromethane–benzene (2:1) under the conditions described above. After work-up, the crude product is chromatographed twice on silicic acid, eluting with 50% ethyl acetate in benzene to afford a colorless syrup (1.8 g) of ethyl 2'-O-(2,3,4,6-tetra-O-acetyl-β-D-glucopyranosyl)-4'-O-acetyl-DL-pantothenate (XIV), R_f (50% ethyl acetate in benzene) = 0.41. A solution of (XIV) (1.0 g) dissolved in 20 ml of anhydrous methanol is treated with 5 ml of 0.2 M barium methoxide at room temperature for 24 hr to remove all protection groups. Working up of the reaction mixture in the same manner gives a hygroscopic powder (490 mg) of 2'-*O*-(β-D-glucopyranosyl)-DL-pantothenic acid (XV), $[\alpha]_D$ + 12.0° (c = 1, H_2O).

Syntheses Utilizing Selective Glucosidation

In order to improve the synthesis of 4'-*O*-(β-D-glucopyranosyl)-D-pantothenic acid (TJF), selective glucosidation of methyl D-pantothenate, which is easily derived from inexpensive calcium D-pantothenate, has been examined. In contrast with acetobromoglucose, tetra-*O*-benzoyl-α-D-glucopyranosyl bromide reacts selectively with methyl D-pantothenate at the position of the C-4' hydroxyl group, and a crystalline product is easily isolated in good yield.

4'-*O*-(β-D-*Glucopyranosyl*)-D-*pantothenic Acid (TJF)*

A mixture of methyl D-pantothenate (3.35 g) and pulverized mercuric cyanide (4.0 g) in 30 ml of dry benzene is gently refluxed with stirring, to which a solution of tetra-*O*-benzoyl-α-D-glucopyranosyl bromide[14] (10 g) in 100 ml of benzene is slowly added over 1.5 hr. The mixture is refluxed with stirring for 2 hr and then filtered after cooling. The filtrate is worked up in the usual manner, and the crude product is dissolved in 100 ml of ethyl ether. Standing of the solution at 0° overnight allows precipitation of 4.3 g of methyl 4'-*O*-(2,3,4,6-tetra-*O*-benzoyl-β-D-glucopyranosyl)-D-pantothenate (XVI), m.p. 114.5°–116°. To 1.6 g of (XVI) dissolved in methylene chloride (5 ml) and anhydrous methanol (15 ml) is added 6 ml of 0.5 M sodium methoxide in methanol; the mixture is kept at room temperature for 24 hr. After addition of water (10 ml) and neutralization with 1 N HCl, the reaction mixture is concentrated *in vacuo,* and then the aqueous solution of the residue is extracted with ethyl ether. The aqueous layer is charged on a column of Dowex 1-X8 (OH^-) and, after thorough washing with water, the column is eluted with 0.5 M acetic acid to afford 700 mg of 4'-*O*-(β-D-glucopyranosyl)-D-pantothenic acid (TJF), $[\alpha]_D$ −9.9° (c = 1, H_2O).

4'-*O*-(β-D-*Glucopyranosyl*)-DL-*pantetheine*

To a boiling mixture of DL-pantothenonitrile[15] (4.0 g), mercuric cyanide (5.0 g) and calcium sulfate (10 g, freshly activated) in 45 ml of nitromethane–benzene (2:1) is slowly added a solution of tetra-*O*-

---

[14] "Methods in Carbohydrate Chemistry,"[9] Vol. II, p. 227.
[15] M. Shimizu, see this series, Vol. 18 [324].

$$\text{HO}-\text{CH}_2-\overset{}{\underset{}{\text{C}}}(\text{CH}_3)-\overset{\overset{\displaystyle \text{OH}}{\vert}}{\text{CH}}-\text{CONH}-\text{CH}_2-\text{CH}_2-\text{CN}$$

benzobromoglucose

$$\text{BzGlu}\overset{\beta}{\underline{\hspace{1cm}}}\text{O}-\text{CH}_2-\text{C}(\text{CH}_3)-\overset{\overset{\displaystyle \text{OH}}{\vert}}{\text{CH}}-\text{CONH}-\text{CH}_2-\text{CH}_2-\text{CN}$$

(XVII)

cysteamine

$$\text{BzGlu}\overset{\beta}{\underline{\hspace{1cm}}}\text{O}-\text{CH}_2-\text{C}(\text{CH}_3)-\overset{\overset{\displaystyle \text{OH}}{\vert}}{\text{CH}}-\text{CONH}-\text{CH}_2-\text{CH}_2-\text{C}\underset{\text{S}}{\overset{\text{N}}{<}}$$

(XVIII)

1. NaOCH$_3$
2. AcOH

$$\text{Glu}\overset{\beta}{\underline{\hspace{1cm}}}\text{O}-\text{CH}_2-\text{C}(\text{CH}_3)-\overset{\overset{\displaystyle \text{OH}}{\vert}}{\text{CH}}-\text{CONH}-\text{CH}_2-\text{CH}_2-\text{CONH}-\text{CH}_2-\text{CH}_2-\text{SH}$$

(XX)

benzoyl-α-D-glucopyranosyl bromide (11.6 g) dissolved in 20 ml of benzene, and the mixture is gently refluxed with stirring for 15 hr. The reaction mixture is worked up in the usual manner and chromatographed on silicic acid, eluting with 3% ethanol in chloroform to afford 4'-O-(2,3,4,6-tetra-O-benzoyl-β-D-glucopyranosyl)-DL-pantothenonitrile (XVII), which is recrystallized from ethyl acetate–ethyl ether (9.7 g), m.p. 167°–171°. A solution of (XVII) (2.0 g) and cysteamine (300 mg) dissolved in 10 ml of ethanol is gently refluxed under nitrogen for 15 hr. The reaction mixture is concentrated *in vacuo* and chromatographed on silicic acid eluting with 4% ethanol in chloroform to give a syrup (1.4 g) of 2-[2-{4-O-(2,3,4,6-tetra-O-benzoyl-β-D-glucopyranosyl)-DL-pantamido }ethyl]-2-thiazoline (XVIII). To a cold solution of (XVIII) (1.4 g) dissolved in 5 ml of methylene chloride is added 21 ml of 50 mM sodium methoxide in methanol under nitrogen and the mixture is stored at room temperature for 24 hr. After neutralization with 2.1 ml of 0.5 M acetic acid, the aqueous solution is concentrated *in vacuo* and the aqueous solution is extracted with ethyl ether. The aqueous layer is applied on a Dowex 50W-X2 (H$^+$) column; after thorough washing with water, the column is eluted with 1 M ammonia. The eluate is lyophilized to a hygroscopic powder (580 mg) of 2-[2-{4-O-(β-D-glucopyranosyl)pantamido }ethyl]-2-thiazoline (XIX). A solution of (MIX) (350 mg) dissolved in 10 ml of 0.1 M acetic acid is kept at room temperature under nitrogen for 24 hr and then lyophilized to afford a white powder (340 mg) of 4'-O-(β-D-glucopyranosyl)-DL-pantetheine (XX). Compound (XX) is easily oxidated by hydrogen peroxide to 4'-O-(β-D-glucopyranosyl)-DL-pantethine (XXI). To a solution of (XX) (300 mg) in 3 ml of water are added ferrous sulfate heptahydrate (1 mg) and 28% ammonia (0.1 ml); then 3% hydrogen peroxide is added dropwise until the color of the solution changes to pale yellow from red purple (about 0.6 ml). The resulting solution is passed through the columns of Dowex 50W-X8 (H$^+$) and Dowex 1-X8 (OH$^-$), and the neutral fraction is lyophilized to afford a hygroscopic powder (270 mg) of (XXI).

[44] Synthesis of Coenzyme A and Its Biosynthetic Intermediates by Microbial Processes[1]

By Sakayu Shimizu, Yoshiki Tani, and Koichi Ogata[2]

CoA has been prepared by extraction from microorganisms and by chemical synthesis.[3] However, these methods are not practical because of their low yields or their complexity. Simple and rapid microbial methods for the preparation of CoA and its biosynthetic intermediates are described below.

Synthesis of CoA[4–6]

CoA is enzymically synthesized from pantothenic acid, cysteine, and ATP by cells of *Brevibacterium ammoniagenes*. The reaction proceeds as follows:

Pantothenic acid $\xrightarrow{1}$ 4'-phosphopantothenic acid $\xrightarrow{2}$ 4'-phosphopantothenoylcysteine $\xrightarrow{3}$

4'-phosphopantetheine $\xrightarrow{4}$ 3'-dephospho-CoA $\xrightarrow{5}$ CoA

Reactions (1) through (5) are catalyzed, respectively, by pantothenate kinase (EC 2.7.1.33), phosphopantothenoylcysteine synthetase (EC

---

[1] This work was supported in part by grants from the Ministry of Education, Japan. We wish to thank Professor H. Yamada, Kyoto University, for his interest and for valuable discussion regarding this work.

[2] K. Ogata is now deceased.

[3] A. Kornberg and E. R. Stadtman, this series, Vol. 3 [131]; M. Shimizu, this series, Vol. 18 [55]. See also K. Ogata, *Adv. Appl. Microbiol.* **19**, 209 (1975).

[4] K. Ogata, S. Shimizu, and Y. Tani, *Agric. Biol. Chem.* **34**, 1757 (1970); **36**, 84 (1972).

[5] S. Shimizu, Y. Tani, and K. Ogata, *Agric. Biol. Chem.* **36**, 370 (1972).

[6] S. Shimizu, K. Miyata, Y. Tani, and K. Ogata, *Biochim. Biophys. Acta* **279**, 583 (1972); *Agric. Biol. Chem.* **37**, 607, 615 (1973).

6.3.2.5), phosphopantothenoyl cysteine decarboxylase (EC 4.1.1.36), dephospho-CoA pyrophosphorylase (EC 2.7.7.3), and dephospho-CoA kinase (EC 2.7.1.24). For 1 mol of CoA 4 mol of ATP are required.[7] The bacterium contains high activities of all the enzymes necessary for the operation of this pathway.[8]

Dried Cell Method for Laboratory Experiments[4,5]

Principle. CoA is synthesized in high yield from pantothenic acid, cysteine, and ATP when these are incubated with dried cells of *Brevibacterium ammoniagenes* as the enzyme.

Reagents

Sodium pantothenate
L-Cysteine
ATP (disodium salt)
$MgSO_4 \cdot 7H_2O$
Potassium phosphate buffer, pH 6.0
Sodium laurylbenzenesulfonate
2-Mercaptoethanol
Dried cells of *Brevibacterium ammoniagenes* IFO 12071[9]
Charcoal for chromatography[10]
DEAE-cellulose (Cl^- form) or Dowex 1-X2 (Cl^- form)

Procedure. PREPARATION OF DRIED CELLS. *B. ammoniagenes* is aerobically grown for 1–2 days at 28° in a medium (1000 ml) containing (in grams) glucose, 10; peptone, 15; K_2HPO_4, 3; NaCl, 2g; yeast extract, 1; $MgSO_4 \cdot 7H_2O$, 0.2. Cells are centrifuged and washed with 0.85% NaCl. The washed cells are spread on a glass plate and dried with an electric fan at room temperature and then with P_2O_5 *in vacuo.* The dried cells can be kept at $-15°$ for at least 2 years without any loss of the activity.[8]

REACTION. The reaction mixture[11] (300 ml) contains pantothenate (3 mmol), cysteine (6 mmol), ATP (6 mmol), $MgSO_4$ (3 mmol), potassium phosphate buffer (45 mmol), sodium laurylbenzenesulfonate (600 mg), and dried cells (30 g). It is incubated for 10 hr at 37°. A typical time course for CoA synthesis is shown in Fig. 1.

---

[7] G. M. Brown, *J. Biol. Chem.* **234**, 370 (1959).

[8] S. Shimizu, K. Kubo, H. Morioka, Y. Tani, and K. Ogata, *Agric. Biol. Chem.* **38**, 1015 (1974).

[9] This strain of *B. ammoniagenes* is available from the authors and also from the American Type Culture Collections as ATCC 6871.

[10] For preparation, see this series, Vol. 18 [24].

[11] A better yield of CoA is obtained by replacing pantothenate with an equimolar amount of pantetheine or by increasing ATP to 9 mmol.

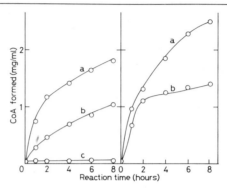

Fig. 1. Syntheses of CoA from pantothenic acid (left) and from pantetheine (right) by dried cells of *Brevibacterium ammoniagenes*. Curve a: complete system (see text); b: sodium laurylbenzenesulfonate omitted; c: cysteine or ATP omitted.

Isolation. The reaction mixture is immersed in a boiling water bath for 10 min, and cells are removed by centrifugation. The cells are resuspended in water (200 ml), boiled for 5 min, centrifuged, and discarded. The combined supernatant liquid (515 ml; CoA content,[12] 720 mg) is adjusted to pH 3.0 with 4 N HCl and applied to a column of charcoal (5.5 × 20 cm). The column is washed with water and the adsorbed nucleotides are eluted with ethanol–28% ammonia water–water (50:5:45, by volume). The eluate is concentrated to about 100 ml under reduced pressure below 35°, and the pH is adjusted to 7.0 with 14% ammonia water. 2-Mercaptoethanol (25 ml) is added, and the mixture is kept at 7° overnight. The solution is then applied to a column of DEAE-cellulose (5.5 × 36 cm).[13] The column is washed with 35 mM LiCl to remove pantothenic acid, AMP, 4′-phosphopantothenic acid, 3′-dephospho-CoA, ADP, etc. Reduced CoA is then eluted with 50 mM LiCl.[14] Both eluents contain 0.1% 2-mercaptoethanol and 0.003 N HCl. This fraction is adjusted to pH 4.5 with 0.5 N LiOH and concentrated to small volume under reduced pressure below 35°. Ethanol is added, and the solution is again concentrated. This is repeated several times to remove most of the water, and the final residue is dissolved in a small volume of methanol. Addition of 20 volumes of acetone precipitates the lithium salt, which is collected, washed repeatedly with methanol–acetone (1:10, by volume), then with acetone–ether (1:1, by volume), and dried over P_2O_5 *in vacuo* to give the

---

[12] CoA is measured by the phosphotransacetylase method. See this series, Vol. 18 [53].

[13] Chromatographic procedure on DEAE-cellulose is essentially the same as that described by J. G. Moffatt and H. G. Khorana, *J. Am. Chem. Soc.* **83**, 663 (1961).

[14] This fraction is essentially free from ATP because it is completely consumed for CoA synthesis. Try linear salt gradient elution if ATP is present.

trilithium salt of reduced CoA (391 mg; purity, 87%[15]; recovery, 47%. Found[16]: C, 27.04; H, 5.05; N, 10.42%; adenosine: P: SH, 1:3.03:0.95. $C_{21}H_{33}O_{16}N_7P_3SLi_3\cdot8H_2O$ requires C, 27.13; H, 5.37; N, 10.54%; adenosine: P: SH, 1:3:1). After washing with 0.1 M LiCl in 0.01N HCl, further elution of the column with 0.2 M LiCl in 0.01N HCl gives oxidized CoA. The pooled fraction is worked up similarly to give the lithium salt (153 mg), which is then converted to reduced CoA by treating with 50% aqueous solution of 2-mercaptoethanol (142 mg; purity, 83%[15]; recovery, 16%, adenosine: P: SH found[16]: 1:2.91:0.85; required: 1:3:1).

A more rapid and simpler procedure for purification is as follows.[17] The combined supernatant obtained as described above (270 ml; CoA content,[12] 213 mg) is treated with charcoal (6.0 g). The adsorbed nucleotides are eluted with 40% acetone containing 0.28% ammonia (350 ml). The eluate is concentrated and neutralized as above. To the concentrate, diluted H_2O_2 solution is added with stirring until the solution becomes SH-free, then Dowex 1-X2 (40 ml) is added. The mixture is stirred for 30 min and filtered. The resin is washed well with 0.2 M LiCl in 0.01 N HCl (four 150-ml portions), then the oxidized CoA is eluted with 0.4 M LiCl in 0.01 N HCl (three 100-ml portions). The concentrated solution is worked up similarly after treatment with 50% aqueous 2-mercaptoethanol to give the lithium salt of reduced CoA (153 mg; purity, 91%[15]; recovery, 65%; adenosine: P, found: 1:3.10; required: 1:3).

Comment. The present method may be very useful for preparation of [14]C-, [3]H-, [35]S-, and ^{32}P-labeled CoA. Preparative synthesis of [^{14}C]CoA by this method has been described.[18]

Fermentative Process for Large-Scale Production[6]

Principle. Pantothenic acid, cysteine, and AMP, when added to cultures of *B. ammoniagenes*, give CoA in a high yield. Glucose, inorganic phosphate, and Mg^{2+} in high concentrations are necessary for generation of ATP from AMP.

Reagents
Calcium pantothenate
L-Cysteine

[15] Determined by the method of G. Michal and H. U. Bergmeyer, *in* "Methods of Enzymatic Analysis" (H. U. Bergmeyer, ed.), 2nd ed., Vol. 4, p. 1972. Academic Press, New York, 1974.

[16] Total phosphate is determined by the method of C. H. Fiske and Y. SubbaRow, *J. Biol. Chem.* **66**, 375 (1925); see also this series, Vol. 3 [115]; and SH by the method of G. L. Ellman, *Arch. Biochem. Biophys.* **82**, 70 (1959).

[17] S. Schimizu, H. Morioka, Y. Tani, and K. Ogata, *J. Ferment. Technol.* **53**, 77 (1975).

[18] K. Hosoki, S. Kurooka, and Y. Yoshimura, *Radioisotopes* **21**, 502 (1972).

AMP (free acid)
Cetylpyridinium chloride
2-Mercaptoethanol
Duolite S-30[19]
Charcoal for chromatography[9]
Dowex 1-X2 (Cl⁻ form)

Procedure. FERMENTATION.[20] *B. ammoniagenes* is cultured for 24 hr at 28° as described above. A 10% inoculum of this culture is added to a liter of pH 7.6 fermentation medium containing (in grams): glucose, 100; urea, 6 (autoclaved separately); yeast extract, 10; K_2HPO_4, 20; $MgSO_4 \cdot 7H_2O$, 10; AMP, 2. Cultivation is carried out at 28° with vigorous aeration for 5–7 days. After 3 days' cultivation, pantothenate (2 g), cysteine (2 g), and cetylpyridinium chloride (1 g) are added, and cultivation is continued for another 2–4 days. A typical time course for the fermentation is shown in Fig. 2.

ISOLATION. The culture broth is boiled for 3 min and centrifuged. The supernatant (1000 ml; CoA content,[12] 2.5 g), in which CoA is present as the disulfide as a result of the vigorous aeration during the fermentation, is passed through a column of Duolite S-30 (3 × 20 cm) to remove brown color, after which the column is washed with water (100 ml). The solution passing through the column is directly applied to a column of charcoal (5.5 × 30 cm), and the substances adsorbed are eluted with 40% acetone containing 0.28% ammonia. The eluate (acidic fraction) containing acetone is directly adsorbed on a column of Dowex 1-X2 (5.5 × 42.5 cm).

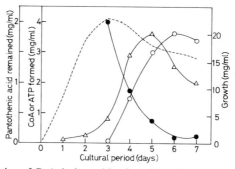

FIG. 2. Accumulation of CoA during cultivation of *Brevibacterium ammoniagenes*. Fermentation was carried out as described in the text accept for the addition of biotin (30 µg/liter). O, CoA; ●, pantothenic acid; △, ATP; ----, cell growth.

---

[19] Diamond Alkali Co., U.S.A. The resin is washed with 5 volumes of 3% NaOH and then with water until the eluate is neutral. This is repeated three times.

[20] Usually fermentation is carried out with 1000 ml of medium in a 3-liter jar fermentor or with 20 ml of medium in a 300 to 500-ml shaking flask.

Washing with 10 mM LiCl in 0.01 N HCl, 40 mM LiCl in 0.01 N HCl, and 0.2 M LiCl in 0.01 N HCl removes AMP, ADP, and ATP, respectively. Washing with 0.4 M LiCl in 0.01 N HCl gives oxidized CoA, which is then worked up as described above to give the lithium salt of reduced CoA (894 mg: purity, 84%[15]; recovery, 30%; adenosine: P: SH: pantothenic acid, found[21]: 1:3.07:0.88:0.93; required: 1:3:1:1).

Other Methods. A similar fermentative process using *Sarcina lutea* IAM 1099 is described by Nishimura *et al.*[22] A rapid purification procedure using affinity chromatographic techniques is also described.[23]

Synthesis of Intermediates of CoA Biosynthesis[6,24]

CoA itself and all the intermediates of CoA biosynthesis, with the exception of 4′-phosphopantothenoylcysteine, are synthesized selectively with high yields by use of the individual reactions involved in the CoA biosynthetic pathway. Both the dried cell method and the fermentative method can be used. The amounts of CoA and its biosynthetic intermediates obtained by these methods are summarized in the table.

SYNTHESIS OF COA AND ITS BIOSYNTHETIC INTERMEDIATES BY
Brevibacterium ammoniagenes[a]

Product	Precursors	Nucleotides	Yield (mg/ml)
Dried cell method			
P-pantothenic acid	Pantothenic acid	ATP	3–4
P-pantetheine	Pantothenic acid + cysteine	ITP + CTP	2–3
	P-pantothenic acid + cysteine	CTP	3–4
	Pantetheine	ITP	2–3
3′-Dephospho-CoA	Pantothenic acid + cysteine	ATP	1–2
CoA	Pantothenic acid + cysteine	ATP	2–3
	Pantetheine	ATP	3–4
Fermentative method			
P-pantothenic acid	Pantothenic acid	AMP	4–5
P-pantetheine	Pantothenic acid + cysteine	GMP + CMP	3–4
CoA	Pantothenic acid + cysteine	AMP	3–5

[a] Abbreviations: P-pantothenic acid; 4′-Phosphopantothenic acid; P-pantetheine; 4′-phosphopantetheine.

[21] Determined by the pantothenic acid release method of Novelli, this series Vol. 3 [132]. For P and SH, see footnote 16.

[22] N. Nishimura, T. Shibatani, T. Kakimoto, and I. Chibata, *Appl. Microbiol.* **28,** 117 (1974).

[23] I. Chibata, T. Tosa, and Y. Matsuo, this series, Vol. 34 [21].

[24] S. Shimizu, S. Satsuma, K. Kubo, Y. Tani, and K. Ogata, *Agric. Biol. Chem.* **37,** 857 (1973).

Synthesis of 4'-Phosphopantothenic Acid[24]

Principle. 4'-Phosphopantothenic acid is produced when cysteine is omitted from the reaction mixture.

Reagents
  Sodium pantothenate
  ATP (disodium salt)
  $MgSO_4 \cdot 7H_2O$
  Potassium phosphate buffer, pH 6.0
  Sodium laurylsulfate
  Dried cells of *B. ammoniagenes* (see previous section)
  Dowex 1-X2 (Cl⁻ form)

Procedure. REACTION. The reaction mixture (300 ml) containing pantothenate (3.4 mmol), ATP (5.1 mmol), $MgSO_4$ (3.4 mmol), potassium phosphate buffer (34 mmol), sodium laurylsulfate (680 mg), and dried cells (20.4 g) is incubated for 8 hr at 37°. More than 90% of the pantothenate added is phosphorylated during the incubation.

ISOLATION. The supernatant obtained by the procedure described above is diluted to 1000 ml with water and applied to a column of Dowex 1-X2 (4.2 × 40 cm). Washings with water and then with 0.007 N HCl remove small amounts of adenine and adenosine and large amounts of AMP. Washing with 10 mM LiCl in 0.01 N HCl gives 4'-phosphopantothenic acid.[25] This is pooled and worked up as described above to give the lithium salt (608 mg; purity, 84–88%[25]; recovery, 58%; complete characterization of a more highly purified sample is described by Shimizu *et al.*[24]).

Other Method. 4'-Phosphopantothenic acid is also obtained from calcium pantothenate and AMP by fermentative process.[6,24]

Synthesis of 4'-Phosphopantetheine[24]

Principle.[26] 4'-Phosphopantothenic acid most rapidly couples with cysteine when CTP is present in place of ATP; GTP, ITP, and UTP are inactive nucleotides. The reaction with CTP gives 4'-phosphopantetheine, while CoA is the main product in the reaction mixture with ATP. 4'-Phosphopantetheine is consumed to form CoA only when incubated with ATP. Other nucleotides lack the ability to couple with this substrate.

---

[25] 4'-Phosphopantothenic acid is detected as pantothenic acid after phosphatase digestion. See this series, Vol. 3 [132]. Purity of this compound is also measured by this method.

[26] S. Shimizu, K. Kubo, S. Satsuma, Y. Tani, and K. Ogata, *J. Ferment. Technol.* **52**, 114 (1974).

Reagents

4'-Phosphopantothenic acid (lithium salt, obtained as described above)

L-Cysteine

CTP (disodium salt)

$MgSO_4 \cdot 7H_2O$

Potassium phosphate buffer, pH 6.5

Sodium laurylsulfate

2-Mercaptoethanol

Washed cells of *B. ammoniagenes* (see previous section)

Dowex 1-X2 (Cl^- form)

DEAE-cellulose (Cl^- form)

Procedure. REACTION. The reaction mixture (30 ml) containing 4'-phosphopantothenic acid (225 μmol), cysteine (450 μmol), CTP (450 μmol), $MgSO_4$ (300 μmol), potassium phosphate buffer (4.5 mmol), sodium laurylsulfate (60 mg), and washed cells (1.8 g, dry weight) is incubated for 5 hr at 37°. The 4'-phosphopantothenic acid added is almost completely converted to 4'-phosphopantetheine.

ISOLATION.[27] The reaction mixture is worked up as above to give supernatant. The resultant supernatant is diluted to 150 ml with water and applied to a column of Dowex 1-X2 (2.5 × 20 cm), and the column is washed with water (600 ml). Washing with 0.005 N HCl removes CMP. 4'-Phosphopantetheine[28] is eluted together with CDP with 10 mM LiCl in 0.01 N HCl. This eluate is adjusted to pH 5.0 and evaporated to dryness. The mixed lithium salts are triturated with methanol briefly, then applied to a column of DEAE-cellulose (2.5 × 30 cm). Elution is carried with an 1100-ml linear salt gradient (0–60 mM LiCl in 0.003 N HCl). Both initial and gradient elutions contain 0.1% 2-mercaptoethanol. 4'-Phosphopantetheine is well separated from the remaining CDP. Appropriate fractions containing 4'-phosphopantetheine are combined, adjusted to pH 5.0 with 0.5 N LiOH, and evaporated to dryness. The mixed lithium salts (4'-phosphopantetheine and LiCl) are triturated with methanol and finally precipitated as the lithium salt by addition of large volume of ether (47 mg; purity, 96–102%[21]; recovery, 53%; P:pantothenic acid:SH, found[21]: 1:1.05:0.97; required, 1:1:1).

Other Methods.[26] Incubation of pantothenic acid and cysteine with CTP and ITP, GTP, or UTP also gives 4'-phosphopantetheine.[26] This is

---

[27] This is a modified procedure based on the work previously published by S. Shimizu, S. Satsuma, K. Kubo, Y. Tani, and K. Ogata, *Agric. Biol. Chem.* **37**, 857 (1973).

[28] Elution of 4'-phosphopantetheine is followed as pantothenic acid. See this series, Vol. 3 [132].

based on the broad specificity of pantothenate kinase for nucleotides[29] and the requirement of ATP by dephospho-CoA pyrophosphorylase.[22] Pantothenate kinase also catalyzes phosphorylation of pantetheine in the presence of ATP, ITP, GTP, or UTP.[29] Therefore, incubation of pantetheine with ITP, GTP, or UTP also gives 4'-phosphopantetheine.[26] These methods are adaptable to the fermentative process. For example, cultivation of the organism with calcium pantothenate, cysteine, GMP, and CMP gives 4'-phosphopantetheine in a high yield.[6,24]

Synthesis of 3'-Dephospho-CoA[26]

Principle. 3'-Dephospho-CoA is obtained by treating the reaction mixture which has accumulated CoA with 3'-nucleotidase of *Bacillus subtilis*.
Reagents
  Sodium pantothenate
  L-Cysteine
  ATP (disodium salt)
  $MgSO_4 \cdot 7H_2O$
  Tris·HCl buffer, pH 7.2
  Sodium laurylsulfate
  2-Mercaptoethanol
  Dried cells of *B. ammoniagenes* (see previous section)
  Crude 3'-nucleotidase solution[30]
  Charcoal for chromatography[9]
  DEAE-cellulose (Cl⁻ form)
Procedure. REACTION. The reaction mixture (40 ml) containing pantothenate (300 μmol), cysteine (400 μmol), ATP (400 μmol), $MgSO_4$ (400 μmol), Tris·HCl buffer (6.2 mmol), sodium laurylsulfate (40 mg), and dried cells (4 g) is incubated for 8 hr at 37°. The mixture is diluted to 80 ml with water and immersed for 5 min in boiling water; the cells are removed by centrifugation. The supernatant (70 ml; CoA content,[12] 45 mg) is combined with a solution of crude 3'-nucleotidase (30 ml), then the mixture is incubated for 6 hr at 37°. More than 80% of the CoA initially present is inactivated by this treatment.

ISOLATION. The reaction mixture is boiled for 5 min, centrifuged and applied to a column of charcoal (1.6 × 7 cm). After elution, concentration, and treatment with 2-mercaptoethanol as above, the resultant solution is adjusted to pH 7.5 with 0.5 N LiOH, diluted to 100 ml with water,

---

[29] S. Shimizu, K. Kubo, Y. Tani, and K. Ogata, *Agric. Biol. Chem.* **37**, 2863 (1973).
[30] Prepared according to the method described by S. Igarashi and A. Kakinuma, *Agric. Biol. Chem.* **26**, 218 (1962).

and applied to a column of DEAE-cellulose (1.8 × 25 cm). Elution is carried out with a 4000 ml linear salt gradient (0–70 mM LiCl in 0.003 N HCl containing 0.1% 2-mercaptoethanol). 3'-Dephospho-CoA[31] is eluted followed by a peak of ADP. Appropriate fractions containing 3'-dephospho-CoA are combined and worked up as described above to give the lithium salt (33 mg; purity, 80%[21]; recovery, 66%; adenosine:P:SH, found: 1:1.88:0.85; required: 1:2:1).

Other Method. 3'-Dephospho-CoA is easily obtained by treating commercial CoA with 3'-nucleotidase of *Bacillus subtilis* IFO 3032.[32]

---

[31] Elution of 3'-dephospho-CoA is followed by absorption at 260 nm and by the pantothenic acid release method of Novelli. See this series Vol. 3 [132].

[32] S. Kurooka, K. Hosoki, and Y. Yoshimura, *Chem. Pharm. Bull.* **15,** 944 (1967).

[45] Phosphopantothenoylcysteine Decarboxylase from Horse Liver[1]

[4'-Phospho-*N*-(D-pantothenoyl)-L-cysteine Carboxy-lyase, EC 4.1.1.36]

By ROBERTO SCANDURRA, MARIO MORIGGI,
VALERIO CONSALVI, and LAURA POLITI

Assay Method

Principle. The method is based on the measurement of $^{14}CO_2$ released from 4'-phosphopantothenoyl-L-[U-^{14}C]cysteine.

Reagents

4'-Phosphopantothenoyl-L-[U-^{14}C]cysteine (Ba salt), 20 mM
Tris·HCl buffer, 0.2 M, pH 8
2-Mercaptoethanol, 1.8 M
KOH, 20% freshly prepared

Procedure. In the main compartment of a Warburg vessel are placed: 0.5 ml of Tris·HCl buffer, 0.25 ml of 2-mercaptoethanol, 0.25 ml of labeled 4'-phosphopantothenoyl-L-cysteine, the enzyme solution, and water to a final volume of 2.5 ml. In the center well 0.2 ml of KOH solution is placed, and in the side arm 0.5 ml of 5 N H_2SO_4. The vessels are tipped and placed for 30 min in a shaking incubator at 38°. At the end of the incubation time, the H_2SO_4 solution is poured from the side arm to the main

---

[1] R. Scandurra, E. Barboni, F. Granata, B. Pensa, and M. Costa, *Eur. J. Biochem.* **49,** 1 (1974).

METHODS IN ENZYMOLOGY, VOL. 62

compartment. After 30 min the KOH solution of the central well is mixed with an automatic pipette, and 0.15 ml is transferred to a liquid scintillation vial filled with 5 ml of absolute ethanol and 10 ml of a scintillation cocktail containing 0.01% of POPOP [2,2'-p-phenylenebis(5-phenyloxazole)] and 0.5% of PPO (2,5-diphenyloxazole) in toluene. Radioactivity is counted in a scintillation beta counter; for each sample a blank is performed with boiled enzyme and the counts subtracted from those obtained in the reaction vessels.

Enzyme Units and Specific Activity. One enzyme unit is defined as the amount of protein that produces 1nmol of CO_2 per minute. This definition has been preferred to the standard enzyme unit because of the low activity of this enzyme. The specific activity is expressed as units per milligram of protein.

Synthesis of 4'-Phosphopantothenoyl-L-[U-^{14}C]cysteine

L-Cysteine (0.54 g, 4.45 mmol) in 15 ml of H_2O are mixed with 0.1 mCi of L-[U-^{14}C]cysteine (The Radiochemical Centre, Amersham, 10–30 mCi/mmol) and brought to pH 9.6 with 1 N NaOH, under a flow of pure N_2: after 30 min the solution is transferred to a 50-ml round flask and rapidly frozen and lyophilized. Methanol, 35 ml, containing 1.26 g (2.7 mmol) of the barium salt of D-pantothenonitrile 4'-phosphate is added, and the solution is refluxed under pure N_2 for 9 hr in a thermoregulated water bath at 50°. Further purification of the compound is achieved as described by Nagase.[2] The solution is dried under vacuum, dissolved in water (18 ml), brought to pH 5.5 with 1 M oxalic acid, and filtered. The clear solution is warmed at 55° for at least 40 min while maximal absorption at 265 nm disappears and the pH drops to 4.2. The solution is filtered through a Dowex 50 H^+ column (10 ml) and washed with 75 ml of H_2O. The solution is brought to pH 7 with 0.1 M Ba(OH)$_2$, and concentrated to 15 ml. Methanol (200 ml) is added under stirring. The mixture is then filtered, and the precipitate collected is reprecipitated from H_2O with methanol to give the barium salt of 4'-phosphopantothenoyl-L-[U-^{14}C]cysteine (MW 623.2), which is then dried under high vacuum. The product (about 1.5 g) should have a specific activity of about 30 nCi/μmol, and purity higher than 98% as tested by paper chromatography on Whatman No. 4 (ethanol–H_2O, 80:20; $R_f = 0.50$) scanned on a radiochromatoscanner. The compound should be more than 90% in the SH form as detected with Ellman's reagent.[3]

[2] O. Nagase, *Chem. Pharmacol. Bull.* **15**, 648 (1967).
[3] G. L. Ellman, *Arch. Biochem. Biophys.* **82**, 70 (1959).

Enzyme Purification

All the operations are carried out at 4°.

Step 1. Extraction. Horse liver obtained from the local slaughterhouse is freed from connective tissues, frozen at −30°, and stored at this temperature for at least one month. Two kilograms of frozen liver are thawed, suspended in 4 liters ice-cold distilled water, and homogenized in the cold room with a Waring blender for 2 min at 20,000 rpm. To the well-stirred homogenate, 10% phosphoric acid is added dropwise to give a pH of 5.5. After standing 30 min at 4°, the homogenate is immersed in a 100° water bath, warmed with continuous stirring to 57°, at which it is kept for 5 min, then rapidly cooled through an immersed coil to 4° and centrifuged at 20,000 g. The precipitate is reextracted with 1 liter of H_2O and centrifuged.

Step 2. Ammonium Sulfate Fractionation. Supernatants are collected and placed in an ice bath; solid ammonium sulfate is added to 30% saturation. After standing 30 min at 4°, the mixture is centrifuged at 20,000 g for 30 min and the supernatant is discarded.

Step 3. DEAE-Cellulose Chromatography. The precipitate, dissolved in 100 ml of 10 mM potassium phosphate buffer, pH 7.6, is dialyzed overnight against 10 liters of the same buffer; it is clarified by centrifugation and applied to a DEAE-23 (Whatman) cellulose column (3 × 50 cm) equilibrated with 70 mM potassium phosphate buffer, pH 7.6. Elution is started with the same buffer at 0.11 M, and the eluate is discarded until a red ring approaches the end of the column; at this point the buffer molarity is increased to 0.13 M and the eluate is collected in 6-ml fractions. Active proteins, located in the tail of the red peak for about 450 ml, are collected by an ammonium sulfate precipitation to 50% saturation and, after 30 min, centrifuged.

PURIFICATION OF PHOSPHOPANTOTHENOYL-L-CYSTEINE DECARBOXYLASE
FROM HORSE LIVER

Step	Volume (ml)	Protein (mg/ml)	Specific activity (units/mg)	Total activity (units)	Recovery (%)
1. Homogenate, pH 5.5, 57°	6270	113.0	0.13	92206	100
	5190	22.6	0.42	49263	53
2. $(NH_4)_2SO_4$, 30% saturated	100	11.8	17	20060	22
3. DEAE-cellulose chromatography	4	76	34	10336	11.2
4. Electrophoresis, pH 8	10	8.3	89	7387	8

Step 4. Column Electrophoresis. The precipitate is dissolved in 20 ml of 10 mM potassium phosphate buffer, pH 8, dialyzed overnight against 10 liters of the same buffer, and then concentrated to 4 ml. A sample containing 50–70 mg of proteins is made 1.3 M in 2-mercaptoethanol and applied to an electrophoresis column (2.5 × 50 cm), filled with CFl cellulose (Whatman) treated with LiBH$_4$,[4] and equilibrated with 10 mM potassium phosphate buffer, pH 8. Electrophoresis is performed for 14.5 hr at 700 V (25 mA) at 4°. The proteins are eluted in 2-ml fractions. The elution pattern shows three peaks, the enzymic activity being found in the second: active fractions are pooled and concentrated. The purification precedure is summarized in the table.

The enzyme can be stored at −20° for at least 2 months. It is free of phosphatase activity, which is eliminated during the first salt fractionation.

Properties of the Enzyme

The enzyme does not contain pyridoxal 5′-phosphate, but contains covalently bound pyruvate involved in the catalytic activity. Pyruvate has been detected as labeled lactate after treatment of the enzyme with borotritide.[5,6]

Substrate Specificity. The enzyme is specific for 4′-phosphopantothenoyl-L-cysteine and does not decarboxylate pantothenoyl-L-cysteine. The substrate is active in the SH form: its S-methyl derivative is not decarboxylated.

Activators and Inhibitors. The enzyme reaction rate is increased more than 20-fold by sulfhydryl compounds, such as dithiothreitol, cysteine, and 2-mercaptoethanol, and is inhibited by iodoacetamide and 4,4′-bis(dimethylaminodiphenylcarbinol) suggesting an essential role for reduced protein —SH groups. Urea (0.2–0.5 M) increases enzyme activity 2.5-fold and reduces to 0.10-fold the amount of 2-mercaptoethanol (0.18 M) that must be added in the incubation mixture for the enzyme to reach maximal velocity. The enzyme is strongly inhibited by pyridoxal phosphate and by carbonyl reagents, such as sodium borohydride, hydroxylamine, and phenylhydrazine.

Kinetic Properties. The optimal pH of the decarboxylase reaction is about 8. The K_m for 4′-phosphopantothenoyl-L-cysteine is 1.43 × 10⁻³ M at pH 8.0.

---

[4] J. Porath and S. Hjertén, *Methods Biochem. Anal.* **9**, 193 (1962).

[5] R. Scandurra, L. Santoro, M. Moriggi, and L. Politi, 2nd National Congress of Italian Biochemical Society, Venezia 1976, Abstract No. 115.

[6] R. Scandurra, L. Politi, L. Santoro, and M. Moriggi, submitted for publication.

[46] Interconversion of Apo- and Holofatty Acid Synthetases of Rat and Pigeon Liver

By DAVID N. BURTON, CAMELLIA SOBHY, ROBERT A. JENIK, and JOHN W. PORTER

Two enzymes, acyl carrier protein synthetase and hydrolase, which are responsible for the interconversion of apo- and holoacyl carrier proteins, through the addition or removal of the prosthetic group, 4′-phosphopantetheine, have been shown to exist in *Escherichia coli*.[1,2] One of these enzymes, 4′-phosphopantetheine transferase (4′-phosphopantetheine:apofatty acid synthetase transferase) has also been reported to be present in rat[3,4] and pigeon[5,6] livers. This enzyme brings about the transfer of the prosthetic group from coenzyme A (CoA) to the acyl carrier protein portion of the apofatty acid synthetase (FAS).

$$\text{Apofatty acid synthetase} + \text{CoA} \xrightarrow[\text{(transferase)}]{\text{Mg}^{2+} + \text{ATP}}$$

holofatty acid synthetase + 3′,5′-adenosine diphosphate

In rats the adaptive synthesis of FAS protein occurs when animals are fed a fat-free diet following a period of fasting.[7,8] The appearance of FAS enzyme activity in rat liver is, however, delayed 3–5 hr after the start of refeeding, even though the synthesis of FAS protein commences immediately on refeeding.[9] The incorporation of pantothenate into FAS parallels the development of enzyme activity, presumably because the apoenzyme is converted to the holoenzyme by the reaction show above.

A crude enzyme preparation derived from livers of fasted rats[4,10] catalyzes the reverse reaction, namely the inactivation of purified FAS.

[1] J. Elovson and P. R. Vagelos, *J. Biol. Chem.* 243, 3603 (1968).

[2] P. R. Vagelos and A. R. Larrabee, *J. Biol. Chem.* 242, 1776 (1967).

[3] H. L. Yu and D. N. Burton, *Biochem. Biophys. Res. Commun.* 61, 433 (1974).

[4] C. Sobhy and J. W. Porter, *J. Biol. Chem.*, submitted for publication (1978).

[5] M. Kim, A. A. Qureshi, R. A. Jenik, F. A. Lornitzo, and J. W. Porter, *Arch. Biochem. Biophys.* 181, 580 (1977).

[6] A. A. Qureshi, M. Kim, F. A. Lornitzo, R. A. Jenik, and J. W. Porter, *Biochem. Biophys. Res. Commun.* 64, 836 (1975).

[7] S. E. Hicks, D. W. Allmann, and D. M. Gibson, *Biochim. Biophys. Acta* 106, 141 (1965).

[8] D. N. Burton, J. M. Collins, A. L. Kennan, and J. W. Porter, *J. Biol. Chem.* 244, 4510 (1969).

[9] H. L. Yu and D. N. Burton, *Arch. Biochem. Biophys.* 161, 297 (1974).

[10] D. A. K. Roncari, *Can. J. Biochem.* 53, 135 (1975).

METHODS IN ENZYMOLOGY, VOL. 62

Sobhy and Porter[4,11,12] have reported the partial purification of this enzyme, 4'-phosphopantetheine hydrolase, (4'-phosphopantetheine: holofatty acid synthetase hydrolase) and the identification of the reaction products as apoFAS and 4'-phosphopantetheine. The rat liver 4'-phosphopantetheine hydrolase reaction occurs as follows[4]:

$$\text{Holofatty acid synthetase} \xrightarrow[\text{(hydrolase)}]{\text{Mg}^{2+}} \text{4'-phosphopantetheine + apofatty acid synthetase}$$

In this chapter methods of assay for rat and pigeon liver 4'-phosphopantetheine transferase and rat liver 4'-phosphopantetheine hydrolase are presented. In addition, procedures for the isolation and partial purification of these enzymes and methods for the isolation and identification of the end products of these reactions are reported.

Preparation of Substrates

Rat Liver Apofatty Acid Synthetase

Male rats weighing approximately 200 g are fasted, with access to water, for 3 days, refed a fat-free diet for 3 hr, then sacrificed; the livers are excised. The livers are homogenized, and the apoFAS is purified by the procedure of Burton et al.[13] The apoFAS eluted in the DEAE-cellulose step of this procedure is precipitated with ammonium sulfate (33% of saturation) and the precipitated protein collected by centrifugation. The enzyme is then dissolved in 0.5 M potassium phosphate buffer, pH 7.0, containing 1 mM dithiothreitol and 1 mM EDTA, and dialyzed against the same buffer. The apoFAS is stored in the presence of 10% glycerol at a protein concentration of 10 mg/ml at $-20°$. The apoFAS prepared by the above procedure contains negligible amounts of overall FAS activity, thereby indicating the presence of only small amounts of the holoenzyme. The apoenzyme can be further purified by chromatography on a column of Sepharose-ε-aminocaproyl pantetheine.[4]

Larger amounts of the apoenzyme may be prepared by treatment of DEAE-cellulose-purified holoFAS with 4'-phosphopantetheine hydrolase, followed by isolation of the apoFAS by chromatography[4] on Sepharose-ε-aminocaproyl pantetheine (see below).

[11] C. Sobhy, Fed. Proc., Fed. Am. Soc. Exp. Biol. **36,** 777 (1977).
[12] C. Sobhy, Fed. Proc. Fed. Am. Soc. Exp. Biol. **37,** 1430 (1978).
[13] D. N. Burton, A. G. Haavik, and J. W. Porter, Arch. Biochem. Biophys. **126,** 141 (1968).

Rat Liver Holofatty Acid Synthetase

Male rats weighing approximately 200 g are fasted for 3 days, refed a fat-free diet for 48 hr, and sacrificed; the livers are excised. The livers are then homogenized and the holoFAS is purified by the method of Burton *et al.*[13] The product obtained by this procedure is primarily the holoenzyme. The traces of apoFAS that may be present are removed by chromatography on a Sepharose-ε-aminocaproyl pantetheine column.

Rat Liver [1-¹⁴C]Pantothenate-Labeled Holofatty Acid Synthetase

Male rats weighing approximately 200 g are fasted for 48 hr and then refed a fat-free diet. Six hours after the start of refeeding, 20 μCi of D-[1-¹⁴C]pantothenic acid are injected intraperitoneally. Eighteen hours later, another 20 μCi are injected. The rats are sacrificed 1 hr after the second injection,[4] and the holoFAS is isolated as mentioned above.[13]

Pigeon Liver Apo- and Holofatty Acid Synthetases

The apoenzyme is present in small amounts, as compared with the quantity of holoenzyme isolated from pigeon liver, under all conditions studied.[5,6] The FAS isolated from pigeon liver is purified by the method of Hsu *et al.*,[14,15] and then the apo- and holoFAS are separated by chromatography on Sepharose-ε-aminocaproyl pantetheine.

*Separation of Rat and Pigeon Liver Apo- and Holofatty Acid
Synthetases on a Sepharose-ε-aminocaproyl
Pantetheine Column*[16,17]

Preparation of Sepharose−ε-Amino-n-caproic Acid.[18,19] Fifty milliliters of Sepharose 4B (measured as a paste) is washed in a beaker with 1 liter of deionized water, and the gel is allowed to settle for 30 min. The fine particles are then decanted off, and the washing procedure is repeated 4 times. The gel is suspended in 50 ml of water in a 250-ml beaker containing a magnetic stirring bar, and cyanogen bromide (10 g) ground with a mortar and pestle in a well-ventilated hood, is slowly added to the suspension with continuous stirring. The suspension is kept at pH 11 by drop-

[14] R. Y. Hsu, G. Wasson, and J. W. Porter, *J. Biol. Chem.* **240**, 3736 (1965).
[15] R. Y. Hsu, P. H. W. Butterworth, and J. W. Porter, this series, Vol. 14 [4].
[16] F. A. Lornitzo, A. A. Qureshi and J. W. Porter, *J. Biol. Chem.* **249**, 1654 (1974).
[17] F. A. Lornitzo, A. A. Qureshi, and J. W. Porter, *J. Biol. Chem.* **250**, 4520 (1975).
[18] R. Axén, J. Porath, and S. Ernbäck, *Nature (London)* **214**, 1302 (1967).
[19] P. Larsson and K. Mosbach, *Biotechnol. Bioeng.* **13**, 393 (1971).

wise addition of cold 5 N KOH, and the temperature is maintained at 18°–20° by the addition of small pieces of ice. The reaction is complete within 20 min after the addition of cyanogen bromide, and little fluctuation in the pH is observed thereafter. The reaction mixture is filtered on a coarse sintered-glass funnel, and the activated gel is washed under suction with 500 ml of cold (4°) 0.1 N NaHCO$_3$, pH 8.5. The activated Sepharose 4B is added to a solution of ε-aminocaproic acid (5 g) in 50 ml of 0.1 N NaHCO$_3$ that has been adjusted to pH 8.5 with 4 M NaOH.[20] The suspension is gently stirred for 16 hr at room temperature before it is transferred to a coarse sintered-glass funnel and washed under suction with 1 liter of ice-cold deionized water. The Sepharose-ε-amino-n-caproic acid is stored as a moist cake at 4° until used.

Preparation of Sepharose –ε-Aminocaproyl Pantetheine. Three milliliters of water and 3 ml of 0.16 M reduced pantetheine[21] are added, with continuous stirring, to 6 g of Sepharose-ε-amino-n-caproic acid. A solution of 1-ethyl-3 (3-dimethylaminopropyl)-carbodiimide hydrochloride (600 mg) in 0.6 ml of water is then added to the mixture, and the pH is immediately adjusted to 4.5–5.0 with 1 N HCl. After stirring for 18 hr at room temperature, the Sepharose-ε-aminocaproyl pantetheine is washed under suction with 1 liter of ice-cold deionized water on a coarse sintered-glass funnel. The gel can be stored as a moist cake for at least 1 week at 4°. The quantity of bound pantetheine is determined by the Ellman method[22] after hydrolysis of the ligand in 0.1 N KOH for 60 min at room temperature. Approximately 1 μmol of pantetheine should be bound per gram of Sepharose. Prior to use, 1 g of the stored Sepharose-ε-aminocaproyl pantetheine gel[23] is washed on a sintered-glass funnel under suction with 500 ml of cold (4°) deionized water, followed by 100 ml of cold 60 mM potassium phosphate, pH 7.0, containing 1 mM dithiothreitol. The gel is immediately packed in a column (0.7 × 5 cm) at 0°.

Procedure for the Separation of the Rat Liver Enzyme

The DEAE-cellulose-purified apoFAS (5 mg of protein) is thawed and then left at room temperature for 1 hr to ensure the complete reassociation of the complex. The protein solution is then poured onto a column of Sepharose–ε-aminocaproyl pantetheine (1 g) containing 0.6–1.0 μmol of thioester. The flow rate during the loading of the enzyme at 0° and the first

---

[20] This solution is prepared before the activation of Sepharose with cyanogen bromide.

[21] Crystalline pantethine is reduced to pantetheine with sodium amalgam.

[22] G. L. Ellman, *Arch. Biochem. Biophys.* **82**, 70 (1959).

[23] One gram of Sepharose-ε-aminocaproyl pantetheine is used to bind 5 (rat) and 5–10 (pigeon) mg of protein.

elution is 1 ml per 15 min. ApoFAS activity is eluted at 0° with 60 mM potassium phosphate buffer, pH 7.0, containing 1 mM dithiothreitol. Twenty 1-ml fractions are collected, and at that time the absorbance at 280 nm of the eluate is less than 0.1. The eluted fractions are immediately adjusted to an ionic strength of 0.5 M with potassium phosphate. Holo-FAS is eluted from the column at room temperature with a mixture of 200 mM potassium phosphate and 200 mM Tris (1:1), pH 8.5, containing 1 mM dithiothreitol, at a flow rate of 1 ml per 5 min. The fractions are immediately adjusted to pH 7.0 and to 0.5 M ion concentration with 1 M monopotassium phosphate. The eluates are assayed for overall FAS, β-ketoacyl thioester reductase, palmitoyl-CoA deacylase, and malonyl- or acetyl-CoA:pantetheine transacylase activities. The fractions containing apo-FAS are pooled, concentrated by ultrafiltration with an Amicon PM-10 ultrafiltration membrane, and dithiothreitol and glycerol are added to final concentrations of 10 mM and 10%, respectively. The fractions containing holoFAS are similarly treated. The holo- and apoenzymes are stable at 4° under these conditions for several days.

The yield of FAS protein from the column is approximately 85%. It should be noted, though, that the amount of enzyme protein relative to the concentration of thioester groups on the gel, the temperature and the flow rate are critical factors in this procedure. If they are not maintained as specified, poor separation and yields will occur. It should be noted also that the ligand is labile at pH 8.5 and therefore cannot be reused.

Procedure for the Separation of the Pigeon Liver Enzymes[5,6]

Ten milligrams of reassociated pigeon liver FAS[14,15,24-27] are loaded onto 1 g of gel at 0° at a flow rate of 1 ml per 15–20 min. ApoFAS is eluted as described for the rat liver apoFAS, except that the ionic strength of the eluted fractions is not adjusted. HoloFAS is eluted at 25° at a flow rate of 1 ml per 5 min with either 0.2 M potassium phosphate, pH 7.0, or 0.1 M potassium phosphate–0.1 M Tris, pH 8.5, containing 1 mM dithiothreitol. If potassium phosphate–Tris buffer, pH 8.5, is used, 1-ml fractions are collected in tubes containing 100 μl of 1 M KH$_2$PO$_4$. The latter is used to bring the pH of the eluate to 7.0. The fractions containing apo- and holo-FAS are concentrated by ultrafiltration and dialyzed overnight at 4° against 125 ml of 0.2 M potassium phosphate, pH 7.0, containing 1 mM

[24] R. A. Muesing and J. W. Porter, this series, Vol. 35 [7].
[25] A BioGel A-1.5-m column has been substituted for Sephadex G-100.
[26] The pigeons are refed Purina pigeon grain for 48 hr following a fast of 2–3 days.
[27] R. A. Muesing, F. A. Lornitzo, S. Kumar, and J. W. Porter, *J. Biol. Chem.* **250**, 1814 (1975).

EDTA and 2 mM dithiothreitol. The holoenzyme is stored at a protein concentration of 3–10 mg/ml in the presence of 10 mM dithiothreitol (10% glycerol may also be added) at −20°. ApoFAS is a minor component (less than 10%) of the apo–holo mixture loaded onto the column, and therefore its concentration at this stage is only 0.5–2 mg/ml. If this preparation is frozen at −20°, all enzyme activity is lost upon thawing. However, in the presence of 10% glycerol and 10 mM dithiothreitol, activity, as measured by the β-ketoacyl thioester reductase assay, is retained for at least 5 days at 4° and 2 months at −20°.

Preparation of the 4′-Phosphopantetheine Transferase and Hydrolase Enzymes

4′-Phosphopantetheine Transferase

Conditioning of Animals. Male rats, Holtzman albino or Long-Evans and weighing 150–200 g, are starved for 3 days with access to water only. They are then refed a fat-free diet for 12 hr prior to sacrifice. Pigeons are fasted for 2–3 days and then refed grain for 12 hr before sacrifice.

Partial Purification of 4′-Phosphopantetheine Transferase from Rat and Pigeon Livers. The following procedure is used for the partial purification of the transferase activity from both rat and pigeon livers.

Preparation of Liver Cytosol. After conditioning, the animals are decapitated, and their livers are excised and placed on ice. Livers are homogenized in a Waring blender in 1.5 volumes of ice-cold bicarbonate–phosphate buffer (70 mM $KHCO_3$, 85 mM K_2HPO_4, 9 mM KH_2PO_4), pH 8.0, containing 2 mM dithiothreitol, and the homogenate is centrifuged at 4° for 20 min at 27,000 g. The supernatant solution is poured through 2 layers of cheesecloth and then centrifuged at 105,000 g at 4° for 90 min. Again, the supernatant solution is poured through 2 layers of cheesecloth, and dithiothreitol is added to a final concentration of 10 mM. If not used immediately, the preparation is stored in aliquots of 5 ml under nitrogen at −20°.

Heat Treatment. The liver cytosol preparation is divided into 1-ml portions and heated at 56°–57° for 2–3 min in a water bath. No FAS activity should be detected in this preparation after removal of denatured protein by centrifugation at 27,000 g for 20 min. This method of elimination of FAS activity from the transferase preparation is preferable to other methods, such as column chromatography on Sepharose-ε-aminocaproyl pantetheine[6] or specific immunoprecipitation.[3]

Ammonium Sulfate Fractionation. The heat-treated supernatant solution is brought to 20% saturation by addition of a saturated solution of ammonium sulfate, pH 7.0, containing 1 mM dithiothreitol. The ammonium sulfate solution is added dropwise with stirring under a stream of nitrogen at room temperature. After 10 min of stirring, precipitated protein is removed by centrifugation and the supernatant solution is brought to 40% (pigeon) or 50% (rat) of saturation with ammonium sulfate. After stirring for 10 min more, the precipitated protein is collected by centrifugation and the supernatant solution is discarded. The pellet is dissolved in 1 ml of potassium phosphate buffer, pH 7.0 (0.2 M for pigeon, 0.5 M for rat), containing 0.1 μM CoA, 1 mM MgCl$_2$, and 5 mM dithiothreitol. The solution is dialyzed against 250 ml of the same buffer for 3 hr at 4° and assayed immediately for 4'-phosphopantetheine transferase activity. Frequent preparation of this fraction is necessary since pigeon liver 4'-phosphopantetheine transferase activity is lost within 1–2 days.[28]

Properties of Rat and Pigeon Liver 4'-Phosphopantetheine Transferases

The 4'-phosphopantetheine transferase catalyzes the incorporation of 4'-phosphopantetheine into apoFAS through the formation of a phosphodiester bond with a serine residue. This enzyme activity requires the presence of Mg^{2+} and ATP, and is stable for several hours in the presence of dithiothreitol. The function of ATP in this reaction is not yet known. The partially purified rat liver transferase enzyme utilizes rat liver apoFAS and either CoA or *E. coli* acyl carrier protein as substrates.[3] The K_m for CoA is approximately 0.6 μM and the pH and temperature optima for the reaction are 7.0 and 37°, respectively.[29]

4'-Phosphopantetheine transferase is an adaptive enzyme that varies with the nutritional state of the animal.[3] However, it does not seem to be involved in the short-term regulation of FAS activity, but instead it appears to function in the posttranslational modification of the FAS.

Rat Liver 4'-Phosphopantetheine Hydrolase

The 4'-phosphopantetheine hydrolase of rat liver is partially purified according to the method of Sobhy and Porter.[4] Male rats are fasted for 3 days and sacrificed; the excised livers are homogenized at 4° in 0.1 M potassium phosphate buffer, pH 7.0, containing 0.25 M sucrose and 5 mM dithiothreitol. The homogenate is centrifuged at 27,000 g for 30 min, and then the supernatant solution is recentrifuged at 109,000 g for 90 min. The

---

[28] The rat liver 4'-phosphopantetheine transferase activity is stable for several hours. The conditions for storage over a longer period of time have not been determined.

[29] H. L. Yu, Ph.D. Thesis, University of Manitoba, 1975.

hydrolase is partially purified through an ammonium sulfate precipitation step of 0 to 75% saturation. The supernatant solution of the 0 to 75% ammonium sulfate saturation is dialyzed against 0.1 M potassium phosphate buffer containing 25 mM sucrose and 5 mM dithiothreitol, concentrated by lyophilization, and stored at $-20°$ in the homogenizing buffer.

Properties of Rat Liver 4'-Phosphopantetheine Hydrolase

Rat liver 4'-phosphopantetheine hydrolase catalyzes the conversion of holoFAS to the apo form by hydrolysis of the prosthetic group, 4'-phosphopantetheine. Enzyme activity requires the presence of Mg^{2+} and dithiothreitol and is optimal at 37° and pH 7.0. The partially purified enzyme is stable for several months at $-20°$ in the presence of dithiothreitol. Rat liver hydrolase is specific for rat liver holoFAS; it does not hydrolyze 4'-phosphopantetheine from pigeon liver FAS or from CoA.

The activity of the rat liver hydrolase varies with the nutritional and hormonal state of an animal.[30]

Assay of Enzyme Activities

Assays for Holofatty Acid Synthetase

Overall FAS activity from rat or pigeon liver is assayed either spectrophotometrically[31] or by measurement of incorporation of ^{14}C-labeled acetyl- or malonyl-CoA into long-chain fatty acids.[14] Details of the spectrophotometric method used for assay of FAS from rat[32] and pigeon[24] livers are given in an earlier volume in this series.

Assay mixtures for the incorporation of radioactivity into fatty acids contain: potassium phosphate buffer, pH 7.0, 0.5 mmol (for rat) *or* 0.2 mmol (for pigeon); acetyl-CoA *or* [1-^{14}C]acetyl-CoA, 15 nmol, 150,000 dpm; malonyl-CoA *or* [2-^{14}C]malonyl-CoA, 60 nmol, 25,000 dpm; NADPH, 100 nmol; EDTA, 1 μmol; dithiothreitol, 1 μmol; and enzyme, 5–10 μg of purified FAS protein in a final volume of 1 ml. The reaction is started by the addition of the substrate. Incubations are carried out at 30° *or* 37° for 5 min, and then the reaction is stopped by the addition of 30 μl of 60% perchloric acid. One milliliter of ethanol is added to each incubation mixture and the ^{14}C-labeled fatty acids are extracted 3 times with 2 ml of petroleum ether. Aliquots of this solution are assayed for radioactivity in a liquid scintillation spectrometer. A unit of FAS activity is that amount

[30] C. Sobhy and J. W. Porter, unpublished observations.
[31] S. Kumar, J. A. Dorsey, R. A. Muesing, and J. W. Porter, *J. Biol. Chem.* **245**, 4732 (1970).
[32] C. M. Nepokroeff, M. R. Lakshmanan, and J. W. Porter, this series Vol. 35 [6].

of enzyme required to catalyze the formation of 1 nmol of palmitate per minute under the conditions of the assay.

Assays for Apofatty Acid Synthetase Activity

Activity of rat and pigeon liver apoFAS is determined by assaying for partial enzyme activities for fatty acid synthesis. These may be the acetyl- or malonyl-CoA: pantetheine transacylase, palmitoyl-CoA deacylase, or β-ketoacyl thioester reductase activities.[24] The β-ketoacyl thioester reductase activity is the most labile and the palmitoyl-CoA deacylase is the most stable of the partial activities. However, addition of glycerol to the enzyme preparation, to a final concentration of 10%, stabilizes β-ketoacyl thioester reductase activity.

Assays for the Conversion of Apo- to Holofatty Acid Synthetase

RAT LIVER 4'-PHOSPHOPANTETHEINE TRANSFERASE

Principle. The conversion of apo-FAS to holoFAS is measured in a 2-step assay procedure.[4] First, the synthesis of the holoenzyme is carried out in the presence of 4'-phosphopantetheine transferase. The amount of synthesized holoenzyme is then determined by spectrophotometric measurement of enzyme activity. If radioactive CoA is included in the reaction mixture, a measurement is made of the amount of radioactivity transferred from CoA to the apoFAS.

A unit of 4'-phosphopantetheine transferase activity is that amount of enzyme required to catalyze the formation of 1 unit of FAS activity per minute under the conditions of the assay.

Procedure. Reaction mixtures contain: potassium phosphate buffer, pH 7.0, 0.5 mmol; CoA plus [³H]CoA (6000 dpm), 3.0 nmol; MgCl$_2$, 1.0 μmol; ATP, 1.5 μmol; dithiothreitol, 1.0 μmol; apoFAS, 1.0 mg of protein; and 4'-phosphopantetheine transferase, 4.0 mg protein in a final volume of 1.5 ml. The reaction is initiated by adding apoFAS, and the incubation is continued for 35 min at 37°. The reaction is stopped by the addition of EDTA to a final concentration of 2 mM.

The reaction mixture is passed through a 1 × 35 cm column of Sephadex G-50 (fine), to separate the synthesized holoFAS from unreacted CoA. The protein eluate from the column is assayed for overall FAS activity,[32] and for radioactivity, to determine the amount of [³H] pantetheine incorporated into the holoenzyme. If desired, the holo-FAS end product may be separated from unreacted apoenzyme by chromatography on Sepharose-ε-aminocaproyl pantetheine as described ear-

TABLE I
CONVERSION OF RAT LIVER APOFAS TO HOLOFAS IN THE PRESENCE OF
4'-PHOSPHOPANTETHEINE TRANSFERASE ACTIVITY

Conditions	Radioactivity incorporated into FAS (cpm)	FAS activity (nmol palmitate formed/min)
Complete system[a]	2625	4.1
− MgCl₂ ATP and CoA	130	1.1
Complete system[b]	3500	4.0
− MgCl₂ ATP and CoA	150	0.0

[a] ApoFAS prepared from liver homogenates of animals as described in text.
[b] ApoFAS prepared in the hydrolase reaction and then isolated by chromatography on Sepharose-ε-aminocaproyl pantetheine.

lier, and the milligrams of protein of holo- and apoFAS may be determined by the immunochemical assay method described below. Table I demonstrates the conversion of apo- to holoFAS in the presence of 4'-phosphopantetheine transferase activity.

Similar results are obtained by the method described by Yu and Burton.[3] In this method a particle-free supernatant solution from livers of rats fed a fat-free diet for 3 hr following starvation is prepared as previously described.[32] The fraction precipitating between 20% and 40% of saturation with ammonium sulfate is dialyzed against 0.15 M potassium phosphate buffer, pH 7.0, containing 1 mM dithiothreitol and then used as the apoFAS preparation. A crude preparation of 4'-phosphopantetheine transferase is obtained from livers of rats fed a fat-free diet for 4–48 hr (maximum activity found at 12 hr) following starvation,[3] exactly as described above for the preparation of crude apoFAS, except that the 20–50% ammonium sulfate fraction is used. This preparation of the transferase enzyme contains holoFAS which must be removed before assay of transferase activity. This can be accomplished by heat treatment as described earlier in this article, or by the addition of rabbit antiserum against FAS in sufficient quantity to precipitate all holoFAS activity.

4'-Phosphopantetheine transferase activity is assayed as described above, and aliquots of the protein eluate from the gel filtration column, used to separate synthesized holoFAS from unreacted CoA, are assayed for overall FAS activity.

PIGEON LIVER 4'-PHOSPHOPANTETHEINE TRANSFERASE

Reaction mixtures for the first stage of the enzyme assay contain: potassium phosphate buffer, pH 7.0, 0.2 mmol; CoA, 3.0 nmol; MgCl₂,

1.0 μmol; ATP, 1.5 μmol; dithiothreitol, 5.0 μmol; apoFAS, 0.5–1.0 mg of protein; and 4'-phosphopantetheine transferase, 4.0 mg of protein in a final volume of 1 ml. The reaction is carried out for 30 min at 30° and then stopped by the addition of 2 mM EDTA. The reaction mixture is dialyzed for 3 hr at room temperature against 250 ml of 0.2 M potassium phosphate buffer, pH 7.0, containing 1 mM EDTA and 5 mM dithiothreitol.

In the second stage of the assay, the FAS activity generated in the first stage is measured by the method involving incorporation of [2-^{14}C]malonyl-CoA into long-chain fatty acids.

When quantitation of the amount of holoenzyme formed is desired, the first stage of this assay is carried out with 1 mg or more of apoFAS. After dialysis the reaction mixture is stored at $-20°$ overnight. Apo- and holo-FAS are then separated as described above for the preparation of substrate, except that the whole reaction mixture is loaded onto the Sepharose-ϵ-aminocaproyl pantetheine column. The apo fraction eluted from this column is freed of other contaminating proteins by DEAE-cellulose chromatography, as described for the purification of FAS.[14,15,24] A column of 1 × 6 cm is used in this procedure.

Assay for the Conversion of Rat Liver Holo- to Apofatty Acid Synthetase

Principle. The rat liver 4'-phosphopantetheine hydrolase deactivates FAS by catalyzing the hydrolysis of the prosthetic group, 4'-phosphopantetheine, from the holoenzyme. The conversion of holoFAS to apoFAS is measured in a two-step assay.[4] In the first step apoFAS is formed in the presence of 4'-phosphopantetheine hydrolase, and in the second step the remaining overall FAS activity and the amount of 4'-phosphopantetheine released are measured.

Procedure. Reaction mixtures contain: potassium phosphate buffer, pH 7.0, 0.1 mmol; sucrose, 0.25 mmol; dithiothreitol, 5 μmol; MgCl$_2$, 2.0 nmol; holoFAS, 100 μg of protein, *or* holoFAS plus [1-^{14}C]pantothenate-labeled holoFAS (3000 dpm), 100 μg of protein; and 4'-phosphopantetheine hydrolase, 20 μg of protein in a final volume of 0.1 ml. The reaction is started by the addition of holoFAS and allowed to proceed for 45 min at 37°. The reaction is stopped by the addition of EDTA (final concentration, 10 mM). In the second step of the assay, an aliquot (0.05 ml) of the reaction mixture is assayed for holoFAS activity as described earlier. To another aliquot (0.05 ml) of the reaction mixture, trichloroacetic acid is added and the precipitated protein is removed by centrifugation. The supernatant solution is then assayed for radioactivity to determine the amount of [^{14}C]4'-phosphopantetheine released. A unit of 4'-phosphopantetheine hydrolase activity is that amount of enzyme required to catalyze the formation of 1 unit of apoFAS, measured as the loss of 1 unit of holoFAS activity per minute under the conditions of the assay.

When large quantities of apoFAS are desired, a larger volume of the above-described reaction mixture, containing 5–10 mg of holofatty acid synthetase, is used. After incubation, the reaction mixture is dialyzed against 60 mM potassium phosphate buffer, pH 7.0, containing 1 mM dithiothreitol, for 1 hr at room temperature. Separation of the apo- and holoenzymes is then effected by column chromatography on Sepharose-ε-aminocaproyl pantetheine. The apoFAS is further purified as described in the section on rat liver apofatty acid synthetase.[13]

Separation and Quantitation of End Products

Separation of the end products of the transferase and hydrolase reactions is accomplished by chromatography on a Sepharose-ε-aminocaproyl pantetheine column as described above. The number of units of apo- and holoFAS present in the eluates from the column may be determined as described earlier in this chapter. Absolute quantities, in terms of milligrams of protein, of apo- and holoFAS are determined immunochemically with rabbit antiserum to FAS prepared as described below.

Table II demonstrates the conversion of holoFAS to apoFAS and the separation of the two forms of enzyme by column chromatography on Sepharose–ε-aminocaproyl pantetheine.

Immunochemical Determination of Apo- and Holofatty Acid Synthetases

The amounts of apo- and holoFAS protein can be determined immunochemically by a quantitative precipitin method[33,34] or by a radial diffusion technique based on the method of Mancini et al.[35]

Preparation of Rabbit Antirat FAS Antisera. DEAE-cellulose or sucrose density gradient purified rat liver fatty acid synthetase is mixed with an equal volume of complete Freund's adjuvant and injected into rabbits subcutaneously over the shoulders at 4 sites (5 mg/rabbit). A second injection, 5 mg FAS in incomplete Freund's adjuvant, is given 10 days later. The rabbits are bled at 2-week intervals after the second injection.

---

[33] E. A. Kabat and M. M. Mayer, "Experimental Immunochemistry," 2nd ed., p. 22. Thomas, Springfield, Illinois, 1961.

[34] R. K. Brown, this series, Vol. 11 [91].

[35] G. Mancini, J. P. Vaerman, A. O. Carbonera, and J. F. Heremans, *in* "Protides of the Biological Fluids," 11th colloquium, (H. Peeters, ed.), p. 370. Elsevier, Amsterdam, 1964.

TABLE II

Conversion of Rat Liver HoloFAS to ApoFAS in the Presence of
4'-Phosphopantetheine Hydrolase Activity

| | HoloFAS | | ApoFAS | |
Conditions	Amount[a] (mg)	FAS activity (units)	Amount[a] (mg)	Palmitoyl-CoA deacylase activity (units)
Complete system				
− hydrolase	1.64	69.8	0	—
+ hydrolase	0.7[b]	28.4	0.79	33.6

[a] Amounts of apo- and holoFAS were determined immunochemically by the quantitative precipitin technique after separation on Sepharose-ε-aminocaproyl pantetheine.

[b] The recovery of holoFAS from the Sepharose-ε-aminocaproyl pantetheine column was estimated to be 82%.

The antiserum is purified through two successive ammonium sulfate precipitation steps of 0 to 50% saturation and a DEAE-cellulose chromatography step, as described by Livingston,[36] except that the DEAE-cellulose column is washed and eluted with 20 mM potassium phosphate, pH 8.0. The purified antibody is concentrated by the addition of ammonium sulfate to 50% saturation. The precipitate is dissolved and dialyzed overnight in 20 mM potassium phosphate, pH 8.0, and the antibody is stored at −20°.

Quantitative Precipitin Method. To obtain a standard curve for the quantitative precipitin reaction between rabbit anti-FAS γ-globulin and FAS, increasing amounts of FAS are added to a constant volume of antibody, 200 μl, in 0.01 M potassium phosphate buffer, pH 7.0, containing 0.9% NaCl and 1% Triton X-100 in a final volume of 1 ml. The reaction mixture is incubated for 1–2 hr at 37°, and then for 24–48 hr at 4° for maximum precipitation. The precipitate is collected by centrifugation and washed twice with the same buffer. The precipitate is dissolved in 0.2 N NaOH, and its protein content is determined by the Lowry method.[37] The supernatant solution is divided into two equal parts; antibody is added to one, and FAS is added to the other to check for the presence of antigen and antibody, respectively. The milligrams of protein of the immunoprecipitant complex are plotted as a function of the milligrams of FAS added to the reaction mixture.

In immunochemical assays of samples containing FAS, three different

[36] D. M. Livingston, this series, Vol. 34 [91].

[37] O. H. Lowry, N. J. Rosebrough, A. L. Farr, and R. J. Randall, *J. Biol. Chem.* **193**, 265 (1951).

aliquots of the sample, each estimated to contain a quantity of FAS close to the equivalence point, are incubated with the rabbit anti-FAS γ-globulin in the same manner described above, and the exact number of milligrams of FAS in the immunoprecipitate is computed from the standard curve.

Radial Diffusion Method. Gels of 2 mm thickness and containing 1% agarose, 1% rabbit antiserum against FAS,[38] 0.5 mM EDTA and 0.25 M potassium phosphate buffer, pH 7.0, are prepared on glass plates measuring 20 × 20 cm. Wells 2.5 mm in diameter are cut in the gel. Aliquots of 5 μl of each solution, containing known or unknown quantities of apo- or holoFAS, are placed in each well, and diffusion is allowed to continue at room temperature for 4 days. The gels are immersed overnight in several liters of 0.15 M NaCl to remove soluble proteins and then are washed with deionized water. The diameter of the precipitin rings formed can be measured immediately, or after staining with Amido Black, 0.1% in 0.55 M acetate buffer, pH 3.75. The diameter of the precipitin ring bears a linear relationship to the logarithm of the amount of apo- or holoFAS placed in the well over a range of about 1–10 μg of protein.

[38] The serum fraction precipitating between 0 and 40% of saturation with ammonium sulfate is used.

[47] Purification and Properties of Pantetheinase from Horse Kidney

By SILVESTRO DUPRÈ and DORIANO CAVALLINI

Pantetheine → pantothenate + cysteamine

Pantetheinase (EC 3.5.1.-, pantetheine hydrolase) hydrolyzes specifically one of the carboamide linkages in pantetheine, with the formation of pantothenate and cysteamine. Cofactors are not required, but a reducing compound (such as cysteine, mercaptoethanol, dithiothreitol) must be present for full activity.

This hydrolytic effect was first postulated by Brown[1] and Novelli,[2] then demonstrated by Cavallini et al.[3,4] The enzyme has been extensively studied by Duprè and co-workers[5-9] and by Kameda and Abiko.[10]

[1] G. M. Brown, J. A. Craig, and E. E. Snell, *Arch. Biochem.* **27**, 473 (1950).

[2] G. D. Novelli, F. J. Schmetz, Jr., and N. O. Kaplan, *J. Biol. Chem.* **206**, 533 (1954).

[3] D. Cavallini, C. De Marco, and C. Crifò, *Boll. Soc. Ital. Biol. Sper.* **40**, 1973 (1964).

[4] D. Cavallini, S. Duprè, M. T. Graziani, and M. G. Tinti, *FEBS Lett.* **1**, 119 (1968).

[5] S. Duprè, M. T. Graziani, M. A. Rosei, A. Fabi, and E. Del Grosso; *Eur. J. Biochem.* **16**, 571 (1970).

The enzyme may be connected to the dissimilative pathway of CoA and of acyl carrier protein, allowing the turnover of the pantothenate moiety. It seems to cover an important role also in metabolic pathways leading to taurine. In connection with pantothenoyl 4'-phosphate L-cysteine decarboxylase (EC 4.1.1.36), pantetheinase functions in the metabolic route that allows the conversion of cysteine to taurine in those tissues (as brain or heart) where the other main route, through cysteine sulfinic acid decarboxylase or cysteic acid decarboxylase, is less efficient or absent.[11–15]

Assay Methods

Principle. The activity is pantetheinase can be assayed by measurement of the amount of pantothenate released from pantetheine,[16] or by titrating reaction products in a pH-stat apparatus.[17] The first assay has been developed for crude extracts or less purified enzyme preparations and requires longer incubation times. The second assay allows the derivation of initial velocities and is better used for more purified enzyme solutions. The details of these methods are described below.

Reagents
Potassium phosphate buffer, 1 M, pH 7.6
Cysteine hydrochloride, 0.3 M

[6] S. Duprè, M. A. Rosei, L. Bellussi, E. Del Grosso, and D. Cavallini, *Eur. J. Biochem.* **40,** 103 (1973).

[7] L. Bellussi, S. Duprè, L. Pecci, and L. Santoro, *Physiol. Chem. Phys.* **6,** 505 (1974).

[8] S. Duprè, M. A. Rosei, L. Bellussi, R. Di Gioacchino, and D. Cavallini, *FEBS, 8th Meeting,* Abstract 457, (1972).

[9] S. Duprè, M. A. Rosei, L. Bellussi, E. Barboni, and R. Scandurra, *Int. Congr. Biochem. 9th,* Abstract 2o28, p. 98 (1973).

[10] K. Kameda and Y. Abiko, unpublished work. Cited in "Metabolism of Sulfur Compounds," p. 13 Vol. VII of "Metabolic Pathways" (D. M. Greenberg, ed.), 3rd ed. Academic Press, New York, 1975.

[11] S. Duprè, F. Granata, L. Santoro, R. Scandurra, G. Federici, and D. Cavallini, *Ital. J. Biochem.* **24,** 369 (1975).

[12] D. Cavallini, R. Scandurra, S. Duprè, G. Federici, L. Santoro, G. Ricci, and D. Barra, *in* "Taurine" (R. Huxtable and A. Barbeau, eds.) p. 59. Raven Press, New York, 1976.

[13] D. Cavallini, R. Scandurra, S. Duprè, L. Santoro, and D. Barra, *Physiol. Chem. Phys.* **8,** 157 (1976).

[14] R. Scandurra, L. Politi, S. Duprè, M. Moriggi, D. Barra, and D. Cavallini, *Bull. Mol. Biol. Med.* **2,** 172 (1977).

[15] R. Scanduzza, G. Federici, S. Duprè, and D. Cavallini, *Bull. Mol. Biol. Med.* **3** (1978) in press.

[16] S. Duprè, M. T. Graziani, and M. A. Rosei, *Ital. J. Biochem.* **19,** 132 (1970).

[17] S. Duprè, A. Antonucci, P. Piergrossi, and M. Aureli, *Ital. J. Biochem.* **25,** 229 (1976).

[^{14}C]Pantethine, 0.1 M in water (8×10^4 dpm/μmol)(Method A)[18]
Pantethine, 25 mM + dithiothreitol, 75 mM (Method B)
Dithiothreitol, 50 mM
Perchloric acid, 4 M

Method A. Reaction mixtures are prepared to contain 0.05 ml of phosphate buffer, 0.05 ml of [^{14}C]pantethine, 0.05 ml of cysteine hydrochloride, enzyme preparation and enough water to make a final volume of 1.0 ml. Incubation is for 60 min or less at 38°. The reaction is stopped with 0.1 ml of HClO$_4$; after centrifugation the solution is neutralized with 2 N KOH and centrifuged again. Deproteinization can be omitted if 3 mg or less of protein are incubated. An aliquot is applied as a narrow band on a 3×31 cm Whatman No. 1 paper strip and submitted to electrophoresis in the cold, in 50 mM acetate buffer, pH 5.0 (400 V, 5 hr). Radioscanning is performed in a gas-flow chromatoscanner. Quantitation is obtained by integrating the area corresponding to labeled pantothenate (which migrates to the anode) and by comparing it to the total radioactivity present on paper.

Method B. The reaction mixture, in a 5-ml thermostated, magnetically stirred reaction vessel, contains: suitable amounts of enzyme solution, dialyzed overnight against deionized water, and 0.2 ml of dithiothreitol. The pH value is adjusted to the desired value with diluted HCl or KOH, water is added to make 1.3 ml, and the reaction is primed by the addition of 0.2 ml of pantethine plus dithiothreitol, previously adjusted to the desired pH. Titration is performed with 1 mM HCl or KOH dispensed from a microburette, at 38°; CO$_2$-free nitrogen is bubbled when the determination is made in the alkaline range. The amount of pantetheine hydrolyzed is calculated from the equivalents of titrating agent, taking into account the dissociation of pantothenic acid ($pK = 4.4$) and cysteamine ($pK = 8.31$).

Definition of Unit and Specific Activity. A unit of enzyme is that amount that hydrolyzes 1 μmol of substrate in 1 min under the conditions described in Method A. Protein concentration is determined by the method of Gornall against bovine serum albumin.[19] Specific activity is in units per milligram of protein.

[18] Pantethine is prepared by the method of Viscontini *et al., Helv. Chim. Acta* **37**, 375 (1954); it is also available commercially. DL-[^{14}C]Pantethine is prepared by the same method, starting with DL-[1-^{14}C]pantolactone (Sorin, Saluggia, Italy). After passage through a Dowex-2 OH$^-$ column, the product is radiochromatographically pure.
[19] A. G. Gornall, G. J. Bardawill and M. H. David, *J. Biol. Chem.* **177**, 751 (1949).

Purification

Step 1. Homogenate. Horse kidney cortex, cut in small pieces, can be stored in the frozen state for several weeks at $-20°$ without significant loss of enzymic activity. It is left to thaw and homogenized for 10 min in a Waring blender at high speed (1000 g with 1000 ml of deionized water).

Step 2. Heating and Ammonium Sulfate Fractionation. The homogenate is brought to pH 5.5, with efficient stirring, by dropwise addition of 10% H_3PO_4 and heated in a boiling water bath to $54°-55°$ (about 10 min are needed to reach this temperature). It is rapidly cooled in an ice bath and centrifuged for 60 min at 3500 g. The supernatant solution is adjusted to pH 7.0 with 1 M KOH (in the cold and with stirring) and then fractioned by ammonium sulfate precipitation. The precipitate from 0.45 to 0.65 saturation (after addition of solid ammonium sulfate, the solution is allowed to stand 30 min with stirring in the cold before the precipitate is spun down by centrifugation) is dissolved in the minimum amount of 10 mM potassium phosphate, pH 7.6. The enzyme solution is dialyzed overnight against 25 liters of water.

Step 3. CM-Cellulose Chromatography. The dialyzed solution, brought in an ice bath to pH 5.0 with 2 N acetic acid and clarified by centrifugation, is passed in the cold through a CM-cellulose column (4 × 60 cm, equilibrated with 20 mM acetate, pH 5.0, capacity 0.7 meq/g). Proteins eluted with the same buffer containing 30 mM NaCl are pooled and precipitated with ammonium sulfate at 0.70 saturation. The precipitate, collected by centrifugation, is dissolved in 10 mM potassium phosphate, pH 7.6, and dialyzed overnight against the same buffer.

Step 4. DEAE-Cellulose Chromatography. The clarified solution is poured in the cold into a DEAE-cellulose column (3 × 30 cm, capacity 1.0 meq/g, equilibrated with 10 mM potassium phosphate, pH 7.6); the bed is washed with the same buffer containing 70 mM NaCl (discarded), and then eluted with the same buffer containing 0.2 M NaCl. The pooled solution is precipitated with ammonium sulfate at 0.70 saturation; the precipitate is collected by centrifugation and dissolved in the minimum amount of 10 mM potassium phosphate, pH 7.6 (about 3 ml).

Step 5. Gel Filtration. The clear solution is fractionated through a Sephadex G-200 2.5 × 90 cm column, equilibrated with 10 mM potassium phosphate, pH 7.6, and eluted backward with the same buffer in 2-ml fractions, by means of a peristaltic pump at a flow rate of about 0.15 ml/min. The elution profile shows invariably four separated peaks at 280 nm, and enzyme activity is located in the third peak. Active fractions are collected and the protein concentrated either by precipitation with ammonium sulfate at 0.70 saturation or by ultrafiltration.[20] This preparation

[20] E. R. Huens and E. M. Shooter, *J. Mol. Biol.* **3,** 257 (1961).

SUMMARY OF PURIFICATION OF PANTETHEINASE FROM KIDNEY

Step	Fraction	Protein (mg)	Specific activity (units/mg)	Yield (%)
1	Homogenate	220,000	0.000025	100
2	Heating and ammonium sulfate fractionation	3,900	0.0012	85
3	CM-cellulose chromatography	400	0.01	72
4[a]	DEAE-cellulose chromatography	190	0.035	61
5	Sephadex G-200 filtration	45	0.1	41
6	Recycling on Sephadex G-75	3.5	0.4	12.8

[a] A pool of two preparations from the previous step was worked up in this step.

is not pure and shows on disc electrophoresis one heavy and four or five faint bands.

Preparations homogeneous on ultracentrifugation and showing one band in disc electrophoresis have been obtained not routinely by repeated recycling[21] on Sephadex G-75, collecting fractions showing only one electrophoretic band.

A summary of the purification is given in the table.

Properties

Stability. The enzyme is quite stable and can be stored in concentrated solutions of step 5 for several months at neutral pH at $-20°$ without significant loss of the activity. Diluted solutions, kept at $0°$, lose activity at a rate of about 5% per week. The stability is good between pH 5.0 and 7.6, and rapidly decreases outside these limits.

pH optimum. The enzyme exhibits optimum activity at pH values between 4.0 and 5.5

Specificity. The enzyme is highly specific for the natural D-(+) isomer of pantetheine in the reduced form.[7] Against D-(+)-pantetheine 4'-phosphate an activity of about 10% is detected. Pantothenoylcysteine, pantothenoyl-4'-phosphate-cysteine, L-(−)-pantetheine, pantetheine thiazoline, CoA, pantothenate, and β-alanylcysteamine are hydrolyzed very poorly or not at all.[6] The products of the reaction have been identified as D-(+)-pantothenate and cysteamine.

Inhibitors and Activators. The reaction is aspecifically activated, to various extent, by the addition of 5 mM thiol. The highest effect is given by mercaptoethanol and dithiothreitol.

[21] H. Determann, "Gel Chromatography," p. 49. Springer-Verlag, Berlin and New York, 1968.

A marked inhibition by product is observed (5 mM cysteamine, 44%; 50 mM pantothenate, 60%); inhibition by excess substrate is also present, but it is observed only at longer incubation times, whereas it is completely absent when the activity is performed with the pH-stat method, and initial velocities are recorded. Fifteen millimolar NaN_3 inhibits 90%; iodoacetate and iodoacetamide are inhibitors at high concentrations when preincubated with the enzyme, and the excess is removed by dialysis before activity determination.[9]

Kinetic Data and Physical Properties. The apparent K_m for pantetheine is 5 mM. A molecular weight of 55,000 has been calculated by ultracentrifugation. The enzyme has an isoelectric point of 4.8 and an extinction coefficient $E_{1\,cm}^{1\%} = 15.2$ (280 nm).[9]

Distribution. The enzyme has been found in liver, kidney, heart, and muscle of horse, cattle, pig, rat, and cow. The highest content is found in pig kidney. Intracellular distribution has been studied in rat liver, where the activity is exclusively located in the microsomal–lysosomal fraction. In kidney the enzyme is found in the soluble fraction.[10]

[48] Pantothenase from *Pseudomonas fluorescens*

By R. KALERVO AIRAS

Pantothenase (EC. 3.5.1.22) is a bacterial amidase that specifically decomposes pantothenic acid to β-alanine and pantoic acid. It is the first enzyme in the degradative metabolism of pantothenate, shown in Scheme 1[1–5]:

The utilization of pantothenic acid as a carbon source of bacteria was reported by Metzger,[6] who found, in soil samples, some bacterial strains growing with pantothenate as the sole source of carbon. All these were pseudomonads. In the classification of Stanier *et al.*[7] nearly all the pseudomonads utilizing pantothenate fall into the biotype C of *Pseudomonas fluorescens*.

[1] P. Mäntsälä, *Acta Chem. Scand.* **26,** 127 (1972).

[2] V. Nurmikko, E. Salo, H. Hakola, K. Mäkinen, and E. E. Snell, *Biochemistry* **5,** 399 (1966).

[3] C. T. Goodhue and E. E. Snell, *Biochemistry* **5,** 403 (1966).

[4] P. T. Magee and E. E. Snell, *Biochemistry* **5,** 409 (1966).

[5] O. Hayaishi, Y. Nishizuka, M. Tatibana, M. Takeshita, and S. Kuno, *J. Biol. Chem.* **236,** 781 (1961).

[6] W. I. Metzger, *J. Bacteriol.* **54,** 135 (1947).

[7] R. Y. Stanier, N. J. Palleroni, and M. Doudoroff, *J. Gen. Microbiol.* **43,** 159 (1966).

METHODS IN ENZYMOLOGY, VOL. 62

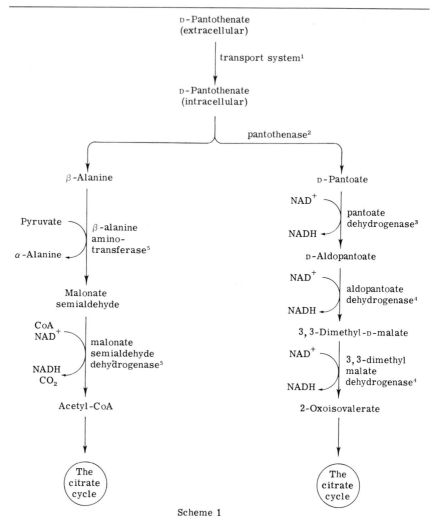

Scheme 1

Assay Methods

Pantothenase activity can be measured either by measuring with ninhydrin the amino group liberated during the reaction or by using [1-^{14}C]pantothenic acid as substrate and measuring the radioactive β-alanine.[8,9] The latter method is much more exact, and only it is described here in detail.

[8] R. K. Airas, *Biochem. J.* **130,** 111 (1972).
[9] R. K. Airas, *Biochem. J.* **157,** 415 (1976).

Principle. The method is based on the reaction:

D-[1-^{14}C]Pantothenate → β-[1-^{14}C]alanine + D-pantoate

The formed β-alanine is separated from the pantothenate with paper chromatography, and the radioactivity of the β-alanine spot is counted.

Reagents

Buffer: 100 mM potassium phosphate, pH 7.4

Substrate: 150 mM pantothenate, potassium salt. From calcium pantothenate, the calcium ion is removed with an ion exchanger. For instance, 6 mmol (2.86 g) of calcium pantothenate are dissolved in 40 ml of water. The solution is passed through a column (2 × 10 cm) of Dowex 50. Fifty milliliters of the solution containing pantothenic acid is collected, and the acid is titrated to pH 7 with 5 M KOH. The final concentration of pantothenate is adjusted to 150 mM according to the amount of KOH used up in titration.

Radioactive substrate: D-[1-^{14}C]pantothenate is diluted in water to give 2 × 10^7 dpm/ml. The background caused by free β-alanine can be remarkably diminished, if the radioactive substrate solution is passed through a small column (1 × 5 cm) of Dowex 50.

The reagents are stored at −20°.

Procedure. The reaction mixture (125 μl) contains 25 μl (or equal amounts) each of buffer, substrate solution, radioactive substrate solution, water, and enzyme solution. At zero time the enzyme solution (or substrate and radioactive substrate) is pipetted to the reaction mixture and the tube is transferred from 0° to a water bath at 20°. After 30 min the reaction is stopped by pipetting a 25-μl sample of the reaction mixture on to Whatman No. 1 chromatography paper located 1 cm above the surface of 30% formic acid. The chromatograms are developed in butan-2-one/methanol/water/36% HCl (60:15:15:1). R_f values: β-alanine, 0.30; pantothenate, 0.85. Strips of the chromatograms from R_f 0.1 to 0.5, containing the β-alanine spots, are cut off and immersed in a toluene-based scintillation liquid and counted for radioactivity. β-Alanine is not soluble in toluene, so the same scintillation liquid can be used repeatedly after removing the counted paper strips.

Nearly any buffer between pH 6.5 and 9 is suitable for the reaction mixture, except borate. Cl$^-$ ions should be avoided because of the inhibiting effect. Low buffer concentrations should be avoided as well, because of the increasing β-alanine–β-[^{14}C]alanine exchange reaction (see later). The exchange reaction can also be avoided if 25 μl of 50 mM β-alanine are added into the reaction mixture instead of water.

Purification Procedure

Cultivation of Pseudomonas fluorescens

Pantothenate is the inducer of pantothenase of *Pseudomonas fluorescens,* and the enzyme is repressible with many carbon sources.[10] The highest yield of pantothenase can be attained if the bacterium is grown on pantothenate as the sole source of carbon and nitrogen. However, a more rapid growth and nearly as high specific activity of pantothenase can be reached if glucose is an additional carbon source, pantothenate being the sole source of nitrogen.

The medium used for growing *Pseudomonas fluorescens* UK-1 is as follows: 10 mM potassium pantothenate, 10 mM glucose, 20 mM potassium phosphate (pH 7.0), 1 mM $MgSO_4$, and 10 μM $FeSO_4$. For 1 liter of the medium 5 mmol of calcium pantothenate (giving 10 mM pantothenate) is dissolved in about 30 ml of water, and 6 mmol of K_2CO_3 is dissolved in another 30 ml of water. The solutions are warmed to about 70° and then mixed together. The formed calcium carbonate precipitate is filtered off, 20 mmol of KH_2PO_4 are dissolved in the filtrate, and the excess CO_2 is removed with suction. The volume is then adjusted to 1 liter, the $MgSO_4$ and $FeSO_4$ are added, and the pH is adjusted to 7 with 2.3 ml of 5 M KOH.

The growth temperature is 25°. Aeration with an air pump or stirrer is necessary. The cells must be gathered during the exponential growth because pantothenase is very rapidly inactivated after cessation of growth.

The samples for pantothenase assays are taken from the bacterial culture as follows: The sample containing 5 to 50 μg of cell dry weight is centrifuged, and the bacterial pellet is suspended in 50 μl of 5 mM potassium phosphate, pH 7.4. The sample is stored at $-20°$. The enzyme is made accessible to substrates by freezing and thawing the samples four times.

Enzyme Isolation

Two methods have been described for the purification of pantothenase.[11,12] The method described here in detail is a modification of one of them.[12] The procedure is summarized in the table.
   Reagents
      Piperazine-HCl buffer, 20 mM, pH 5.9
      Potassium phosphate buffer, 40 mM, pH 6.8
      DEAE-Sephadex A-50 (Pharmacia)

[10] P. Mäntsälä and V. Nurmikko, *Suom. Kemistilehti* **B43,** 414 (1970).
[11] J. Puisto and V. Nurmikko, *Suom. Kemistilehti* **B43,** 44 (1970).
[12] R. K. Airas, E. A. Hietanen, V. T. Nurmikko, *Biochem. J.* **157,** 409 (1976).

PURIFICATION OF PANTOTHENASE FROM *Pseudomonas fluorescens*

Purification step	Total protein (mg)	Total activity (U)	Yield (%)	Specific activity (U/mg)
Crude extract	1040	247	100	0.238
Ammonium sulfate fractionation	561	158	64	0.282
DEAE-Sephadex	77	87	35	1.13
Affinity column	8.22	19.0	7.7	4.33
Hydroxyapatite	1.07	15.8	6.4	14.8

Hydroxyapatite (Sigma)
Affinity gel: 3-aminobenzene boronic acid bound to succinyl aminoethyl Sepharose 4B

Ethylene diamine is coupled to Sepharose 4B (Pharmacia) through cyanogen bromide activation.[13] The succinyl group is then coupled to the aminoethyl Sepharose 4B,[14] and finally 3-aminobenzene boronic acid is coupled with carbodiimide reaction[15] [1-ethyl-(3-dimethylaminopropyl)carbodiimide (EDC) as the coupling reagent].

Step 1. Ammonium Sulfate Precipitation. The entire procedure is carried out at 0–6°. The centrifuged cell mass (about 50 g) is suspended in two volumes of piperazine buffer, and the suspension is pressed through a French press. The resulting solution is made $1 M$ with ammonium sulfate and after 30 min is centrifuged at 15,000 g for 15 min. The supernatant is then made 1.8 M with ammonium sulfate and let stand for 1 hr. The precipitate is gathered by centrifugation at 15,000 g for 15 min, dissolved in 100 ml of the piperazine buffer, and then dialyzed overnight against 3 liters of 20 mM KCl.

Step 2. Ion-Exchange Chromatography on DEAE-Sephadex. A column containing 4 g of DEAE-Sephadex A-50 is equilibrated with the piperazine buffer. The dialyzed enzyme solution is applied to the column, and the first, turbid protein peak is eluted from the column with about 400 ml of the piperazine buffer. Thereafter elution with a linear gradient of

[13] I. Parikh, S. March, and P. Cuatrecasas, this series, Vol. 34, p. 79.
[14] P. Cuatrecasas, *J. Biol. Chem.* **245,** 1327 (1970).
[15] H. F. Hixson and A. H. Nishikawa, this series, Vol. 34, p. 446.

ammonium sulfate is started. The mixing chamber contains 300 ml of the piperazine buffer, and 300 ml of 0.5 M (NH$_4$)$_2$SO$_4$ in the piperazine buffer is led into it. Pantothenase will be eluted at about 0.2–0.25 M ammonium sulfate. The fractions containing pantothenase are pooled, and pantothenase is precipitated by making the solution 2.2 M in ammonium sulfate. After standing overnight the precipitate is collected by centrifugation at 15,000 g for 10 min. The precipitate is then dissolved in 50 ml of 40 mM potassium phosphate, pH 6.8.

Step 3. Affinity Chromatography. About 30 ml of the affinity gel is washed with 1 liter of 1 M NaCl, with 2 liters of water, and finally with 500 ml of 40 mM potassium phosphate, pH 6.8, before it is packed into a column. The pantothenase sample from step 2 is applied to the column, and elution is carried out with the potassium phosphate buffer until the first protein peak has come out from the column (about 150 ml of the phosphate buffer). Pantothenase is then eluted from the column with 50 mM NaCl in 40 mM potassium phosphate, pH 6.8.

Step 4. Hydroxyapatite Chromatography. Hydroxyapatite is packed into a column (3 × 1.5 cm), and 30 ml of 40 mM potassium phosphate, pH 6.8, is passed through the column. The pooled pantothenase-containing fractions from step 3 are applied. Pantothenase passes through the column without attaching to the gel. The pantothenase-containing fractions are collected, and the enzyme is stored at −20°.

Properties

Purity. The purified pantothenase exhibits a single band in disc-gel electrophoresis, sodium dodecyl sulfate–acrylamide electrophoresis and electrofocusing on acrylamide gel. The disc-gel band of thermally (at 40°) inactivated pantothenase is identical with that of the native enzyme, and thus the final purified pantothenase may contain an unknown amount of inactivated forms of the enzyme.

Physical Properties. Using gel filtration on Ultrogel AcA 34, a molecular weight of 97,400 ± 4800 has been obtained.[12] The subunit molecular weight obtained by means of sodium dodecyl sulfate–acrylamide gel electrophoresis was 50,000 ± 700. Pantothenase thus seems to be a dimer having a molecular weight of about 100,000.

The isoelectric point determined by means of electrofocusing in polyacrylamide gels[12] is 4.7 at 10°.

Stability. Pantothenase is stable between pH 5.5 and 9.5. The enzyme loses some activity on freezing; in 20 mM potassium phosphate (pH 7) the decrease of activity is about 20% per round of freezing and thawing. Pantothenate and oxalate (an inhibitor) protect against inactivation. Frozen enzyme retains its activity at −20° for at least 2 years.

Above 30° the enzyme is rapidly inactivated. At 35° in 20 mM potassium phosphate buffer, the inactivation rate is 7.2% per minute.[16] The activation energy of inactivation is 220 kJ/mol. Owing to the thermal inactivation, the assay temperature of pantothenase has normally been chosen as low as 20°.

Many carboxylic acids, e.g., oxamate, 3-oxoglutarate, 2-oxomalonate, oxalate, and oxaloacetate, protect the enzyme against thermal inactivation.[16] The protection constants of these are generally slightly lower than the inhibition constants of the same compounds. The activation energy for inactivation of the enzyme is somewhat higher in the presence of these compounds than without them.

Specificity. Pantothenase specifically decomposes D-pantothenate.[17] In addition, it slowly decomposes hydroxypantothenate with an additional OH group in the pantoate part, and pantoyl-γ-aminobutyrate with γ-aminobutyrate instead of β-alanine. A number of other structural relatives of pantothenate remain undecomposed.[17]

Kinetic Data. The K_m value of pantothenase for pantothenate in phosphate buffer at pH 7.0 for 20° is 15 mM[9]; this is dependent on pH. The pK value of the ionizable group at the substrate binding site is 7.0, the basic form of this group being active enzymically. The pH-independent (theoretical) K_m is 7.6 mM. Both the K_m and the pK are dependent on temperature, the van't Hoff enthalpy of substrate binding being -14 kJ/mol, and that of H$^+$ ion binding being -50 kJ/mol.[9]

Pantothenase is inhibited by a series of carboxylic acids.[16] The inhibition with oxalate and oxaloacetate is of the noncompetitive type and depends on pH.[9] The pH dependence suggests an ionizable group with a pK of 6.4 at the inhibitor binding site, the inhibitor binding to the acidic form of the ionizable group. The theoretical, pH-independent K_i of the oxalate inhibition is 0.36 mM, and the corresponding K_i at pH 7.0 is 1.8 mM.

When the substrate binding, inhibitor binding, and ionizations at the substrate and inhibitor binding sites are taken into account, a reaction velocity equation can be formulated for pantothenase:

$$v = \frac{V}{\left[1 + \dfrac{K_m(1 + \text{H}/K_1)}{s}\right]\left[1 + \dfrac{i}{K_i(1 + K_2/\text{H})}\right]}$$

where K_1 is the H$^+$ ion dissociation constant at the substrate binding site and K_2 is the corresponding constant at the inhibitor binding site. This equation explains quite well the velocity dependences of the enzyme when v vs. s, v vs. pH, K_m vs. pH, v vs. i or K_i vs. pH are measured between pH 6 and 8.

[16] R. K. Airas, *Biochim. Biophys. Acta* **452**, 193 (1976).
[17] P. Mäntsälä and V. Nurmikko, *Suom. Kemistilehti* **B43**, 47 (1970).

Reaction Mechanism. The following reaction sequence is suggested for pantothenase:

$$E + S \rightleftharpoons ES \rightleftharpoons E'PB \underset{B}{\longleftrightarrow} E'P \longrightarrow E + P$$

where B represents β-alanine, and P represents pantoate. The scheme assumes the existence of acyl enzyme (E'P), and the irreversible stage is assumed to be the hydrolysis of the acyl enzyme. The reaction sequence is proved by the existence of a β-alanine–β-[^{14}C]alanine exchange reaction: In the presence of substrate—and only in the presence of substrate—radioactive β-alanine is transferred into pantothenate. The backward reaction expresses itself also as a curved reaction velocity curve. When the formed β-alanine goes with increasing rate back into pantothenate, the measured reaction becomes lower.

The backward reaction of β-alanine is affected by the buffer concentration in the reaction mixture, or, more exactly, by the amount of that form of the buffering agent with bound hydrogen ion (AH form). With high AH concentrations, the β-alanine exchange reaction is diminished.

The β-alanine exchange reaction disturbs also the normal pantothenase assay by causing nonlinear reaction velocity curves. This can be avoided by using high buffer concentrations, rather on the acidic side of the buffer pK value. The disturbance can also be avoided by adding nonradioactive β-alanine into the reaction mixture. The formed radioactive β-alanine is then diluted by the nonradioactive compound, and it is returned in smaller amounts. The β-alanine exchange reaction offers an opportunity of preparing radioactively labeled pantothenate from nonradioactive pantothenate and labeled β-alanine.

Reactivation after Thermal Inactivation. A special property of pantothenase is a nearly total reactivation after thermal inactivation *in vivo*.[8] The reactivation occurs when the heat-treated (1 hr at 40°) bacterial cells, grown on pantothenate, are grown at 25° on good carbon sources, other than pantothenate. Chloramphenicol does not prevent the reactivation. Some reactivation occurs also *in vitro,* but not at all so complete as *in vivo*.[18]

Similarities to Other Enzymes. Pantothenase has some common properties with other amidases, especially with the serine proteases. The pH dependences of the enzymic reactions are similar, the activities being affected by an ionizable group with a pK value near 7. Phenylmethane sulfonyl fluoride, an inhibitor of serine proteases, is also capable of inactivating pantothenase. Pantothenase is competitively inhibited by *m*-

[18] R. K. Airas, *Biochim. Biophys. Acta* **452,** 201 (1976).

aminobenzene boronic acid with a K_i value of 0.1 mM at pH 7. For chymotrypsin[19] and subtilisin[20] the corresponding K_i values are of same order of magnitude.

[19] K. A. Koehler and G. E. Lienhard, *Biochemistry* **10,** 2477 (1971).
[20] R. N. Lindqvist and C. Terry, *Arch. Biochem. Biophys.* **160,** 135 (1974)

Section V

Biotin and Derivatives

[49] Isotopic Dilution Assay for Biotin: Use of [14C]Biotin

By ROSS L. HOOD

Principle. Biotin binds stoichiometrically to avidin to form a stable biotin–avidin complex that can be precipitated, leaving any excess biotin in the supernatant. The isotope dilution assay described is based on the competition between a known quantity of radioactive biotin and an unknown quantity of nonradioactive biotin for the binding site on avidin.[1,2] After precipitation of the avidin–biotin complex, the amount of excess [14C]biotin remaining in the supernatant will depend on the dilution of [14C]biotin by the nonradioactive biotin. The unknown quantity of biotin is then calculated by linear regression analysis, a technique that incorporates a correction factor for quenching in liquid scintillation counting systems and a level of confidence for the reliability of the assay.

Sensitivity. Using [14C]biotin (45 mCi/mmol), the assay is sensitive to 1 ng of biotin per assay. The same procedure can be used with [3H]biotin (2.5 Ci/mmol); the method is then sensitive to 0.05 ng of biotin per assay.

Reagents

d-[carbonyl-14C]Biotin (45 mCi/mmol)
d-Biotin
Avidin (0.54 unit/mg)
NaCl, 0.9%
H_2SO_4, 1 M
NaOH, 20%
$ZnSO_4$, 2%
NaOH, 0.1 M
Scintillation solution suitable for measuring radioactivity in aqueous solutions

[14C]*Biotin Calibration.* A solution of d-[carbonyl-14C]biotin (0.02 μg of biotin/100 μl) is prepared in 0.9% NaCl. Final calibration is achieved by preparing a nonradioactive biotin standard (0.025 μg/100 μl) and checking the concentration of the [14C]biotin against the standard, using the assay procedure described below.

Avidin Preparation. A solution of avidin in 0.9% NaCl (avidin solution 5, Table I) is prepared such that the avidin in it binds 80–90% of the [14C]biotin, when 50 μl of the [14C]biotin standard is added to 100 μl of solution 5. Avidin solutions 0 to 4 are prepared by quantitative dilution of

[1] R. L. Hood, *J. Sci. Food Agric.* **26**, 1847 (1975).

[2] R. L. Hood, *Anal. Biochem.* **79**, 635 (1977).

METHODS IN ENZYMOLOGY, VOL. 62

TABLE I
REAGENTS FOR A TYPICAL BIOTIN ASSAY[a]

	Blank					Assay			
Centrifuge tube	[14C]Biotin (μl)	0.9% NaCl (μl)	Avidin solution[b] No.	μl	Centrifuge tube	[14C]Biotin (μl)	Biotin unknown (μl)	Avidin solution[b] No.	μl
B_0	50	50	0	100	A_0	50	50	0	100
B_1	50	50	1	100	A_1	50	50	1	100
B_2	50	50	2	100	A_2	50	50	2	100
B_3	50	50	3	100	A_3	50	50	3	100
B_4	50	50	4	100	A_4	50	50	4	100
B_5	50	50	5	100	A_5	50	50	5	100

[a] From R. L. Hood, Anal. Biochem. **79**, 635 (1977).
[b] The procedure for the preparation of the avidin solutions is described in the text.

avidin solution 5 with 0.9% NaCl to yield concentrations that are comparable to that in solution 5; e.g., the concentration of avidin in solution 5 is
five times that of solution 1.

Acid Hydrolysis of Biological Material. Tissue samples (2 g) are homogenized in 10 ml of 1 M H_2SO_4, using an Ultra Turrax homogenizer,
and then hydrolyzed by autoclaving at 15 psi for 1 hr. Animal feeds and
feed ingredients are ground in a Wiley mill fitted with a 1-mm screen, and
10 g of the feed are hydrolyzed in 80 ml of 1 M H_2SO_4.

Assay Procedure. The reagents that are added to a series of blank and
assay tubes (15-ml disposable centrifuge tubes) are listed in Table I. The
avidin solutions are added only after the two biotin solutions are
thoroughly mixed. In this example (Table I), 50 μl of a nonradioactive
biotin standard is added to tubes A_0 to A_5. However, in the assay of biological samples with a low biotin content, this volume can be increased,
and an equal volume of 0.9% NaCl is then added to tubes B_0 to B_5. The
final volume in each series of tubes is not important, although it should be
kept to a minimum for convenience. The tubes are stoppered and incubated with shaking for 10 min in a water bath at 37°. Two hundred microliters of 2% Zn SO_4 are added, and the solution is mixed, then 200 μl of 0.1
M NaOH are added. The tubes are centrifuged for 5 min and equal aliquots (maximum volume without disturbing the precipitate) of the supernatant are transferred from each tube to a scintillation vial for quantitation
of the [^{14}C]biotin which remains unbound in solution. Sufficient commercial scintillator to solubilize the aqueous supernatant is added to each vial,
and the radioactivity is determined with a liquid scintillation spectrometer.

Calculations. The quantity of biotin in an unknown sample is calculated, using Eqs. (1) and (2), from the slope and intercept of the straight
line determined by linear regression analysis on the relative avidin concentration (X) and the counts per minute in the supernatant (Y) (Table II).

$$\text{Amount (ng) of unknown biotin} = C \times (1 - Z)/Z \qquad (1)$$

where

$$Z = [(A_s)(B_I)]/[(B_s)(A_I)] \qquad (2)$$

B_s and A_s are the slopes of the regression lines for the blank series (B_0 to
B_5) and assay series (A_0 to A_5) of tubes, respectively; B_I and A_I are the in-
tercepts calculated for the respective regression lines when the relative
avidin concentration is zero; C is the amount (ng) of [^{14}C]biotin added to
each tube. Examples of the calculations are shown in Table II.

Comments. In the biotin assay, the amount of [^{14}C]biotin in the supernatant, after precipitation of the avidin–biotin complex, is dependent on

TABLE II

RESULTS FROM AN ISOTOPE DILUTION ASSAY FOR BIOTIN[a]

Concentration of avidin[b]	Radioactivity in supernatant (cpm $\times$ 10^{-3})			
	Blank	Standard biotin	Standard biotin quenched sample[c]	Standard biotin: 20% error in one determination
0	2.939	2.990	2.634	2.990
1	2.509	2.752	2.425	2.752
2	2.061	2.575	2.269	2.060
3	1.487	2.339	2.061	2.339
4	1.049	2.157	1.900	2.157
5	0.595	1.951	1.718	1.951

Mathematical parameters from above results

Correlation coefficient	−0.999	−0.999	−0.999	−0.870
Slope	−0.476	−0.206	−0.182	−0.191
Intercept, $x = 0$	2.964	2.976	2.622	2.854
Significance	0.0001	0.0001	0.0001	0.05
[14C]Biotin added (ng)	9.5	9.5	9.5	9.5
z uncorrected[d]		0.433	0.382	0.401
z quench corrected[e]		0.431	0.432	0.416
Biotin (ng)		12.5	12.5	13.3

[a] From R. L. Hood, *Anal. Biochem.* **79**, 635 (1977).
[b] The units are arbitrary, and each solution was prepared by dilution from a common solution.
[c] Color quenched with 500 μl of liver digest. The biotin was removed from the digest by treatment with avidin.
[d] z uncorrected = A_s/B_s.
[e] z quench corrected is calculated from Eq. (2).

the dilution of the [14C]biotin by the biotin unknown. This dilution and the linear binding relationship between avidin and biotin are shown in Fig. 1.

The effects of quenching on the regression analysis are also shown in Fig. 1 and Table II. Quenching diminishes the slope of the regression line and causes an error in the calculation of Z (Table II), unless corrected for by taking into account the value of the intercept when the avidin concentration is zero. This correction (B_I/A_I) adjusts the counting efficiency to that obtained from the blank series of assay tubes, and then the results for tubes B_0-B_5 become directly comparable to those for tubes A_0-A_5. If this correction had not been made in the results of the quenched biotin standard in Table II, a result of 15.4 ng of biotin, rather than 12.5 ng, would have been obtained.

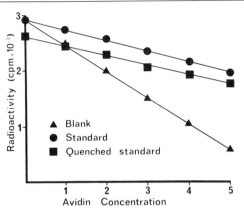

FIG. 1. Effect of quenching on an isotope dilution assay for biotin [R. L. Hood, *Anal. Biochem.* **79**, 635 (1977)].

Variation in the avidin concentration within a series of six determinations provides a built-in assessment of the reliability of the values used in the regression analysis. By using six avidin concentrations, correlation coefficients better than -0.99 are routinely obtained. If the correlation coefficient is not better than that routinely obtained, the regression analysis should be repeated after eliminating a value that is considered incorrect or has been calculated by a statistical procedure to fall outside desired confidence limits. If, after elimination of the value, the correlation coefficient is not better than that routinely obtained, the assay requires repeating. An example of the results expected after a 20% error was simulated in one determination is shown in Table II.

A recovery of 98.1% $\pm$ 2.1% was obtained when 10 μg of biotin was added to six tubes before hydrolysis.[1] Results obtained from biological material are more reproducible from the same digest than from separate digests of the same material. This is particularly true for animal feeds and is considered to be a problem of sampling, not of the actual assay procedure. A widely used microbiological assay using *Lactobacillus plantarum* as the test organism gave similar results to the radiochemical assay, whether the assays were performed with the same digest or different digests.[1]

When [³H]biotin of high specific activity is substituted for [¹⁴C]biotin in the assay procedure, the method is suitable for the determination of biotin levels in blood serum or plasma.

[50] Isotope Dilution Assay for Biotin: Use of [³H]Biotin

By K. Dakshinamurti and L. Allan

Avidin binds to biotin forming a complex. The specificity and stability of this binding has long been recognized.[1] The stability of the avidin–biotin complex is increased in a medium of high salt concentration. Both avidin and the avidin–biotin complex can be adsorbed on bentonite.[2] If radioactive biotin is allowed to compete with nonradioactive biotin for binding to avidin of known specific activity, the radioactivity of the avidin–biotin complex will depend on the extent of dilution of radioactive biotin by cold biotin. Based on this principle, a method for the assay of biotin, sensitive in the assay range up to 20 ng per assay tube, was developed.[3] The specific activity of the available [carbonyl-¹⁴C]biotin was the limiting factor in regard to the sensitivity of the method. The procedure described uses tritiated biotin of higher specific activity and a concentration step with activated charcoal to increase the sensitivity.

Method

Preparation of Bentonite Suspension. The method is essentially the same as that of Fraenkel-Conrat.[1] Bentonite (20 g) is suspended in 500 ml of distilled water and stirred vigorously for 1–2 hr until a homogeneous suspension is obtained. This is centrifuged at 3000 g for 15 min, and the supernatant is further centrifuged at 10,000 g for 15 min. The pellet is collected and suspended in 20 ml of 0.2 M ammonium carbonate. Bentonite concentration is determined by drying an aliqot of the suspension at 60° overnight and weighing the residue. Suitable dilution is made to give a suspension containing 10 mg of bentonite per milliliter. The suspension can be kept for several weeks and does not precipitate.

Standardization of Avidin. The dye-binding method of Green[4] is used to determine the specific activity of the avidin preparation. The working standard solution of avidin is made to a concentration of 0.033 unit/ml.

Determination of the Equivalence between Avidin and d-[³H]Biotin

---

[1] H. Fraenkel-Conrat, B. Singer, and A. Tsugita, *Virology* **14,** 54 (1961).

[2] R. D. Wei and L. D. Wright, *Proc. Soc. Exp. Biol. Med.* **117,** 17 (1964).

[3] K. Dakshinamurti, A. D. Landman, L. Ramamurti, and R. J. Constable, *Anal. Biochem.* **61,** 225 (1974).

[4] N. M. Green, *Biochem. J.* **94,** 23C (1965).

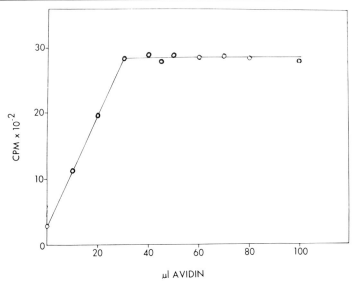

FIG. 1. Determination of the equivalence between avidin and [³H]biotin.

Solution. In a series of Eppendorf centrifuge tubes are placed 1.0 ml of 0.2 M ammonium carbonate solution, 25 μliters of [³H]biotin (2.4 Ci/mmol) solution (4200 cpm, 1.0 ng), and 200 μliters of 8 M sodium sulfate. Varying volumes (0–100 μliters) of the standard avidin solution are added to the tubes and allowed to mix for 5 min in an Eppendorf shaker at room temperature. After this, 0.3 ml of the bentonite suspension is added; the tubes are kept in the shaker for another 10 min and filtered on a Millipore filter (GSWPO2500) GS with a pore sixe of 0.22 μm. The filter is washed with 5 ml of 0.2 M ammonium carbonate, dried, wetted with a drop of 95% ethanol, and transferred to a scintillation vial containing 15 ml of the scintillator[5] containing 0.25% 2,5-diphenyloxazole, 0.009% 1,4-bis-2-(5-phenyloxazolyl)benzene, and 31% ethanol in toluene. Radioactivity is determined using a Beckman LS 250 Liquid Scintillation Spectrometer. Figure 1 is a representative curve used to determine the equivalence between avidin and biotin. From the curve it is seen that 30 μl of the avidin solution (1.0 munit) completely complexes with the [³H]biotin in 25 μl of the solution added. From this the concentration of [³H]biotin is calculated to be 40 ng/ml.

Biotin Calibration Plot. To a series of Eppendorf centrifuge tubes are added 1.0 ml of 0.2 M ammonium carbonate, 200 μl of [³H]biotin (4200 cpm, 1 ng), and increasing amounts (0–2 ng) of cold biotin in a volume of

[5] K. Dakshinamurti and P. R. Desjardins, *Can. J. Biochem.* **46**, 1261 (1968).

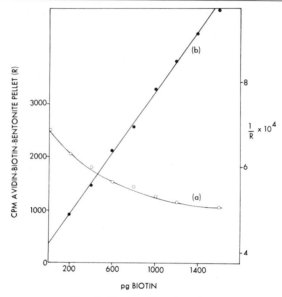

FIG. 2. Biotin calibration plot.

50 μl. After mixing, 30 μl of the avidin working standard (which will bind 1 ng of biotin) are added. After mixing in a shaker for 5 min, 0.3 ml of the bentonite suspension is added and mixed for a further 10 min. The precipitate of the avidin–biotin complex adsorbed on the bentonite is filtered through the Millipore filter as described earlier and washed; the radioactivity is determined in a liquid scintillation counter.

The biotin calibration curve is given in Fig. 2; curve (a) is a plot of the radioactivity of the avidin–biotin–bentonite pellet against the biotin content of the assay tube. According to the isotope dilution principle $1/R = 1/R_0 + 1/b \cdot R_0 X$ where R_0 is the total radioactive biotin (cpm) bound by avidin in the absence of any cold biotin, R is the radioactivity (cpm) bound by the same amount of avidin in the presence of X ng of cold biotin, and b is the amount (ng) of [³H]biotin used in the assay. Curve (b) is a plot of $1/R$ vs. X. The calibration curve varies very little from day to day. The isotope dilution relationship can also be used to calculate b from the values for R, R_0, and X. This serves as an internal cross check of the concentration of radioactive biotin.

Hydrolysis of Tissue to Release Bound Biotin. Varying amounts (150–200 mg) of freeze-dried tissue or 1.0 ml of serum containing up to 5 ng of biotin are hydrolyzed with 1.0 ml of 4.5 N H$_2$SO$_4$ by autoclaving at 15 psi for 1 hr. The hydrolyzate is neutralized with 1 ml of 4.5 N NaOH, made up to 5.0 ml and filtered on a coarse-grade filter paper.

In case of material having a lower biotin content, a larger amount of the starting material is used. This is followed by concentration of the biotin in the hydrolyzate. Up to 10 g of the original material can be used and the filtered hydrolyzate diluted to 100 ml with water. To this 1 g of activated charcoal (Norit A) is added. The mixture is shaken for 2 hr at room temperature and centrifuged. The pelleted charcoal is washed by resuspension in 8 ml of water, mixed for 30 min, and centrifuged. The washed pellet is extracted sequentially with six 2-ml portions of a mixture of ethanol–water–ammonia (10:10:1). The extracts are pooled and the biotin content is determined.

Limit of Determination. This method can be applied to the determination of biotin in the range 0–1.0 ng per assay tube, which is comparable to the range of the microbiological assay with *Lactobacillus plantarum* ATCC 8014.

Application. The isotope dilution procedure is sensitive enough to be used for the assay of biotin in biological material such as tissues and serum.

[51] A Rapid, Sensitive Fluorometric Assay for Avidin and Biotin

By HENRY J. LIN and JACK F. KIRSCH

Principle. The binding of biotin to avidin is accompanied by a decrease in the fluorescence intensity of the protein.[1,2] These observations, combined with the known high affinity of avidin for biotin, led to the development of the assay described. Although not designed for biological fluids containing fluorescent impurities, the present method provides the most rapid and sensitive assay for samples containing relatively pure material. It is particularly convenient in that commercial avidin and biotin can be used directly. No radioisotopic or chemical syntheses are required.[3]

Reagents

Sodium phosphate buffer, 0.15 M, pH 7.0

Biotin standard (Sigma), ca. 20 μg/ml (80 μM)

Avidin standard (Worthington), ca. 250 μg/ml (3.7 μM)

Procedure. All solutions are prepared in 0.15 M sodium phosphate buffer at pH 7, ionic strength = 0.35. The fluorescence measurements

[1] N. M. Green, *Biochem J.* **90,** 564 (1964).

[2] C. F. Chignell, D. K. Starkweather, and B. K. Sinha, *J. Biol. Chem.* **250,** 5622 (1975).

[3] H. J. Lin and J. F. Kirsch, *Anal. Biochem.* **81,** 442 (1977).

METHODS IN ENZYMOLOGY, VOL. 62

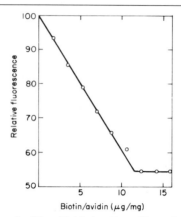

Biotin/avidin (μg/mg)

FIG. 1. Typical titration of avidin with biotin. About 25 μg of avidin in 2.0 ml (180 nM) of 0.15 M sodium phosphate buffer, pH 7, was titrated with 2.0-μl aliquots of a 90 μM (22 μg/ml) solution of biotin. Fluorescence was measured at 350 nm with excitation at 290 nm. Each avidin molecule is tetrameric and binds four molecules of biotin [N. M. Green, *Biochem. J.* **92**, 16C (1964)]. Data of H. J. Lin and J. F. Kirsch, *Anal. Biochem.* **81**, 442 (1977).

described were made with a Hitachi Perkin–Elmer spectrofluorometer (Model MPF-2A) whose emission monochromator was calibrated with indole (Aldrich Chemical Company, m.p. 52–53°).[4] All measurements were made in the ratio recording mode with the temperature between 20 and 24°, the excitation monochromator set at 290 nm using a bandwidth of 4 nm, and the fluorescence monitored after each addition of titrant at 350 nm with a bandwidth of 16 nm.

Assay of Avidin. A standard solution of biotin is prepared by dissolving 5 mg in 5.0 ml of buffer followed by dilution to 20 μg/ml. Two milliliters of approximately 10 μg/ml of avidin solution are pipetted into a 4-ml fluorometer cell and titrated with successive 2.0-μl aliquots of the standard 20-μg/ml biotin solution using a syringe microburete (e.g., Micrometric Instrument Co., Model SB2). Additions are made until no further decrease in fluorescence is observed, indicating the attainment of the equivalence point (Fig. 1).

Assay of Biotin. Typically, a solution of avidin standardized with biotin as described above is diluted to 250 μg/ml and is used to titrate 2 ml of a solution that contains an unknown amount of biotin in the fluorometer cell. Relative fluorescence is measured after each addition of 2.0-μl aliquots of the avidin solution from a syringe microburette, and a plot is made of fluorescence versus volume of avidin solution. The point at which the slope undergoes a sharp increase indicates the equivalence

[4] I. B. Berlman, "Handbook of Fluorescence Spectra of Aromatic Molecules," 1st ed., p. 101. Academic Press, New York, 1965.

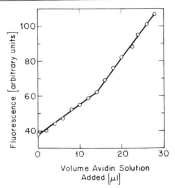

FIG. 2. Typical titration of biotin and avidin. Biotin, 33 ng in 2.0 ml (68 nM) of 0.15 M sodium phosphate buffer, pH 7, was titrated with 2.0-μl aliquots of a solution of 250 μg of avidin (3.7 μM) per milliliter. Fluorescence was measured at 350 nm with excitation at 290 nm. Data of H. J. Lin and J. F. Kirsch *Anal. Biochem.* **81**, 442 (1977).

point, and the quantity of biotin is determined from the specific activity of the avidin solution (Fig. 2).

Notes. The useful range of the biotin assay is 20 to 120 ng. Between these limits, dilution upon addition of avidin will be less than 5%. Use of a wider emission slit may improve sensitivity. Ionic strength is not critical, since identical results have been obtained with 0.25 M sodium phosphate buffer, in which ionic strength = 0.6.[3] The general applicability of these assays may be limited by the need to obtain samples nearly free of other fluorescent species, such as protein and free tryptophan. To achieve maximum sensitivity, the tryptophan concentration, for example, should not greatly exceed 1–2 μM. A few other proteins are known to bind biotin, such as streptavidin[5] and a recently described egg yolk protein,[6] but avidin is the only biotin-binding protein that has been found in egg white. For samples from which interfering substances are absent, the assay is rapid and extremely parsimonious, since only microgram quantities of avidin are required to bind 20 ng of biotin.

[5] L. Chaiet and F. J. Wolf, *Arch. Biochem. Biophys.* **106**, 1 (1964).
[6] H. G. White, B. A. Dennison, M. A. Della Fera, C. J. Whitney, J. C. McGuire, H. W. Meslar, and P. H. Sammelwitz, *Biochem. J.* *157*, 395 (1976).

[52] Microassay of Avidin

By HEIKKI A. ELO and PENTTI J. TUOHIMAA

Avidin is a biotin-binding protein first detected in the egg white of the hen[1] and later in the avian and amphibian oviduct.[2] The study of avidin induction has been extensive since avidin induction by progesterone in the chick oviduct is one of the best models for study of the mechanism of progesterone action.[3]

Many methods for avidin assay have been developed.[1,4-9] Korenman and O'Malley[7] introduced a rapid assay for chick oviduct avidin in which the 105,000 g supernatant solution was used and the avidin–[14C]biotin complex was adsorbed onto bentonite. After filtering and rinsing the avidin–[14C]biotin–bentonite complex on a Millipore filter, the radioactivity was counted in a liquid scintillation counter. We give here two modified bentonite adsorption methods that are extremely rapid[9,10] or sensitive.[11] The modified bentonite adsorption method used in the routine avidin assay allows more than 30 tissue samples to be determined daily in duplicate. By means of the bentonite adsorption method, we have studied progesterone-independent avidin induction in oviductal and nonoviductal chicken tissues caused by tissue injury and acute inflammation.[10,12] The bentonite adsorption method can be used for assay of avidin in most nonoviductal chick tissues except for the liver and kidney.[9,12]

Routine Bentonite Adsorption Method[9,10]

Principle. The assay is based on the specific binding of ^{14}C-labeled biotin to avidin, adsorption of the avidin–[14C]biotin complex onto ben-

[1] R. E. Eakin, E. E. Snell, and R. J. Williams, *J. Biol. Chem.* **140,** 535 (1941).
[2] R. Hertz and W. H. Sebrell, *Science* **96,** 257 (1942).
[3] B. W. O'Malley, W. L. McGuire, P. O. Kohler, and S. G. Korenman, *Recent Prog. Hormone Res.* **25,** 105 (1969).
[4] N. M. Green, *Biochem. J.* **89,** 585 (1963).
[5] N. M. Green, *Biochem. J.* **94,** 23C (1965).
[6] R. D. Wei and L. D. Wright, *Proc. Soc. Exp. Biol. Med.* **117,** 17 (1964).
[7] S. G. Korenman and B. W. O'Malley, *Biochim. Biophys. Acta* **140,** 174 (1967).
[8] B. W. O'Malley and S. G. Korenman, *Life Sci.* **6,** 1953 (1967).
[9] M. S. Kulomaa, H. A. Elo, and P. J. Tuohimaa, *Biochem. J.* **175,** 685 (1978).
[10] H. Elo, P. Tuohimaa, and O. Jänne, *Mol. Cell. Endocrinol.* **2,** 203 (1975).
[11] H. Elo and P. Tuohimaa, *Biochem. J.* **140,** 115 (1974).
[12] H. A. Elo, M. S. Kulomaa, and P. J. Tuohimaa, *Comp. Biochem. Physiol.,* in press (1978).

tonite, washing of radioactive bentonite in the test tube, and counting the radioactivity in a liquid scintillation counter.

Reagents

Homogenization buffer: 70 mM KCl, 4 mM MgCl$_2$, 70 mM NaCl, 20 mM sodium phosphate, pH 7.1

Ammonium carbonate, 0.2 M

d-[*carbonyl*-^{14}C]Biotin (38–57 mCi/mmol)

Bentonite

Procedure. Tissue is homogenized in 6–40 volumes of buffer using a motor-driven Teflon pestle, and the homogenate is centrifuged at 2500 g for 25 min. The assay is processed in duplicate from the aliquots of the supernatant solution; normally this corresponds to 25 mg and 50 mg of tissue in 2 ml of homogenization buffer. Note that the amount of avidin in the aliquot of the supernatant solution can not exceed 4 μg.

All procedures are made in the test tube at room temperature. [^{14}C]Biotin, 0.3 nmol in 0.5 ml of 0.2 M ammonium carbonate, is mixed with an aliquot of the supernatant solution. After a 15-min incubation, 40 mg of bentonite in 2 ml of 0.2 M ammonium carbonate is added, and the tube is shaken several times over a 15-min period. The tube is then centrifuged at 300 g for 5 min, and the bentonite sediment containing the avidin–[^{14}C]biotin complex is washed in the tube four times with 3 ml of 0.2 M ammonium carbonate. The bentonite sediment is transferred into a scintillation vial three times with 3.5 ml of scintillation solution, and the radioactivity is counted in a liquid scintillation counter. The sensitivity of the method is 30–50 ng of avidin.

Bentonite Adsorption Micromethod[11]

Principle. The principle is the same as in the routine bentonite adsorption method, but ^{3}H-labeled biotin is used and the radioactivity is counted in a tritium counter with high efficiency and low background.

Reagents

(+)-[^{3}H]Biotin, 114 mCi/mmol (Roche Products Ltd., Basel, Switzerland)

Mixture of metallic zinc powder and Na$_2$CO$_3$ (4:1, w/w)

Other reagents as in the routine bentonite adsorption method.

Procedure. Two hundred microliters of the 105,000 g (or 2500 g) supernatant solution (3–4 mg or less of tissue) and 1 nCi of [^{3}H]biotin in 50 μl of 0.2 M ammonium carbonate are incubated for 15 min in a small tube at room temperature. Then 3 mg of bentonite in 150 μl of 0.2 M ammonium carbonate is added, and the mixture is shaken several times over a 10–15-min period. After centrifugation at 300–500 g for 5 min, the bentonite sediment is washed in the tube five times with 1.5 ml of 0.2 M ammonium carbonate.

After washing, the bentonite sediment is transferred into the counting ampoule with 0.15 ml and 0.2 ml of 0.2 M ammonium carbonate. The ampoule is centrifuged and the ammonium carbonate solution is removed by aspiration. The radioactive bentonite sediment is then lyophilized overnight. Thereafter 40 mg of the mixture of metallic zinc powder and Na_2CO_3 (4:1, w/w) is added; the ampoule is closed, then the radioactivity of [^{3}H]biotin is combusted into tritium gas by heating the ampoule at 640° for 1 hr.[13]

Tritium activity is counted in a gas counter with high efficiency for tritium (60–70%) and low background (0.5–1 cpm for pure methane carrier gas; 3–4 cpm for the bentonite micromethod). An automatic apparatus with ten proportional counting tubes is manufactured by LKB-Wallac, Turku, Finland. The sensitivity of the bentonite adsorption micromethod with the presently available [^{3}H]biotin of low specific activity is 1–3 ng of avidin. If 5 mg of bentonite is used, 5–7 mg of tissue can be assayed with the same sensitivity.

[13] M. Parvinen, E. Soini, and P. Tuohimaa, *Anal. Biochem.* **55**, 193 (1973).

[53] A Radioimmunoassay for Chicken Avidin[1]

By Markku S. Kulomaa, Heikki A. Elo,
and Pentti J. Tuohimaa

There are a number of assay methods for avidin.[2–9] Most of these are based on the specific biotin binding property of avidin, but are hampered by the variable concentrations of endogenous tissue biotin. In addition, unknown biotin-binding factors of the tissue may affect results.[10] In our studies we have shown the induction of avidin in a number of nonovi-

[1] Supported by Grant No. 760-0526 from the Ford Foundation, New York.
[2] R. E. Eakin, E. E. Snell, and R. J. Williams, *J. Biol. Chem.* **140**, 535 (1941).
[3] N. M. Green, *Biochem. J.* **94**, 23C (1965).
[4] N. M. Green, *Biochem. J.* **89**, 585 (1963).
[5] R. D. Wei and L. D. Wright, *Proc. Soc. Exp. Biol. Med.* **117**, 17 (1964).
[6] S. G. Korenman and B. W. O'Malley, *Biochim. Biophys. Acta* **140**, 174, (1967).
[7] H. Elo, P. Tuohimaa, and O. Jänne, *Mol. Cell. Endocrinol.* **2**, 203 (1975).
[8] H. Elo and P. Tuohimaa, *Biochem. J.* **140**, 115 (1974).
[9] B. W. O'Malley and S. G. Korenman, *Life Sci.* **6**, 1953 (1967).
[10] H. A. Elo, M. S. Kulomaa, and P. J. Tuohimaa, *Comp. Biochem. Physiol.*, in press (1978).

METHODS IN ENZYMOLOGY, VOL. 62

ductal chick tissues. These studies indicate that the ^{14}C-labeled biotin–bentonite method is not a valid method for assay of avidin in the liver and kidney.[10,11] Here we present a radioimmunoassay for chicken avidin[11] under conditions where endogenous tissue biotin may vary.

Principle. Avidin radioimmunoassay (RIA) is based on the competition of ^{125}I-labeled and unlabeled avidin to the same binding site of the antibody. The bound and free avidin are then separated with a second antibody bound to solid matrix (DASP) and centrifuged. Radioactivity of the pellet is counted in a gamma counter.

Reagents

Freund's complete adjuvant (Difco)

Na[125]I (The Radiochemical Centre, Amersham; 100 mCi/ml; carrier free, for iodination)

Chloramine-T and sodium metabisulfite (Merck)

Avidin (Sigma, 11.2 units/mg)

Sephadex G-50 (Pharmacia, medium)

Sheep anti-(rabbit γ-globulin) immunosorbent or DASP (Organon)

Sodium phosphate buffers, 0.5 M and 50 mM, pH 7.5[12]

PB buffer: 50 mM sodium phosphate buffer (pH 7.5) containing 0.15 M NaCl, 10 mM EDTA, and 0.01% (w/v) merthiolate (Thiomersal)[12]

Bovine serum albumin (Sigma, 96–99% purity)

Preparation of Antiserum. Avidin antiserum is prepared by subcutaneous injections of avidin (1 mg) in Freund's complete adjuvant into the rabbit at 2-week intervals for 4–6 months. Blood is taken by cardiac puncture, and serum is separated by centrifugation (15 min, 3000 g, 4°). Antiserum is stored in small fractions at $-80°$.

Preparation of Labeled Avidin.[13] For avidin labeling 1 mCi of Na[125]I (10 μl), 0.5 M sodium phosphate buffer, pH 7.5 (10 μl), 20 μg of avidin (10 μl), and 50 μg of chloramine-T (10 μl) in 50 mM sodium phosphate buffer (pH 7.5) are mixed in a plastic tube. After 40 sec the reaction is stopped with 100 μg of sodium metabisulfite (100 μl) in 50 mM sodium phosphate buffer (pH 7.5).

[125]I-Labeled avidin is separated from free radioiodide and damaged proteins on a Sephadex G-50 column (0.8 × 25 cm) which has been previously equilibrated with PB buffer (50 ml) containing 0.1% (w/v) bovine serum albumin (BSA). Fractions containing [^{125}I]avidin are pooled, diluted (1:5) with the PB buffer containing 0.5% (w/v) BSA and stored at $-20°$ until used in the assay. The purity of the product is checked on a

[11] M. S. Kulomaa, H. A. Elo, and P. J. Tuohimaa, *Biochem. J.* **175**, 685 (1978).

[12] Deionized water was further purified by distillation in a quartz-glass apparature.

[13] F. C. Greenwood, W. M. Hunter, and J. S. Glover, *Biochem. J.* **89**, 114 (1963).

second Sephadex column. The labeled avidin remains immunoactive at least for 6 weeks.

Assay Procedure[11]

All dilutions are made with PB buffer containing 0.5% BSA. Tissue samples are homogenized and centrifuged (25 min, 2500 g, at room temperature) as described in this volume [52]. Four hundred microliters of sample or standard, 100 μl of antiserum (diluted $1-2 \times 10^{-5}$) and 50 μl of [^{125}I]avidin (10–20 ng) are mixed in a plastic tube and incubated overnight (16–20 hr) at room temperature (22°–24°). Thereafter, 1 ml of sheep anti-(rabbit γ-globulin) immunosorbent (DASP, diluted 1:20) is added and the tube is rotated slowly end-over-end for 2 hr at room temperature. The visible precipitate is separated by centrifugation in a swing-out rotor (2–3 min, 1000 g, at room temperature) and washed twice with 2 ml of PB buffer containing 0.5% BSA. The radioactivity of the pellet is counted in a gamma counter.

The maximal binding control (B_0) contains 400 μl of buffer instead of sample and the background control 500 μl instead of sample and antiserum. The standard curve ranges from 1 to 500 ng of avidin per milliliter. The standard avidin (250 μg) is dissolved in 1 ml of the oviduct homogenate supernatant, then diluted 1:10 with PB buffer containing 0.5% BSA and stored at −80°. The final dilution is made immediately before the assay.

Comments

Specific activity of [^{125}I]avidin is 20–60 μCi/μg. Antiserum is diluted so that 20–30% of the total radioactivity is bound to antibody in the absence of unlabeled avidin (B_0). The standard curve is linear from 1–2 ng to 100–200 ng of avidin per milliliter in the logit-log scale. The cross-reaction for ovalbumin is smaller than 0.015%. An excess of biotin affects the avidin RIA by decreasing values about 15%. Recoveries of RIA are 64–131% depending on the tissue. One technician can assay 30–60 supernatant samples and avidin standards in a day.

[54] The Synthesis and Use of Spin-Labeled Analogs of Biotin in the Study of Avidin

By Birandra K. Sinha and Colin F. Chignell

Spin labels are stable free radicals that can be used as probes for biologically important macromolecules, e.g., proteins, nucleic acids, membranes.[1,2] The most commonly employed spin labels utilize the nitroxide free radical, since this radical is very stable in aqueous solutions at physiological pH values.[1,2]

Avidin is a tetrameric protein (mass 68,000 daltons), found in raw egg white, which binds four molecules of biotin [(I) vitamin H].[3-5] The biotin binding sites, one per subunit, are grouped in two pairs at opposite ends of the short axis of the avidin molecule.[6] Since the carboxyl group of biotin is not necessary for binding to avidin,[6] analogs of biotin [(II)–(V)] containing the nitroxide free radical were synthesized to study the binding of biotin to avidin and the topography of the biotin binding sites.

Synthesis of the Spin-Labeled Biotin Analogs

4-{[5-(Hexahydro-2-oxo-1H-thieno[3,4-d]imidazol-4-yl)-1-oxopentyl]-amino}2,2,6,6-tetramethyl-1-piperidinyloxy (II) (Method 1). To a solution of 0.28 g (1.64 mmol) of 4-amino-2,2,6,6-tetramethyl-1-piperidinyloxy (VI) (Aldrich, Eastman) in 25 ml of absolute EtOH, 0.4 g (1.64 mmol) of biotin and 0.45 g (10% excess) of *N*-ethoxycarbonyl-2-ethoxy-1,2-dihydroquinoline (EEDQ, Aldrich) are added. The mixture is stirred at room temperature for 48 hr and filtered, and the alcohol solution is concentrated under reduced pressure. The residual semisolid is dissolved in CHCl$_3$, and the CHCl$_3$ solution is washed with bicarbonate solution and H$_2$O and dried (Na$_2$SO$_4$). Removal of CHCl$_3$ affords a pink-orange solid (0.42 g, 65%). Crystallization from CHCl$_3$-*n*-hexane gives pure (II), m.p. 197.5–198.5°.

3-{[5-(Hexahydro-2-oxo-1H-thieno[3,4-d]imidazol-4-yl)-1-oxopentyl]amino} 2,2,5,5-tetramethyl-1-pyrrolidinyloxy (III). This is prepared by

---

[1] L. J. Berliner (ed.), "Spin Labeling: Theory and Applications." Academic Press, New York, 1975.

[2] P. Jost and O. H. Griffith, *in* "Methods in Pharmacology," (C. F. Chignell, ed.), Vol. 2, p. 223. Appleton, New York, 1972.

[3] N. M. Green, *Biochem. J.* **89**, 585 (1963).

[4] N. M. Green, *Biochem. J.* **89**, 609 (1963).

[5] N. M. Green, *Biochem. J.* **92**, 16C (1964).

[6] N. M. Green, L. Konieczny, E. J. Toms, and R. C. Valentine, *Biochem. J.* **125**, 781 (1971).

METHODS IN ENZYMOLOGY, VOL. 62

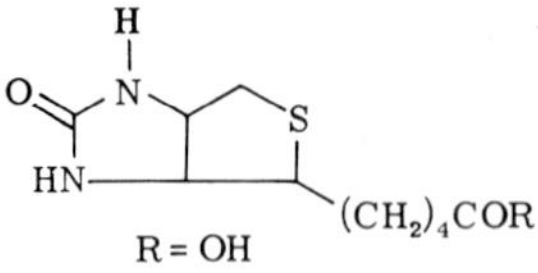

(I) 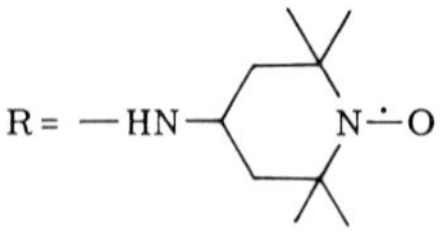   (II) 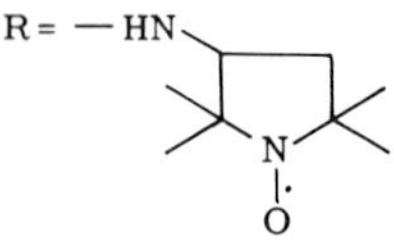   (III)

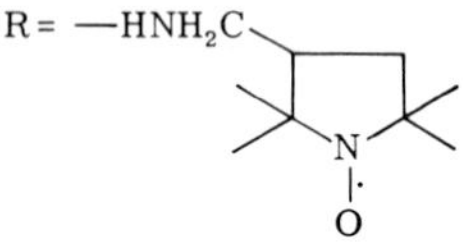

R = OH

R = —HNH$_2$C (IV)

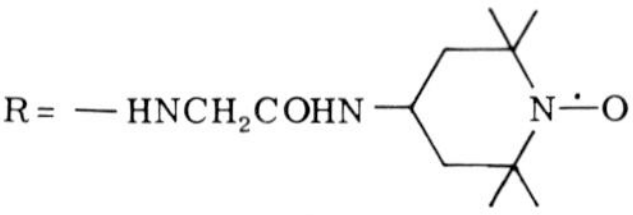

R = —HNCH$_2$COHN (V)

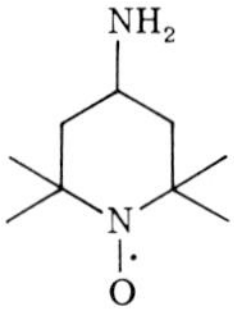

(VI) 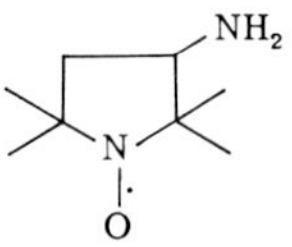   (VII) 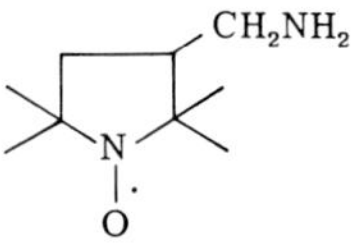   (VIII)

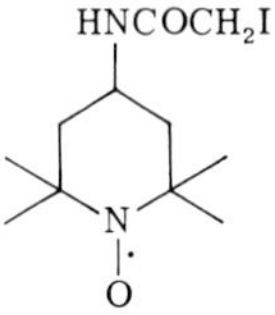

(IX) 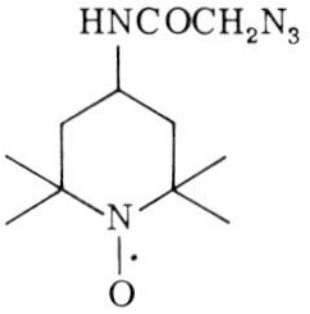   (X) 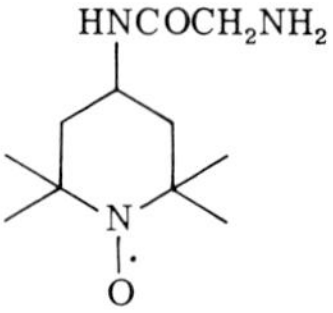   (XI)

method 1 from (VII) (Eastman) and biotin in MeOH at 40°–50° for 2 hr, and then for 48 hr at room temperature, in 77% yield. Crystallization from CHCl$_3$-Et$_2$O gives (III) as a yellow solid, m.p. 139°–140°.

3-⟨{[5-(Hexahydro-2-oxo-1 H-thieno[3,4-d]imidazol-4-yl)-1-oxopentyl] amino}methyl⟩-2,2,5,5-tetramethyl-1-pyrrolidinyloxy (IV). This was prepared from (VIII) and biotin by method 1. Crystallization from CHCl$_3$-n-hexane gives (IV) as yellow solid, m.p. 181°–183°.

4-(2-Azidoacetamido)-2,2,6,6-tetramethyl-1-piperidinyloxy (X). To a solution of 0.85 g (2.5 mmol) of (IX) in 15 ml of dry DMF, 1.6 g (24.6 mmol) of sodium azide is added. The reaction mixture is stirred at room temperature for 18 hr, diluted with water, and saturated with NaCl. The aqueous solution is extracted with ether (3 × 25 ml). The ether solution is washed with saturated NaCl solution, dried (Na_2SO_4), filtered, and concentrated under reduced pressure. This affords 0.63 g (99%) of a red liquid; IR ($CHCl_3$) shows the characteristic band for $-N_3$ at 2240 cm^{-1} and for $>C = 0$ at 1720 cm^{-1}.

4-(2-Aminoacetamido)-2,2,6,6-tetramethyl-1-piperidinyloxy (XI). This is prepared from (X) by $NaBH_4$ reduction of the azido group using 2-propanol as solvent. A solution of 0.6 g (2.3 mmol) of (X) in 10 ml of dry 2-propanol is added to 0.21 g (5.5 mmol) of $NaBH_4$ in 10 ml of dry *i*-PrOH with stirring. The reaction mixture is refluxed for 18 hr and cooled; the solvent is removed under reduced pressure. The residual oil is dissolved in $CHCl_3$, and the $CHCl_3$ solution is extracted with dilute HCl. The acid solution is made basic with dilute NaOH and extracted with $CHCl_3$. The $CHCl_3$ solution is dried (Na_2SO_4), and the solvent is removed to give 0.52 g (98%) of pure (XI) (GC-MS, IR) as a red oil.

4-(({[5-(Hexahydro-2-oxo-1H-thieno[3,4-d]imidazol-4-yl)-1-oxopentyl] amino }acetyl)amino)-2,2,6,6-tetramethyl-1-piperidinyloxy (V). Compound (V) is obtained by condensation of (XI) with biotin according to method 1, described earlier, in 35–40% yield. Crystallization from $CHCl_3$-Et_2O affords (V) as a pink solid, m.p. 132–135°.

Purification of Avidin

Avidin is purchased from a commercial source (Worthington Biochemical Corporation) and purified by column chromatography over carboxymethyl cellulose according to the procedure of Green and Toms.[7] The concentration of avidin can be estimated either from the optical absorption of the sample at 280 nm ($E_1^{1 \, cm} = 15.4$)[7] or by titration with 2-(4-hydroxyazobenzene)benzoic acid (HABA).[8]

Evidence That the Spin-Labeled Biotin Analogs (II–V) Occupy the Same Binding Sites on Avidin As Does Biotin

Principle. Since the spin-labeled biotin analogs differ considerably in their molecular structure from biotin, it is necessary to establish that they occupy the same binding sites of avidin. This is most easily done on the

---

[7] N. M. Green and E. J. Toms, *Biochem. J.* **118**, 67 (1970).

[8] Abbreviations used: HABA, 2-(4'-hydroxyazobenzene)benzoic acid; TEMPO, 2,2,6,6-tetramethylpiperidine-1-oxyl; ESR, electron spin resonance.

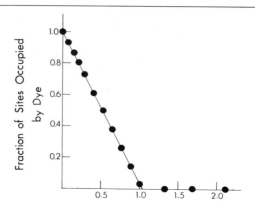

Molecules of Spin Label / Binding Site

FIG. 1. Titration of a mixture containing avidin (5 μM) and HABA (90 μM) in 50 mM sodium phosphate buffer, pH 7.4, with spin label (II). Reproduced from C. F. Chignell, D. K. Starkweather, and B. K. Sinha, *J. Biol. Chem.* **250**, 5622 (1975), with permission.

basis of competitive binding experiments and overall stoichiometry. Green has shown[9] that when HABA binds to avidin the absorption maximum of the dye shifts from 348 to 480 nm. Each avidin molecule has four binding sites for HABA from which the dye can be displaced by biotin. The addition of spin label II to the HABA–avidin complex results in the displacement of an equivalent amount of dye (Fig. 1). Similar results are obtained with labels (III)–(V).

Procedure. Aliquots of an ethanol solution of the spin labels (0.5 mM) are added via a Hamilton syringe with a constant-delivery adaptor to a cuvette containing a solution (2 ml) of avidin (5 μM) and HABA (90 μM) in 50 mM sodium phosphate buffer (pH 7.4). After each addition, the optical density is measured at 500 nm. The concentration of bound HABA is calculated from the known molar extinction coefficients for free ($\epsilon = 600$) and bound ($\epsilon = 34,500$) dye.

Principle. The binding of biotin to avidin produces a shift in the tryptophan fluorescence emission maximum of the protein to shorter wavelengths and a concomitant decrease in the quantum yield of avidin fluorescence.[10] All the biotin spin labels produce changes in the tryptophan fluorescence of avidin similar to that seen with biotin (Fig. 2 and Table I). Spin labels (II)–(IV) are more efficient quenchers of avidin fluorescence; label (V) produces about the same quenching as biotin

[9] N. M. Green, *Biochem. J.* **94**, 23C (1965).
[10] N. M. Green, *Biochem. J.* **90**, 564 (1964).

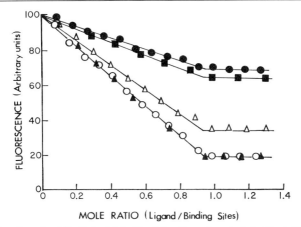

FIG. 2. Titration of avidin (5.5 μM) in 50 mM sodium phosphate buffer (pH 7.4) with spin labels (II) (○——○), (III) (▲——▲), (IV) (△——△), (V) (■——■), and biotin (●——●). Reproduced from C. F. Chignell, D. K. Starkweather, and B. K. Sinha, *J. Biol. Chem.* **250,** 5622 (1975), with permission.

(Table I). The fluorescence titration experiments confirm that avidin binds four molecules of each spin-labeled biotin analog.

Procedure. Aliquots of an ethanol solution of the spin labels (0.5 mM) are added via a microsyringe to a solution of avidin (5.5 μM) in 50 mM sodium phosphate buffer (pH 7.4). After each addition the tryptophan fluorescence is measured in a spectrophotofluorometer using activation and emission wavelengths of 290 nm and 336 nm, respectively. The solutions are exposed to the activating light just long enough to obtain the reading to avoid photodecomposition. The relative quantum yields of

TABLE I

FLUORESCENCE EMISSION MAXIMA AND RELATIVE QUANTUM YIELDS OF COMPLEXES
BETWEEN AVIDIN AND BIOTIN AND BIOTIN SPIN LABELS

	Fluorescence	
Compound added to avidin	Emission maximum (nm)	Relative quantum yield
None	336	100
Biotin	326	68.8
(II)	326	18.2
(III)	325	16.0
(IV)	325	32.3
(V)	330	62.4

avidin and its complexes with biotin and the spin labels are calculated from the areas under their respective emission curves. If the spectrophotofluorometer does not display corrected spectra, then the emission curves should be corrected for response of detector by standard procedures.[11]

ESR Spectral Evidence That the Biotin Spin Label (II) Binds to Avidin

Principle. The ESR spectrum of spin label (II) (Fig. 3) in dilute aqueous solution consists of three sharp lines of approximately equal intensity with a splitting of about 17 G between adjacent lines. The ESR spectrum of any nitroxide radical is dependent upon several parameters including (a) the polarity of the microenvironment, (b) molecular motion, (c) orientation with respect to the magnetic field, and (d) presence of other paramagnetic species.[2] The effect of a decrease in molecular motion on the ESR spectrum of a nitroxide radical is shown in Fig. 4. In this experiment 2,2,6,6-tetramethyl piperidine-1-oxyl has been dissolved in glycerol and its ESR spectrum recorded as a function of temperature. At 43° the ESR spectrum of TEMPO is similar to that of biotin label (II) in aqueous

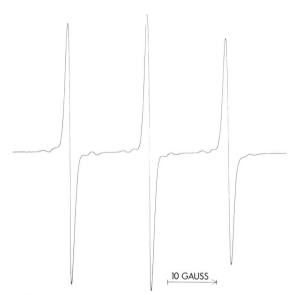

10 GAUSS

FIG. 3. The ESR spectrum of spin label (II) (10 μM) dissolved in 50 mM sodium phosphate buffer (pH 7.4).

[11] R. F. Chen, *Anal. Biochem.* **20**, 339 (1967).

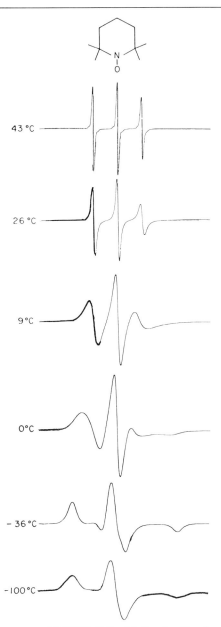

FIG. 4. The ESR spectrum of TEMPO (0.5 mM) dissolved in glycerol. Adapted from P. Jost, A. S. Waggoner, and O. H. Griffith, *in* "Structure and Function of Biological Membranes" (L. I. Rothfield, ed.), p. 83. Academic Press, New York, 1971.

solution at room temperature (Fig. 3). As the temperature of the glycerol solution is decreased, the lines begin to broaden until at $-36°$ a broad asymmetric spectrum is obtained. This ESR spectrum, often referred to as a "rigid glass" spectrum, is obtained whenever a spin label is randomly oriented and molecular motion is either absent or very slow on the ESR time scale, i.e., when $\tau \geqslant 10^{-8}$ sec where τ is the rotational correlation time.[2]

When less than two equivalents of biotin label II are added to avidin, the ESR spectrum of the radical is characteristic of a highly immobilized nitroxide (Fig. 5, curve A). This spectrum provides direct evidence for the binding of spin label (II) to avidin.

Procedure. ESR spectra are recorded with a spectrometer operating at 9.5 GHz. Quartz aqueous sample cells (e.g., Varian E-248; Scanco S-812) are used to obtain spectra at ambient temperature. For all other temperatures, it is necessary to use a quartz microcell (e.g., Varian E248-1; Scanco S808) in conjunction with a suitable variable temperature accessory. Temperatures may be measured with a telethermometer (e.g., Yellow Springs Instrument Company, Model 42SC). Spectra are obtained using a modulation amplitude of 1 G, a power setting of 20 mW, a 100-G sweep width, and a scan time of 8 min. A detailed description of the prac-

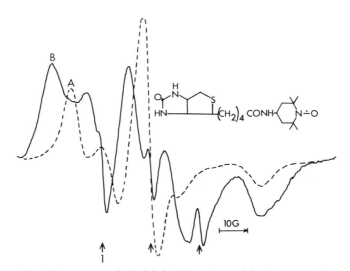

Fig. 5. The ESR spectrum of spin label (II) bound to avidin (80.0 μM) in 50 mM sodium phosphate buffer (pH 7.4). The concentration of spin label was 0.1 mM in A and 0.32 mM in B. The arrows mark the position of the sharp three-line spectrum of a small amount of unbound spin label. Spectrum A was recorded 3 hr after the addition of spin label II. Reproduced from C. F. Chignell, D. K. Starkweather, and B. K. Sinha, *J. Biol. Chem.* **250**, 5622 (1975), with permission."

tical aspects of recording the ESR spectra of nitroxide radicals is outside the scope of this article but may be found in an excellent monograph by Jost and Griffith.[2]

Measurement of the Distance Between Adjacent Spin Labels Bound to Avidin

Principle. The ESR spectrum of avidin containing four bound equivalents of label (II) is shown in Fig. 5, Curve B. For reasons that will not be discussed here, it appears that this spectrum is a composite of two spectra, one of which is similar to that observed when only two equivalents of label (II) are bound to avidin (Fig. 5, curve A). If it is assumed that 30% of spectrum B (Fig. 5) is due to spectrum A (Fig. 5) then subtraction results in the spectrum shown in Fig. 6. A comparison of the subtracted spectrum shown in Fig. 6 with spectrum A in Fig. 5 indicates that each of the hyperfine lines has been split into a doublet. This type of spectrum is characteristic of highly immobilized nitroxide radicals undergoing electron–electron dipolar interactions.[12] Green and co-workers[6] in a series of elegant experiments have shown that the biotin-binding sites on avidin are arranged in two pairs at opposite ends of the short axis of the

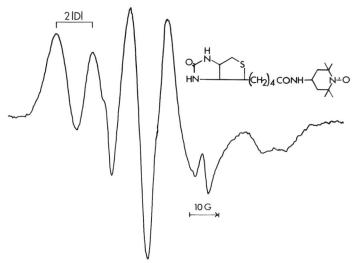

FIG. 6. The ESR spectrum of spin label (II) bound to avidin derived from Fig. 5, B spectrum by computer subtraction of Fig. 5, spectrum A. Reproduced from C. F. Chignell, D. K. Starkweather, and B. K. Sinha, *J. Biol. Chem.* **250,** 5622 (1975), with permission.

[12] P. Ferruti, D. Gill, M. P. Klein, H. H. Wang, G. Entine, and M. Calvin, *J. Am. Chem. Soc.* **92,** 3704 (1970)

TABLE II
ESR Parameters of Complexes between the Spin-Labeled Biotin
Analogs and Avidin

Compound	Maximum hyperfine splitting of 2:1 complex (G)	Dipolar splitting of 4:1 complex (G)	Calculated distance between adjacent nitroxides (A)
(II)	69.0	13.8	16.0
(III)	64.9	11.9	16.7
(IV)	63.5	14.2	15.7
(V)	62.0	—	—

avidin molecule. Thus, it seems reasonable to assume that the spectrum in Fig. 6 results from an interaction between spin labels bound to adjacent nitroxides. The distance between adjacent bound nitroxides (r_{12}) can be calculated from the equation

$$r_{12} = [(5.56 \times 10^4)/2D]^{1/3}$$

where $2D$ is the maximal splitting of the individual hyperfine lines in gauss (Fig. 6). The calculated distance for spin label (II) was 16 Å. Similar results were also obtained for labels(III) and (IV) (Table II). By contrast, the ESR spectrum of a 4:1 complex of spin label (V) and avidin showed no evidence for interaction between labels bound to adjacent sites (Fig. 7).

Procedure. Subtraction of ESR spectra is best achieved with the aid of a digital computer. The spectrum shown in Fig. 6 was obtained by first collecting the spectra (Fig. 5,A and B) in a Nicolet Instrument Corpora-

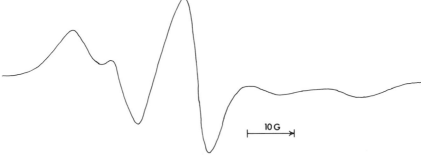

Fig. 7. The ESR spectrum of spin label (V) (60.0 μM) in the presence of avidin (15.0 μM) and 50 mM sodium phosphate buffer (pH 7.4). Reproduced from C. F. Chignell, D. K. Starkweather, and B. K. Sinha, *J. Biol. Chem.* **250,** 5622 (1975), with permission.

tion Model 1072 computer, punching them onto paper tapes and then reading the tape into a Digital Equipment Corporation PDP10 computer. The spectra were then subtracted in the form of matrices using the MLAB program.[13] An excellent discussion of the use of dedicated computers in ESR has been written by Klopfenstein *et al.*[14]

Redistribution of Spin Label (II) among Available Binding Sites

Principle. When less than two equivalents of spin label (II) are added to avidin, the ESR spectrum of the complex is time dependent (Fig. 8). Examination of Fig. 8 reveals that, although some of the bound spin labels are undergoing dipolar interactions immediately after mixing, no such interactions are present 2.5 hr later. This finding indicates that on mixing some of the labels are bound to sites that are adjacent to other occupied

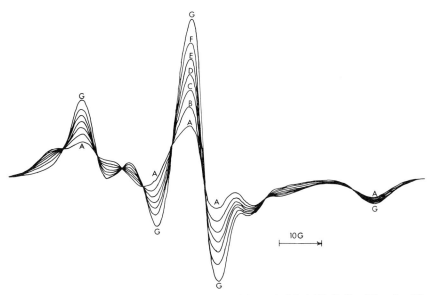

Fig. 8. The ESR spectrum of a solution containing spin label (II) (0.12 mM) and avidin (69.0 μM) in 50 mM sodium phosphate buffer (pH 7.4). Spectra were recorded at the following minutes after mixing: A, 3.5; B, 11.5; C, 20.5; D, 30.5; E, 46.5; F, 75.5; G, 156.5. Scan time was 4 min. Reproduced from C. F. Chignell, D. K. Starkweather, and B. K. Sinha, *J. Biol. Chem.* **250,** 5622 (1975), with permission.

[13] G. D. Knott and D. K. Reece, *Proc. ONLINE 1972 Conf.* Vol. 1, p. 497, Brunel University, England, 1972.

[14] C. Klopfenstein, P. Jost, and O. H. Griffith *in* "Computers in Chemical and Biochemical Research" (C. E. Klopfenstein and C. L. Wilkins, eds.), Vol. 1, p. 175. Academic Press, New York, 1972.

sites, and the remainder are bound to sites that are adjacent to empty sites. The time dependence of the ESR spectra suggests that, after mixing, the spin labels redistribute among the available site in such a way that each bound label is adjacent to an empty site.

Procedure. For this experiment an ethanol solution (12 μl) of spin label II (10 mM) was added at time zero to a solution (1.0 ml) of avidin (69 μM) dissolved in 50 mM in a quartz aqueous sample cell. Mixing was achieved with the aid of a disposable plastic syringe connected to the bottom of the cell by means of an 18-gauge needle embedded in a Teflon stopper. The spectra were recorded using 4-min scan times.

Measurement of the Rate of Dissociation of Biotin Spin Label (II) from Avidin

Principle. The rate of dissociation of spin label (II) from a 4:1 complex with avidin can readily be followed by measuring the concentration of the unbound label from the amplitude of its low field line (arrow 1 in Fig. 5, spectrum B) after the addition of a large excess of biotin. The results shown in Fig. 9 indicate that the rate of dissociation of spin label II from 4:1 complex with avidin is biphasic. Approximately half of the labels dissociate from avidin quite rapidly ($k_{diss} = 2.51 \times 10^{-4}$ sec^{-1}), and the remainder dissociate at a much slower rate ($k_{diss} = 1.22 \times 10^{-5}$ sec^{-1}). This finding suggests that when spin label (II) binds to a site adjacent to an occupied site, then it dissociates more rapidly than when it is bound to a site adjacent to an empty one.

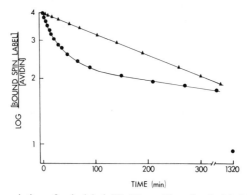

FIG. 9. The dissociation of spin label (II) (●——●) and spin label (V) (▲——▲) from avidin. The following concentrations were employed: avidin, 12.5 μM; spin label, 50 μM; 50 mM sodium phosphate buffer (pH 7.4). The experiment was initiated by the addition of biotin (2.5 mM). Reproduced from C. F. Chignell, D. K. Starkweather, and B. K. Sinha, *J. Biol. Chem.* **250**, 5622 (1975), with permission.

Procedure. A solution of biotin is added at zero time to a solution (1.0 ml) containing avidin (12.5 μM), spin label (II) (50.0 μM), and 50 mM sodium phosphate buffer. The amplitude of the low-field line (Fig. 5, spectrum B, arrow 1) is then monitored as a function of time. After the experiment is completed, the reaction mixture is replaced with a series of standard solutions of spin label (5–25 μM), and the concentration of spin label displaced by the biotin is calculated from a standard curve.

Binding Site Topography of Avidin Based on Experiments with Spin Labels

A model based on the spin label studies described here and elsewhere is shown in Fig. 10. The main features of the model are as follows: (a) the arrangement of binding sites in pairs at each end of the short axis of the avidin molecule; (b) the location of the binding sites in a depression on the

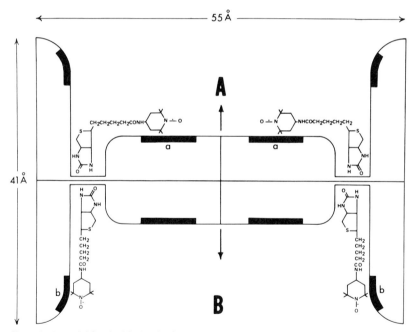

FIG. 10. A model for the biotin binding sites of avidin. In this figure, we have followed the convention of N. M. Green, L. Konieczny, E. J. Toms, and R. C. Valentine, *Biochem. J.* **125**, 781 (1971) and ascribed a 222 symmetry to the avidin tetramer. The figure itself represents a diagrammatic projection of the biotin binding sites onto a plane containing the 2-fold axis indicated by the arrows. A, spin label (II) bound to adjacent accessory sites (a); B, spin label (II) bound to nonadjacent accessory sites (b). Reproduced from C. F. Chignell, D. K. Starkweather, and B. K. Sinha, *J. Biol. Chem.* **250**, 5622 (1975), with permission.

protein surface; (c) the existence of two accessory sites (Fig. 10a and b), which interact specifically with the piperidine bearing nitroxide moieties; (d) the distance between the nitroxide groups of label II bound to adjacent sites in 16 Å; (e) the existence of tryptophan groups at or near accessory sites a. A more detailed discussion of this model may be found elsewhere.[15]

[15] C. F. Chignell, D. K. Starkweather, and B. K. Sinha, *J. Biol. Chem.* **250**, 5622 (1975).

[55] The Avidin–Biotin Complex in Affinity Cytochemistry

By Edward A. Bayer, Ehud Skutelsky, and Meir Wilchek

Principle

The use of affinity methods for the localization, visualization, and subsequent evaluation of specific cellular components has been termed affinity cytochemistry.[1,2] In general, the technique is based upon the preparation of a mixed conjugate, comprising a biologically active molecule attached chemically to a potentially perceptible probe (e.g., fluorescein, ferritin, peroxidases, hemocyanin, etc.), whereby the resultant product retains both detectability and biological activity. A wide spectrum of biologically active molecules have been coupled to the above probes, including antibodies,[3-5] lectins,[6-9] hormones,[10] lipoproteins,[11] vitamins,[12] sugars,[13] cations,[14] and anions,[15] for use in light, fluorescent, and electron microscopical studies. Since the defined electron opacity of the

[1] E. A. Bayer, M. Wilchek, and E. Skutelsky, *FEBS Lett.* **68**, 240 (1976).
[2] E. Skutelsky and E. A. Bayer, *Isr. J. Med. Sci.* **12**, 1355 (1976).
[3] S. J. Singer and A. F. Schick, *J. Biophys. Biochem. Cytol.* **9**, 519 (1961).
[4] S. Avrameas, *Immunochemistry* **6**, 43 (1969).
[5] M. C. Raff, *Sci. Am.* **234**, 30 (1976).
[6] G. L. Nicholson and S. J. Singer, *Proc. Natl. Acad. Sci. U.S.A.* **68**, 942 (1971).
[7] R. T. Parmley, B. J. Martin, and S. S. Spicer, *J. Histochem. Cytochem.* **21**, 912 (1973).
[8] N. K. Gonatas and S. Avrameas, *J. Cell Biol.* **59**, 436 (1973).
[9] J. F. Ash and S. J. Singer, *Proc. Natl. Acad. Sci. U.S.A.* **73**, 4575 (1976).
[10] L. Jarrett and R. M. Smith, *Proc. Natl. Acad. Sci. U.S.A.* **72**, 3526 (1975).
[11] R. G. W. Anderson, J. L. Goldstein, and M. S. Brown, *Proc. Natl. Acad. Sci. U.S.A.* **73**, 2434 (1976).
[12] E. A. Bayer, E. Skutelsky, T. Viswanatha, and M. Wilchek, *Mol. Cell. Biochem.* **19**, 23 (1978).
[13] M. Monsigny, C. Kieda, D. Gros, and J. Schrevel, *Proc. Eur. Congr. Electron Microsc.* *6th*, p. 39 (1976).

ferritin iron core affords superior resolution qualities, ferritin represents the electron microscopic marker of choice.

The use of the high-affinity avidin–biotin complex has been shown to circumvent some of the problems relating to ferritin–protein conjugation.[16] In addition, this method may be employed in order to unify and facilitate certain aspects of affinity cytochemical techniques.[1,2,17]

The following steps are involved in this approach: (a) Biotin is attached via an appropriate reactive derivative either directly to cell surface functional groups (sugars, amino acids, etc.) or to a biologically active molecule (antibody, lectin, etc.). (b) In the latter case, the biotinylated conjugate is incubated with an appropriate target (intact cells or tissue, enzymically or chemically treated cells, membrane preparations, subcellular fractions, defined macromolecules). (c) Subsequent incubation with ferritin–avidin conjugates enables the ultrastructural visualization of the given cell surface receptor. (d) Proper controls using nonbiotinylated preparations and/or unconjugated ferritin should always be implemented.

An alternative method, which constitutes a permutation of the above method, has also been attempted in our laboratory. Biotinylated mem-

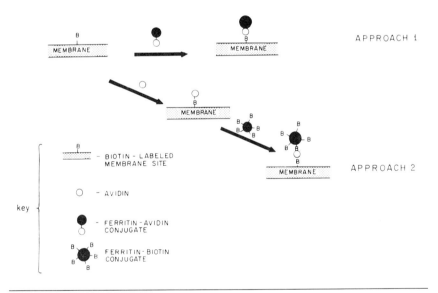

[14] D. Danon, L. Goldstein, Y. Marikovsky, and E. Skutelsky, *J. Ultrastruct. Res.* **38,** 500 (1972).

[15] E. Skutelsky and E. A. Bayer, unpublished results (1977).

[16] E. A. Bayer, E. Skutelsky, D. Wynne, and M. Wilchek, *J. Histochem. Cytochem.* **24,** 933 (1976).

[17] H. Heitzmann and F. M. Richards, *Proc. Natl. Acad. Sci. U.S.A.* **71,** 3537 (1974).

brane sites are saturated with free avidin. Since the latter is a tetramer, subsequent interaction with biotin-conjugated ferritin also results in specific labeling of cell surfaces. This method is somewhat less tedious than that involving ferritin–avidin conjugates, since the preparation and analysis of protein–protein conjugates are precluded. However, the resultant cell surface label is much less uniform, and consequently the method is less reliable. Both approaches are summarized in the accompanying scheme (p. 309).

Preparation of Ferritin–Avidin Conjugates

In the following account we describe the preparation of ferritin–avidin conjugates both by reductive alkylation[16] and by the conventional glutaraldehyde method. The former has been shown to yield unit-paired conjugates.

Reagents

Ferritin, 6 times recrystallized (Pentex, Inc.)

Avidin, 12.3 units/mg protein (Sigma Chem. Co.)

Sodium *meta*-periodate (Merck)

Sodium borohydride (Merck)

Glutaraldehyde (70% solution, Ladd Research Industries)

Ammonium bicarbonate

ABS; acetate-buffered saline, pH 4.5: sodium chloride, 9 g/liter, is added to a solution of acetate buffer, pH 4.5[18]

BBS; borate-buffered saline, pH 8.5: sodium chloride, 9 g/liter, is added to a solution of borate buffer, pH 8.5[19]

PBS; phosphate-buffered saline, pH 7.0: sodium chloride, 9 g/liter, is added to a solution of phosphate buffer, pH 7.0.[20]

Biotin–Sepharose affinity columns[21]

Reductive-Alkylation Method. Commercial avidin (15 mg in 5 ml of ABS) is added to ferritin (100 mg, 1 ml). Sodium *meta*-periodate (Merck, 0.66 ml, 0.1 M solution) is added to a final concentration of 10 mM. The mixture is stirred for 3 hr in ice, dialyzed for 6 hr against ABS at 4°, followed by a second dialysis overnight at 4° against BBS. A fresh solution of sodium borohydride (10 mg/ml in 10 mM NaOH) is prepared, and 0.5 ml is added to the ferritin–avidin conjugates in an ice bath. After 1 hr the solution is dialyzed against PBS. The conjugates are washed twice by centrifugation (100,000 g, 3 hr) and resuspended to 1 mg/ml ferritin.

[18] G. Gomori, this series, Vol. 1, p. 140.

[19] G. Gomori, this series, Vol. 1, p. 145.

[20] G. Gomori, this series, Vol. 1, p. 143.

[21] E. Bayer and M. Wilchek, this series, Vol. 34, p. 265.

Glutaraldehyde Method. Commercial avidin (15 mg in 3 ml of PBS) is added to a commercial solution of ferritin (1 ml, 100 mg) and stirred. Glutaraldehyde (440 μl, 0.5% solution) is added slowly to a final concentration of 0.05%. The reaction is allowed to proceed for 1 hr at room temperature and is then quenched with 0.1 M ammonium bicarbonate. The conjugates are dialyzed overnight against PBS. Large aggregates are removed by centrifugation at 10,000 g for 30 min. The supernatant is centrifuged at 100,000 g for 3 hr. The pellet, consisting of free ferritin and ferritin–avidin conjugates, is resuspended in PBS, and the 100,000 g centrifugation step is repeated. (The supernatant fractions, comprising free avidin, may be saved for future preparations of ferritin–avidin conjugates.) The washed conjugates are resuspended in PBS to a final concentration of 1 mg/ml ferritin (A_{440} = 1.1).

Alternatively, in order to achieve unidirectionality, ferritin can be treated with an excess of glutaraldehyde (15%) and subsequently interacted with avidin following gel filtration on Sephadex G-25 to remove free glutaraldehyde.[22,23]

Analysis of Conjugates. The relative size of ferritin–avidin conjugates can be tested by gel filtration on a Sepharose 6B column. The extent of ferritin conjugation is assayed by affinity chromatography on a biotin-containing affinity column. The difference in absorbance (A_{440} applied-effluent) represents the amount of active ferritin–avidin conjugate.

Preparation of Ferritin–Biotin Conjugates[12]

Reagents
  Ferritin
  BNHS, biotinyl-N-hydroxysuccinimide ester, prepared as described
    below
  Sodium bicarbonate
  PBS, phosphate-buffered saline, pH 7.3, prepared as described
    above
  Avidin–Sepharose affinity column.[24] Sepharose 4B (10 g) is washed
    well with distilled water and resuspended in 10 ml of distilled
    water. The suspension is stirred constantly with a magnetic stirrer.
    Cyanogen bromide (1.25 g) is added and the pH maintained
    between 10.0 and 11.0 by the dropwise addition of 2 N sodium hydroxide. After 10 min, the activated gel is filtered and washed
    extensively with cold 0.1 M sodium bicarbonate. Avidin (20 mg in

[22] H. Otto, H. Takamiya, and A. Vogt, *J. Immunol. Methods* **3**, 137 (1973).
[23] Y. Kishida, B. R. Olsen, R. A. Berg, and D. J. Prockop, *J. Cell Biol.* **64**, 331 (1975).
[24] A. Bodanszky and M. Bodanszky, *Experientia* **26**, 237 (1970).

20 ml of sodium bicarbonate) is added, and the suspension is stirred overnight at 4°. The gel is washed, then stored suspended in water with the addition of a few crystals of sodium azide to prevent microbial growth.

Procedure. A solution (1 ml) of ferritin (50 mg) is brought to pH 8.5 with 0.1 *M* sodium bicarbonate. BNHS (3 mg) is added. The reaction is stirred for 1 hr at room temperature and overnight at 4°. The conjugate is dialyzed exhaustively against PBS and sterilized through a Millipore filter. The extent of biotinylation is determined by passage of the derivatized ferritin through an avidin–Sepharose affinity column.

Preparation of Reactive Biotinyl Derivatives

Biotinyl-N-Hydroxysuccinimide Ester (BNHS)

    Reagents
      Dicyclohexylcarbodiimide (DCC) (Fluka)
      Biotin (Sigma Chemical Co.)
      N-Hydroxysuccinimide
      Dimethylformamide (DMF)
      Ether
      Isopropanol

Procedure. DCC (0.8 g) is added to a DMF solution (12 ml) containing biotin (1 g) and N-hydroxysuccinimide (0.6 g). The suspension is stirred overnight at room temperature. The precipitate is filtered and then the filtrate is evaporated under reduced pressure. The residue is washed well with ether, and the product is recrystallized from isopropanol. Yield: 1 g (70%).

Biotin Hydrazide (BHZ)

    Reagents
      Thionyl chloride (Fluka)
      Biotin
      Methanol
      Hydrazine hydrate
      Dimethylformamide

Procedure. Thionyl chloride (1 ml) is added slowly to a chilled solution (10 ml) of methanol (in an ice–saline bath). To this solution, biotin (1 g) is added and left overnight at room temperature. The solvent is evaporated to dryness. Methanol (10 ml) is added, and the solvent is again evaporated to dryness. The residue is redissolved in 5 ml methanol, hydrazine hydrate (1 ml) is added, and the reaction proceeds overnight at room temperature. The precipitate (biotin hydrazide) is filtered and washed with ether. A second crop may be obtained upon concentration of the filtrate. The samples are recrystallized from dimethylformamide. Total yield: 80%.

Direct Biotinylation of Functional Groups

Sialic Acid Residues.[25] Cells (10^8/ml) are washed and resuspended in PBS (1 ml). Sodium *meta*-periodate is added to a final concentration of 1 mM, and the treatment is allowed to proceed for 30 min in an ice bath. The cells are then washed twice with PBS and resuspended in a solution of biotin hydrazide (2.5 mg/ml). After 1 hr at room temperature, the cells are washed three times in PBS and fixed in 2% glutaraldehyde (1 ml in PBS).

Galactose and N-Acetylgalactosamine Residues.[17] The primary hydroxyl groups of these sugars are treated with galactose oxidase,[26] and the resultant aldehydes are interacted with biotin hydrazide. Cells (approximately 10^8) are washed twice with PBS and mixed with sodium borohydride (2 mM in PBS) in order to quench endogenous oxidized membrane components. The cells are washed twice in buffer and biotin hydrazide (2.5 mg/ml PBS) is added. The suspension is treated with galactose oxidase (10 units, Sigma Chem. Co.) for 3 hr at 37°, washed twice with buffer, and fixed in glutaraldehyde.

Biotinylation of Biologically Active Proteins

For relatively stable, biologically active proteins, such as antibodies, lectins, and polypeptide hormones, the following biotinylation procedure has proved effective.[1,27]

Biotinyl-*N*-hydroxysuccinimide ester (BNHS) dissolved in dimethylformamide (DMF), is added to a solution, pH 7.0 or higher, of the desired protein in a 1:50 v/v and 5:1 mol/mol ratio. For example, an aliquot (0.1 ml) containing 0.5 μmol BNHS (1.7 mg/ml DMF) is added to a solution containing goat antirabbit IgG antibodies (16 mg protein in 5 ml of PBS). The solution is kept at room temperature for 4 hr and dialyzed overnight

[25] E. Skutelsky, D. Danon, M. Wilchek, and E. A. Bayer, *J. Ultrastruct. Res.* **61,** 325 (1977).

[26] C. G. Gahmberg and S.-I. Hakomori, *J. Biol. Chem.* **248,** 4311 (1973).

[27] E. Skutelsky and E. A. Bayer, *Proc. Eur. Congr. Electron Microsc. 6th,* p. 198 (1976).

at 4° against PBS with one buffer change. Biotinylated antibodies or lectins may be stored at −20°. Whole antiserum or unpurified lectins may be biotinylated in the above manner and used for affinity cytochemical studies in their unrefined state. The reaction with BNHS can be restricted mostly to α-amines by performing the reaction between pH 5.0 and 6.0.

Relatively unstable proteins (or those subject to loss of biological activity upon chemical modification) may require additional or alternative treatment, e.g., modification of cysteines, tyrosines, etc., or separation of biotinylated proteins from underivatized material after mild biotinylation.

Affinity Cytochemistry of Cell Surfaces

Viable cells (10^8) or cells fixed with 2% glutaraldehyde for 30 min are washed and incubated with an appropriate solution of biotinylated lectins or antibodies (0.5–1.0 mg protein/ml PBS) for 30 min at room temperature. Normal or optimal conditions of interaction should be used with any other protein type. Controls comprise labeling with underivatized protein samples. After the previous step, cells are washed, fixed with glutaraldehyde, and treated subsequently with 2% bovine serum albumin. Further treatment with ferritin–avidin conjugates (1 mg of protein per milliliter) results in specific labeling of biotin-tagged sites. Cells, the functional groups of which have been directly biotinylated, are labeled with ferritin–avidin conjugates by similar treatment. Samples prepared in this manner are then processed for electron microscopy.[28]

Comments

The above-described procedure is appropriate for analysis of labeled material in thin sections, freeze-etched replicas, shadow casting or negatively stained samples by transmission electron microscopy. Additionally the system can be easily adopted for scanning electron microscopy. The use of the avidin–biotin complex in this context is appealing for a variety of reasons: (a) Only one conjugate (ferritin–avidin, fluorescent–avidin, etc.) need be prepared and characterized for all affinity systems; (b) biotin can be attached to small ligands and macromolecules efficiently and under very mild conditions; (c) the size, physical characteristics, and biological activity of biotin-derivatized proteins, are, in most cases, only nominally affected; (d) the avidin–biotin complex is of exceptionally high affinity and stability; and (e) both avidin and biotin are commercially available in large quantities.

[28] J. H. Luft, *J. Biophys. Biochem. Cytol.* **9**, 409 (1961).

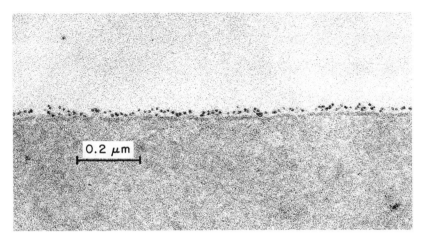

FIG. 1. Perpendicular section of human erythrocyte cell surface membrane specifically labeled for sialic acid residues. Note the distance separating the ferritin from the outer surface of the membrane.

An example of the ferritin–avidin labeling pattern, achieved by direct biotinylation of sialic acid residues on cell surfaces of intact human erythrocytes, is shown in Fig. 1.

In addition to the use of the avidin–biotin complex for the localization and visualization of receptors on cell membranes (affinity cytochemistry), this complex can also be used for the isolation of such receptors by affinity chromatography[12,29] or affinity partitioning.[30] For this purpose a sandwich technique is applicable, similar to the antibody approach,[31] namely, Sepharose–avidin:biotinylated binding protein:receptor. The avidin–biotin complex has also been used for the inhibition of bacteriophages[32] and for gene localization and enrichment studies.[33,34]

[29] E. A. Bayer and M. Wilchek, this volume [63].

[30] S. D. Flanagan and S. H. Barondes, J. Biol. Chem. 250, 1484 (1975).

[31] M. Wilchek and M. Gorecki, FEBS Lett. 31, 149 (1973).

[32] J. M. Becker and M. Wilchek, Biochim. Biophys. Acta 264, 165 (1972).

[33] L. Angerer, N. Davidson, W. Murphy, D. Lynch, and G. Attardi, Cell 9, 81 (1976).

[34] J. Manning, M. Pellegrini, and N. Davidson, Biochemistry 16, 1364 (1977).

[56] Egg Yolk Biotin-Binding Protein: Assay and Purification

By Harry W. Meslar and Harold B. White, III

Assay Method

Principle. The assay is based on the temperature-accelerated exchange of bound biotin for free [^{14}C]biotin.[1] The procedure is, therefore, useful for biotin-binding proteins with no free binding sites.[2]

Reagents

Potassium phosphate, 50 mM, pH 7.2

[^{14}C]Biotin, 15 μg/ml, 20 μCi/mg in pH 7.2 buffer[3]

Bentonite, 10 mg/ml suspension in pH 7.2 buffer

Bray's solution, liquid scintillation fluid[4]

Definition of Unit. One unit is defined as the amount of protein that will bind 1 μg of d-biotin.

Procedure. A sample, containing 2.6×10^{-4} to 2.6×10^{-3} units (2–20 μg of egg yolk biotin-binding protein or avidin) is pipetted into a test tube. A 50-μl portion of [^{14}C]biotin is added, and the mixture is incubated for 15 min at the appropriate temperature (65° for yolk biotin-binding protein, 85° for avidin). The sample is cooled on ice, and 300 μl of the bentonite suspension is added to adsorb the protein. The slurry is poured into a vacuum-sampling manifold equipped with nitrocellulose filters and washed 3 times with 1 ml of cold 50 mM potassium phosphate buffer, pH 7.2. After the last wash has dried, the filters are dissolved in scintillation fluid and the bound radioactivity is measured in a liquid scintillation counter. Blank values are obtained by running a series of assays at different protein concentrations and extrapolating to zero protein. The filtering of materials of high lipid content (e.g., egg yolk) is greatly aided by replacing the first buffer wash with 1 ml of water-saturated 1-butanol. The butanol dissolves the lipids without damaging the protein.

---

[1] H. B. White, B. A. Dennison, M. A. DellaFera, C. J. Whitney, J. C. McGuire, H. W. Meslar, and P. H. Sammelwitz, *Biochem. J.* **157**, 396 (1976).

[2] For assay of avidin with available binding sites, see N. M. Green, this series, Vol. 18A, p. 414; R. Wei, this series, Vol. 18A, p. 424; S. G. Korenman and B. W. O'Malley, this series, Vol. 18A, p. 427.

[3] In our experience 30–50% of the radioactivity in commercially prepared [^{14}C]biotin will not bind to avidin. The specific activity of samples diluted with unlabeled biotin must be adjusted to compensate for these impurities.

[4] G. A. Bray, *Anal. Biochem.* **1**, 279 (1960).

Purification Procedure[5]

General. All buffers used are potassium phosphate, pH 7.2, at the concentration given. The procedure outlined is for 5 dozen eggs.

Lipid Removal.[6] The eggs are cracked and the white and yolks are separated. The yolks are broken and 1.5 times the yolk volume in distilled water is added. The diluted yolk is well stirred. This solution is centrifuged at 13,200 g for 30 min. The supernatant is collected, and solid NaCl is added to make the solution 1 M in salt. After the salt has completely dissolved, the solution is poured into centrifuge bottles, two-thirds the solution volume in 1-butanol is added, and the mixture is shaken vigorously.[7] Centrifugation of the resulting emulsion at 13,200 g for 15 min causes a lipid-free aqueous phase and a lipid-rich organic phase to form, with a solid layer of precipitated protein at the interface. The two liquids are poured into a separatory funnel and the lower aqueous phase is removed. The aqueous phase is dialyzed overnight against running water at 4° to remove NaCl and butanol.

Phosphocellulose Ion-Exchange Chromatography. To the dialyzed lipid-free yolk solution is added 150 ml of settled phosphocellulose, equilibrated with 50 mM potassium phosphate, pH 7.2, and the suspension is stirred. The phosphocellulose is allowed to settle, and the supernatant is decanted. The phosphocellulose is poured into a column (2.5 × 50 cm) attached to a peristaltic pump and a UV column monitor. The column is washed with 50 mM potassium phosphate until the OD_{280} of the eluate has regained baseline. The flow is discontinued, and the 50 mM buffer is replaced by 0.5 M potassium phosphate. The pump is started, and the sharp peak of protein eluted is collected and pooled.

Affinity Chromatography. The phosphocellulose eluate is placed in a pressure concentrator equipped with an ultrafiltering membrane (0.45 μm) along with 10 ml of biotinyl-Sepharose.[8] The total volume must not exceed 150 ml in a 400-ml concentrator. The concentrator is lowered into a 55° water bath equipped for magnetic stirring[9] and pressurized with nitrogen, taking care not to increase the volume in the concentrator. The af-

[5] H. W. Meslar, S. A. Camper, and H. B. White, III, *J. Biol. Chem.* **253,** 6979 (1978).

[6] H. W. Meslar and H. B. White, III, *Anal. Biochem.* **91,** 75 (1978).

[7] R. K. Morton, this series, Vol. 1, p. 40.

[8] Biotinyl-Sepharose is prepared by water-soluble carbodiimide coupling of biotin to hexamethylene-diamine substituted Sepharose 4BCL. The coupling must take place at pH 5.7–6.0. Below pH 5.7 the biotin becomes insoluble, and pH 6.0 is the upper limit for successful carbodiimide coupling. See I. Parikh, S. March, and P. Cuatrecasas, this series, Vol. 34B, p. 77.

[9] An aluminum pot on top of a heater-stirrer has been used quite successfully in our laboratory for this purpose.

Fraction	Total units	Specific activity (units/mg)[a]
Diluted yolk	378	6.82×10^{-3}
Phosphocellulose eluate	58.6	1.42
Gel filtration eluate[b]	5.25	12–13

[a] Protein determinations were made using the Bio-Rad dye binding assay kit, Bio-Rad Laboratories, Richmond, California.

[b] Activity of the affinity gel eluate is not measured because of the excess unlabeled biotin present at that stage.

finity gel-protein slurry is washed with 2.0–2.3 liters of 50 mM potassium phosphate and depressurized. The gel is poured into a water-jacketed column (1.0 × 15 cm) and washed with 0.3 M potassium phosphate until the OD_{280} of the eluate reaches baseline. The water jacket is filled with water at 65°, and the column is filled with 10 ml of 40 mM biotin, dissolved in 0.2 M potassium phosphate. The system is allowed to heat and then is slowly eluted (0.6 ml/min). The entire biotin-containing peak is pooled.

Gel Filtration. The eluted material is concentrated to a volume of 2 ml, made dense with solid sucrose, and applied to a column of Ultrogel AcA 44,[10] equilibrated with 0.2 M potassium phosphate, and eluted. Fractions of constant specific activity are pooled and lyophilized.

Summary of the Purification. The procedure is summarized in the table. This procedure has been successfully repeated 10 times. The protein shows a single band on gel electrophoresis, shows a single N-terminal, and gives a single precipitin band with antiserum raised against the mixture of proteins eluted from phosphocellulose.

Properties

Stability. The protein is unstable in dilute solution at low ionic strength (50 mM potassium phosphate). At higher ionic strength, it is stable for months when frozen. The protein denatures at temperatures above 70°.

Molecular Structure. Egg yolk biotin-binding protein appears to be a tetramer of identical subunits. The molecular weight of the tetramer is about 74,300.

pH Optimum. Biotin is tightly bound from pH 5 to above pH 10. Below pH 5.0, the biotin-binding activity decreases and at pH 2.3 the protein is completely inactive.

[10] LKB Instruments, Rockville, Maryland.

Temperature-Induced Biotin Release. Yolk biotin-binding protein has a maximum rate of biotin release at 70°, above which temperature the protein begins to denature.

Comparison with Egg White Avidin.[5] The properties of yolk biotin-binding protein are very different than those reported for avidin.[11] The pI is about pH 4.6, as opposed to about pH 10 for avidin. The yolk protein, although heavier (74,300 as opposed to 68,000), contains fewer residues of lysine, threonine, isoleucine, and aspartic acid. Biotin release is more than 1000-fold faster from the egg yolk protein than from the egg white protein. Neither protein cross-reacts with antibodies directed against the other. Cyanogen bromide cleavage generates peptides containing N-terminal leucine from egg yolk biotin-binding protein and N-terminal threonine and tryptophan from avidin.

[11] N. M. Green, *Adv. Protein Chem.* **29**, 85 (1975).

[57] Antibodies That Bind Biotin and Inhibit Biotin-Containing Enzymes[1]

By MELVIN BERGER

Antibodies with affinity for biotin are produced following immunization of rabbits with antigens consisting of biotin linked covalently as a hapten to a carrier protein such as bovine serum albumin or bovine γ-globulin.[2] In recent years considerable technology has been developed for the use of specific antibodies as biochemical reagents.[3,4] Antibodies that bind biotin thus could be very useful in the purification and analysis of biotin-containing enzymes. In particular, the Fab fragments of these antibodies could be used as specific, monovalent, biotin-binding proteins, offering an advantage over the tetravalent avidin molecule in some cases. The avidity of these antibodies for enzyme-bound biotin allows enzyme inhibition to be used as a simple assay for the production of biotin-binding antibody. Enzyme inhibition can also serve as a convenient measure of unsaturated binding sites in studies of the interaction of biotin or its derivatives with the antibodies.

[1] This work was supported by National Institutes of Health Grants Nos. 5T01-GM-00035 and AM-12245 and was performed in the laboratory of Dr. Harland G. Wood in the Department of Biochemistry, Case Western Reserve University.
[2] M. Berger, *Biochemistry* **14**, 2338 (1975).
[3] J. B. Robbins and R. Schneerson, this series, Vol. 34, p. 703.
[4] D. M. Livingston this series, Vol. 34, p. 723.

Preparation of Biotin–Protein Conjugates

Principle. A water-soluble carbodiimide is used to activate the carboxyl group on the valeric acid side chain of biotin. Amide linkages are then formed with free amino groups in the carrier protein. The use of ^{14}C as a tracer allows quantitation of the incorporation of biotin into the conjugates.

Reagents
  d-Biotin
  [2'-^{14}C]Biotin, 58 Ci/mol (Amersham-Searle)
  1-Ethyl-3-(3-dimethylaminopropyl) carbodiimide hydrochloride
  Bovine γ-globulin (Cohn fraction II)
  Crystalline bovine serum albumin
  Tris·HCl, 50 mM, pH 9.0
  Potassium phosphate buffer, 0.1 M, pH 6.8
  Aqueous pyridine, 50%
  NaCl, 0.15 M (0.9%)

Procedure for Biotinyl Bovine Serum Albumin

Biotin (500 mg; 2 mmol) and 15 × 10^6 cpm of [^{14}C]biotin are suspended in 7.5 ml of 50% aqueous pyridine with constant stirring at room temperature. Carbodiimide (2.5 g; 13 mmol) is dissolved in 12.5 ml of 50% aqueous pyridine, and this mixture is added dropwise to the biotin suspension. Stirring is continued for an additional 30 min at room temperature; during this time a clear solution is formed. Albumin (250 mg; 3.7 μmol) in 6.25 ml of distilled H$_2$O is added dropwise to the biotin–carbodiimide reaction mixture, and stirring is continued at room temperature for 4.5 hr. Following this reaction, the mixture is dialyzed for 24 hr against 4 changes of 4 liters each of 0.9% NaCl. Determination of the radioactivity in aliquots of the complete reaction mixture and of the product which is retained on dialysis indicates that 30 μmol of biotin is bound to protein in the average preparation. This corresponds to approximately 8 mol of biotin per mole of serum albumin. The dialyzed protein conjugate may be stored at −5°.

Procedure for Biotinyl Bovine γ-Globulin

Biotin (50 mg; 0.2 mmol) is dissolved in 10 ml of 0.018 N NaOH. The pH is adjusted to 5.5 by the addition of 0.14 ml of 0.2 N HCl in small increments, and 6.8 × 10^6 cpm of [^{14}C]biotin are added to the solution, which is kept at room temperature. One milliliter of the biotin solution is used to dissolve 250 mg (1.3 mmol) of the carbodiimide, and the reaction is allowed to proceed at room temperature. The pH rises to 7 within 5–6 min, and 0.02 ml of 2 N HCl is added to return the pH to 5. The reaction mixture is held at room temperature for another 6–7 min and then added to a solution of 100 mg (0.67 μmol) of γ-globulin in 6 ml of 50 mM Tris·HCl (pH 9.0) with gentle stirring at room temperature which is continued for 4.5 hr. The reaction mixture is extensively dialyzed against 0.1 M potassium phosphate buffer (pH 6.8) at room temperature for 24 hr. Slight turbidity may develop during this time. Determination of the radioactivity present in the biotin solution and in the dialyzed product indicates that 1.6 μmol of biotin is bound to protein corresponding to 2.6 mol of biotin per mole of γ-globulin.

Immunization and Collection of Antisera

New Zealand white rabbits weighing approximately 3 kg are maintained on a standard laboratory chow diet. Protein solutions are diluted to 2 mg/ml using 0.9% NaCl. Equal volumes of protein solution and Complete Freund's Adjuvant (Bacto No. 0638-59, obtained from Difco, Inc.) are emulsified by repeated pumping through a two-hub, 18-gauge needle using 10-ml glass syringes.[5]

Either biotin–protein conjugate may be used as the antigen, but a higher titer of biotin-binding antibodies is obtained if the more highly substituted biotinyl-bovine serum albumin is used to immunize the rabbits. For the initial immunization, 1 ml of the emulsion is injected into each of the hind foot pads of each rabbit. Booster injections of 2 mg of the antigen in 1 ml of 0.9% NaCl are given in the peripheral ear vein at 10–14-day intervals. Nonimmune control sera are drawn before the initial immunization, and test bleedings are obtained from the central artery of the ear immediately before each booster immunization. Terminal bleeding by cardiac puncture is performed 7–9 days after the last immunization. In all cases, the blood is allowed to clot at 37° for 1 hr and then stored at 0° overnight. The antiserum is separated from the clots by centrifugation and stored at −10° in the presence of 0.02% sodium azide.

Assays for Biotin-Binding Antibodies

Precipitation

The simplest test for production of antibodies against the antigen is the precipitin ring test. Serial dilutions of the sera to be tested are prepared

---

[5] R. B. Brown, this series, Vol. 11, p. 917.

TABLE I
PRECIPITIN RING TESTS WITH ANTI-BIOTINYL γ-GLOBULIN[a]

Protein	Highest serum dilution forming visible precipitate
1. Biotinyl γ-globulin	256
2. γ-Globulin	128
3. Biotinyl serum albumin	8
4. Serum albumin	None

[a] Antiserum was drawn 10 days after the fourth immunization. This table is reproduced from M. Berger, *Biochemistry* **14**, 2338 (1975) with permission of the American Chemical Society.

using 0.9% NaCl or nonimmune (preimmunization or commercially available normal rabbit serum) rabbit serum. Forty microliters of diluted serum is placed at the bottom of a 6×50 mm tube and $40 \mu l$ of the antigen solution, diluted to 1 mg/ml in 0.9% NaCl, is carefully layered over the serum. The tubes are held vertically and incubated at room temperature for 30 min, during which time a ring of precipitation occurs at the interface between the serum and the antigen. Results are expressed as the highest serum dilution giving a visible precipitin ring. Precipitation of the antigen is caused by antibodies against biotin, but also by antibodies against the carrier protein itself. Independent estimation of these two types of antibodies is possible if nonbiotinylated carrier protein and a different biotinylated protein are also used in the precipitin tests, since the latter will be precipitated only by antibodies binding to the biotins, as shown in Table I.

In contrast, the biotin-binding antibodies do not precipitate transcarboxylase or pyruvate carboxylase, possibly because of the inability of a sufficient number of IgG molecules to span the distance between biotins on different enzyme molecules or because the binding sites on the antibody are saturated by adjacent biotins of a single enzyme molecule.[2] The antibodies do, however, form stable high-molecular-weight complexes with biotin enzymes,[2,6] and the complexes may be separated from uncomplexed antibody and enzyme by gel filtration or density gradient centrifugation.[6] Stable complexes are also formed between the antibodies and isolated biotin carboxyl carrier proteins or biotinyl peptides derived from biotin enzymes.[7]

[6] F. R. Harmon, M. Berger, and H. G. Wood, unpublished results, 1976.
[7] M. Berger and H. G. Wood, *J. Biol. Chem.* **251**, 7021 (1976).

Enzyme Inhibition

Reagents

Transcarboxylase (methylmalonyl-CoA:pyruvate carboxytransferase, EC 2.1.3.1.), prepared from *Propionibacterium shermanii* and assayed by the method of Wood *et al.*[8] The following reagents are necessary for the enzyme assay:

Methylmalonyl-CoA, 3 mM

NADH, 3 mM

Assay mixture: 0.1 M sodium pyruvate, 2.0 ml; 1.0 M potassium phosphate, pH 6.8, 7.0 ml; malate dehydrogenase (40 units/ml), 1.0 ml

Potassium phosphate buffer, 0.3 M pH 6.8

Normal rabbit serum (Gibco)

Procedure. Transcarboxylase (2.8 μg, approximately 0.1 unit) in 0.15 ml of 0.3 M potassium phosphate buffer (pH 6.8) and 0.10 ml of a variable combination of normal rabbit serum and antiserum are incubated for 30 min at room temperature. Aliquots of 0.02 ml are assayed for transcarboxylase activity by measuring the rate of formation of oxaloacetate from methylmalonyl-CoA and pyruvate, using malate dehydrogenase and following the rate of NADH oxidation in a Gilford Model 2000 recording spectrophotometer.[8] A unit is defined as 1 μmol of oxaloacetate produced per minute. Control assays with the methylmalonyl-CoA omitted are used to correct for the amount of NADH oxidase and lactic dehydrogenase present in the serum. Units are expressed on the basis of the total incubation system. As shown in Fig. 1, the activity decreases linearly as the amount of antiserum increases until about two-thirds of the enzyme activity is lost. The slope of the linear portion of this curve is used to calculate the number of units of enzyme that would be inhibited by 1.0 ml of antiserum.

A similar procedure is used for chicken liver pyruvate carboxylase [pyruvate:CO$_2$ ligase (ADP), EC 6.4.1.1.] which is assayed by following the formation of oxaloacetate from pyruvate, ATP, and HCO$_3^-$ in the presence of 15 mM acetyl-CoA using a linked assay system similar to that used for transcarboxylase.[9] This procedure could be adapted for use with any other biotin enzyme with appropriate conditions for maintaining the stability of the enzyme during the incubation with the antiserum.

Results. Time studies of the course of inhibition of the enzyme by the antiserum indicate that over 50% of the enzyme activity is lost within 1.5 min of the addition of the antiserum, and about 85% of the activity is lost

---

[8] H. G. Wood, B. Jacobson, B. I. Gerwin, and D. B. Nortrop, this series, Vol. **13**, 215.

[9] M. C. Scrutton, M. R. Olmsted, M. F. Utter, this series, Vol. 13, p. 235.

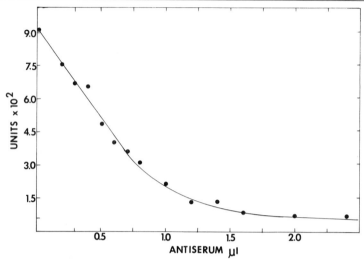

Fig. 1. Inhibition of transcarboxylase by antibiotinyl serum albumin. Procedure as described in text. Reproduced from M. Berger, *Biochemistry* **14**, 2338 (1975) with permission of the American Chemical Society.

within 15 min with very little further loss of enzyme activity for over 60 min.[2] The rapidity of the effect of the antiserum on transcarboxylase and the ease with which the enzyme is assayed makes enzyme inhibition a convenient tool to quantitate the production of biotin-binding antibodies during the course of immunization. Two of three rabbits immunized with biotinyl–bovine serum albumin gave very similar responses, as shown in Table II, whereas the third rabbit produced antiserum with the capacity to inhibit only 8.7 units of transcarboxylase per milliliter.

TABLE II

TRANSCARBOXYLASE INHIBITION BY ANTISERA DURING THE COURSE OF IMMUNIZATION WITH BIOTINYL SERUM ALBUMIN

Number of immunizations	Days since initial immunization	Units of transcarboxylase inhibited by 1 ml of antiserum
1	17	0.52
2	28	0.86
3	42	3.0
4	52	12.6
8	77	53.5
16	170	79.4

Binding of Free Biotin

  Reagents
  d-Biotin
  Antiserum (diluted 50-fold in normal rabbit serum)
  Potassium phosphate buffer, 0.3 *M* pH 7.0
  Potassium phosphate buffer, 0.25 *M* pH 6.5
  Transcarboxylase and assay reagents as above

Principle. Enzyme inhibition is used to detect unsaturated binding sites on the antibodies after the antiserum is incubated with biotin. The amounts of antiserum and enzyme to be used are chosen to be in the linear range of inhibition as shown on the curve of Fig. 1—about 70% inhibition of enzyme activity is suitable. Previous studies with this system indicate that little free antibody remains in the presence of excess enzyme and that little or no dissociation of biotin–antibody complex occurs after addition of the enzyme.[2]

Procedure. A variable amount of biotin or biocytin (0–40 ng) in 0.32 ml of 0.3 *M* potassium phosphate buffer (pH 7.0) is incubated for 30 min at room temperature with 0.08 ml of antiserum diluted 50-fold with nonimmune rabbit serum. This corresponds to 1.6 μl of undiluted antiserum with capacity to inhibit 70–80 units transcarboxylase per milliliter. Transcarboxylase, 0.10 ml (5.2 μg, 0.18 unit) in 0.25 *M* potassium phosphate buffer (pH 6.5) is added, and the incubation is continued for 30 min at room temperature. Aliquots (0.02 ml) are assayed for transcarboxylase

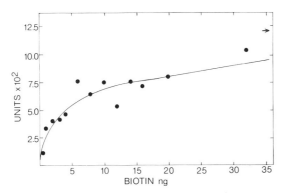

Fig. 2. Relief of immune inhibition of transcarboxylase by addition of free biotin. Procedure as described in text. The arrow at the right indicates the transcarboxylase activity in a control incubation containing normal serum in place of antiserum. Note that each incubation contains 0.18 unit of transcarboxylase, but the amount of antiserum is chosen to provide only 65–70% inhibition. Thus the control with antiserum present but no biotin added contains about 0.06 units of transcarboxylase activity, but this value is subtracted from every other incubation.

activity in the standard assay for the enzyme. Controls with no biotin added and no methylmalonyl-CoA in the assay are also run; the results are subtracted from those obtained with the complete mixture. In addition, controls with only nonimmune serum (no antiserum) may be run to check on the stability of the enzyme. The results of a typical experiment with biotin are shown in Fig. 2. As the shape of the curve resembles that of a classical enzyme–substrate interaction, the values could be replotted on double-reciprocal plots allowing an estimate of the apparent binding constant for biotin to antibody, which in this case was found to be 5.0×10^{-8} M.[2] Similar results were obtained with biocytin.

Purification of Biotin-Binding Antibodies

The immunoglobulin fraction of the serum may be purified by standard techniques including ammonium sulfate precipitation and DEAE-cellulose chromatography as outlined by Livingston.[4]

Biotinyl–Sepharose may be used to quantitatively remove the biotin-binding antibodies from the antiserum.[10] The antibodies can then be recovered in approximately 40% yield by elution with 1.5 M acetic acid.[11] More recently, we have found that the biotin-binding antibodies may be absorbed onto lipoyl–Sepharose and eluted using 0.1 M acetic acid with recovery of 70% of the enzyme-inhibiting antibody.[11]

[10] M. Berger, Ph.D. thesis, Case Western Reserve University, 1976.
[11] F. R. Harmon and H. G. Wood, unpublished results, 1977.

[58] Microbiological Biosynthesis of Biotin[1]

By Yoshikazu Izumi, Yoshiki Tani, and Koichi Ogata[2]

In a previous volume of this series,[3] one of the authors described the biosynthesis of DTB and biotin using growing microorganisms. Recent studies have used cell-free systems to determine the enzymic steps from pimelic acid to DTB as shown in Fig. 1.[4] The final step from DTB to biotin

[1] Abbreviations used: KAPA, 7-keto-8-aminopelargonic acid; DAPA, 7,8-diaminopelargonic acid; DTB, dethiobiotin; PLP, pyridoxal 5'-phosphate; PMP, pyridoxamine 5'-phosphate; CoA, coenzyme A.
[2] K. Ogata is now deceased.
[3] K. Ogata, this series, Vol. 13, p. 390.
[4] Y. Izumi and K. Ogata, Adv. Appl. Microbiol. 22, 145 (1977).

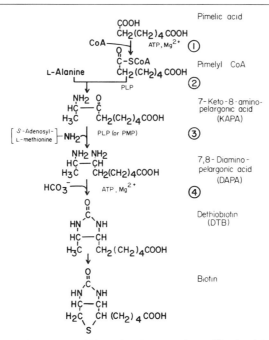

FIG. 1. Biosynthetic pathway of biotin in microorganisms. ①, Pimelyl-CoA synthetase; ②, KAPA synthetase; ③, DAPA aminotransferase; ④, DTB synthetase.

has not yet been enzymically resolved. Therefore, the following sections deal with the methodology concerning the enzymic studies of the biosynthesis of DTB from pimelic acid using various bacteria.

I. Pimelyl-CoA Synthetase[5,6]

Assay Method

Principle. Pimelyl-CoA synthetase catalyzes the synthesis of pimelyl-CoA from pimelic acid, CoA, and ATP with Mg^{2+} as a cofactor. Because too small an amount of pimelyl-CoA is formed to be detectable directly by the hydroxamic acid method, the assay of pimelyl-CoA synthetase is performed by measuring the amount of KAPA formed in a coupled reaction with the use of KAPA synthetase, which catalyzes the

[5] Y. Izumi, H. Morita, Y. Tani, and K. Ogata, *Agric. Biol. Chem.* **38**, 2257 (1974).
[6] Y. Izumi, H. Morita, Y. Tani, and K. Ogata, *Biochim. Biophys. Acta* **264**, 210 (1972).

synthesis of KAPA from pimelyl-CoA and L-alanine with PLP as the coenzyme. The KAPA is quantitatively determined by a microbiological assay (paper disc plate method) with *Saccharomyces cerevisiae*[7] with KAPA as the standard. The reactions for the enzyme assay are carried out using the procedures A and B described below.

Reagents

Pimelic acid, 10 mM and 0.1 M (neutralized with NaOH)

CoA, 1 mM and 2 mM

ATP disodium salt, 0.1 M and 10 mM

MgCl$_2$, 0.1 M and 50 mM

Potassium phosphate buffer, 1 M, pH 7.0

Tris·HCl buffer, 1 M, pH 8.0

PLP, 1 mM

KAPA synthetase from *Bacillus sphaericus* IFO 35625, 50 units/ml (see unit definition in Section II)

Procedure A. The assay system consists of 1 μmol of pimelic acid, 10 nmol of CoA, 1 μmol of ATP, 1 μmol of MgCl$_2$, 10 μmol of potassium phosphate buffer, 5 μmol of L-alanine, 10 nmol of PLP, 1 unit of KAPA synthetase, and enzyme solution in a final volume of 120 μl. The reaction is carried out for 2 hr at 37°, and is stopped by heating the mixture in boiling water for 2 min.

Procedure B. In this procedure, the two reactions are separately carried out. The first reaction mixture, a pimelyl-CoA synthetase system, contains 0.1 μmol of pimelic acid, 20 nmol of CoA, 0.1 μmol of ATP, 0.5 μmol of MgCl$_2$, 20 μmol of Tris·HCl buffer, and enzyme solution in a final volume of 70 μl. The reaction is carried out for 1 hr at 37°, and is stopped by heating the mixture in boiling water for 2 min. The mixture for the KAPA synthetase reaction containing 5 μmol of L-alanine, 10 μmol of PLP, 1 unit of KAPA synthetase, and 10 μmol of potassium phosphate buffer in a total volume of 50 μl is added to this first reaction mixture. The reaction is carried out for another 1 hr at 37°, then is stopped by heating the whole in boiling water for 2 min.

Enzyme Unit. One unit of enzyme activity is defined as the amount of enzyme necessary to synthesize 1 nmol of KAPA under the conditions described. Specific activity is expressed as units per milligram of protein. Protein is determined by the method of Lowry *et al.*[8] and by measuring the absorbancy at 280 nm.

[7] E. E. Snell, R. E. Eakin, and R. J. Williams, *J. Am. Chem. Soc.* **62**, 175 (1940).

[8] O. H. Lowry, N. J. Rosebrough, A. L. Farr, and R. J. Randall, *J. Biol. Chem.* **193**, 265 (1951).

Purification Procedure

All operations are carried out around 5° unless otherwise stated. The buffer used throughout the purification is potassium phosphate buffer, pH 7.0, containing 5 mM 2-mercaptoethanol.

Step 1. Preparation of Cell-Free Extract. *Bacillus megaterium* NIHB 12 is cultivated at 28° for 24 hr with shaking in 500 ml of medium containing 50 g of soluble starch, 30 g of peptone, 5 g of meat extract, 1 g of K_2HPO_4, 1 g of KCl, and 0.5 g of $MgSO_4 \cdot 7H_2O$ in 1 liter of tap water, pH 7.0. The cells grown in 10 liters of the culture medium are harvested, washed 3 times with 0.85% NaCl solution, and resuspended in 10 mM buffer. The cells are disrupted by the action of a Kaijo-Denki 19-kHz ultrasonic oscillator below 15° for 30 min. The cells and debris are removed by centrifugation at 10,000 g for 20 min. The resultant supernatant solution is dialyzed overnight against 10 liters of 50 mM buffer.

Step 2. First Ammonium Sulfate Fractionation. Solid ammonium sulfate is added to the dialyzed cell-free extract to 0.70 saturation. The precipitate is removed by centrifugation at 10,000 g for 20 min. The supernatant solution is dialyzed overnight against two changes of 10 liters of the same buffer.

Step 3. DEAE-Cellulose Column Chromatography. The dialyzed enzyme solution is applied to a DEAE-cellulose column (4.5 × 34 cm) equilibrated with 10 mM buffer. After the column has been washed with 0.1 M buffer containing 0.1 M NaCl, the enzyme is eluted with 0.1 M buffer containing 0.2 M NaCl, and the active fractions are pooled.

Step 4. Second Ammonium Sulfate Fractionation. Solid ammonium sulfate is added to the enzyme solution to 0.30 saturation. After standing for 30 min, the precipitate is removed by centrifugation and discarded. The ammonium sulfate concentration is then increased to 0.50 saturation. After standing for 30 min, the resulting precipitate is collected by centrifu-

TABLE I

PURIFICATION OF PIMELYL-CoA SYNTHETASE FROM *Bacillus megaterium*

Step	Total protein (mg)	Total activity[a] (units)	Specific activity[a]
Cell-free extract	2642	4884	1.9
Ammonium sulfate	2028	7739	3.8
DEAE-cellulose	353	6369	18.0
Ammonium sulfate	63	3876	62.0

[a] Assayed by procedure A.

gation and dissolved in 50 mM buffer. The solution is then dialyzed overnight against the same buffer.

The results of the purification procedure are summarized in Table I.

Properties

Effect of Temperature and pH. The optimum temperature is 32°. The pH optimum is around 8.0.

Specificity. Dicarboxylic acids other than pimelic acid, such as oxalic acid, succinic acid, glutaric acid, and azelaic acid, do not serve as substrate. ADP is 85% as effective as ATP, whereas AMP and the other nucleotides do not replace ATP at all. Mn^{2+} has a remarkable effectiveness, which is 50% higher than Mg^{2+}.

Inhibitors. A remarkable inhibition is observed with such metal-chelating agents as α,α'-dipyridyl, o-phenanthroline, and EDTA, and with Fe^{3+}.

Distribution

Enzyme activities are observed in *Escherichia coli* AKU[9] 0007, *Aerobacter aerogenes* IFO 12010, *Micrococcus roseus* IFO 3764, *Pseudomonas riboflavina* IFO 3140, and *P. fluorescens* AKU 0821, as well as *B. megaterium*.

II. KAPA Synthetase[10-12]

Assay Method

Principle. KAPA synthetase catalyzes the synthesis of KAPA from pimelyl-CoA and L-alanine with PLP as a coenzyme. The KAPA synthesized is determined by bioassay.

Reagents

Potassium phosphate buffer, 1 M, pH 7.0

Pimelyl-CoA (synthesized by the method of Eisenberg and Star[13]), 1.5 mM

[9] Stock cultures preserved in the Laboratory of Fermentation Physiology, Department of Agricultural Chemistry, Kyoto University, Kyoto, Japan.

[10] Y. Izumi, H. Morita, Y. Tani, and K. Ogata, *Agric. Biol. Chem.* **37,** 1327 (1973).

[11] Y. Izumi, H. Morita, Y. Tani, and K. Ogata, *Agric. Biol. Chem.* **36,** 510 (1972).

[12] Y. Izumi, K. Sato, Y. Tani, and K. Ogata, *Agric. Biol. Chem.* **37,** 1335 (1973).

[13] M. A. Eisenberg and C. Star, *J. Bacteriol.* **96,** 1291 (1968).

L-Alanine, 0.5 M
PLP, 1 mM

Procedure. The enzyme assay is performed in a standard reaction mixture containing 10 μmol of potassium phosphate buffer, 15 nmol of pimelyl-CoA, 5 μmol of L-alanine, 10 nmol of PLP, and enzyme in a final volume of 100 μl. Reaction is carried out for 1 hr at 37° and stopped by heating the mixture for 2 min in a boiling water bath. The KAPA synthesized is microbiologically determined by the paper disc plate method using *Saccharomyces cerevisiae*.

Enzyme Unit. One unit of enzyme activity is defined as the amount of enzyme that synthesizes 1 nmol of KAPA under the assay conditions. Specific activity is defined as units per miligram of protein per hour. Protein is determined as described in Section I.

Purification Procedure

All operations are carried out around 5° unless otherwise stated. The buffer used throughout the purification is potassium phosphate buffer, pH 7.0, containing 5 mM 2-mercaptoethanol.

Step 1. Preparation of Cell-Free Extract. *Bacillus sphaericus* IFO 3525 is grown at 28° for 24 hr in a 20-liter jar-fermentor containing 15 liters of medium consisted of 20 g of glycerol, 20 g of peptone, 5 g of casamino acid, 1 g of K_2HPO_4, 1 g of KCl, 0.5 g of $MgSO_4 \cdot 7H_2O$, 0.01 g of $MnSO_4 \cdot 4 - 6H_2O$, 0.01 g of $FeSO_4 \cdot 7H_2O$, 2 mg of thiamine·HCl, and 2 g of Antifoam AF emulsion (Dow Corning Silicone) in 1 liter of tap water, pH 7.2. The cells are harvested, washed with 0.85% NaCl solution, and resuspended in 10 mM buffer. The suspension is subjected to the action of a Kaijo-Denki 19-kHz ultrasonic oscillator below 10° for 10 min. The cells and debris are removed by centrifugation. The resultant supernatant is dialyzed overnight against 10 liters of the same buffer.

Step 2. Ammonium Sulfate Fractionation. To 580 ml of the dialyzed cell-free extract is added solid ammonium sulfate to 0.40 saturation, adjusting the pH to 7.0. After standing for 30 min, the precipitate is removed by centrifugation and discarded. The ammonium sulfate concentration is then increased to 0.70 saturation. After standing for 30 min, the precipitate is collected by centrifugation and dissolved in 10 mM buffer. The solution is then dialyzed overnight against 15 liters of the same buffer.

Step 3. Protamine Sulfate Treatment. One-tenth volume of 5.5% protamine sulfate solution at pH 7.0 is added to the dialyzate and the precipitate formed is removed by centrifugation.

Step 4. DEAE-Cellulose Column Chromatography. The enzyme solution is divided into four portions, each of which is subjected to DEAE-

TABLE II
PURIFICATION OF KAPA SYNTHETASE FROM *Bacillus sphaericus*

Step	Total protein (mg)	Total activity (units)	Specific activity
Cell-free extract	11,190	44,430	4.0
Ammonium sulfate	3,192	21,610	6.8
Protamine sulfate	1,439	17,330	12.0
DEAE-cellulose	327	4,416	13.5

cellulose column chromatography. The enzyme solution is applied to the column (2.5 × 20 cm) equilibrated with 10 mM buffer. The column is washed with 30 mM buffer, then the enzyme is eluted with 50 mM buffer. The active fractions are pooled and concentrated with the addition of ammonium sulfate to 0.70 saturation. The resultant precipitate is dissolved and dialyzed against 50 mM buffer.

The results of the purification procedure are summarized in Table II.

Properties

Effect of Temperature and pH. The optima of temperature and pH are 60° and 7.0, respectively.

Stability. The enzyme preparation can be preserved at −20° over 5 months in 10 mM potassium phosphate buffer, pH 7.0, containing 5 mM 2-mercaptoethanol, with little loss of activity. The enzyme is quite stable up to 60°.

Specificity. Among amino acids, only L-alanine is active as a substrate, condensing with pimelyl-CoA.

Inhibitors. Typical inhibitors of PLP enzymes such as phenylhydrazine, semicarbazide, hydroxylamine, D- and L-penicillamine, show marked inhibition to the reaction. Such metal ions as Ag$^+$, Cu^{2+}, Cd^{2+}, Zn^{2+}, and Co^{2+} significantly inhibit the reaction. Various amino acids such as L-cysteine, L-serine, D-alanine, glycine, D-histidine, and L-histidine inhibit the enzyme activity. The inhibition is competitive with L-alanine. Complete repression occurs when the cells grow in the presence of 0.1 μg per milliliter of biotin.

Distribution

Escherichia coli AKU 0007, *Micrococcus roseus* IFO 3764, *Bacillus subtilis* IAM 1193, *Pseudomonas fluorescens* AKU 0821, as well as *B. sphaericus* have relatively high activity.

III. DAPA Aminotransferase[14,15]

Assay Method

Principle. DAPA aminotransferase catalyzes the synthesis of DAPA from KAPA and *S*-adenosyl-L-methionine with PLP or PMP as a coenzyme. The assay of the DAPA synthesized in the reaction mixture is performed by measuring the amount of DTB formed in a coupled reaction using DTB synthetase, which catalyzes the synthesis of DTB from DAPA and HCO_3^- in the presence of ATP and Mg^{2+}. The DTB is quantitatively determined by a microbiological assay (paper disc plate method) with *Bacillus subtilis* AKU 0236,[16] a biotin-requiring bacterium, the growth of which can be supported well by DTB but very weakly by DAPA and KAPA. The reaction for the enzyme assay is performed using the following two procedures.

Reagents

KAPA, 0.1 mM

S-Adenosyl-L-methionine (hydrogen sulfate), 10 mM

PLP, 1 mM

Potassium phosphate buffer, 1 M, pH 8.0

DTB synthetase from *Pseudomonas graveolens* IFO 3406, 125 units/ml (see unit definition in Section IV)

$NaHCO_3$ (freshly prepared before use), 0.5 mM

ATP disodium salt, 0.1 mM

$MgCl_2$, 0.1 mM

Procedure A. The standard reaction mixture contains 1 nmol of KAPA, 0.1 μmol of *S*-adenosyl-L-methionine, 10 nmol of PLP, 10 μmol of potassium phosphate buffer, 2.5 units of DTB synthetase, 5 μmol of $NaHCO_3$, 1 μmol of ATP, 1 μmol of $MgCl_2$, and enzyme solution in a final volume of 120 μl. The reaction is carried out for 2 hr at 37° and is stopped by heating the mixture for 2 min in boiling water.

Procedure B. In this procedure, the two reactions are separately carried out. The first reaction mixture, a KAPA–DAPA transamination system, contains 1 nmol of KAPA, 0.1 μmol of *S*-adenosyl-L-methionine, 10 nmol of PLP, 10 μmol of potassium phosphate buffer, and enzyme solution in a final volume of 70 μl. The reaction is carried out for 2 hr at 37°, and is stopped by heating the mixture for 2 min in boiling water. To the heat-treated reaction mixture is added the following mixture for DTB synthetase reaction: the mixture contains 5 μmol of $NaHCO_3$, 1 μmol of

[14] Y. Izumi, K. Sato, Y. Tani, and K. Ogata, *Agric. Biol. Chem.* **39,** 175 (1975).

[15] Y. Izumi, K. Sato, Y. Tani, and K. Ogata, *Agric. Biol. Chem.* **37,** 2683 (1973).

[16] S. Iwahara, S. Takasawa, T. Tochikura, and K. Ogata, *Agric. Biol. Chem.* **30,** 1069 (1966).

ATP, 1 μmol of $MgCl_2$, and 2.5 units of DTB synthetase in a total volume of 50 μl. The reaction is carried out for another 2 hr at 37°, and is stopped by heating the whole for 2 min in boiling water.

Enzyme Unit. One unit of enzyme activity is defined as the amount of enzyme synthesizing 1 nmol of DTB for 2 hr in the assay systems described above. Specific activity is expressed as units per milligram of protein. Protein is determined as described in Section I.

Purification Procedure

All operations are carried out around 5° unless otherwise stated. The buffer used throughout purification is potassium phosphate buffer, pH 7.0, containing 5 mM 2-mercaptoethanol.

Step 1. Preparation of Cell-Free Extract. Brevibacterium divaricatum NRRL 2311 is cultivated at 28° for 48 hr in a 30-liter jar-fermentor containing 15 liters of medium consisting of 20 g of glycerol, 80 g of corn steep liquor, 1 g of K_2HPO_4, 1 g of KCl, and 0.01 g of $MnSO_4 \cdot 4 - 6H_2O$ in 1 liter of tap water, pH 7.0. The cells are harvested by centrifugation, washed with tap water, and suspended in about 1 liter of 10 mM buffer. The cell suspension is subjected to the action of a Kaijo-Denki 19-kHz ultrasonic oscillator for 4 hr. The cells and debris are removed by centrifugation. The supernatant solution is used as the crude cell-free extract.

Step 2. Ammonium Sulfate Fractionation. Solid ammonium sulfate (3.9 kg) is added to the crude cell-free extract (10.3 liters) from 150 liters of the culture medium to bring to 0.50 saturation. After stirring it for 30 min, the resulting precipitate is collected by centrifugation, and dissolved in 10 mM buffer. The solution is then dialyzed overnight against 15 liters of the same buffer.

Step 3. Acetone Treatment. Acetone is added to the dialyzed enzyme solution below 0° to 70% (v/v), with continuous stirring. After standing overnight at 5°, the resulting precipitate is collected by centrifugation and dissolved in 50 mM buffer. The enzyme solution is then dialyzed overnight against 15 liters of the same buffer.

Step 4. DEAE-Cellulose Column Chromatography. The enzyme solution is applied to a DEAE-cellulose column (6 × 70 cm) equilibrated with 50 mM buffer. The protein is eluted stepwise with 0.1 M and 0.2 M buffer. The active fractions, which are eluted with 0.2 M buffer, are combined and concentrated by salting out with ammonium sulfate at 0.60 saturation. The resulting precipitate is dissolved in 10 mM buffer, and dialyzed overnight against the same buffer.

Step 5. Hydroxyapatite Column Chromatography. The enzyme solution is applied to a hydroxyapatite column (6 × 18 cm) equilibrated with 10 mM buffer. The protein is eluted stepwise with 30 mM, 50 mM, and 0.1

TABLE III

PURIFICATION OF DAPA AMINOTRANSFERASE FROM *Brevibacterium divaricatum*

Step	Total protein (mg)	Total activity[a] (units)	Specific activity[a]
Cell-free extract	56,800	26,100	0.46
Ammonium sulfate	39,100	21,100	0.54
Acetone	11,400	65,900	5.8
DEAE-cellulose	650	62,600	96.3
Hydroxyapatite	43	32,400	753
Sephadex G-150 (I)	15	19,500	1300
Sephadex G-150 (II)	9.5	22,700	2390

[a] Assayed by procedure A.

M buffer. The active fractions, which are eluted with 0.1 M buffer, are combined and concentrated by the salting-out described above.

Step 6. First Sephadex G-150 Gel Filtration. The resulting precipitate is dissolved in 3.0 ml of 10 mM buffer, and filtered on a Sephadex G-150 column (2 × 100 cm) equilibrated with the same buffer. The buffer is allowed to flow at a rate of 20 ml/hr, and 3-ml fractions are collected. The active fractions are combined and concentrated by salting-out.

Step 7. Second Sephadex G-150 Gel Filtration. The resulting precipitate is dissolved in a minimum volume of 10 mM buffer, and filtered through the same Sephadex column by the procedure described above. The active fractions are combined and concentrated by salting-out. The resulting precipitate is dissolved in a minimum volume of 10 mM buffer. The enzyme solution is dialyzed overnight against 10 mM buffer.

The purified enzyme preparation is homogeneous on polyacrylamide gel disc electrophoresis, and it sediments as a single component in the ultracentrifuge. A summary of the purification is shown in Table III.

Properties

Effect of Temperature and pH. The optima of temperature and pH are 37° and 8.5, respectively. The enzyme is stable in a range of 7 to 60° and at a pH range of 7.0 to 11.0.

Specificity. Only S-adenosyl-L-methionine shows activity as an amino donor. 7-Amino-8-ketopelargonic acid is only one-hundredth as active as KAPA as an amino acceptor. Both PLP and PMP are effective as cofactors.

Inhibitors. Enzyme activity is extremely inhibited by phenylhydrazine, semicarbazide, isoniazid, and hydroxylamine. Inhibitors of thiol enzymes, such as p-chloromercuribenzoate, $HgCl_2$, and iodoacetate, also strongly inhibit enzyme activity. Among metal ions, Co^{2+} and As^{5+}

remarkably inhibit activity. A complete repression of enzyme synthesis is observed on the addition of only 0.1 μg of biotin per milliliter to the culture medium.

Distribution

Salmonella typhimurium AJ 2635, *S. typhimurium* AJ 2636, *Bacillus roseus* IAM 1257, *Micrococcus roseus* IFO 3764, and *Escherichia coli* AKU 0015, as well as *B. divaricatum,* have relatively high activity.

IV. DTB Synthetase[17]

Assay Method

Principle. DTB synthetase catalyzes the synthesis of DTB from DAPA, HCO_3^-, ATP, and Mg^{2+}. The DTB synthesized is determined by bioassay with *Bacillus subtilis.*

    Reagents
      Potassium phosphate buffer, 1 M, pH 7.2
      DAPA·2HCl, 0.1 mM
      $NaHCO_3$ 0.5 mM (freshly prepared before use)
      $MgCl_2$, 0.1 mM
      ATP disodium salt, 0.1 mM

Procedure. The standard reaction mixture contains 100 μmol of potassium phosphate buffer, 0.1 μmol of DAPA, 50 μmol of $NaHCO_3$, 10 μmol of ATP, and enzyme solution in a total volume of 1.2 ml. Incubation is carried out for 2 hr at 37°. The reaction is stopped by heating the mixture for 2 min in boiling water. The DTB is determined quantitatively by a microbiological method (paper disc plate method) with *B. subtilis.*[16]

Enzyme Unit. One unit of enzyme activity is expressed as the amount of enzyme that synthesizes 1 nmol of DTB under the assay conditions. Specific activity is defined as units per milligram of protein per 2 hr. Protein is determined as described in Section I.

Purification Procedure

All operations are carried out at 5° unless otherwise stated. The buffer used throughout the purification is potassium phosphate buffer, pH 7.0, containing 5 mM 2-mercaptoethanol.

---

[17] K. Ogata, Y. Izumi, K. Aoike, and Y. Tani, *Agric. Biol. Chem.,* **37**, 1093 (1973).

Step 1. Preparation of Cell-Free Extract. Pseudomonas graveolens IFO 3406 is cultivated at 28° for 24 hr in a 20-liter jar-fermentor containing 15 liters of medium consisted of 20 g of glycelol, 20 g of peptone, 1 g of K_2HPO_4, and 0.5 g of $MgSO_4 \cdot 7H_2O$ in 1 liter of tap water, pH 7.2. The cells from 100 liters of culture broth are harvested by centrifugation, washed with 50 mM buffer, and suspended in the same buffer. The suspension is subjected to the action of a Kaijo-Denki 19-kHz ultrasonic oscillator for 30 min. The cells and debris are removed by centrifugation. The resultant supernatant is dialyzed overnight against two changes of a total of 70 liters of the same buffer.

Step 2. Ammonium Sulfate Fractionation. To 2300 ml of the supernatant solution is added 349.6 g of solid ammonium sulfate to 0.20 saturation. After standing overnight, the precipitate is removed by centrifugation and discarded. The ammonium sulfate concentration is then increased to 0.50 saturation by the addition of 485.3 g of solid ammonium sulfate. After standing overnight, the precipitate is collected by centrifugation and suspended in 50 mM buffer. The solution is then dialyzed for 10 hr against two changes of a total of 20 liters of the same buffer. The inactive precipitate formed is removed by centrifugation.

Step 3. First DEAE-Cellulose Column Chromatography. The dialyzed enzyme solution (1960 ml) is applied to a DEAE-cellulose column (6.5 × 90 cm) equilibrated with 10 mM buffer. After the column is washed with the same buffer, the enzyme is eluted with 0.1 M buffer containing 0.4 M NaCl. The active fractions are pooled to give 2560 ml, concentrated by the addition of ammonium sulfate to 0.50 saturation. The resultant precipitate is dissolved and dialyzed against 50 mM buffer.

Step 4. Second DEAE-Cellulose Column Chromatography. The dialyzed enzyme solution (175 ml) is applied to a second DEAE-cellulose column (3 × 45 cm) equilibrated with 0.1 M buffer and subsequently with the same buffer containing 0.1 M NaCl. The enzyme is eluted with an NaCl concentration gradient, using 0.1 M buffer. The final concentration of NaCl is 0.4 M. The active fractions are combined and concentrated with the addition of ammonium sulfate to 0.50 saturation. The resultant precipitate is dissolved and dialyzed against 50 mM buffer.

Step 5. Hydroxyapatite Column Chromatography. The dialyzed enzyme solution (52 ml) is chromatographed on a hydroxyapatite column (4 × 5 cm) equilibrated with 10 mM buffer. After the column is washed with the same buffer, the enzyme is eluted with 30 mM buffer. The active fractions are combined to give 1250 ml and concentrated by the addition of ammonium sulfate to 0.50 saturation.

Step 6. Sephadex G-100 Gel Filtration. The precipitate is dissolved in a minimum volume of 25 mM buffer. The enzyme solution is filtered on a

TABLE IV

PURIFICATION OF DTB SYNTHETASE FROM *Pseudomonas graveolens*

Step	Total protein (mg)	Total activity (units)	Specific activity
Cell-free extract	114,880	165,000	1.44
Ammonium sulfate	97,440	81,500	0.84
DEAE-cellulose (I)	3,328	71,800	21.5
DEAE-cellulose (II)	2,835	28,800	9.75
Hydroxyapatite	149.5	25,800	158.0
Sephadex G-100	12.0	3,100	259.0

Sephadex G-100 column (2.4 × 88 cm) equilibrated with the same buffer. The buffer is allowed to flow at a rate of 18 ml/hr, and 3 ml fractions are collected.

A summary of the purification procedure is presented in Table IV.

Properties

Effect of Temperature and pH. The optimum temperature and pH are about 50–55° and 7.0–8.0, respectively.

Stability. The enzyme precipitated by salting-out with ammonium sulfate can be stored in ammonium sulfate solution at 5° over a few years without any loss of activity.

Specificity. The activity of biotin diaminocarboxylic acid as substrate is about one-tenth that of DAPA. CTP, UTP, GTP, and ITP are 10–20% as effective as ATP. But ADP and AMP cannot replace ATP. Mn^{2+} shows 95–136% and Fe^{2+} 71–91%, of the activity of Mg^{2+}.

Inhibitors. The enzyme reaction is strongly inhibited by such chelating agents as EDTA, α,α'-dipyridyl, and o-phenanthroline. ADP, which is formed from ATP in the reaction, shows competitive inhibition toward ATP. The substrates DAPA and biotindiaminocarboxylic acid are competitive with each other.

Acknowledgment

We wish to thank Professor H. Yamada, Kyoto University, for his interest in this work and for valuable discussion.

[59] Preparation of Pimeloyl-Coenzyme A

By Max A. Eisenberg

Principle. Pimeloyl-CoA is prepared by the ester exchange method of Kloss and Dickinson.[1] However, pimelic acid forms a polymeric rather than a cyclic anhydride; thus, the usual procedures for thioester formation produce both the mono- and dithioesters, which must be separated prior to the exchange reaction. Separation is based on the differential solubility of the monoester in dilute base.

Preparation of the α-Anhydride of Pimelic Acid

The polymeric anhydride is prepared by the method of Hill and Carothers.[2] Dry pimelic acid, 10 g (62.5 mmol) and 27.4 ml (290 mmol) of acetic anhydride are added to a 250-ml, round-bottom flask fitted with a reflux condenser and a calcium chloride drying tube. The reaction mixture is heated under reflux for 5 hr and then permitted to cool. Volatile material is distilled off under vacuum (15 mm) at 30°–34°, and the yellowish oily residue is dissolved in hot benzene and filtered while hot. Petroleum either (30°–60°) is then slowly added until the solution becomes turbid. On refrigeration, a white solid separates.

Preparation of Mono-*o*-thiocresol Pimelate

The solvent is removed, and the white solid pimelic anhydride is transferred to a three-neck, 250-ml round-bottom flask fitted with a condenser, drying tube, and an addition funnel. Anhydrous ethyl ether (75 ml) is added to the reaction flask and the mixture is heated until the anhydride dissolves. *o*-Thiocresol (1.6 ml) in 50 ml of dry ethyl ether is added dropwise to the reaction mixture over a 20-min period with stirring. The reaction mixture is heated under reflux and stirred overnight. Thin-layer chromatography on silica gel in benzene–acetone (9:1 v/v) reveals two purple spots with the hydroxamic acid test.[3] The spots of R_f values of 0.43 and 0.81 represent the monothioester and dithioester, respectively.

The mixture is poured into a separatory funnel and extracted six times with 25 ml of 2% sodium bicarbonate. The combined aqueous extracts are acidified with 2 N HCl (Congo Red) and extracted with ether until the water layer becomes clear. By repeating the extraction procedure and

---

[1] R. A. Kloss and J. E. Dickinson, *Biochim. Biophys. Acta* **70**, 90 (1963).

[2] J. W. Hill and W. H. Carothers, *J. Am. Chem. Soc.* **55**, 5023 (1933).

[3] See this series, Vol. 3 [137].

PAPER CHROMATOGRAPHY (R_f) OF SYNTHETIC PIMELOYL-CoA[a]

Sample	Solvent system[b]							
	1				2			
	Ultraviolet light	Hydroxamic acid	Nitroprusside	Nitroprusside + methanolic KOH	Ultraviolet light	Hydroxamic acid	Nitroprusside	Nitroprusside + methanolic KOH
CoA	0.01	-	-	-	0.17	-	-	-
Pimelic o-thiocresol ester	0.10	-	++	++	0.45	-	++	++
	0.91	++	-	++	0.88	++	-	++
Reaction mixture	0.04	-	-	-	0.25	-	-	-
	0.09	-	-	-	0.50	++	-	++
	0.18	++	-	++	0.75	-	-	+
	0.93	+	-	+	0.89	+	-	+

[a] M. A. Eisenberg and C. Star, *J. Bacteriol.* **96**, 1291 (1968).

[b] Solvent system 1: *n*-butyl alcohol–acetic acid–water (60:15:25); solvent system 2: ethyl alcohol and sodium acetate, pH 4.5 (1:1). Symbols: ++, intense; +, light; −, negative.

monitoring the extent of resolution by thin-layer chromatography, an ether solution is finally obtained which is free of the dithioester. A comparison of the chromatographic behavior of the hydroxamic acid derivatives from purified monothioester and monomethyl ester on silica gel thin-layer chromatography in butanol–acetic acid–H_2O (60: 15: 25) shows the same R_f value of 0.86. After drying the final ether extract over anhydrous sodium sulfate, the ether is removed with a stream of dry nitrogen leaving a yellow viscous oil.

Preparation of Pimeloyl-Coenzyme A

Pimeloyl-CoA is prepared from the purified mono-o-thiocresol ester of pimelic acid and coenzyme A according to the procedure of Kloss and Dickinson.[1] Purified mono-o-thiocresol pimelate, 26 mg (97 μmol) and 6.0 mg of coenzyme A (7.7 μmol) are added to 2.5 ml of distilled water in a 12-ml, glass-stoppered centriguge tube. The tube is shaken until the monothioester dissolves. $KHCO_3$ (20 mg) is then added, and the reaction mixture is allowed to stand at room temperature for 1 hr, at which time the solution becomes cloudy owing to the formation of insoluble o-thiocresol. The pH is adjusted to 4 with dry Dowex 50 (H^+)-4X, and the reaction mixture is extracted with ether until the water layer becomes clear. The sample is kept cold during the extraction procedure. The water layer is then separated from the resin, which is washed with a small amount of water. The combined aqueous solutions are adjusted to pH 5.0 with 0.1 N KOH, the dissolved ether is removed by bubbling nitrogen through the solution, and the solution is frozen at $-20°$. The yield ranges from 37 to 52% of the theoretical value.

Chromatographic analysis of the pimeloyl-CoA on Whatman 3 mm paper is shown in the table. One major and 3 minor bands are observed under ultraviolet light. The nitroprusside test[3] for free thiol is negative unless the paper is subsequently treated with methanolic KOH; then 3 pink spots appear on the chromatogram from solvent system 2, and 2 spots from solvent system 1. The hydroxamic acid test for thioester is positive for only two of the components. One of the components corresponds to pimelic mono-o-thiocresol ester and the other, with R_f values of 0.18 and 0.50 in solvent systems 1 and 2, respectively, corresponds to pimeloyl-CoA.

The above preparation can be used as such for the assay of 7-keto-8-aminopelargonic acid synthetase.[4] The pimeloyl-CoA can be purified by DEAE-cellulose chromatography according to the procedure of Gregolin et al.[5]

[4] This volume (7-keto-8-aminopelargonic acid synthetase).
[5] C. Gregolin, E. Ryder, and M. D. Lane, *J. Biol. Chem.* **243**, 4227 (1968).

[60] 7,8-Diaminopelargonic Acid Aminotransferase

By MAX A. EISENBERG and GERALD L. STONER

$$\underset{\substack{\text{7-Keto-8-aminopelargonic}\\\text{acid (7KAP)}}}{\overset{\substack{\text{H}_2\text{N}\quad\text{O}\\|\quad\ \ ||}}{\text{CH}_3-\text{CHC}-(\text{CH}_2)_5\text{COOH}}} \xrightarrow[\text{PLP}]{\text{Ado-Met}} \underset{\substack{\text{7,8-Diaminopelargonic}\\\text{acid (DAPA)}}}{\overset{\substack{\text{H}_2\text{N}\quad\ \text{NH}_2\\|\qquad\ \ |}}{\text{CH}_3-\text{CH}-\text{CH}(\text{CH}_2)_5\text{COOH}}}$$

This enzyme is unique among aminotransferase enzymes in its absolute requirement for S-adenosyl-L-methionine (Ado-Met) as the amino donor. The enzyme has been demonstrated in resting cells and in cell-free extracts of a variety of organisms[1-3] and purified from *Escherichia coli*[4] and *Brevibacterium devaricatum*.[3]

Assay Method

Principle. The conversion of 7KAP to DAPA is estimated by the agar plate bioassay procedure[5] utilizing a biotin auxotroph that is defective in the aminotransferase enzyme. This mutant will utilize DAPA, but not 7KAP, as a growth factor. The assay strain, *bio*A-109 was selected for its ability to grow on concentrations of DAPA as low as 5 ng/ml. Most of the mutants in this class require 25–50 times as much DAPA for growth.

Reagents

Tris·HCl buffer, 0.75 M, pH 8.5
7-Keto-8-aminopelargonic acid,[6] 0.1 mM, prepared in 10 mM HCl
S-Adenosyl-L-methionine, 25 mM. Purity of commercial preparation should be checked by chromatography
7,8-Diaminopelargonic acid sulfate standards (0.05 to 2.5 μM)
Pyridoxal phosphate, 0.002 M
Dithiothreitol, 0.05 M
Trichloroacetic acid, 12%
Protein solution
Bioassay medium: Vogel–Bonner[7] minimal agar with 0.01% L-

[1] M. A. Eisenberg and G. L. Stoner, *J. Bacteriol.* **108,** 1135 (1971).
[2] C. H. Pai, *J. Bacteriol.* **105,** 793 (1971).
[3] Y. Izumi, K. Sato, Y. Tami, and K. Ogata, *Agric. Biol. Chem.* **39,** 175 (1975).
[4] G. L. Stoner and M. A. Eisenberg, *J. Biol. Chem.* **250,** 4029 (1975).
[5] D. S. Genghof, C. W. H. Partridge, and F. H. Carpenter, *Arch. Biochem. Biophys.* **17,** 413 (1948).
[6] T. Suyama and S. Kaneo, Japanese patent No. 19,716. *Biol. Abstr.* **60,** 4013 f (1964).
[7] H. J. Vogel and D. M. Bonner, *J. Biol. Chem.* **218,** 97 (1956).

leucine·HCl, 0.005% thiamine·HCl, and 0.007% 2,3,5-Triphenyl-tetrazolium chloride

Procedure. The total volume of the reaction mixture is 0.50 ml. The enzyme preparation (0.05–0.10 ml) is preincubated for 10 min at 37° with 0.05 ml of Tris buffer, 0.05 ml of dithiothreitol, and sufficient distilled water to bring the volume to 0.30 ml. Equal volumes of both substrates, pyridoxal phosphate, and Tris buffer are mixed in a separate tube, and 0.2 ml of the mixture is added to start the reaction. The incubation is continued for 30 min without shaking, and the reaction is terminated by the addition of 0.25 ml of 12% trichloroacetic acid. The precipitate is removed by centrifugation, and 20 μl of the supernatant are spotted on 6-mm filter paper penicillin-assay disks which are placed onto the agar surface of the minimal medium seeded with *bio*A-109. Disks containing 20 μl of the appropriate standards of DAPA are also included on the plate. The plates are incubated for 18 hr at 37°. The concentration of DAPA is estimated from the diameter of growth.[5]

Definition of Unit. One unit of DAPA aminotransferase activity is that amount which catalyzes the formation of 1 μmol of DAPA per minute under the standard conditions of assay. The specific activity is expressed as units per milligram of protein as determined by the Lowry[8] procedure with crystalline bovine serum albumin as the standard or at low concentrations of protein by a modification of the light-scattering method of Tappan.[9]

Purification

In the procedure described below, all buffers are prepared with glass-distilled water. Determinations of buffer pH are made at ambient temperature. All procedures are carried out at 0–4°, except where noted.

Growth of Cells. Regulatory mutants (*bio*R) were used for the enzyme purification because of the derepressed levels of the biotin-synthesizing enzymes. Cells are grown in 50 liters of a tryptone–maltose medium in a 100-liter capacity fermentor (Model F-130, New Brunswick Scientific Co.). The tryptone broth contained in 48 liters of deionized water: 500 g of Bacto-tryptone, 250 g NaCl, 125 g MgSO$_4$·7H$_2$O and 25 ml of 10% NaOH. After sterilization in the fermentor, 500 ml of filter-sterilized 20% maltose is added aseptically. The inoculum is 2 liters of an overnight culture grown in the same medium. The cells are grown to late log phase (200 Klett units, 66 filter) with yields of 5–7 g/liter. The cells are used immedi-

[8] O. H. Lowry, N. J. Rosebrough, A. L. Farr, and R. J. Randall, *J. Biol. Chem.* **193**, 265 (1951).

[9] D. V. Tappan, *Anal. Biochem.* **14**, 171 (1966).

ately or spread in a thin layer and frozen at $-20°$. The enzyme is stable at this temperature for one year.

Step 1. Preparation of Cell-Free Extract. The enzyme is prepared from 75 g of *E. coli* K12 strain Y_{10-1}, *bio*R-10 previously stored at $-20°$. The cells are thawed and suspended in buffer A: 50 mM potassium phosphate, pH 7.0, containing 10 mM 2-mercapoethanol, 1 mM EDTA, and 50 μM pyridoxal phosphate (4 ml of buffer per gram of cells). The cells are disrupted by intermittent sonication (Bronson, Model W 140D), and the sonic extract is centrifuged at 30,000 rpm for 1 hr (30 rotor, Beckman, Model L).

Step 2. Protamine Sulfate Precipitation of Nucleic Acids. A 2% solution of protamine sulfate prepared in buffer A is added dropwise (0.25 ml per milliliter of extract) while the extract is continuously stirred. Stirring is continued for another 20 min, and the mixture is centrifuged at 12,000 rpm for 30 min (GSA rotor. Sorvall RC2-B).

Step 3. Ammonium Sulfate Fractionation. The protein fraction precipitating between 35 and 55% ammonium sulfate saturation is prepared as follows: Powdered ammonium sulfate (62.7 g) is slowly added to the protamine-treated supernatant (300 ml) while the mixture is continuously stirred. Stirring is continued for 20 min; after centrifugation as in step 2, the precipitate is discarded. Ammonium sulfate (42.6 g) is added to the supernatant (325 ml) as before. The precipitate is collected by centrifugation, dissolved in 70 ml of buffer B (10 mM potassium phosphate buffer, pH 7.0, containing 10 mM 2-mercaptoethanol and 10 mM pyridoxal phosphate), and dialyzed overnight against 3 liters of the same buffer.

Step 4. Acid Treatment. Acetic acid (50 mM) is added to the dialyzed enzyme solution (79 ml) until the pH falls to 5.1 and the solution is allowed to stand undisturbed for 1 hr. After centrifugation at 18,000 rpm for 40 min (SS-34 rotor, Sorvall RC2-B), the phosphate concentration of the supernatant is raised to 50 mM by the addition of 1 M potassium phosphate, pH 7.0, and the pH is adjusted to 7.0 by the addition of 0.1 N NaOH.

Step 5. Heat Treatment. The enzyme solution, contained in two 125-ml flasks, is heated with gentle swirling in a 65° water bath to a temperature of 59°–60°, and then transferred to a 60°–61° bath for an additional 10 min. After rapid cooling in ice water, the precipitated protein is removed by centrifugation for 60 min as in step 4, and the supernatant is frozen at $-20°$.

Step 6. DEAE-Cellulose Column Chromatography. The heat-treated enzyme is thawed and centrifuged at 12,000 rpm for 20 min (GSA rotor, Sorvall RC2-B). The enzyme solution is then placed in a 50° water bath for

20 min. This treatment has been found to activate the enzyme and thus eliminates a shoulder on the main activity peak eluted from the DEAE-cellulose column. After chilling to 0°, NaCl is added to a final concentration of 50 mM and the solution is applied to a DEAE-cellulose column (2.6 × 60 cm) which has been equilibrated with buffer C; 50 mM potassium phosphate buffer, pH 7.0, containing 10 mM 2-mercaptoethanol, 10 mM PLP, and 50 mM NaCl. The enzyme solution (82 ml) is applied to the column at a flow rate of 2 ml/min and then eluted with a salt gradient consisting of 650 ml of the equilibrating buffer in the mixing chamber and 650 ml of buffer C containing 0.30 M NaCl in the feeding reservoir. Fractions of 15 ml are collected at a flow rate of 1.5 ml/min. The absorbance of each fraction at 280 nm is determined, and even-numbered fractions are assayed for DAPA aminotransferase activity. Fractions 64–80, which contain 81% of the enzyme activity, are combined (250 ml), and the protein is precipitated by the addition of ammonium sulfate to 60% saturation. The suspension is centrifuged at 12,000 rpm (GSA rotor, Sorvall RC2-B) for 1 hr. The precipitated protein is dissolved in 10 ml of buffer D (10 mM potassium phosphate buffer, pH 6.0, containing 10 mM 2-mercaptoethanol, 1 mM EDTA, and 0.01 mM pyridoxal phosphate) and dialyzed for 5 hr against 2 liters of the same buffer. The dialyzed enzyme is centrifuged for 60 min as in step 4, and the supernatant is refrigerated overnight.

Step 7. Phosphocellulose Column Chromatography. The dialyzed enzyme solution (11.5 ml) from the DEAE-cellulose column is applied to a column (1.6 × 20 cm) of phosphocellulose (Whatman P-11) previously equilibrated with 250 ml buffer D, and is eluted with the same buffer. Fractions of 3 ml are collected at a flow rate of 0.5 ml/min. The enzyme is eluted in the nonadsorbed protein fractions. Fractions showing biological activity are combined, made 20 mM in phosphate by the addition of 0.5 M potassium phosphate, pH 6.8, and then concentrated to 5.5 ml using a PM-30 membrane in a Diaflo ultrafiltration cell (Model 402) under nitrogen pressure (25 lb).

Step 8. Hydroxyapatite Column Chromatography. The concentrated eluate from the phosphocellulose column is placed in a 50° water bath for 20 min as in Step 7. After chilling to 0°, the solution is applied to a column (3.2 × 4 cm) of hydroxyapatite (Bio-Rad HTP) which has been equilibrated with 20 mM potassium phosphate buffer, pH 6.8, containing 10 mM 2-mercaptoethanol and 10 mM pyridoxal phosphate. The column is washed first with 1 column volume (35 ml) of the equilibrating buffer, then eluted stepwise with 1 column volume of the same buffer containing 50 mM phosphate, followed by 1.5 column volumes of 0.10 M phosphate

SUMMARY OF PURIFICATION

Step	Volume (ml)	Total protein (mg)	Units ($\times 10^3$)	Specific activity	Yield (%)
1. Cell-free extract	244	6560	653	0.10	100
2. Protamine sulfate	300	4920	640	0.13	98
3. $(NH_4)_2SO_4$	79	2730	410	0.15	63
4. Acid treatment	98	1310	407	0.31	62
5. Heat treatment	90	530	380	0.70	58
6. DEAE-cellulose	250	69	218	3.2	33
7. Phosphocellulose	33	30	178	5.9	27
8. Hydroxyapatite	40	9.6	91	9.5	14
9. 1st Sephadex G-100	32	1.1	63	57	10
10. 2nd Sephadex G-100	13	0.33	32	98	5

buffer and finally 1.5 column volumes of 0.50 M phosphate buffer. Fractions of about 3 ml are collected at a flow rate of 0.8 ml/min. The enzyme is eluted by 0.10 M phosphate buffer. The contents of tubes showing biological activity are pooled and concentrated to 2.2 ml by ultrafiltration.

Steps 9 and 10. Sephadex G-100 Column Chromatography and 2-Mercaptoethanol Treatment. Two Sephadex G-100 columns (2.6 $\times$ 100 cm and 1.4 $\times$ 100 cm) are washed with buffer C until the beds are stabilized. The concentrated enzyme solution is made 10% in glycerol, applied to the larger column, and eluted with buffer C. Fractions of 4 ml are collected at a flow rate of 0.5 ml/min. The contents of tubes showing biological activity are combined and concentrated to 1.4 ml by ultrafiltration. 2-Mercaptoethanol is added to a final concentration of 10% and the mixture incubated for 30 min at 37°. The denatured protein is removed by centrifugation at 35,000 rpm for 30 min (40 rotor, Beckman Model L).

Glycerol is added to the supernatant to a final concentration of 10%, and the 2-mercaptoethanol, which interferes with the bioassay for DAPA, is removed by chromatography on the smaller Sephadex column. The enzyme is eluted from the column with buffer C and fractions of 2.2 ml are collected at a flow rate of 0.2 ml/min. A single protein peak which coincided with the enzyme activity is observed. Each fraction is made 20% in glycerol and stored frozen at $-20°$.

A summary of the purification procedure shown in the table indicates an overall purification of nearly 1000-fold with a 5% yield. After the second Sephadex G-100 column only a single major protein band with several minor components is seen on polyacrylamide electrophoresis. The enzyme band constitutes 86% of the total protein as estimated from the area under the protein peaks on a gel scan.

Enzyme Properties

Products of the Reaction. While DAPA can be readily identified by chromatography and quantitated by bioassay, the expected keto product, *S*-adenosyl-2-oxo-4-methylthiobutyric acid, canot be detected. Studies with *S*-adenosyl-L-[2-^{14}C]methionine indicate the product to be labile and suggest that it decomposes nonenzymically to form *S*-methylthioadenosine and 2-oxo-3-butenoic acid.

Molecular Weight. The molecular weight determination by gel filtration and by sucrose gradient centrifugation is 84,000 and 104,000, respectively. The average of 94,000 is taken as an estimate of the molecular weight of the holoenzyme. Sodium dodecyl sulfate disc gel electrophoresis shows one band with a molecular weight of 47,000, indicating 2 subunits. The enzyme can be resolved by preincubation with Ado-Met; the apoenzyme which is thus formed elutes later than the holoenzyme on Sephadex G-200. The position corresponds to a molecular weight of 47,000, indicating that in the absence of the cofactor the enzyme readily disassociates into its subunits. The isoelectric point of the enzyme is 4.7, as determined by isoelectric focusing.

Substrate Specificity. Two analogs of 7KAP, 8-keto-7-aminopelargonic acid and 7,8-diketopelargonic acid can be transaminated by the enzyme to form DAPA. The enzyme is highly specific for *S*-adenosyl-L-methionine.

Kinetic Properties.[10] The K_m values for Ado-Met, 7KAP, pyridoxamine phosphate, and pyridoxal phosphate are 0.20 μM, 1.2 μM, 21 μM, and 32 μM, respectively. The V_{max} for the reaction is 0.158 $\mu mol/mg$ per minute, which represents a turnover number of 17. 7KAP shows strong substrate inhibition with both the resting cells and the purified enzyme. The 7KAP inhibition is competitive with Ado-Met with a K_i of 25 μM. The analog, 8-keto-7-aminopelargonic acid is not inhibitory and has a K_m of 1.0 mM. The V_{max} is about one-sixth that attained with 7KAP.

Inhibitors. *S*-Adenosyl-L-(2 hydroxyl-4-methylthio)butyric acid and adenosine are competitive with Ado-Met and noncompetitive with 7KAP, whereas *S*-adenosyl-L-ethionine, adenine, and 8-keto-7-aminopelargonic acid are noncompetitive with either substrate.

[10] G. L. Stoner and M. A. Eisenberg, *J. Biol. Chem.* **250**, 4037 (1975).

[61] Dethiobiotin Synthetase

By MAX A. EISENBERG and KENNETH KRELL

$$
\begin{array}{c}
H_2N \quad NH_2 \\
| \quad\quad | \\
HC-CH \\
| \quad\quad | \\
H_3C \quad CH_2(CH_2)_4COOH
\end{array}
+ ATP + CO_2 \xrightarrow{Mg^{2+}}
\begin{array}{c}
O \\
\| \\
HN^{\diagup C \diagdown}NH \\
| \quad\quad | \\
HC \quad\quad CH \\
| \quad\quad | \\
H_3C \quad\quad CH_2(CH_2)COOH
\end{array}
+ ADP + P_i
$$

7, 8-Diaminopelargonic   Dethiobiotin
    acid (DAPA)   (DTB)

      The interrelationship between 7,8-diaminopelargonic acid and dethio-biotin was first established when both vitamers were found to satisfy the biotin requirement of certain biotin auxotrophs.[1,2] A precursor–product relationship was suggested from a genetic and biochemical analysis of biotin auxotrophs of *Escherichia coli*,[3,4] and this was experimentally confirmed by the demonstration of the formation of dethiobiotin from DAPA, first in resting cells[5] and then in cell-free extracts of *E. coli*.[6,7] The enzyme has been purified from *E. coli*[8,9] and *Pseudomonas graveolens*.[10]

Assay

      Principle. The enzyme preparation is incubated with DAPA, ATP, Mg^{2+}, and HCO_3^-, and the reaction mixture is deproteinized. The supernatant fluid is analyzed for the product, dethiobiotin, by the agar plate bioassay procedure. The assay organism, a biotin auxotroph (*bio*D-302) which has a defective dethiobiotin synthetase enzyme, can utilize both DTB and biotin as growth factors, but not DAPA.

      Reagents
      Tris·HCl buffer 0.5 *M*, pH 7.7
      ATP, 0.2 *M*, + MgCl₂, 0.2 *M*. Both reagents are combined in one solution, and the pH is adjusted to 6.0.

[1] G. Pontecorvo, *Adv. Genet.* **5,** 141 (1953).
[2] S. Okumura, T. Tsugawa, T. Tsunoda, and S. Motozaki, *Nippon Nogei Kagaku Kaishi,* **36,** 203 (1962).
[3] A. DelCampello-Campbell, G. Kayajanian, and A. Campbell, *J. Bacteriol.* **94,** 2065 (1967).
[4] B. Rolfe and M. A. Eisenberg, *J. Bacteriol.* **96,** 515 (1968).
[5] M. A. Eisenberg and K. Krell, *J. Bacteriol.* **98,** 1227 (1969).
[6] M. A. Eisenberg and K. Krell, *J. Biol. Chem.* **244,** 5503 (1969).
[7] C. H. Pai, *J. Bacteriol.* **99,** 696 (1969).
[8] K. Krell and M. A. Eisenberg, *J. Biol. Chem.* **245,** 6558 (1970).
[9] P. Cheeseman and C. H. Pai, *J. Bacteriol.* **104,** 726 (1970).
[10] K. Ogata, Y. Izumi, K. Oike, and Y. Tani, *J. Agric. Biol. Chem. Jpn.* **37,** 1093 (1973).

7,8-Diaminopelargonic acid sulfate, 1.0 mM. Prepared from either d- or dl-dethiobiotin[11] and purified by column chromatography on Bio-Rad Ag 1-X8 (200–400 mesh) to remove dethiobiotin.
d-Dethiobiotin standards (0.05 to 2.5 μM)
NaHCO$_3$, 0.4 M, prepared fresh
Trichloroacetic acid, 12%
Bioassay medium[12]

Procedure. The reaction mixture in a volume of 1.0 ml contains the following final concentrations: 0.15 M Tris·HCl buffer, pH 7.7; 10 mM ATP; 10 mM MgCl$_2$; 0.125 mM DAPA; and 40 mM NaHCO$_3$. The mixture is equilibrated at 37°, and the reaction is started by the addition of the enzyme preparation. The incubation is continued for 10 min, and the reaction is terminated by the addition of 0.50 ml of 12% trichloroacetic acid. The precipitated protein is removed by centrifugation, and 20 μl of the supernatant fluid are spotted onto 6-mm filter paper penicillin assay disks, which are placed onto the agar surface of a plate seeded with *bio*D-302. Disks containing 20 μl of the appropriate standards of DTB are included on the plate. The plates are incubated for 18 hr. The concentration of DTB is estimated from the diameter of growth.[12]

Dethiobiotin synthase activity in disc gel electropherograms can be readily measured without eluting the enzyme from the gel. The gel is cut into slices of approximately 2 mm, sandwiched between two filter disks, and placed onto the agar surface of the bioassay plate. The uppermost disk is moistened with 20 μl of the assay mixture described above, and the plates are incubated overnight at 37°. The mean diameter of growth around each disk is taken as a measure of dethiobiotin activity.

Definition of Unit. An enzyme unit is that amount which will catalyze the synthesis of 1 nmol of dethiobiotin in 10 min under the assay conditions indicated above. The specific activity is the units per milligram of protein determined by the Lowry method[13] using crystalline bovine serum albumin as the standard.

Purification

Enzyme Sources. It is possible to increase the enzyme levels 3-fold over the wild-type strain of *E. coli* by derepressing a biotin auxotroph or by using a *bio*R mutant that has a defective repressor and is therefore derepressed for all the enzymes in the biotin pathway. A more convenient

[11] V. du Vigneaud, D. B. Melville, K. Folkers, D. E. Wolf, R. Mozingo, J. C. Keresztesy, and S. A. Harris, *J. Biol. Chem.* **146**, 475 (1942).
[12] M. A. Eisenberg and G. Stoner, this volume [61].
[13] O. H. Lowry, N. J. Rosebrough, A. L. Farr, and R. J. Randall, *J. Biol. Chem.* **193**, 265 (1951).

source for greater quantities of enzyme is a λ lysogen carrying the gene for dethiobiotin synthetase (*bio*D). λ Phage BLS7 strain N821, carries a portion of the *E. coli* biotin operon, including the *bio*D, gene which is under λ control. Heat induction of the lysogen leads to replication, transcription, and extended periods of protein synthesis resulting in a 3-fold increase of the enzyme over the depressed levels.

For large-scale production, the lysogen is grown at 34° in 50 liters of Tryptone–maltose medium[12] to mid-log phase (100 Klett units, 66 filter). The temperature is rapidly raised to 41°, and the cells are kept at this temperature with vigorous stirring and aeration for 30 min. The temperature is lowered to 34°, and incubation is continued for 3–5 hr. The cells, harvested by continous-flow centrifugation, are spread in a thin layer and frozen at −20°. The yield is 190–240 g. The enzyme is stable at −20° for at least 8 months.

Step 1. Crude Extract. All steps in the purification are carried out at 0° in the presence of 3 mM 2-mercaptoethanol except where indicated. The cells are thawed and suspended in a sufficient volume of 10 mM phosphate buffer, pH 7.5, to give a final volume of 300 ml. The cells are disrupted by intermittent sonication (Bronson, Model W 140D) for a total of 7 min while cooled in an ice–alcohol bath. The mixture is diluted to 500 ml with the same buffer and the solution stirred until the viscosity is reduced. The addition of pancreatic DNase (0.1 μg/ml) will reduce the viscosity more rapidly. The mixture is centrifuged for 60 min at 11,500 rpm (Sorvall-GSA rotor, RC2-B), and the supernatant is removed. The sediment is resuspended in 100 ml of buffer and recentrifuged. The combined supernatants are dialyzed overnight in 3 liters of 50 mM Tris·HCl buffer, pH 7.0, with one change of buffer.

Step 2. First Ammonium Sulfate Fractionation. The dialyzed crude extract is centrifuged, and sufficient powdered ammonium sulfate for 30% saturation is added to the supernatant solution (490 ml) with continuous stirring. Stirring is continued for another 30 min, and the suspension is centrifuged for 1 hr as described above. More ammonium sulfate is added to the supernatant solution (517 ml) to give a 45% saturated solution, and stirring is continued for another 30 min. The precipitate is separated by centrifugation as above, and the supernatant fluid is discarded. The precipitate is dissolved in 200 ml of 0.15 M Tris·HCl buffer, pH 7.0, and dialyzed for 2 hr in 3 liters of the same buffer with one change.

Step 3. Protamine Sulfate Precipitation. A sufficient volume of protamine sulfate (16 mg/ml in 50 mM Tris·HCl buffer pH 7.0) is slowly added to the 30–45% ammonium sulfate fraction (230 ml) with stirring to give a final concentration of 1 mg of protamine sulfate to 6 mg of protein. The mixture is stirred continuously for another 20 min, then centrifuged for 60 min; the sediment is discarded.

Step 4. Second Ammonium Sulfate Fractionation. The supernatant solution from the protamine sulfate step (250 ml) is subjected to a second ammonium sulfate fractionation exactly as in step 2. The resulting precipitate is dissolved in 35 ml of 50 mM Tris·HCl buffer, pH 7.0 containing 0.1 M NaCl and dialyzed against the same buffer for 4 hr with one change.

Step 5. First DEAE-Cellulose Column Chromatography. The dialyzed fraction is chromatographed on a column (2.5 × 54.5 cm) of DEAE-cellulose (Whatman DE-52) which has been previously equilibrated with the dialysis buffer. The column is eluted with a linear gradient of 0.1 M to 0.3 M NaCl in a total volume of 3 liters of 50 mM Tris chloride buffer, pH 7.0. Fractions of 10 ml are collected at a flow rate of 1 ml/min and assayed for biological activity. Only one component with biological activity is observed emerging in the range of 0.11 to 0.13 M NaCl. The fractions (28–42) containing enzyme activity are pooled and the protein is precipitated with ammonium sulfate at 45% saturation. The precipitate is dissolved in the Tris·HCl buffer without NaCl and dialyzed overnight against the same buffer with two buffer changes.

Step 6. Second DEAE-Cellulose Column Chromatography. The dialyzate from step 5 is rechromatographed in the same manner described above except that a shallower gradient of 0 to 0.2 M NaCl in a total volume of 2 liters of buffer is used. Fractions of 15 ml are collected at a flow rate of 1 ml/min. The contents of the tubes (40–60) showing biological activity are pooled, and the protein is precipitated by the addition of ammonium sulfate to 45% saturation. The precipitate is dissolved in 5 mM phosphate buffer, pH 7.5, containing 6 mM 2-mercaptoethanol and dialyzed against 3 liters of the same buffer with 2 buffer changes.

Step 7. Hydroxyapatite Column Chromatography. The dialyzed sample is applied to a column (1.5 × 10 cm) of hydroxyapatite (Bio-Rad

SUMMARY OF PURIFICATION

| | | | Dethiobiotin synthetase | | |
Step	Volume (ml)	Total protein (mg)	Total (units)	Specific activity (units/mg)	Yield (%)
Dialyzed cell-free extract of induced strain N821 cells	490	16,370	63,500	5.	100
(NH$_4$)$_2$SO$_4$, 1st	230	5,060	76,820	15	121
Protamine sulfate	256	3,710	67,840	18	107
(NH$_4$)$_2$SO$_4$, 2nd	35	3,120	68,075	22	107
DEAE-cellulose column, 1st	25	860	50,000	59.	79
DEAE-cellulose column, 2nd	7.8	50	12,012	244	19
Hydroxyapatite	6.3	15	1,329	864	2

HT), which has been previously equilibrated with the dialysis buffer. The column, which is run at ambient temperature, is eluted with the same buffer at a flow rate of 7 ml/hr. The major peak shows the A_{280} and biological activity to be coincidental.

A summary of the purification is shown in the table. The procedure results in an 190-fold enrichment of the enzyme with a 2% yield. Disc gel electrophoresis of the most purified fraction shows two minor bands comprising less than 5% of the total protein.

Enzyme Properties

Stability. The enzyme is stable at $-20°$ for 2–3 weeks through the second DEAE-cellulose chromatography step. However, after step 7 over 90% of the activity is lost when stored for 24 hr at $-20°$. Storage at $0°$ decreases the rate of inactivation, and at $4°$ the enzyme is stable for about 2 weeks.

Molecular Weight. The molecular weight of the enzyme determined by gel filtration on BioGel P-150 (100–200 mesh) is 42,000. On the basis of SDS disc gel electrophoresis, the molecular weight is estimated to be 24,500, indicating that the native enzyme is a dimer.

Substrate Specificity. CTP is only half as active as ATP for the synthesis of dethiobiotin, whereas ITP, UTP, and GTP are less than 10% as active as ATP.

Diaminobiotin, which can be considered an analog of DPA with an intact tetrahydrothiophene ring, is 37% as active as DAPA under conditions that are optimal for the natural substrate. The product of the reaction was identified as biotin.

CO_2 rather than HCO_3^- is the actual substrate in the dethiobiotin synthetase reaction as determined by the carbonic anhydrase procedure of Cooper *et al.*[14]

Kinetic Parameters. The apparent K_m values for the reactants in the dethiobiotin synthetase reaction are: DAPA, $1.3 + 0.5$ μM; ATP, $5.0 + 2.0$ μM; Mg^{2+}, 0.6 ± 0.1 mM; and HCO_3^-, 3.4 ± 0.7 mM. Appropriate corrections of the latter value by the carbonic anhydrase procedure gave an apparent K_m value of 40 μM for CO_2.

Inhibition. None of the substrates are inhibitory at concentrations up to ten times their respective K_m values. However, ADP is competitive with ATP, with a K_i value of 0.23 mM. Dethiobiotin concentrations up to 1 mM are not inhibitory, but at 10 mM there is 30% inhibition of the enzyme activity.

[14] T. G. Cooper, D. Filmer, M. Wishnick, and M. D. Lane, *J. Biol. Chem.* **244**, 1081 (1968).

[62] Synthesis and Use of Specifically Tritiated Dethiobiotin in the Study of Biotin Biosynthesis by *Aspergillus niger*

By RONALD J. PARRY and M. G. KUNITANI

I. Introduction

The vitamin (+)-biotin (**1**) is widely distributed in plant and animal tissues, where it functions as the cofactor for a variety of enzymic carboxylation reactions.[1] A number of fungi and bacteria synthesize biotin from pimelic acid via a metabolic sequence whose final step is the conversion of (+)-dethiobiotin (**2**) into (+)-biotin[2-5] [Eq. (1)]. The conversion of

(+)-dethiobiotin into (+)-biotin is an unusual biosynthetic process since it involves the introduction of sulfur at two saturated carbon atoms. At the time when we began the work described in detail below, very little was known concerning the nature of the reaction(s) responsible for sulfur introduction. Li *et al.*[6] had reported an experiment in which a mixture of carbon-14-labeled dethiobiotin and "randomly" tritiated dethiobiotin was administered to *Aspergillus niger* and doubly labeled biotin was isolated. The tritium to carbon-14 ratio of the biotin led the authors to suggest that approximately four hydrogen atoms are removed from carbon atoms 1-4 of **2** is a result of its conversion to biotin; it was also suggested that sulfur introduction may proceed with the introduction of unsaturation of C-2 and/or C-3 of dethiobiotin. Unfortunately, these experimental results are attended by considerable uncertainty due to the use of "randomly" tri-

[1] F. Lynen, *Biochem. J.* **102**, 381 (1967).

[2] S. Okumura, R. Tsugawa, T. Tsunada, and S. Motozaki, *Nippon Nogei Kagaka Kaishi* **36**, 599, 605 (1962).

[3] M. A. Eisenberg, *J. Bacteriol.* **86**, 673 (1963).

[4] S. Iwahara, M. Kikuchi, T. Tochikura, and K. Ogata, *Agric. Biol. Chem.* **30**, 304 (1966).

[5] M. Eisenberg, R. Maseda, and C. Star, *Fed. Proc., Fed. Am. Soc. Exp. Biol.* **27**, 762 (1968).

[6] H. C. Li, D. A. McCormick, and L. D. Wright, *J. Biol. Chem.* **243**, 6442 (1968).

METHODS IN ENZYMOLOGY, VOL. 62

tiated dethiobiotin. Because of this fact and because of the interesting nature of the transformation of **2** into **1**, we chose to reexamine the mechanism of the conversion by using specifically tritiated forms of dethiobiotin.[7,8]

II. Syntheses of Specifically Tritiated Forms of (±)-Dethiobiotin

In order to examine the mechanism of sulfur introduction in biotin biosynthesis, it was decided to synthesize (±)-dethiobiotin labeled specifically with tritium at C − 1, − 2, − 3, or − 4. Attempts to utilize published syntheses[9–12] of (±)-dethiobiotin for this purpose proved unsatisfactory. It therefore was necessary to devise a new synthesis of (±)-dethiobiotin that would allow the introduction of specific tritium labels.[7]

Synthesis of [1-³H](±)-Dethiobiotin (22) (Fig. 1)

Synthesis of Acetal Ester 5

In a dry 3-liter, three-neck flask equipped with a reflux condenser, a mechanical stirrer, and a fritted gas inlet tube are placed 225 g of fused, anhydrous sodium acetate and 20 g of 10% palladium on carbon. Two liters of benzene are introduced, the mixture is stirred vigorously, and hydrogen gas is bubbled through the mixture at a rapid rate. 5-Ethoxycarbonyl-*n*-pentanoyl chloride (**3**)[13] (99.7 g, 0.52 mol) is added to the mixture, and the progress of the reduction is monitored by gas-liquid chromatography (GLC) (6 ft, 10% UCW 98). Upon completion of the reduction (ca. 1–1.5 hr), the mixture is washed with 3 × 500 ml of 1 *N* sodium bicarbonate, dried by addition of 10 g of anhydrous magnesium sulfate, and suction-filtered through a bed of Celite. Ethylene glycol (75 ml) and *p*-toluenesulfonic acid (1 g) are added to the filtrate and the mixture is refluxed using a Dean–Stark trap to remove the water. Gas chromatographic analysis (6 ft, 10% UCW 98) reveals that the acetalization is complete within 30 min. The hot benzene solution is poured into a mixture of 500 ml 1 *N* bicarbonate and 500 g of ice. The benzene layer is sepa-

[7] R. J. Parry, M. G. Kunitani, and O. Viele, III, *J. Chem. Soc., Chem. Commun.*, p. 321 (1975).
[8] R. J. Parry and M. G. Kunitani, *J. Am. Chem. Soc.* **98**, 4024 (1976).
[9] J.-B. Bourquin, O. Schnider, and A. Grüssner, *Helv. Chim. Acta* **28**, 528 (1945).
[10] J. L. Wood and V. du Vigneaud, *J. Am. Chem. Soc.* **67**, 210 (1945).
[11] R. Duschinsky and L. A. Dolan, *J. Am. Chem. Soc.* **67**, 2079 (1945).
[12] G. Swain, *J. Chem. Soc., London,* 1552 (1948).
[13] G. B. Brown, M. D. Armstrong, A. W. Moyer, W. P. Anslow, B. R. Baker, M. V. Querry, S. Berstein, and S. R. Safir, *J. Org. Chem.* **12**, 160 (1947).

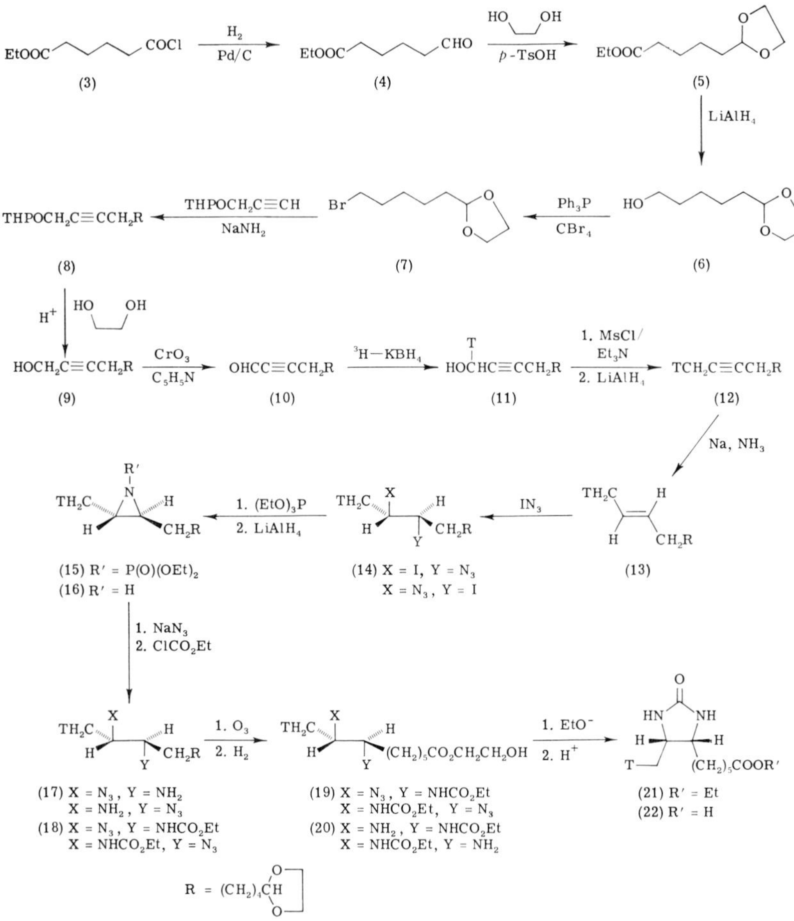

EtOOC~~~COCl $\xrightarrow[\text{Pd/C}]{\text{H}_2}$ EtOOC~~~CHO $\xrightarrow[p\text{-TsOH}]{\text{HO} \quad \text{OH}}$ EtOOC~~~⟨acetal⟩

(3)   (4)   (5)

$\downarrow$ LiAlH$_4$

THPOCH$_2$C≡CCH$_2$R $\xleftarrow[\text{NaNH}_2]{\text{THPOCH}_2\text{C≡CH}}$ Br~~~⟨acetal⟩ $\xleftarrow[\text{CBr}_4]{\text{Ph}_3\text{P}}$ HO~~~⟨acetal⟩

(8)   (7)   (6)

$\downarrow$ H$^+$, HO OH

HOCH$_2$C≡CCH$_2$R $\xrightarrow[\text{C}_5\text{H}_5\text{N}]{\text{CrO}_3}$ OHCC≡CCH$_2$R $\xrightarrow{^3\text{H−KBH}_4}$ $\overset{\text{T}}{\text{HO}\overset{|}{\text{C}}\text{HC≡CCH}_2\text{R}}$ $\xrightarrow[\text{2. LiAlH}_4]{\text{1. MsCl/Et}_3\text{N}}$ TCH$_2$C≡CCH$_2$R

(9)   (10)   (11)   (12)

$\searrow$ Na, NH$_3$

TH$_2$C⋯N(R')⋯H / H⋯CH$_2$R $\xleftarrow[\text{2. LiAlH}_4]{\text{1. (EtO)}_3\text{P}}$ TH$_2$C⋯X⋯H / H⋯Y⋯CH$_2$R $\xleftarrow{\text{IN}_3}$ TH$_2$C⋯H / H⋯CH$_2$R

(15) R' = P(O)(OEt)$_2$ (14) X = I, Y = N$_3$ (13)
(16) R' = H   X = N$_3$, Y = I

$\downarrow$ 1. NaN$_3$ 2. ClCO$_2$Et

TH$_2$C⋯X⋯H / H⋯Y⋯CH$_2$R $\xrightarrow[\text{2. H}_2]{\text{1. O}_3}$ TH$_2$C⋯X⋯H / H⋯Y⋯(CH$_2$)$_3$CO$_2$CH$_2$CH$_2$OH $\xrightarrow[\text{2. H}^+]{\text{1. EtO}^-}$ (ureido ring)

(17) X = N$_3$, Y = NH$_2$ (19) X = N$_3$, Y = NHCO$_2$Et   (21) R' = Et
     X = NH$_2$, Y = N$_3$ X = NHCO$_2$Et, Y = N$_3$ (22) R' = H
(18) X = N$_3$, Y = NHCO$_2$Et   (20) X = NH$_2$, Y = NHCO$_2$Et
     X = NHCO$_2$Et, Y = N$_3$ X = NHCO$_2$Et, Y = NH$_2$

R = (CH$_2$)$_4$CH⟨O⟩

FIGURE 1

rated, washed with 500 ml of saturated brine, dried over magnesium sulfate, filtered, and evaporated. The residue is distilled *in vacuo* yielding 73.8 g (71%) of the ester acetal **5**, b.p. 75°–80°/0.03 mm. The purity by GLC is ca. 90%.

Reduction of Ester 5 to Acetal Alcohol 6

About 700 ml of dry THF and 40 g (1.08 mol) of LiAlH$_4$ are mechanically stirred in a 3-liter, 3-neck flask. The mixture is chilled to −78° in a Dry-Ice acetone bath, and ester **5** (147.5 g, 0.722 mol) is added dropwise over 15 min. After stirring overnight at room temperature, the reaction is quenched by careful addition of ca. 30 ml of 40% aqueous KOH. The mixture is stirred overnight, and 2 liters of ether are added. Additional aqueous KOH solution is then added with stirring over a 2-day period until the salts are white and granular. About 20 g of anhydrous magnesium sulfate is added, and the mixture is filtered. The filter cake is washed with ether, and the combined filtrate and washings are evaporated to dryness. The residue is distilled *in vacuo* to yield 108 g (94%) of the acetal alcohol **6**, b.p. 95°–100°/0.4 mm. The purity of the product is ca. 95% by GLC analysis (6 ft, UCW-98).

Conversion of Alcohol 6 to Bromide 7

Approximately 600 ml of dry THF and 48.0 g (0.30 mol) of alcohol **6** are magnetically stirred in a 2-liter round-bottom flask, and the solution is cooled to −10° in an ice–salt bath. Carbon tetrabromide (129 g, 0.39 mol) is added, followed by the addition of 102 g (0.39 mol) of triphenylphosphine over a period of 5 min. After 1 hr of stirring, the ice–salt bath is removed and the mixture is stirred for an additional 45 min at room temperature. The reaction mixture is filtered, and the filtrate is concentrated *in vacuo* to a volume of ca. 300 ml. One liter of petroleum ether is added slowly, with swirling, to the concentrated filtrate. The precipitated solid is filtered off, washed with petroleum ether, and the combined filtrate and washings concentrated *in vacuo* to a thick oil. About 1 liter of petroleum ether and 5 g of MgSO$_4$ are added to the residue and the mixture chilled in the freezer (−15°) overnight. The precipitated solid is removed by filtration and the filtrate concentrated *in vacuo*. The residual oil is distilled *in vacuo* to give 33.9 g (51%) of the bromide **7**, b.p. 65°–73°/0.1 mm. Purity by GLC is about 90% (6 ft, UCW-98).

Alkylation with Bromide 7 to give Tetrahydropyranyl Ether 8

Five hundred milliliters of liquid ammonia are distilled into a three-liter, three-neck flask equipped with a Dry-Ice condenser, a mechanical stirrer, and a dropping funnel. After addition of a small fragment of ferric nitrate to the liquid ammonia, freshly cut sodium (5.34 g, 0.255 mol) is added slowly with stirring to generate gray sodium amide. The formation of sodium amide is complete after 1–2 hr. The mixture is chilled to −78° in a Dry-Ice acetone bath, and the THP ether of propargyl alcohol[14] (40.8 g, 0.291 mol) is added dropwise with stirring. The mixture is refluxed for 0.5 hr (−33°) with stirring. The reaction mixture is again chilled to −78°,

[14] G. R. Owen and C. B. Reese, *J. Chem. Soc.*, *C*, 2401 (1970).

and 33.9 g (0.146 mol) of the bromide 7 is added dropwise with stirring. The resulting mixture is stirred and refluxed at $-33°$ for 4 hr. The ammonia is then allowed to evaporate. Ammonium chloride (10 g) and 300 ml of water are added to the residue. The aqueous solution is extracted with 3×100 ml of ether. The combined ether extracts are then washed with water, dried with magnesium sulfate, treated with 0.1 g of Norit, filtered, and evaporated to give the crude product. Vacuum distillation gives a forerun of unreacted THP ether of propargyl alcohol followed by 25.6 g (62%) of the acetylenic acetal 8, b.p. $140°-150°/0.03$ mm. The purity of the product determined by GLC analysis (2 ft, 10% OV-17) is about 95%.

Preparation of the Alcohol Acetal 9

The acetylenic acetal 8 (4.24 g, 0.015 mol) is stirred for 3 hr at room temperature in a mixture of 500 ml of glacial acetic acid, 170 ml of water, and 170 ml of ethylene glycol. The water and acetic acid are removed *in vacuo*, and the residue is neutralized with aqueous KOH solution. The mixture is then made basic with 1 *N* sodium bicarbonate. About 600 ml of water are added, and the solution is extracted with 4×250 ml of ether. The combined extracts are washed with 100 ml of 1 *N* bicarbonate, 200 ml of brine, and then dried with magnesium sulfate. Filtration and removal of the solvent gives the crude propargyl alcohol 9. The alcohol 9 is somewhat unstable, and, consequently, it is normally used without purification. The alcohol can be distilled, with care to avoid overheating, to give 2.40 g (80%), b.p. $80-120°/0.05$ mm. The purity of the distilled compound is about 90% by GLC (2 ft, 10% OV-17).

Oxidation of Alcohol Acetal 9 to Aldehyde 10 and Reduction with [³H]Potassium Borohydride to the Alcohol 11

The procedure used for the oxidation is the Collins method as modified by Ratcliffe and Rodehorst.[15]

A mixture of dry pyridine (20 ml) and dry methylene chloride (300 ml) is stirred at room temperature, and anhydrous chromium trioxide (12.0 g, 0.12 mol) is added. The resulting deep-red solution is stirred at room temperature for 15 min. The alcohol acetal 9 (3.96 g, 0.02 mol) is added to the stirred solution, whereupon a dark, tarry precipitate immediately forms. The reaction mixture is stirred for an additional 15 min, the supernatant is decanted, and the insoluble tar washed with ether. The combined supernatant and ether washings are concentrated *in vacuo* and 600 ml of ether

[15] R. Ratcliffe and R. Rodehorst, *J. Org. Chem.* **35**, 4000 (1970).

added to the concentrate. The ethereal solution is suction-filtered, and the filtrate is washed successively with *cold* 1 *N* HCl (2 × 100 ml), 1 *N* sodium bicarbonate (200 ml), and brine (50 ml). The solution is then dried over sodium sulfate, filtered, and evaporated to give the crude aldehyde **10** (2.37 g, 60%).

The crude aldehyde **10** (1.96 g, 0.01 mol) is dissolved in 50 ml of absolute ethanol and ca. 5 mg of sodium borohydride are added. The solution is stirred for 10 min, and [^{3}H]potassium borohydride (6 mg, 500 mCi) is added. The resulting mixture is stirred at room temperature for 2 hr. Excess sodium borohydride (380 mg, 0.01 mol) is added, and the reaction is stirred overnight. The mixture is evaporated to dryness, 5 ml of 40% aqueous KOH solution and 150 ml of water are added, and the alkaline solution is extracted with 4 × 60 ml of ether. The combined ether extracts are washed with water and brine and dried over magnesium sulfate. Removal of the ether gives a residue that is distilled bulb-to-bulb in a Kugelrohr apparatus to yield the tritiated alcohol **11**, 1.02 g (53%), b.p. 80°–100°/0.03 mm (210 mCi).

Conversion of Alcohol 11 to Alkyne 12

The labeled alcohol **11** (1.01 g, 5 mmol) is dissolved in 30 ml of dry THF, and dry triethylamine (1.14 ml, 8.1 mmol) is added. The solution is cooled to −20°, and distilled methanesulfonyl chloride (0.43 ml, 5.6 mmol) is added dropwise with stirring. A white precipitate of triethylamine hydrochloride forms almost immediately. Stirring at −20 to −15° is continued for 10 min after completion of the addition, and the mixture is then cooled to −78°. Lithium aluminum hydride (0.78 g, 20 mmol) is added cautiously. After addition of the hydride, the reaction is stirred at −78° for 10 min, the cooling bath removed, and stirring continued for 1.5 hr. At the end of this time, the reaction mixture is diluted with 50 ml of ether and several milliliters of 40% aqueous KOH are added slowly with stirring. The white, granular precipitate is removed by filtration, and the filter cake is washed with ether. The combined filtrate and washings are evaporated to dryness, the residue is redissolved in ether, and the solution is dried with magnesium sulfate. The dried ether solution is concentrated *in vacuo*, and the residue is distilled using a Kugelrohr apparatus to yield 0.72 g (76%) of the labeled alkyne **12** (b.p. 62°–64°/0.03 mm) (193 mCi).

Reduction of Alkyne 12 to trans-Alkene 13

Liquid ammonia (60 ml) is distilled into a dry 250-ml three-neck flask equipped with a nitrogen inlet and a Dry-Ice condenser after flushing the flask with dry nitrogen. The tritiated alkyne (0.57 g, 3.1 mmol) is added to the liquid ammonia, and the solution is stirred magnetically. The mixture is cooled to $-78°$, and sodium metal (0.36 g, 15.6 mmol) is added in small pieces with stirring while the flask is flushed with nitrogen. When the addition of the sodium is complete, the cooling bath is removed and the mixture is refluxed ($-33°$) with stirring for 2 hr. At the end of this period, 1.5 g of solid ammonium nitrate is added to the reaction followed by 30 ml of water. The Dry-Ice condenser is removed from the flask, and the ammonia is allowed to evaporate. The aqueous residue is extracted with ether (3 × 25 ml) and the combined extracts are washed with 2 × 20 ml of cold 1 N HCl. The ethereal layer is then washed with water (25 ml), 1 N bicarbonate (25 ml), and brine. After drying of the ether solution with magnesium sulfate, removal of the ether gives a residue that is distilled on a Kugelrohr to yield 0.47 g (84%) of the tritiated *trans*-alkene **13** (b.p. 47°–49°/0.01 mm) (122 mCi).

Conversion of Tritiated Alkene *13* to Tritiated Aziridine *16*

Sodium azide (0.42 g, 6.5 mmol) is suspended in 5 ml of dry acetonitrile, and the mixture is cooled to $-10°$ in an ice–salt bath. Iodine monochloride (0.15 ml, 3 mmol) is added dropwise with stirring, and residual ICl is rinsed in with a little dry acetonitrile. After 15 min stirring, the tritiated *trans*-olefin **13** (0.47 g, 2.6 mmol) is added and the mixture is stirred at $-10°$ for 15 min. The mixture is then allowed to warm to room temperature with stirring, and stirring is continued for 12 hr. Water (70 ml) is added, and the mixture is extracted with 3 × 30 ml of ether. The combined ether extracts are washed with 10 ml of 5% $Na_2S_2O_3$ solution, with water, and finally, with brine. The ether solution is then dried with magnesium sulfate, filtered, and evaporated to give the crude, labeled iodoazide **14**, which is used without purification.

The crude iodoazide **14** is dissolved in 10 ml of ether, and a boiling chip is added. Distilled triethylphosphite (0.45 g, 2.5 mmol) is added, and the solution is allowed to stand overnight. The ether is then removed from the reaction and the residual oil checked by IR spectroscopy to see if azide absorption remains (2100 cm^{-1}). If the oil exhibits azide absorption, several drops of triethyl phosphite are added and the mixture is allowed to stand for 24 hr. The reaction mixture is then examined by IR to determine if any azide remains. If azide absorption is still present, more triethyl phosphite is added, and the reaction is allowed to continue.

When the reaction mixture no longer exhibits azide absorption, indicating that formation of the phosphonoaziridine **15** is complete, the crude product is added dropwise to a stirred suspension of lithium aluminum hydride (0.25 g, 6.5 mm) in 20 ml of ether. The resulting mixture is stirred at room temperature for 3 hr and then quenched by the slow addition of enough 40% KOH to produce a white, granular precipitate. Magnesium sulfate (1 g) is added, and the solution is filtered. The filtrate is concentrated *in vacuo*, and the residual crude aziridine **16** is distilled bulb to bulb to give 0.31 g (60%) of labeled aziridine (**16**) (b.p. 75°/0.02 mm) (70 mCi).

Conversion of Tritiated Aziridine 16 to [1-³H-](±)-Dethiobiotin (22)

Tritiated aziridine **16** (0.31 g, 1.6 mmol, 73 mCi), sodium azide (0.51 g, 78 mmol), and ammonium chloride (0.42 g, 78 mmol) are dissolved in 30 ml of 8:2 ethanol–water, and the mixture is stirred and refluxed for 10 hr at 80°. The mixture is concentrated to a small volume (ca. 5 ml) and 50 ml of water are added. Aqueous ammonium hydroxide is added until the solution is basic and the basified mixture is extracted with ether (4 × 25 ml). The combined ethereal extracts are washed with water and brine, and dried over sodium sulfate. Evaporation of the ether gives the crude, tritiated aminoazide **17** (0.40 g) which is used without purification.

The labeled aminoazide **17** (0.40 g) is dissolved in 25 ml of ether, and 0.3 ml of dry triethylamine is added (2.3 mmol). The solution is cooled to −10° in an ice–salt bath, and distilled ethyl chloroformate (0.22 ml, 2.3 mmol) is added dropwise with stirring. A white precipitate of triethylamine hydrochloride forms almost immediately. The mixture is stirred at −10° to 0° for an additional 40 min after completion of the addition, magnesium sulfate (0.5 g) is added, and the mixture is filtered. The filtrate is washed with water, dried over MgSO₄, and evaporated to give the crude tritiated azidourethane **18** (0.35 g) (73% from aziridine).

The crude azidourethane is dissolved in 20 ml of methanol, the solution is cooled to −78°, and ozone is bubbled through the solution until a permanent blue color is formed. Oxygen is then bubbled through the solution for 10 min, whereupon the solution becomes colorless. Four drops of dimethyl sulfide are added to the mixture with swirling, and the temperature is allowed to rise to that of the room. Evaporation of the methanol gives the crude azidourethane ester **19** (0.33 g).

The azidourethane ester **19** is dissolved in 20 ml of absolute ethanol, and 75 mg of 10% palladium on carbon catalyst are added. Hydrogenation at atmospheric pressure for 12 hr is followed by removal of the catalyst and evaporation of the solvent to yield the crude aminourethan ester **20** (0.33 g). The crude product exhibits no azide absorption in the infrared at 2100 cm⁻¹.

The crude aminourethane ester **20** is stirred with a mixture of 11 ml of 0.1 *M* sodium ethoxide and 15 ml of absolute ethanol under an N₂ atmosphere at 80° for 4 days. The solution is concentrated *in vacuo* to near dryness, and 25 ml of water are added to the residue. The pH of the resulting solution is adjusted to 7 with solid ammonium chloride and the aqueous solution extracted with 6 × 30 ml of chloroform. The combined chloroform extracts are washed with brine and dried over MgSO$_4$. Evaporation of the chloroform gives a residue which is purified by preparative thin-layer chromatography on three 1-mm HF$_{254}$ silica plates (20 × 20 cm) using 10% methanol in ethyl acetate as the solvent. The main band, R_f 0.35–0.55, is removed and eluted with hot methanol (3 × 30 ml); the combined eluents are concentrated *in vacuo* to give a thick yellow oil, which is the tritiated ethyl ester of (±)-dethiobiotin (**21**, 160 mg) (47%).

The dethiobiotin ethyl ester (160 mg) is dissolved in 7 ml of 0.5 N HCl, and the solution is allowed to stand for 5 days at room temperature. Yellow crystals of (±)-dethiobiotin slowly separate. The liquid is decanted and the crystals are dissolved in 2 ml of hot acetic acid. The hot solution is treated with Norit (50 mg), filtered, and evaporated to dryness *in vacuo*. The white residue is recrystallized from a minimum amount of hot water to yield 42 mg (12%) of fine white needles of [1-^{3}H](±)-dethiobiotin (**22**) (2.5 mCi), m.p. 163°–164°.

Synthesis of [3-^{3}H](±)-Dethiobiotin (27) (Fig. 2)

Reduction of Propargyl Alcohol 9 to the [3-^{3}H]trans-Allylic Alcohol 23

The reduction procedure is a modification of that described by Grant and Djerassi.[16]

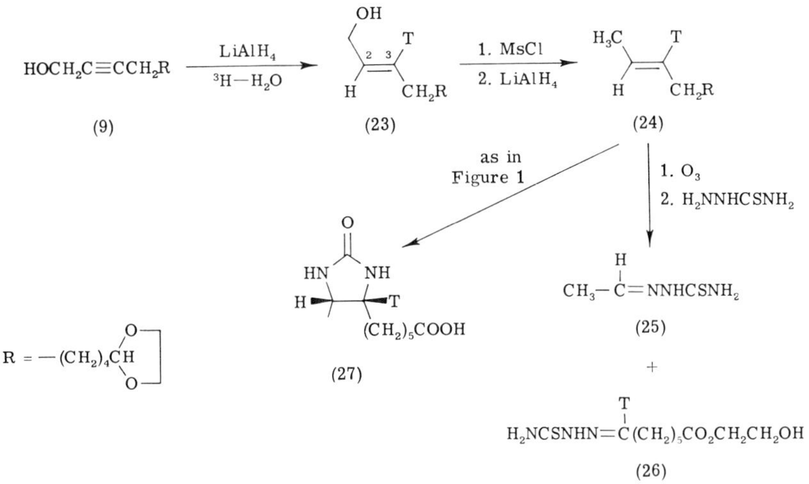

FIGURE 2

[16] B. Grant and C. Djerassi, *J. Org. Chem.* **39**, 968 (1974).

The propargyl alcohol 9 (1.4 g, 7.1 mmol) is added slowly to a stirred solution of 270 mg (7.1 mmol) of lithium aluminum hydride in 20 ml of dry THF. After 3.5 hr of stirring, the reaction is quenched with 0.40 ml (2 Ci) of tritiated water, added slowly over a 1 hr period. The reaction is stirred overnight and filtered; the filtrate is evaporated to dryness. The residue is dissolved in 200 ml of ether, and the ethereal solution is washed with 50 ml of 1 N NaOH, 50 ml of water, and 50 ml of brine. The solution is dried over MgSO$_4$, filtered, concentrated, and the residue is vacuum distilled to yield 1.155 g (77%) of the [3-^{3}H]$trans$-allylic alcohol 23, b.p. 60°–80°/0.03 mm. The radiochemical yield is ca. 2% (41.8 mCi).

Conversion of the Tritiated Allylic Alcohol 23 to the Tritiated Alkene 24

The conversion of the [3-^{3}H]$trans$-allylic alcohol 23 to the [3-^{3}H]alkene 24 is accomplished by formation of the mesylate and reduction with lithium aluminum hydride in a manner analogous to that described above for the conversion of propargyl alcohol 11 to alkyne 12.

Ozonolysis of Tritiated Alkene 24 and Derivatization of the Products as Thiosemicarbazones 25 and 26

The lithium aluminum hydride–[^{3}H]H$_2$O reduction of propargyl alcohol 9 leads to labeling of both C-2 and C-3 of the resulting alkene 24. The distribution of the label between these two carbon atoms shows some variation each time the reaction is carried out. The relative amounts of radioactivity at C-2 and C-3 are determined by ozonolysis of the alkene, derivatization of the products with thiosemicarbazide, and measurement of radioactivity contained in the two thiosemicarbazones 25 and 26.

A small quantity of tritiated $trans$-alkene 24 (ca. 10 mg) is diluted with radioinactive $trans$-alkene to give a total of 368 mg (2 mmol). The diluted compound is dissolved in 20 ml of methanol, the solution cooled to −78°, and ozone bubbled through the mixture until the blue color of ozone persists for 20 min. Oxygen is then bubbled through the solution until the blue color vanishes, and 10 drops of dimethyl sulfide are added. The solution is then allowed to warm to room temperature. Sodium acetate (0.95 g, 7 mmol) and thiosemicarbazide (0.41 g, 4.5 mmol) are dissolved in a hot mixture of methanol (3 ml) and water (7 ml). The hot solution is added to the ozonolysis mixture, and the resulting solution is stirred overnight at room temperature. The solution is concentrated in $vacuo$ to remove methanol and then diluted with 100 ml of water. The aqueous solution is extracted with chloroform (7 × 60 ml) and the combined chloroform extracts washed with water, dried with MgSO$_4$, and evaporated to yield ca. 0.60 g of a thick yellow oil. This material is chromatographed on six 1-mm silica gel GF-254 preparative plates (20 × 20 cm) using 1 : 1 benzene:ethyl acetate as solvent. The two main bands (R_f = 0.45, 0.15) are removed and eluted with hot methanol (2 × 100 ml). Evaporation of the eluent of the upper band gives a residue that is crystallized from water to yield 66 mg of acetaldehyde thiosemicarbazone (25), m.p. 146°–147° (28%). The eluent of the lower band gives a residue which is crystallized from water to yield 10 mg of thiosemicarbazone 26, m.p. 74°–76° (19%). The sample of [3-³H]trans-alkene 24 used in this experiment gave thiosemicarbazones 25 and 26, which carried 17% and 83%, respectively, of the total radioactivity.

Conversion of the Tritiated Alkene 24 into [3-³H](±)-Dethiobiotin (27)

The tritiated alkene 24 is converted into [3-³H](±)-dethiobiotin (27) by the methods outlined for the conversion of the tritiated alkene 13 to [1-³H](±)-dethiobiotin (22).

Synthesis of [2,3-³H](±)-Dethiobiotin (32) (Fig. 3)

Reduction of Propargyl Alcohol 9 to [2,3-³H]trans-Allylic Alcohol 28

Propargyl alcohol 9 (1.98 g, 10 mmol) is added dropwise with stirring to a mixture of 40 ml of dry THF, 350 mg of lithium aluminum hydride (9.3 mmol), and 28.6 mg of [³H]lithium aluminum hydride (0.74 mmol, 100

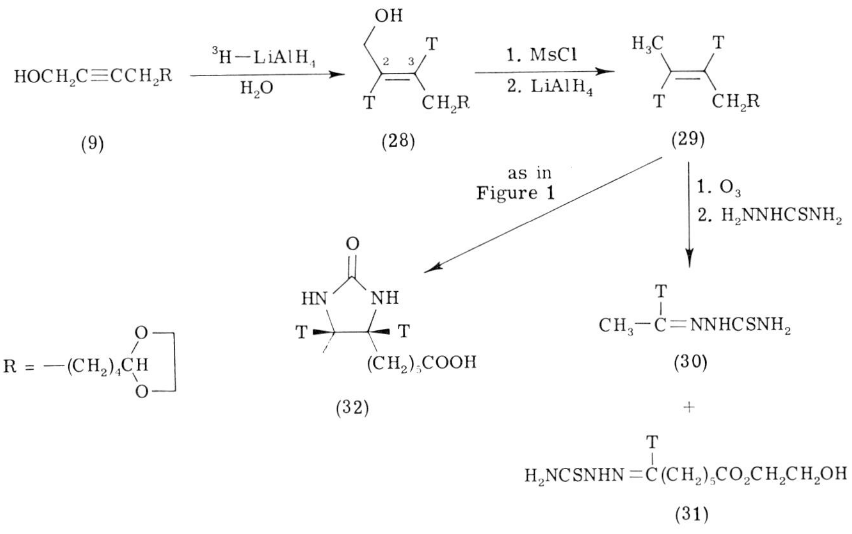

FIGURE 3

mCi). The resulting mixture is stirred overnight at room temperature and worked up in the manner described for the synthesis of the [3-³H]*trans*-allylic alcohol **23**. The yield of distilled [2,3-³H]*trans*-allylic alcohol **28** is 1.70 g (85%). The radiochemical yield is 45 mCi (45%).

Reduction of [2,3-³H]trans-Allylic Alcohol 28 into [2,3-³H]trans-Alkene 29

The reduction of the labeled allylic alcohol **28** to the corresponding labeled alkene **29** is carried out in precisely analogous fashion to the reduction of propargyl alcohol **11** to alkyne **12**.

Ozonolysis of Tritiated Alkene 29 and Derivatization of The Products as Thiosemicarbazones 30 and 31

The ozonolysis and derivatization are carried out in the manner described for the [3-³H]alkene **24**. The acetaldehyde thiosemicarbazone (**30**) derived from **29** was found to carry 58% of the radioactivity of the alkene, while the thiosemicarbazone **31** carried 42% of the radioactivity.

Conversion of Tritiated Alkene 29 into [2,3-³H](±)-Dethiobiotin (32)

The [2,3-³H]alkene **29** is transformed into [2,3-³H](±)-dethiobiotin (**32**) by the methods outlined for the conversion of the [1-³H]alkene **13** into [1-³H](±)-dethiobiotin (**22**).

Synthesis of 4 (RS)-[³H](±)Dethiobiotin (38) (Fig. 4)

Oxidation of Alcohol 6 to Aldehyde 33

A mixture of dry pyridine (9.5 ml) and dry methylene chloride (150 ml) is stirred magnetically at room temperature, and dry chromium trioxide (6.0 g, 0.06 mol) is added. The deep red solution is stirred at room temperature for 15 min. The alcohol **6** (1.36 g, 0.01 mol) is added and stirring continued for 15 min. At the end of that time, the supernatant liquid is decanted from the tarry material at the bottom of the flask, and the insoluble tar washed with ether. The combined supernatant and ether washings are concentrated *in vacuo* and the residue is mixed with 300 ml of ether. The ethereal solution is suction filtered, and the filtrate is washed successively with *cold* 1 N HCl (2 × 50 ml), 1 N sodium bicarbonate (2 × 50 ml), and brine (25 ml). The ethereal solution is dried over MgSO₄ and evaporated to give the crude aldehyde, which is distilled, b.p. 70°–80°/0.1 mm, to yield 1.15 g (72%) of **33**, which is 95% pure by GLC (6 ft, 10% UCW-98).

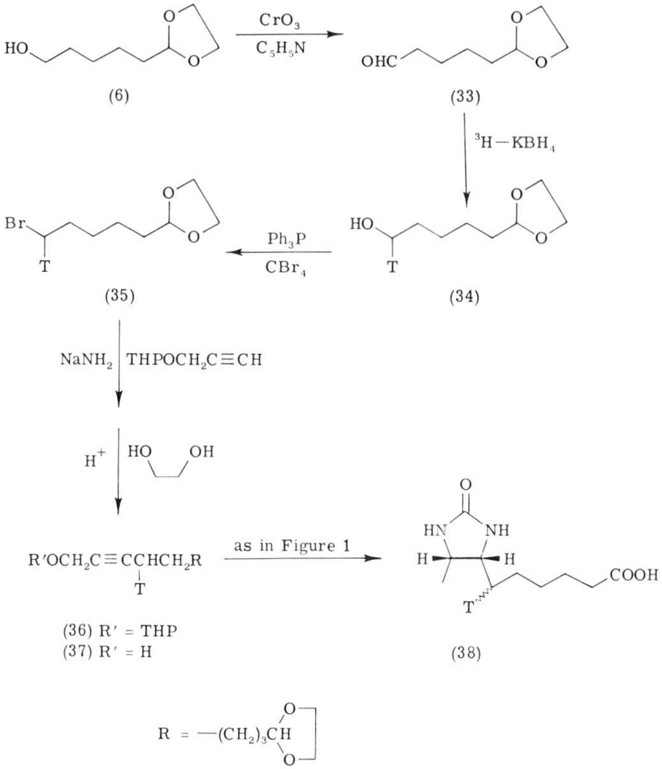

FIGURE 4

Reduction of Aldehyde 33 to Tritiated Alcohol 34

Aldehyde **33** (1.2 g, 7.0 mmol) is dissolved in 30 ml of absolute ethanol, and a few milligrams of sodium borohydride are added with stirring. After about 5 min, [³H]potassium borohydride (50 mg, 1.06 mmol, 500 mCi) is added. The mixture is stirred for 45 min and radioinactive sodium borohydride (0.280 g, 7.4 mmol) is then added. The resultant mixture is stirred overnight. Work-up is similar to the preparation of the labeled alcohol **11**. Distillation of the crude product gives 0.67 g (57%) of the tritiated alcohol **34**. The radiochemical yield is 240 mCi (48%).

Conversion of the Tritiated Alcohol 34 to the Tritiated Bromide 35

The procedure utilized to convert the labeled alcohol **34** into the corresponding bromide **35** is precisely analogous to that used to transform the unlabeled alcohol **6** into the unlabeled bromide **7**.

Alkylation of the Tetrahydropyranyl Ether of Propargyl Alcohol with
  Tritiated Bromide 35: Synthesis of 4 (RS)-[³H]Alkyne 36

The labeled bromide **35** is converted into the acetylene **36** in the manner described for the conversion of the unlabeled bromide **7** into the unlabeled alkyne **8**.

Conversion of 4(RS)-[³H]Alkyne 36 into
  4 (RS)-[³H]Propargyl Alcohol 37

Removal of the tetrahydropyranyl ether group from **36** is accomplished as described above for the removal of the THP ether group from the unlabeled THP ether **8**.

Transformation of the 4 (RS)-[³H]Propargyl Alcohol 37 into
  4 (RS)-[³H](±)-Dethiobiotin (38)

The conversion of the 4 (RS)-[³H]alcohol **37** into 4 (RS)-[³H](±)-dethiobiotin (**38**) is carried out in the same fashion as the conversion of the [1-³H]alcohol **11** into [1-³H](±)-dethiobiotin (**22**) recorded above.

III. Synthesis of [10-¹⁴C](±)-Dethiobiotin (40) (Fig. 5)

10-¹⁴C(±)-Dethiobiotin (**40**) is prepared by utilizing reactions reported in the literature (Fig. 5).[17]

Preparation of (±)-7,8-Diaminopelargonic Acid Dihydrochloride (39)

(±)-Dethiobiotin (50 mg, 0.24 mmol) is dissolved in 2 ml of concentrated hydrochloric acid, and the solution is refluxed (140°) for 6 hr. The resulting pale yellow solution is evaporated to dryness, and the residue is triturated with 0.5 ml of absolute ethanol. The resulting solid is recrystalized twice from methanol–ether to give white crystals, 50 mg (83%), m.p. 202°–203°.

Conversion of (±)-7,8-Diaminopelargonic Acid Dihydrochloride (39) to
  [10-¹⁴C](±)-Dethiobiotin (40)

[¹⁴C]Phosgene (3.3 mg, 0.03 mmol, 0.5 mCi) is vacuum-transferred to 0.8 ml of a 3.2% solution of radioinactive phosgene in xylene (ca. 22 mg of unlabeled phosgene, 0.22 mmol). 7,8-Diaminopelargonic acid dihydro-

[17] D. Melville, K. Hofmann, and V. du Vigneaud, *Science* **94**, 308 (1941).

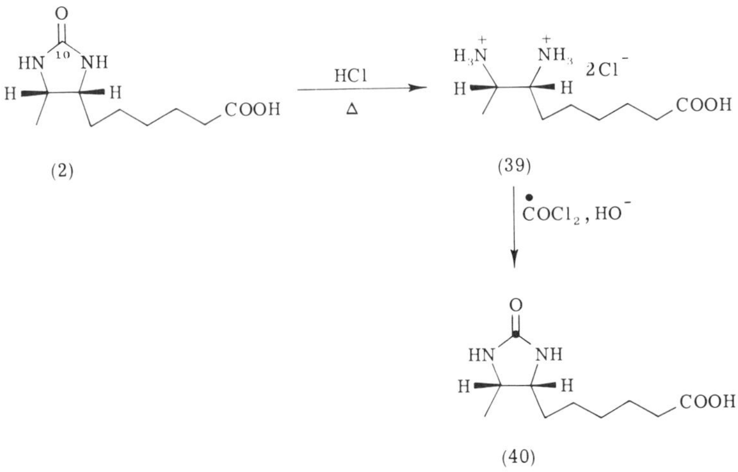

FIGURE 5

chloride (39) (38 mg, 0.15 mmol) is dissolved in 1.5 ml of water, and 1 N NaOH is added to pH 8. The basic aqueous solution is stirred, and the phosgene solution is added dropwise; the pH of the aqueous phase is checked periodically and readjusted to 8 with 1 N NaOH solution. After all the phosgene solution has been added, the mixture is stirred for 15 min with occasional monitoring of the pH. If the pH is 8 after 15 min, the reaction is considered complete. Otherwise, the pH is brought to 8 with 1 N NaOH, and stirring is continued. The xylene is then separated from the aqueous phase, traces of xylene removed from the aqueous phase by ether extraction, and the aqueous phase is then acidified to pH 4. The acidified, aqueous phase deposits crystals on standing. These are removed by filtration and dried to give [10-^{14}C]($\pm$)-dethiobiotin, 21 mg (68%), (0.33 mCi; radiochemical yield, 66%), m.p. 163°–164°.

IV. Administration of Doubly Labeled Dethiobiotin to *Aspergillus niger*

Appropriate quantities of [10-^{14}C]($\pm$)-dethiobiotin (40) and a specifically tritiated form of ($\pm$)-dethiobiotin are weighed out, mixed, and recrystallized from water until a constant specific radioactivity and a constant tritium to carbon-14 ratio are reached.

The growth medium is prepared as described by Wright and Cresson[18]: 1.5 liters of medium contain 45 g of sucrose, 3 g of sodium nitrate, 1.5 g of dipotassium hydrogen phosphate, 0.75 g of magnesium sulfate heptahy-

[18] L. Wright and E. Cresson, *J. Am. Chem. Soc.* **74**, 4156 (1954).

drate, 0.75 g of potassium chloride, 0.015 g of ferrous sulfate heptahy-drate, and about 2 mg of pimelic acid. About 15 mg of zinc sulfate hepta-hydrate was added to the culture medium in the experiment involving ad-ministration of $[2,3\text{-}^3H\text{-}10\text{-}^{14}C](\pm)$-dethiobiotin. The presence of zinc ions appeared to promote mycelium growth and to raise the level of incorpo-ration of dethiobiotin into biotin.

The medium is distributed between three 1-liter flasks, and, after steril-ization, a doubly-labeled form of dethiobiotin (ca. 2–3 mg) dissolved in 5 ml of hot water is syringed into the flasks through a 0.45-μm Millipore filter. The flasks are inoculated from slants of *Aspergillus niger* (ATCC 1004) and incubated with rotary shaking at 30° for 5–6 days. Isolation of the biotin and biotin vitamers is accomplished by modification of a proce-dure of Tepper *et al.*[19] The mycelium is removed by filtration through cloth toweling and the pH of the yellow culture broth adjusted to 2 with conc. HCl. Norit (30 g) is added, the mixture is stirred for a few minutes, and the Norit is collected by gravity filtration. The Norit is washed with distilled water (2 × 1 liter) and suction-filtered. The Norit is then eluted with a solution of water, ethanol, and ammonium hydroxide (10:10:1, v/v/v) (2 × 600 ml) and the combined eluents concentrated *in vacuo* to a yellow oil. Radioinactive (+)-biotin (120–150 mg) is added as carrier and the mixture is dissolved in 110 ml of glacial acetic acid. Hydrogen perox-ide (30%, 30 ml) is added, and the mixture is stirred overnight. Palladium black (0.25 g) is then added to destroy excess peroxide, and the mixture is again stirred overnight at room temperature. When starch-iodide paper in-dicates the absence of peroxides, the solution is filtered and the filtrate is taken to dryness. The yellow residue is dissolved in 5 ml of water, and the solution is neutralized with dilute aqueous NaOH. The solution is applied to a 1.5 × 90 cm column of Dowex 1-X8 (200 mesh, formate form), and elution is carried out with a linear gradient from 2 liters of water to 2 liters of 0.5 N ammonium formate. Fractions of ca. 22 ml are collected, and 0.5 ml is pipetted out of alternate fractions and counted in 10–15 ml of Bray's scintillator. The fractions believed to contain the labeled biotin sulfone are pooled and concentrated *in vacuo;* the concentrate is passed through a 3.5 × 25 cm column of Dowex 50-X8 (50 mesh, H$^+$ form). The column is eluted with 1 liter of 1:1 ethanol–water and the eluent taken to dryness. If crystallization proves possible, the crude biotin sulfone is recrystallized repeatedly from hot water to a constant specific radioactivity and con-stant tritium to carbon-14 ratio. However, in most cases, further purifica-tion of the crude biotin sulfone by chromatography is required. The crude biotin sulfone is esterified by refluxing in an excess of saturated metha-

[19] J. P. Tepper, D. B. McCormick, and L. D. Wright, *J. Biol. Chem.* **241,** 5734 (1966).

TABLE: INCORPORATION OF DOUBLY-LABELED ($\pm$)-DETHIOBIOTIN INTO BIOTIN

Expt. No.	Labeling pattern of administered dethiobiotin	^{3}H/^{14}C for dethiobiotin	^{3}H/^{14}C for biotin sulfone methyl ester	Percent ^{3}H retention
1	3-^{3}H, 10-^{14}C[a]	2.89	3.04	105
2	2, 3-^{3}H, 10-^{14}C[b]	6.05	5.74	95
3	1-^{3}H, 10-^{14}C	6.88	4.81	70
4	4(RS)-^{3}H, 10-^{14}C	5.88	3.10	53

[a] Dethiobiotin had 83% ^{3}H at C-3, 17% ^{3}H at C-2.
[b] Dethiobiotin had 58% ^{3}H at C-2, 42% ^{3}H at C-3.

nolic HCl overnight. The methanolic HCl is removed *in vacuo,* and the residue is purified by preparative thin-layer chromatography on 4 silica GF-254 plates (20 × 20 cm) using 77% ethyl acetate, 20% methanol, and 3% acetic acid as the eluent. The biotin sulfone methyl ester (R_f 0.2–0.5) is eluted from the silica with hot methanol and recrystallized repeatedly from methanol–water (100 : 1) to give material of constant specific radio-activity and constant tritium to carbon-14 ratio, m.p. 160°.

V. Results

The results obtained when (±)-dethiobiotin labeled with tritium at C − 1, −2, −3, or −4 is transformed into (+)-biotin by *A. niger* are summarized in the table. A number of conclusions can be drawn from the data in the table. Experiments 1 and 2 clearly demonstrate that the introduction of sulfur at C − 1 and C − 4 of the dethiobiotin skeleton proceeds without loss of hydrogen from C − 2 or C − 3. It therefore seems unlikely that unsaturation is introduced at C − 2 or C − 3 as a consequence of the sulfur introduction processes; however, the possibility of enzymic removal of hydrogen from C − 2 or C − 3 followed by replacement of the hydrogen without exchange cannot be ruled out. Guillerm, Frappier, and co-workers have recently reported[20] that the incorporation of [2,3-³H]dethiobiotin into biotin by *Escherichia coli* also proceeds without tritium loss. Experiment 3 shows that the incorporation of [1-³H](±)-dethiobiotin into biotin proceeds with about 30% tritium loss. The nature of the reaction associated with the oxidation of the methyl group of dethiobiotin is unknown, but the amount of tritium loss is consistent with the removal of one hydrogen atom from the methyl group by a process exhibiting little or no isotope effect. Experiment 4 reveals that 4 (*RS*)-[³H]1 ±)-dethiobiotin is incorporated into (+)-biotin with about 47% tritium loss. This figure falls within experimental error of that expected (50%) for the stereospecific removal of one hydrogen atom from C − 4 of dethiobiotin as the result of sulfur introduction. Therefore, it appears that the conversion of dethiobiotin into biotin proceeds with the removal of two hydrogen atoms from the former substance.

---

[20] G. Guillerm, F. Frappier, M. Gaudry, and A. Marquet, *Biochimie* **59**, 119 (1977).

[63] The Biotin Transport System in Yeast

By EDWARD A. BAYER and MEIR WILCHEK

A number of early investigations have reported accumulation of biotin in bacteria and yeast under certain growth conditions.[1,2] Studies to elucidate the mechanism and control of this transport were initiated by Lichstein and his associates first with cells of *Lactobacillus arabinosus*[3-6] and subsequently with the yeast *Saccharomyces cerevisiae*.[7] These investigations revealed that cells grown under normal conditions can accumulate biotin intracellularly in unaltered form to at least three orders of magnitude over that present in the external medium. The transport is pH and temperature dependent, stimulated by glucose, and inhibited by biotin analogs—features consistent with the phenomenon being an active transport process. Acute control of biotin transport is accomplished by an "overshoot" phenomenon,[8] whereas protracted regulation of transport is determined by the biotin content of the growth medium and is apparently controlled by a repression of the synthesis of transport components.[9] Since spheroplasts retain biotin uptake ability,[10,11] transport components are evidently distinct constituents of the cell plasma membrane.

In our studies on the biotin transport system in yeast, we have used various affinity methods, including affinity labeling, affinity cytochemistry and affinity chromatography, in order to obtain (i) chemical and structural information relating to transport components; (ii) localization and quantification data, and ultimately (iii) the isolation of transport components. The overall scheme of these studies is shown in Fig. 1.

Affinity labeling studies have been carried out with a series of chemically active, biotin-containing compounds based on the reference compound biotinyl-*p*-nitrophenyl ester (*p*BNP).[12-14] The latter compound

[1] K. K. Kreuger and W. H. Peterson, *J. Bacteriol.* **55**, 693 (1948).
[2] H. Massock and I. L. Baldwin, *J. Bacteriol.* **45**, 34 (1943).
[3] H. C. Lichstein and R. B. Ferguson, *J. Biol. Chem.* **233**, 243 (1958).
[4] H. C. Lichstein and J. R. Waller, *J. Bacteriol.* **81**, 65 (1961).
[5] J. R. Waller and H. C. Lichstein, *J. Bacteriol.* **90**, 843 (1965).
[6] J. R. Waller and H. C. Lichstein, *J. Bacteriol.* **90**, 853 (1965).
[7] T. O. Rogers and H. C. Lichstein, *J. Bacteriol.* **100**, 557 (1969).
[8] J. M. Becker and H. C. Lichstein, *Biochim. Biophys. Acta* **282**, 409 (1972).
[9] T. O. Rogers and H. C. Lichstein, *J. Bacteriol.* **100**, 565 (1969).
[10] J. F. Cicmanec and H. C. Lichstein, *J. Bacteriol.* **119**, 718 (1974).
[11] T. Viswanatha, E. Bayer, and M. Wilchek, *Biochim. Biophys. Acta* **401**, 152 (1975).
[12] J. M. Becker, M. Wilchek, and E. Katchalski, *Proc. Natl. Acad. Sci. U.S.A.* **68**, 2604 (1971).
[13] E. A. Bayer, T. Viswanatha, and M. Wilchek, *FEBS Lett.* **60**, 309 (1975).
[14] E. A. Bayer and M. Wilchek, this series, Vol. 46, p. 613 (1977).

METHODS IN ENZYMOLOGY, VOL. 62

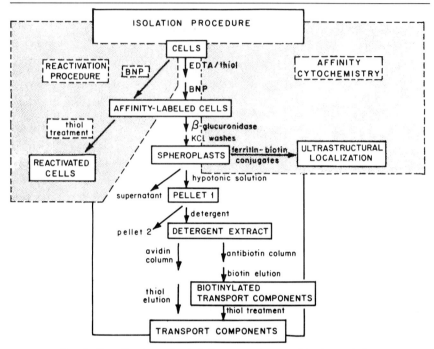

FIG. 1. Overall scheme for isolation procedure used in these studies.

causes the covalent attachment of biotin to transport components, thus inactivating biotin uptake into the cells. Simple treatment with thiols chemically removes the biotin label from transport components and results in total restoration of biotin transport.[11] This phenomenon provides (a) chemical information as to the nature of the bond between the affinity label and the transport protein; (b) a rapid method for the localization, quantification, and identification of transport components during various stages of purification; (c) a potential method for the ultimate regain of binding activity in the purified transport component(s); and (d) a method for switching on and off biological transport systems in whole cells.

The amount of biotin receptors per cell can be determined by two complementary methods.[15] One approach is based on measuring the amount of radioactivity released by thiol treatment of radioactive pBNP-affinity-labeled membrane extracts. The second method involves the use of a biotin-derivatized, impermeant, electron-dense, affinity-cytochemical label (ferritin–biotin conjugates) for subsequent visualization by electron microscopy.

[15] E. A. Bayer, E. Skutelsky, T. Viswanatha, and M. Wilchek, *Mol. Cell. Biochem.* **19,** 23 (1978).

Model isolation experiments have been performed, which take advantage of the specific affinity characteristics and chemical properties of the biotin transport system. Biotinylated components in membrane extracts of affinity-labeled cells can be specifically adsorbed to avidin–Sepharose affinity columns and subsequently eluted with mercaptoethanol. Elution of the latter material should be accompanied by a regain in biological activity. Alternatively, biotinylated transport components can be adsorbed to antibiotin antibody columns and released by competition with free biotin. Treatment of the latter *in vitro* with thiols should also yield viable transport components.

Saccharomyces cerevisiae

Growth Medium. Fleischman strain 139 (ATCC 9896) is grown in Vogel's medium,[16] containing the following additives: inositol (36 μg/ml), calcium pantothenate (2 μg/ml), pyridoxine hydrochloride and thiamine hydrochloride (4 μg/ml each), and biotin (0.25 ng/ml). Glucose (1%) is substituted for sucrose. The above-described medium constitutes a "biotin-sufficient" medium, conducive to maximum production of biotin transport components.[9]

Preparation of Resting Cell Suspension. Cells are grown at 30° with reciprocating shaking. The culture is harvested during the late exponen-

---

[16] H. J. Vogel, *Microbiol. Gen. Bull.* **13**, 42 (1956). Yeasts are maintained on maximal medium slants containing yeast extract (1%), peptone (2%), dextrose (2%), and agar (2%) at 4°, transferred to an inoculum tube containing Vogel's minimal medium for overnight growth, and then grown (1% inoculum) on Vogel's minimal medium. The latter is prepared at 50 times the normal concentration in the following manner: The following chemicals (of the highest analytical grade) are dissolved successively with stirring at room temperature in 750 ml of double-distilled water: tri-sodium citrate dihydrate (125 g); potassium dihydrogen orthophosphate, anhydrous (250 g); ammonium nitrate, anhydrous (100 g); magnesium sulfate septahydrate (10 g), calcium chloride dihydrate (5 g) and trace element solution (5 ml; preparation described below). Chloroform (2 ml) is added as a preservative. The 50-strength Vogel's medium is stored at room temperature and is stable over a period of 4–6 months. The medium is diluted 50-fold with double-distilled water plus the given additives and is sterilized by autoclaving; a solution of glucose is autoclaved separately and then added (to a final concentration of 10 g/1) to the medium. The trace element solution is prepared as follows: The following chemicals (highest purity) are dissolved in distilled water with stirring at room temperature: Citric acid monohydrate (5 g as solubilizing agent); zinc sulfate (5 g); ferrous ammonium sulfate sexahydrate (1 g); cupric sulfate pentahydrate (250 mg); manganous sulfate monohydrate (50 mg); boric acid, anhydrous (50 mg); sodium molybdate dihydrate (50 mg). Chloroform (1 ml) is added as a preservative. The trace element solution is stored at room temperature and is stable over a period of 4–6 months.

tial growth phase, washed twice with distilled water, and resuspended to 1 mg of dry weight per milliliter[17] in 50 mM phosphate solution (pH 4.0).[18]

Affinity Labeling and Reactivation Studies

Preparation of Biotinyl-p-Nitrophenyl Ester (pBNP)[12–14]

Reagents
Biotin (Sigma)
p-Nitrophenol (Fluka)
Dicyclohexyl carbodiimide (Fluka)
Methylene chloride
Diethyl ether
Isopropanol

Procedure. To biotin (250 mg), suspended in methylene chloride (3 ml), are added *p*-nitrophenol (200 mg) and dicyclohexylcarbodiimide (220 mg). The container with the above mixture is stoppered, and the suspension is stirred for 24 hr at room temperature. The precipitate is filtered off, and the filtrate is taken to dryness under reduced pressure. The yellow gummy residue is washed several times with absolute ether. The flakes are dissolved in a minimal amount of boiling isopropanol, and the product is allowed to crystallize overnight at room temperature. The crystals of *p*BNP are collected by filtration and washed with ether. Yield: 120 mg, 33%; m.p. 156°–158°.

Analyses. The product is analyzed by thin-layer chromatography (TLC), R_f = 0.45 (chloroform–methanol, 37:3 v/v). The purity of the compound is established by determination of *p*-nitrophenol (A_{420}) upon treatment of a weighed sample of *p*BNP with 0.2 N NaOH.

Preparation of Radioactive pBNP[11]

[*carbonyl*-^{14}C]*d*-Biotin (0.27 mg, 50 μCi, Radiochemical Centre, Amersham) is treated with 100 μl of methylene chloride containing *p*-nitrophenol (0.27 mg) and DCC (0.23 mg). The reaction takes place for 24 hr; then a weighed amount (25 mg) of pure, nonradioactive *p*BNP is added. The product is recrystallized from isopropanol as described above. Analysis is carried out by combined TLC and autoradiography. Specific activity: 4.4 × 10^6 cpm/mg *p*BNP.

---

[17] An aliquot of a cell suspension is diluted (1:10) with water, and the turbidity is read on a Klett–Summerson colorimeter. A reading of 30 Klett units (420 blue filter) is equivalent to 1 mg of cells, dry weight, per milliliter (about 4 × 10^7 cells) in the original suspension.

[18] A solution of potassium dihydrogen orthophosphate is brought to pH 4.0 with the dropwise addition of phosphoric acid.

Affinity Labeling of Intact Yeast Cells. A resting cell suspension (9 ml) is treated with a solution (1 ml) of *p*BNP (3.7 mg in 100 ml of ethanol)[19] to obtain the desired final concentration (in this case 10 μM). The treatment is allowed to proceed for 30 min at 30°; the cells are then washed three times with distilled water and resuspended to their original volume.

Reactivation Conditions. A washed, *p*BNP-treated, cell suspension (90 ml) is packed by centrifugation, and the pellet is resuspended in 4 ml of phosphate solution, pH 4.0.[18] 2-Mercaptoethanol (40 μl) is added to achieve a final concentration of 0.15 M. The suspension is kept at 30° for 30 min under constant agitation. The treated cells are then washed three times and resuspended in the same phosphate solution to a concentration of 1 mg cells (dry weight) milliliter.[17]

Preparation of Spheroplasts

The conversion of yeast cells to their spheroplasts is accomplished using glusulase.[11,20,21] In a typical experiment, washed cells are subjected to pretreatment with a medium (4.2 ml/g cells, wet weight) containing EDTA (40 mM) and 2-mercaptoethanol (0.1 M). After incubation for 30 min at 30°, the cells are collected by centrifugation, then washed once with distilled water and twice with 0.6 M KCl.

Pretreated cells are suspended in 0.6 M KCl solution (6.4 ml/g cells wet weight) containing EDTA (1.25 mM), potassium phosphate buffer (pH 6.1, 50 mM). Glusulase (0.4 ml; 60,000 units; Endo Laboratories, Inc.) is added, and the reaction is carried out with continuous gentle shaking at 30°. Owing to the fragility of spheroplasts in osmotic shock, the rate of spheroplast formation can be measured in the following manner: Aliquots (50 μl) of the reaction mixture are removed every 10–15 min and diluted with 5 ml of either distilled water or 0.6 M KCl solution (the latter serving as a control). Digestion is normally terminated when there is no further decrease in turbidity at 420 nm. The spheroplasts are collected by centrifugation, washed twice with 0.6 M KCl, and resuspended in the same medium.

Biotin Uptake

Procedure. A resting suspension of cells or spheroplasts (1 mg dry weight)[17] after appropriate chemical treatment, is mixed with an equal

---

[19] Yeasts thrive in media containing up to 10% ethanol. In this case, the affinity label is readily soluble in ethanol; the latter also serves as an energy source, activating the transport system, thus permitting specific interaction with *p*BNP.

[20] E. Cabib, this series, Vol. 22, p. 120.

[21] J. R. Villaneuva and I. A. Acha, *Methods Microbiol.* **4**, 665 (1971).

volume of 0.1 M potassium phosphate solution (pH 4.0) containing glucose (2%); 0.6 M KCl being present in experiments with spheroplasts. After 20 min of incubation, a solution of [^{14}C]biotin (20 μM diluted 1:100 v/v to a final concentration of 0.2 μM) is added to glucose-pretreated cell suspensions.[22] Biotin uptake is measured either by the Millipore filtration or the centrifugation technique as described below.

Millipore Filtration Technique.[7] Aliquots (1 ml) of the reaction mixture are taken at desired time intervals and filtered rapidly on Millipore membrane filters (0.45 μm pore, type HA). After washing with water or KCl solution, the filters are transferred to vials, to which scintillation fluid (10 ml) is added,[23] and the samples are counted in a scintillation spectrometer. This method is inappropriate for measuring biotin uptake in spheroplasts.

Centrifugation Technique.[11] Aliquots (1 ml) are drawn at various time intervals and transferred to a suspension of washed lyophilized yeast cells,[24] which serve as carriers. The suspension is then centrifuged (2500 rpm) on a clinical centrifuge, and the pellet is washed three times with 0.6 M KCl solution. The washed pellet is suspended in 1 ml of water, and the entire suspension is transferred to scintillation fluid for counting.[23] Biotin uptake of both cells and/or spheroplasts may be performed with this method.

Affinity Cytochemistry

A solution (2.5 ml) containing glucose (2%), 0.6 M KCl, and 50 mM potassium phosphate (pH 4.0) is added to an isotonic, resting suspension of spheroplasts (10^8 cells/2.5 ml). The suspension is treated for 15 min at 30° with gentle shaking. A solution of ferritin–biotin conjugates (200 μg of ferritin per milliliter, final concentration)[25] is added, and the suspension is incubated an additional 15 min. The suspension is centrifuged once (500 rpm) and resuspended in an isotonic solution of glutaraldehyde (2% final concentration) for 5 min at room temperature. The spheroplasts are washed twice in 0.6 M KCl and fixed in 2% glutaraldehyde for 30 min at room temperature. After fixation, spheroplasts are washed twice and processed for electron microscopy.[26]

[22] In order to calculate apparent K_m of biotin transport, varying concentrations of biotin may be used. The initial rate of uptake is measured and the reciprocal plotted against that of biotin concentration. The inverse of the X-intercept provides the apparent K_m.

[23] G. A. Bray, *Anal. Biochem.* **1**, 279 (1960).

[24] The amount of cells per sample used for the uptake assay is too low for collection and washing by centrifugation. Lyophilized cells, which are inert with respect to biotin transport, may therefore be used as carriers. For this purpose, about 10 mg of lyophilized cells may be used per sample.

[25] E. A. Bayer, E. Skutelsky, and M. Wilchek, this volume [55].

[26] J. H. Luft, *J. Biophys. Biochem. Cytol.* **9**, 409 (1961).

Isolation Experiments

Preparation of Affinity-Labeled Spheroplasts. The scheme for isolation experiments of this nature is shown in Fig. 1. It should be observed that, in this case, mercaptoethanol treatment precedes incubation of cells with *p*BNP, since the former reagent releases biotin from affinity-labeled transport components. Thus, a resting cell suspension, grown from 6 liters of "biotin-sufficient" medium (yield: 4 g dry weight = 1.6×10^{11} cells)[17] is incubated (30°, 30 min) in a solution (50 ml) containing 0.15 M mercaptoethanol and 40 mM EDTA. The cells are collected by centrifugation and washed twice with distilled water. After treatment (30°, 30 min) with 10 μM [^{14}C]*p*BNP, the cells are washed once with distilled water and twice with 0.6 M KCl. Affinity-labeled cells are then suspended in a 0.6 M KCl solution (40 ml) containing EDTA (1.25 mM) and potassium phosphate buffer (50 mM, pH 6.1), to which β-glucuronidase (1.8 g; 500,000 units; type H-1; Sigma Chemical Co.) is added. The reaction proceeds at 30° for approximately 1 hr. The spheroplasts are washed twice with 0.6 M KCl.

Subcellular Fractionation. Washed, packed, affinity-labeled spheroplasts are lysed in a hypotonic medium (20 mM potassium phosphate, pH 6.8, 20 ml) and stored overnight at $-18°$. After centrifugation for 30 min at 20,000 g, the supernatant is carefully decanted. The pellet is washed three times with the same buffer and treated with an appropriate detergent or chemical agent. Treatment with 0.1% Triton X-100, 0.1% deoxycholate or a butanol/water mixture (1:1) is equally effective.

Purification Procedures. The detergent extract is applied to an avidin–Sepharose affinity column.[25] The amount of biotin-tagged components which are bound to the affinity column is determined by differential radioactivity measurements of applied versus effluent fractions. Supposed transport components are released from the affinity column upon treatment with 0.15 M mercaptoethanol. The eluent is dialyzed against water, and the solute is concentrated to 100 μl.

Alternatively, the detergent extract may be applied to an antibiotin antibody–Sepharose affinity column.[27] Specific release of biotin-tagged components can be achieved by passing a solution of biotin (1 mM) over the column. The radioactive eluent fractions are pooled. Biotin may now be released by mercaptoethanol treatment. The reaction mixture is dialyzed exhaustively, and the solution is concentrated to 100 μl.

---

[27] Antibodies to biotin-conjugated bovine serum albumin are produced in rabbits according to M. Berger [*Biochemistry* **14**, 2338 (1975)]. The antibiotin antibodies are isolated from antiserum by affinity chromatography on biotin-Sepharose (E. A. Bayer and M. Wilchek, this series, Vol. **34**, p. 265) and attached to cyanogen bromide-activated Sepharose in a procedure similar to the immobilization of avidin [see footnote 25].

Comments

Affinity Labeling. A homologous series of affinity-labeling reagents, containing both the biologically active ureido moiety and the chemically reactive *p*-nitrophenyl ester group, has been synthesized.[13,14] The inhibitory and reactivation capacities were examined, and the information obtained suggests the existence of two nucleophilic amino acid residues adjacent to the biotin binding site. In this manner, topographical information may be acquired in systems inaccessible to X-ray crystallographic analysis.

Other thiols, e.g., dithiothreitol, have been shown to be equally effective in restoration of biotin transport in *p*BNP-inhibited cells.[28] In addition thiol reagents, e.g., iodoacetic acid,[9] *N*-ethylmaleimide, *p*-hydroxymercuribenzoate, and other mercurials,[28] also inhibit biotin transport. Inhibition by mercurials, in fact, are subject to reactivation by thiols. It is especially interesting to note that another active-ester derivative of biotin, biotinyl-*N*-hydroxysuccinimide ester, which is chemically more reactive to ϵ-amino groups of lysine, failed to inhibit appreciably biotin uptake levels under the same conditons.[12] Hence the mode of *p*BNP action in its inhibition of biotin transport apparently comprises the specified labeling of an essential cysteine residue on a given component of the transport system.

Sites per Cell. The use of affinity cytochemistry in specific localization and quantification of receptor sites offers an alternative and complementary method to label tracing determinations. For the biotin transport system, values of 4000 sites per cell were visualized by ferritin–biotin conjugates versus 8000 sites/cell obtained by measuring the radioactivity (derived from [^{14}C]*p*BNP) released by thiolysis of membrane extracts.[15] The use of affinity cytochemistry is especially appealing for studies of this nature and may easily be applied to other transport systems, membrane-bound binding proteins or membrane-bound enzymes. The preparation of ligand-conjugated ferritin is relatively quick and easy, and cell fractionation studies are unnecessary.

Isolation Studies. Sodium dodecyl sulfate–gel electrophoresis of the mercaptoethanol eluent from avidin–Sepharose yielded a Coomassie Blue-positive major band, corresponding to an apparent molecular weight of 16,000. No conditions have yet been found for the restoration of biotin binding in the isolated material. Thus, the connection between these supposed transport components with the biotin transport system can only be based upon their initial affinity for *p*BNP and the correlation between reactivation and their release via thiols from affinity columns.

[28] Unpublished results.

[64] Enzymic Reduction of d-Biotin d-Sulfoxide to d-Biotin

By ALICE DEL CAMPILLO-CAMPBELL, DANIEL DYKHUIZEN, and P. PATRICK CLEARY

Biotin sulfoxides are ubiquitous in nature, present in culture filtrates of various microorganisms and in cows' milk, and the d-isomer is an oxidative product contaminating most preparations of authentic biotin.[1] In an aerobic environment biotin d-sulfoxide may be one of the major chemical forms of biotin and, therefore, an important source of biotin for organisms able to utilize this oxidized vitamer. One example is *Escherichia coli*, which has a biotin sulfoxide reduction system separate from the biotin biosynthetic pathway.[2] The natural function of sulfoxide reduction may be either to scavenge biotin sulfoxide from the environment or else to reduce bound intracellular biotin that happens to become oxidized. The reduction of biotin d-sulfoxide to biotin is a complex reaction. At least three proteins appear to be required for this conversion. *E. coli* strains with mutations in genes ABFCD located near 17 min in the *E. coli* chromosome[3] are unable to synthesize biotin, but (except for one mutant, bioC18) they are able to grow with biotin sulfoxide. Enzyme extracts from these mutants, except bioC18, catalyze the conversion of biotin sulfoxide to biotin.[2] Starting with a strain carrying the bioA24 mutation, four genes have been identified which are required for biotin sulfoxide utilization, bisA, bisB, bisC, and bisD.[4] Two of these genes map near the biotin gene cluster, and the other two are in other parts of the *E. coli* chromosome. *Bio*C18 is a double mutant, $bio^- bis^-$. Enzyme extracts from the bis mutants do not catalyze the conversion of biotin sulfoxide to biotin. However, all the extracts contain a heat-stable protein, which is required for the reaction.

The reaction catalyzed by biotin sulfoxide reductase may be written as follows:

$$\text{Biotin sulfoxide} + \text{protein(SH)}_2 \rightarrow \text{biotin} + \text{protein-S}_2 + \text{H}_2\text{O}$$
$$\text{(heat stable)}$$

Protein-S$_2$ may be reduced either chemically with dithiothreitol (DTT) or enzymically in the presence of NADPH (reduced nicotinamide adenine dinucleotide phosphate) and a heat-labile protein, which is stimulated by

[1] L. D. Wright, E. Cresson, and C. A. Driscoll, *Proc. Soc. Exp. Biol. Med.* **86,** 480 (1954).

[2] P. P. Cleary and D. Dykhuizen, *Biochem. Biophys. Res. Commun.* **56,** 629 (1974).

[3] A. L. Taylor and C. D. Trotter, *Bacteriol. Rev.* **36,** 504 (1972).

[4] D. Dykhuizen, *J. Bacteriol.* **115,** 662 (1973).

METHODS IN ENZYMOLOGY, VOL. 62

FAD (flavin adenine dinucleotide). This latter enzyme is present in *bio*A24 and *bio*C18 extracts.

Protein-S_2 does not appear to be either thioredoxin[5] or glutaredoxin.[6] Purified yeast thioredoxin kindly supplied by A. Holmgren has very little activity in the above reaction. Enzyme extracts from a thioredoxin-deficient mutant, *E. coli* B tsn C7004,[7] and a thioredoxin reductase-deficient mutant, *E. coli* KK1048[8] catalyze the conversion of biotin sulfoxide to biotin.[9] Glutathione, glutathione reductase, and NADPH are required for glutaredoxin activity.[6] Enzyme extracts from the thioredoxin-deficient mutant and the partially purified protein-S_2 do not catalyze the conversion of biotin sulfoxide to biotin in the presence of the glutathione system.[9]

Genetic studies have shown that the *bis* gene products are not needed for the reduction of methionine sulfoxide to methionine,[4] a reaction known to require thioredoxin.[10]

Assay

Principle. The assay depends on the formation of *d*-biotin from *d*-biotin *d*-sulfoxide in the presence of an enzyme extract from a wild-type (*bis*⁺) strain, *bio*A-24, or in the presence of partially purified enzyme fractions from *bio*A24 and *bio*C18 *bis*A strains. The biotin formed is assayed by the disk microbiological method.[11] The assay organism, strain SA291,[4] is missing the entire biotin operon and therefore needs biotin for growth.

Reagents

   d-Biotin *d*-sulfoxide (BDS) is synthesized and purified according to the methods described by D. B. Melville[12] and modified by H. Ruis *et al.*[13] Solutions of BDS contain less than 0.01% of biotin. No detectable biotin is observed in the assay plate with the biotin sulfoxide concentrations employed in the assay.

   Other reagents are commercial preparations.

Procedure. The reaction mixture contains in 0.25 ml, 25 μmol of ammonium acetate, pH 9.5, 6 μmol of dithiothreitol (DTT), 4.8 nmol of *d*-biotin *d*-sulfoxide (BDS) and the enzyme solution, 0.1–10 mg of protein.

[5] P. G. Porqué, A. Baldesten, and P. Reichard, *J. Biol. Chem.* **245**, 2363 (1970).
[6] A. Holmgren, *Proc. Natl. Acad. Sci. U.S.A.* **73**, 2275 (1976).
[7] M. Chamberlin, *J. Virol.* **14**, 509 (1974).
[8] J. Fuchs, *J. Bacteriol.* **129**, 967 (1977).
[9] A. del Campillo-Campbell, unpublished results.
[10] P. G. Porqué, A. Baldesten, and P. Reichard, *J. Biol. Chem.* **245**, 2371 (1970).
[11] K. Krell and M. Eisenberg, *J. Biol. Chem.* **245**, 6558 (1970).
[12] D. B. Melville, *J. Biol. Chem.* **208**, 495 (1954).
[13] H. Ruis, D. B. McCormick, and L. D. Wright, *J. Org. Chem.* **32**, 2010 (1967).

The reaction is started by adding BDS after preincubation of the other ingredients for 10 min at 37°. After 15 min at 37°, the reaction is stopped by adding 0.2 mmol of trichloroacetic acid. The acid precipitate is removed by centrifugation, and the supernatant, usually 20 μl, is assayed for biotin. The assay plate contains 300 ml of 1.4% agar in synthetic medium[14] supplemented with 0.01% 2,3,5-triphenyl-2*H*-tetrazolium chloride, 0.16% dextrose, 10 mg L-histidine, and 1 ml of a fresh culture of strain SA291.

NADPH (0.125 μmol) can be substituted for DTT in the assay, but only 10–20% of the activity is obtained. A NADPH-generating system composed of 0.06 μmol of NADP$^+$, 1 μmol of glucose 6-phosphate, and 0.02 unit of glucose-6-phosphate dehydrogenase is as effective as DTT. With the purified biotin sulfoxide reductase and the heat-stable protein, NADPH and the NADPH-generating system are inactive.

Although the rate of biotin synthesis is nonlinear with crude extracts, linearity can be achieved employing low concentrations of purified biotin sulfoxide reductase (20–200 μg) and large amounts of crude or purified heat-stable protein (2–9 mg).

Units and Specific Activity. Enzyme units are defined as micrograms of biotin formed per 15 min. Specific activity is the number of units per milligram of protein.

Preparation of Enzymes

Growth of Cells

Cells are grown to stationary phase at 37° with shaking in either tryptone broth or synthetic medium[14] containing 0.4% dextrose, 4×10^{-2} mg/ml L-histidine, and 3×10^{-3} μg/ml *d*-biotin. The cells are harvested in the cold at 6000 rpm (Sorvall, GSA rotor) and washed twice with cold 0.9% NaCl. The washed cells are usually frozen ($-20°$) before extraction.

Preparation of Extracts

All procedures are carried out at 0–4° unless otherwise indicated. Phosphate buffer contains 50 m*M* potassium phosphate (pH 7.0) and 3 m*M* 2-mercaptoethanol.

The frozen cells are thawed and resuspended at 125 times the original concentration in cold phosphate buffer. The cell suspension is disrupted with a Bronson sonifier, Model W140. After centrifugation for 20 min at

[14] A. Campbell, *Virology* **14**, 22 (1961).

14,000 rpm (Sorvall SS-34 rotor), the supernatant is dialyzed overnight against phosphate buffer. The protein concentration of the extract is determined by the method of Warburg and Christian.[15] The extracts contain between 30 and 50 mg/ml.

Purification of Biotin Sulfoxide Reductase

Extracts from *E. coli* K12 *bio*A24, strain R879 (16), are used to prepare biotin sulfoxide reductase.

Streptomycin Sulfate Treatment. A 20% solution of streptomycin sulfate is added to the dialyzed extract to precipitate nucleic acids. About 0.3 mg of streptomycin sulfate is added per milligram of protein. After stirring for 10 min, the precipitate is removed by centrifugation at 12,000 rpm and discarded.

Ammonium Sulfate Fractionation. Solid ammonium sulfate is added to the streptomycin sulfate supernatant to 40% saturation. After stirring for 20 min, the precipitate is centrifuged down at 14,000 rpm, dissolved in phosphate buffer, and dialyzed overnight against the same buffer. The buffer is changed in the morning, and the fraction is dialyzed for another 4 hr. This fraction contains between 30 and 50 mg of protein per milliliter.

Ca$_3$(PO$_4$)$_2$ Gel Adsorption. Aged Ca$_3$(PO$_4$)$_2$ gel (Nutritional Biochemicals Corporation) is added to the 0–40% dialyzed ammonium sulfate fraction so that 82–85% of the protein is adsorbed on the gel. The amount of gel required is determined for each batch of Ca$_3$(PO$_4$)$_2$ gel with small trial runs. The mixture is stirred for 10 min and centrifuged for 10 min at 12,000 rpm. The supernatant contains between 4 and 6 mg of protein per milliliter and most of the biotin sulfoxide reductase.

Sephadex G-50 Chromatography. The gel supernatant is applied to a Sephadex G-50 column (Pharmacia, Sweden, medium; 3 × 45 cm) in 10 mM ammonium acetate, pH 8.6, containing 1 mM ethylenediaminetetraacetate (EDTA) and 3 mM 2-mercaptoethanol. The enzyme is eluted with

TABLE I
PURIFICATION OF BIOTIN SULFOXIDE REDUCTASE

Fraction	Volume (ml)	Protein (mg)	Specific activity (units/mg protein)	Total units	Yield (%)
1. Crude extract	115	3350	0.015	50	100
2. Streptomycin supernatant	115	2090	0.029	60	120
3. 0–40% (NH$_4$)$_2$SO$_4$	18	1060	0.045	48	96
4. Ca$_3$(PO$_4$)$_2$ gel supernatant	24	153	0.23	36	72
5. Sephadex fractions	3.9	42	0.71	30	60

[15] O. Warburg and W. Christian, *Biochem. Z.* **310**, 384 (1941).

the same buffer at a flow rate of 20 ml/hr, 4 ml per tube. Tubes 15 to 23 and 24 to 40 are pooled, and the fractions are brought to 70% saturation with solid ammonium sulfate. The precipitates are collected by centrifugation at 14,000 rpm after stirring for 20 min, dissolved in phosphate buffer, and dialyzed overnight against 1 liter of phosphate buffer. The activity of the two fractions is about the same and represents close to a 50-fold purification.

The purification procedure is summarized in Table I.

Purification of Heat-Stable Protein-S_2

Extracts from *E. coli* K12 *bio*C18, strain R876[16] are used. This strain is also *bis*−.[4]

The same steps used in the purification of the biotin sulfoxide reductase are followed up to the $Ca_3(PO_4)_2$ gel step. The protein is eluted from the gel after washing the gel with cold water. The gel is eluted first with 0.1 M potassium phosphate, pH 7, containing 3 mM 2-mercaptoethanol. After 0.5 hr, the mixture is centrifuged for 10 min at 12,000 rpm. The supernatant is saved. The gel is eluted again for another 0.5 hr with 0.1 M potassium phosphate, pH 7.7, containing 3 mM 2-mercaptoethanol. After centrifugation, the two eluates are combined and brought to 80% saturation with solid ammonium sulfate. The precipitate is collected by centrifugation at 14,000 rpm, dissolved in phosphate buffer, and dialyzed overnight against the same buffer. The purification at this stage is about 6-fold.

This fraction is heated in 4-ml batches in a 90°–95° bath for about 1 min, chilled, and centrifuged. The supernatant contains 30–50% of the protein and most of the activity, a 2- to 3-fold purification. Several batches of the heated fractions are combined, and 47 ml (202 mg) are filtered through a PM-10 Amicon ultrafilter. The filtrate contains 9% of the protein with a 5.5-fold purification, and this is concentrated by passing the filtrate through an UM$_2$ Amicon ultrafilter. The overall purification is about 70-fold. The apparent molecular weight of the active protein is approximately 10,000, based on sodium dodecyl sulfate (SDS) acrylamide gel electrophoresis.

Partial Separation of the Heat-Labile Protein-S_2 Reductase

Extracts from *E. coli* K12 *bio*C18, strain R876[16] are the source of this enzyme.

Solid ammonium sulfate is added to a dialyzed extract to 40% satura-

---

[16] A. del Campillo-Campbell, G. Kayajanian, A. Campbell, and S. Adhya, *J. Bacteriol.* **94**, 2065 (1967).

tion. After stirring for 15 min, the precipitate is collected by centrifugation at 14,000 rpm. The supernatant is brought to 80% saturation, and after stirring for 15 min the precipitate is collected by centrifugation. The precipitates, 0–40% and 40–80%, are dissolved in 50 mM Tris·HCl pH 7.6 [tris(hydroxymethyl)aminomethane chloride] containing 1 mM EDTA, and dialyzed overnight against the same buffer. Preliminary attempts at further purification of this enzyme have not been very successful. The results presented in Table II show that the 0–40% fraction contains the protein-S_2 reductase as indicated by the activity in the presence of the NADPH generating system.

Table II shows the requirement for biotin sulfoxide reductase, the heat-stable fraction, and the protein-S_2 reductase in the conversion of biotin sulfoxide to biotin.

TABLE II

REQUIREMENT FOR THREE PROTEIN FRACTIONS IN THE CONVERSION OF BIOTIN
SULFOXIDE (BDS) TO BIOTIN

	Biotin (μg)	
Fractions present[a]	DTT added[b]	NADPH added[b]
1. Purified BDS reductase	0	0
2. Purified protein-S_2	0	0
3. Crude protein-S_2 reductase (R876, 0–40%)	0	0
4. Purified BDS reductase + purified protein-S_2	6.4×10^{-3}	1.6×10^{-4}
5. Purified BDS reductase + crude protein-S_2 reductase (R876, 0–40%)		7.7×10^{-3}
6. Purified BDS reductase + purified protein-S_2 + crude protein-S_2 reductase (R876, 0–40%)		2.3×10^{-2}
7. Crude BDS reductase + crude protein-S_2 reductase (R876, 0–40%)	4.1×10^{-3}	4.5×10^{-4}
8. Crude BDS reductase + R876 (40–80%)	2.2×10^{-3}	0
9. Crude BDS reductase + crude protein-S_2 reductase (R876, 0–40%) + R876 (40–80%)	4.7×10^{-3}	2.6×10^{-3}
10. Crude BDS reductase + crude protein-S_2 reductase (R876, 0–40%), heated 92°, + R876 (40–80%)	5.5×10^{-3}	2.6×10^{-4}

[a] Purified BDS reductase is the 50-fold purified Sephadex fraction, 16.5 μg added. Purified protein-S_2 is the 70-fold purified fraction, 0.19 mg added. Crude protein-S_2 reductase (R876, 0–40%) and (R876, 40–80%) are the fractions described in the text. 3.8 mg, 3.4 mg, and 0.8 mg (heated) added, respectively. Crude BDS reductase is the dialyzed extract, 0.24 mg added.

[b] DTT-dithiothreitol; NADPH generating system. In experiments 1–6, 8 μM FAD is present; in 7–10, no FAD is present.

Properties

pH Optimum and K_m. The overall reaction has a pH optimum of 9.5 in ammonium acetate buffer. There is 31% activity at pH 10, 65% at pH 8.6, 18% at pH 7.0, and 4% at pH 6.0. The apparent K_m for d-biotin d-sulfoxide is 5.7 μM at pH 9.5.

Specificity. NADH is only 34% as active as NADPH in the overall conversion of biotin sulfoxide to biotin.

Inhibitors and Cofactors. The reaction is strongly inhibited by thiol and vicinal thiol inhibitors. Sodium arsenite (1 mM) inhibits 76%; sodium iodoacetate (5 mM) inhibits 43%; manganous chloride (10 mM) and copper sulfate (10 mM) inhibit 79% and 89%, respectively; and EDTA (10 mM) inhibits 87% of the total activity.

Quinacrine, 0.8 mM, inhibits 88%. This inhibition is reversed by flavin adenine dinucleotide (FAD). FAD, 8 μM, enhances the activity 1.5- to 2-fold, while flavin mononucleotide (FMN), 0.8 mM, enhances the activity 1.5-fold.

Stability. Biotin sulfoxide reductase preparations lose about 60% of their activity after 1.5 min at 60°, and the activity is completely destroyed above 90°. The low-molecular-weight protein is stable at 60°, and it loses very little activity above 90°. The protein-S_2 reductase is inactivated above 90°. All the enzymes are stable frozen for long periods of time. They lose some activity on repeated freezing and thawing.

[65] Isolation and Characterization of d-Allobisnorbiotin

By Wha Bin Im, Donald B. McCormick, and Lemuel D. Wright

Isolation

Principle. A pseudomonad is grown aerobically for 5 days on an essential-salts medium containing d-biotin as a sole source of carbon, nitrogen, and sulfur.[1] Double-labeled biotin, the thiolane ring, and the side chain with ^3H, and the ureido carbonyl with ^{14}C are provided in the medium, since the catabolite loses a two-carbon fragment from the side chain and most of the original ureido carbon.[2] The culture filtrates are

[1] R. N. Brady, L. F. Li, D. B. McCormick, and L. D. Wright, *Biochem. Biophys. Res. Commun.* **19**, 777 (1965).

[2] W. B. Im, D. B. McCormick, and L. D. Wright, *J. Biol. Chem.* **248**, 7798 (1973).

concentrated and freed from most of the soluble proteins by mixing with 95% alcohol. The catabolite is separated by two anion-exchange column chromatographies of the culture filtrates on Dowex 1 (formate) using ammonium formate as an eluent in the first chromatography and formic acid in the second. Ammonium ion in the effluents is removed on Dowex 50 (H^+).

Procedure. Two liters of the culture medium are prepared to contain, per liter, 3 g of [^3H,^{14}C]biotin and an inorganic salts mixture.[1] This medium is inoculated with 4 ml of a stationary-phase culture of the pseudomonad and incubated at 37° with continuous shaking for 5 days. Volume ratio of the culture and air during incubation is about 1:4. The culture filtrates are obtained by centrifugation at 10,000 g for 15 min. Supernatants are concentrated to a small volume under reduced pressure and mixed with an equal volume of 95% ethanol. The precipitate is removed by filtration. The filtrate is evaporated to dryness and dissolved in a minimum volume of distilled water. The solution is poured over a Dowex 1-X2 formate anion-exchange column (1.5 × 90 cm). The column is first eluted with 1 liter of water and then a linear gradient formed from 2 liters of water to 2 liters of 0.3 M ammonium formate. Fractions of 15 ml are collected. Radioactivity is determined in 0.2-ml aliquots with 10 ml of Bray's solution[3] in a liquid scintillation spectrometer using appropriate channel selections.

Figure 1 shows a typical elution pattern from chromatography of the culture filtrates containing [^3H,^{14}C]biotin and catabolites over a column

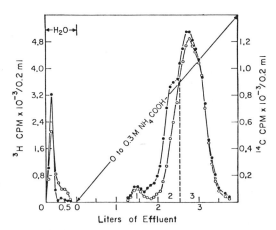

Fɪɢ. 1. Elution pattern from chromatography of [^3H,^{14}C]biotin culture filtrate on a column (1.5 × 90 cm) of Dowex 1-X2 formate. ●, ^3H; ○, ^{14}C.

[3] G. A. Bray, *Anal. Biochem.* **1**, 279 (1960).

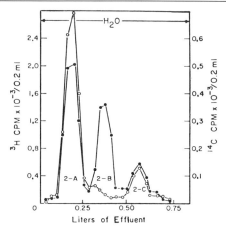

FIG. 2. Elution pattern from rechromatography of the second acidic fraction, 2, on a column (1 × 45 cm) of Dowex 50W (H⁺). ●, ³H; ○, ¹⁴C.

of Dowex 1-X2 formate. The second acidic fraction, the ^3H content of which is much higher as compared to its ^{14}C label, is evaporated to a small volume and poured over a column (1 × 45 cm) of Dowex 50W (H⁺). The column is eluted with water. Figure 2 shows the elution pattern from rechromatography of the second acidic fraction over the column of Dowex 50W (H⁺). Again the second fraction, 2-B, contains the catabolite and *d*-bisnorbiotin. This fraction is concentrated and poured over a column (0.8 × 80 cm) of Dowex AG 1-X2 formate. The column is eluted with a linear gradient from 2 liters of water to 2 liters of 0.03 N formic acid. The catabolite, *d*-allobisnorbiotin, is eluted as the first acidic fraction from the column followed by a minor fraction containing *d*-bisnorbiotin. Upon complete dryness of the first fraction, about 15 mg of *d*-allobisnorbiotin can be obtained as a white powder.

Comments. The catabolite, *d*-allobisnorbiotin, has a *trans*-ureido ring, which is very unstable in acidic solutions.[4] Formic acid in the effluents containing the catabolite should be removed completely under reduced temperature. Yield of the catabolite can be somewhat improved by keeping the culture under nonshaken conditions for an additional 2 days.[2]

Characterization

The structure of the catabolite,[2] *d*-allobisnorbiotin, is shown in Fig. 3. The ^3H specific activity of protons on the thiolane ring and the side chain

⁴ S. A. Harris, R. Mozingo, D. Z. Wolf, A. N. Wilson, and K. Folkers, *J. Am. Chem. Soc.* **67**, 2102 (1945).

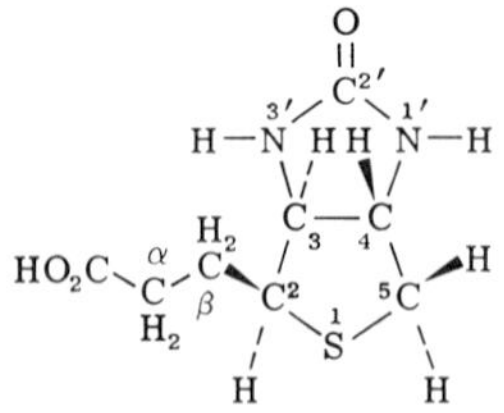

$$
\begin{array}{c}
\text{O} \\
\| \\
\overset{3'}{\text{C}}^{2'} \overset{1'}{} \\
\text{H}-\text{N} \quad \text{H H} \quad \text{N}-\text{H} \\
\text{C}-\text{C} \\
\text{H}_2 \quad {}_3 \quad {}_4 \\
\text{HO}_2\text{C} \underset{\text{H}_2}{\overset{\alpha}{\text{C}}} \underset{\beta}{\overset{}{\text{C}}} \overset{}{\text{C}}^2 \quad {}^5\text{C} \\
\text{H} \quad \text{S} \quad \text{H}
\end{array}
$$

FIG. 3. Structure of *d*-allobisnorbiotin.

of the catabolite is the same as that of *d*-bisnorbiotin, a stereoisomer having the original *cis*-ureido ring. The ^{14}C-specific activity of the carbonyl carbon of the ureido ring of the catabolite, however, is only about one seventh of that of *d*-bisnorbiotin. The molecular weight of the methyl ester of the catabolite is 230.0806 ($C_9H_{14}N_2O_3S$) as determined by high-resolution mass spectra. The R_f values of the catabolite are 0.71 in a solvent system of butanol–formic acid–water (4:1:1, v/v/v) and 0.75 in butanol–acetic acid–water (2:1:1, v/v/v). The avidin–catabolite complex[5] shows $E_{233 \text{ nm}} = 0.9 \times 10^4$, whereas the avidin–*d*-bisnorbiotin complex has $E_{233 \text{ nm}} = 1.7 \times 10^4$. A Schiff base formed between the catabolite and *p*-dimethylaminocinnamaldehyde[6] has $E_{533} = 2.0 \times 10^4$. The optical rotation at 20° (sodium line, C = 1.0) is + 200 for the catabolite and + 130°

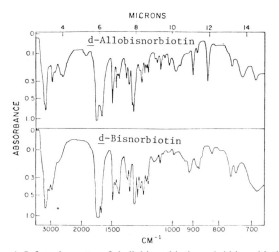

FIG. 4. Infrared spectra of *d*-allobisnorbiotin and *d*-bisnorbiotin.

[5] N. M. Green, *Biochem. J.* **89**, 599 (1963).
[6] D. B. McCormick and J. A. Roth, *Anal. Biochem.* **34**, 226 (1970).

NUCLEAR MAGNETIC RESONANCE CHARACTERISTICS OF *d*-ALLOBISNORBIOTIN AND *d*-BISNORBIOTIN

| Compound | Methylene | Methine | | | Vinyl | | Carboxyl | |
	C-5	C-2	C-3[b]	C-4[b]	α	β	N-1, 3[b]	-COOH[b]
d-Allobisnorbiotin	3.03	3.41	4.05	4.47	2.64	2.08	6.32	9.5
d-Bisnorbiotin	2.95	3.35	4.38	4.38	2.35	2.03	6.40	10.0

Chemical shift for protons as δ (ppm)[a]

[a] Carbon positions in the thiophane ring are numbered counterclockwise from sulfur as 1. See Fig. 3.
[b] Chemical shifts of C-3,4, N-1,3, and the carboxyl proton were determined in D_6-dimethyl sulfoxide with tetramethylsilane as reference; the rest in D_2O, pD 5.5.

for d-bisnorbiotin (in H_2O). Infrared spectra of d-allobisnorbiotin and d-bisnorbiotin are compared in Fig. 4. The proton nuclear magnetic resonance characteristics of the catabolite and d-bisnorbiotin are summarized in the table. The coupling constants between the proton on C_3, a bridgehead carbon, and adjacent protons of d-allobisnorbiotin are 7.5 and 3.0 cps for $J_{4,3}$ and $J_{2,3}$, respectively.

[66] Analysis of Microbial Biotin Proteins

By R. Ray Fall

It is now well established that the biotin-dependent enzymes catalyze analogous two-step reactions that involve a covalently bound biotin residue serving as a "carboxyl carrier" (see references cited in footnotes 1–4 for a review):

The biotin-dependent carboxylases, which include acetyl-CoA, propionyl-CoA, 3-methylcrotonyl-CoA, geranyl-CoA,[5] pyruvate, and urea carboxylases, utilize bicarbonate as the carboxyl donor and require ATP to drive the formation of the carboxybiotin enzyme. The remaining known biotin enzymes, which include methylmalonyl-CoA: pyruvate transcarboxylase, methylmalonyl-CoA decarboxylase, and oxaloacetate decarboxylase, catalyze ATP-independent carboxylation of their respective biotin prosthetic groups. For transcarboxylase methylmalonyl-CoA is the carboxyl donor, and for the decarboxylases the carboxyl transfer to the biotinyl group is followed by decomposition to free biotinyl-enzyme and CO_2.

The distinct nature of the carboxyl donor, biotin, and carboxyl acceptor subsites of these enzymes has been confirmed by the findings that these three subsites are localized on three separate subunits in *Es-*

[1] A. W. Alberts and P. R. Vagelos, *Enzymes* **6**, 37 (1972).
[2] J. Moss and M. D. Lane, *Adv. Enzymol.* **35**, 321 (1971).
[3] M. Obermayer and F. Lynen, *Trends Biochem. Sci.* **1**, 169 (1976).
[4] H. G. Wood and R. E. Barden, *Annu. Rev. Biochem.* **46**, 385 (1977).
[5] As shown by Seubert *et al.* [*Biochem. Z.* **338**, 265 (1963)] and confirmed by us, the substrate for geranyl-CoA carboxylase is the 2Z isomer, *cis*-geranyl-CoA, not the 2E isomer, *trans*-geranyl-CoA. The systematic name for the enzyme (EC 6.4.1.5) should probably be (2Z)-3,7-dimethyl-2,6-octadienyl-CoA:carbon-dioxide ligase (ADP-forming).

cherichia coli,[6,7] Pseudomonas citronellolis,[8] and spinach chloroplast[9] acetyl-CoA carboxylases, and Propionibacterium shermanii[10] transcarboxylase. Several other biotin-dependent carboxylases contain two or more of these subsites fused into one or two polypeptide chains (see Obermeyer and Lynen[3] for a review; several examples are cited in Table I).

The biotin prosthetic group of these enzymes is covalently linked via amide linkage to an ε-amino group of a lysine residue.[2] This makes it feasible to covalently label the biotinyl site with a radioactive label, and subsequently monitor and/or analyze the biotinyl polypeptide(s) present in an organism or purified enzyme preparation. This communication describes techniques we have used for the in vivo radioactive labeling of microbial biotin enzymes, and analysis of the resulting labeled biotinyl polypeptides.

Radiolabeling of Whole-Cell Biotin Proteins

Several groups have successfully labeled biotin proteins in growing microbial cultures by addition of radioactive biotin to the culture medium. Early studies took advantage of available biotin-requiring strains of E. coli[7,11] and yeast[12] to incorporate [14C]biotin from the medium. A major limitation to the use of [14C]biotin for this purpose is that the commercially available material is of relatively low specific activity (20–40 mCi/mmol). For several years we have used [3H]biotin of much higher specific activity (1–3 Ci/mmol), with the result that it is possible to achieve sufficient labeling of biotin proteins even with wild-type strains. For example, a typical addition of a [3H]biotin preparation (7–15 μg/liter; 3.2 Ci/mmol) to wild-type E. coli K12 cultures is sufficient to repress biotin synthesis,[13] and yields [3H]biotin incorporation at a level comparable to that obtained with biotin auxotrophs of E. coli (grown with [3H]biotin; see also Fall and Vagelos[14]). The feasibility of achieving sufficient labeling of a variety of bacteria with [3H]biotin so that whole-cell

[6] A. W. Alberts and P. R. Vagelos, Proc. Natl. Acad. Sci. U.S.A. 59, 561 (1968).

[7] A. W. Alberts, A. M. Nervi, and P. R. Vagelos, Proc. Natl. Acad. Sci. U.S.A. 63, 1319 (1969).

[8] R. R. Fall, Biochim. Biophys. Acta 450, 475 (1976).

[9] C. G. Kannangara and P. K. Stumpf, Arch. Biochem. Biophys. 152, 83 (1972).

[10] H. G. Wood and G. Zwolinski, Crit. Rev. Biochem. 4, 47 (1976).

[11] P. Dimroth, R. B. Guchhait, E. Stoll, and M. D. Lane, Proc. Natl. Acad. Sci. U.S.A. 67, 1353 (1970).

[12] M. Sumper and C. Riepertinger, Eur. J. Biochem. 29, 237 (1972).

[13] C. H. Pai and H. C. Lichstein, Biochim. Biophys. Acta 100, 28 (1965).

[14] R. R. Fall and P. R. Vagelos, J. Biol. Chem. 248, 2078 (1973).

TABLE I. GROWTH MEDIA SUITABLE FOR INCORPORATION OF [³H]BIOTIN INTO MICROBIAL BIOTIN PROTEINS

Strain	Biotin-deficient Medium[a]	[³H]Biotin enzymes detected[b]
a. Bacteria		
Aerobacter aerogenes (NCTC 418)	Stern citrate medium[e] (semi-anaerobic growth)	A, O
	Medium 56, 0.2% glycerol	A
Escherichia coli (several, including K12, B, and W strains)	Medium E[f] or medium 56[g] plus 0.5% glucose or glycerol	A
Enterobacter aerogenes (ATCC 13048)	Medium 56, 0.2% glucose	Yes
Proteus vulgaris	Medium E, 0.2% glucose, 1 mg/ml niacin	Yes
Pseudomonas citronellolis (ATCC 13674)	T medium,[h] various carbon sources	A, G, M, P, Py (depending on inducing conditions)
P. aeruginosa, P. putida (various strains)	T medium, various carbon sources	Yes
Bacillus cereus, B. megaterium, B. subtilis	Medium 56, 0.2% glucose, 0.2% Casamino acids (vitamin free, Difco)	Yes
Lactobacillus plantarum (ATCC 8014)	Biotin assay medium (Difco; anaerobic growth)	Yes
Mycobacterium phlei (ATCC 356)	Medium M[c]	P
Propionibacterium shermanii (ATCC 31673)	Delwiche medium[i] with casein hydrolyzate replaced by Casamino acids (vitamin free, Difco)	Yes
b. Fungi and Yeasts		
Lipomyces starkeyi (ATCC 12659)	Yeast medium[d]	A
Neurospora crassa (sl strain, FGSC 1118)	Scarborough ψVM medium,[j] with biotin omitted, 1% glucose	Yes
Saccharomyces cerevisiae (ATCC 9896)	Yeast medium[d]	A, Py
	Yeast medium[d] with urea (0.2%) as sole N source	A, Py, U

c. Protozoa

Acanthamoeba castellanii (ATCC 30010)	ATCC medium 354 minus biotin	Yes
Crithidia fasiculata (ATCC 11745)	Defined hemoflagellate medium[k] minus biotin	A
Cryptothecodinium cohnii (ATCC 30021)	ATCC medium 460 minus biotin	Yes
Leptomonas sp. (ATCC 30250)	Defined hemoflagellate medium minus biotin	Yes
Tetrahymena pyriformis (mating type II)	Cox semisynthetic medium[l] minus biotin	Yes

[a] In each case the media were supplemented with [^{3}H]biotin (1.35–3.2 Ci/mmol; 5–20 µg per liter of medium).

[b] In some cases, we have purified individual biotin enzymes from the [^{3}H]biotin-labeled cells and shown that they contain the [^{3}H]biotin label. For these enzymes the abbreviations are: A, acetyl-CoA carboxylase; G, geranyl-CoA carboxylase; M, 3-methylcrotonyl-CoA carboxylase; O, oxaloacetate decarboxylase; P, propionyl-CoA carboxylase; Py, pyruvate carboxylase; U, urea carboxylase. In other cases we have demonstrated the feasibility of labeling cellular biotin proteins (e.g., >50,000 cpm [^{3}H]biotin protein/mg protein in crude extracts) using the medium indicated, but have not isolated and identified individual biotin enzymes; these are identified by the word, Yes. It is important to note that, where particular biotin enzymes are indicated, it does not necessarily imply that other biotin enzymes are absent in those cells, although this is true in some instances (e.g., we have been able to detect only one native biotinyl polypeptide, the BCCP subunit of acetyl-CoA carboxylase, in E. coli).

[c] Medium M (g/liter): glycerol, 20 g; ammonium chloride, 2.2 g; Tween-80, 2.0 g; Casamino acids (vitamin free, Difco), 2.0 g; KH_2PO_4, 1.0 g; succinic acid, 1.0 g; $MgSO_4,7H_2O$, 0.5 g; ferric ammonium citrate, 0.5 g; pH adjusted to 7.0 with KOH.

[d] Yeast medium is a modification of the ψVM medium of Scarborough[j]: biotin and NH_4NO_3 are omitted from the 50X stock solution; added (per liter) are 2 g of ammonium sulfate, 2 mg m-inositol, and 400 µg each of calcium pantothenate, pyridoxamine-HCl, and thiamine-HCl (the latter four from a concentrated stock solution). One percent glucose is used as the carbon source. For induction of urea carboxylase, the ammonium sulfate is omitted, and 5 ml of a filter-sterilized 20% (w/v) solution of urea is added.

[e] J. R. Stern, Biochemistry **6**, 3545 (1967).

[f] H. J. Vogel and D. M. Bonner, J. Biol. Chem. **218**, 97 (1956).

[g] J. Monod, G. Cohen-Bazire, and M. Cohen, Biochim. Biophys. Acta **7**, 585 (1951).

[h] M. L. Hector and R. R. Fall, Biochemistry **15**, 3465 (1976).

[i] E. A. Delwiche, J. Bacteriol. **58**, 395 (1949).

[j] T. H. Schulte and G. A. Scarborough, J. Bacteriol. **122**, 1076 (1975).

[k] C. J. Bacchi, C. Lambros, B. Goldberg, S. H. Hutner, and G. D. F. de Carvalho, Antimicrob. Agents Chemother. **6**, 785 (1974).

[l] D. Cox, J. Protozool. **17**, 150 (1970).

[³H]biotinyl polypeptides could be analyzed directly was demonstrated by Fall et al., [15] and we have more recently extended the use of this procedure to several other microorganisms, as described below.

[³H]Biotin is prepared commercially by subjecting an authentic d-biotin sample to catalytic exchange labeling. [16] The resulting material contains many degradation products and should be purified before use. Purification of [³H]biotin is easily achieved by chromatography on Dowex-1-X2 formate as described by Ogata [17] and in detail elsewhere. [14] The specific activity of the purified material has been measured by bioassay with *Lactobacillus arabinosus* [18] or by an avidin binding assay [19] with equivalent results. The purified, concentrated [³H]biotin can be stored for prolonged periods as a sterile solution in ethanol, and then added directly to sterile media for labeling purposes.

In order to incorporate exogenous radioactive biotin efficiently into various microorganisms, it is necessary to use a defined or semidefined growth medium that is relatively free of nonradioactive biotin. Media recipes containing moderate to high levels of supplements (e.g., yeast and liver extracts, peptones) usually also contain significant amounts of free biotin, [20] which will dilute the added radioactive biotin and correspondingly decrease the incorporation of the isotope into cellular biotinyl polypeptides. Many of the biotin-deficient media that we have found to be suitable for the labeling of several different groups of microorganisms are summarized in Table I. In each case the growth and [³H]biotin labeling conditions described produced sufficient labeling of biotinyl polypeptides ($> 50,000$ cpm/mg cell extract protein) so that it is possible to analyze directly whole-cell [³H]biotin polypeptides by sodium dodecyl sulfate (SDS) polyacrylamide gel electrophoresis as described below. It should be emphasized that Table I summarizes only the suitable media that we have tested; it does not represent a systematic attempt to optimize growth or [³H]biotinylation conditions. The information should be useful as a general guideline for constructing particular media for various microorganisms, including those not listed.

Table I also indicates that in some instances we have isolated and identified particular [³H]biotin-labeled enzymes. In these cases, the enzyme noted has been purified free of other known biotin enzymes in the

[15] R. R. Fall, A. W. Alberts, and P. R. Vagelos, *Biochim. Biophys. Acta* **379**, 496 (1975).

[16] For catalytic exchange labeling, samples of d-biotin (25 mg) were submitted to New England Nuclear and treated with 10 Ci of tritium.

[17] K. Ogata, this series, Vol. 18A, p. 390.

[18] L. D. Wright and H. R. Skeggs, *Proc. Soc. Exp. Biol. Med.* **56**, 95 (1944).

[19] D. B. Rylatt, D. B. Keech, and J. C. Wallace, *Arch. Biochem. Biophys.* **183**, 113 (1977).

[20] The biotin content of a variety of commercially available peptones and yeast extracts is presented in the BBL Catalog (p. 163, 5th ed., 1970).

organism, and the [³H]biotin label has been shown to copurify with the particular enzyme.

Analysis of Radiolabeled Biotinyl Polypeptides

The ability to introduce a specific, covalently bound radioactive biotin residue into the relatively few types of biotin proteins present in microbial cells provides a convenient means for detection and/or analysis of cellular biotinyl polypeptides.

Fall *et al.* [15] obtained information on the number and molecular weights of [³H]biotinyl polypeptides present in SDS extracts of whole [³H]biotin-labeled bacteria by analysis of such extracts on SDS polyacrylamide gels. Various bacteria labeled by growth in [³H]biotin-supplemented media, as described above, were disrupted by sonication directly in an SDS buffer.[21] More recently, to avoid proteolysis that may occur under these conditions, we have incorporated 1 m*M* *o*-phenanthroline and phenylmethanesulfonyl fluoride in the disruption buffer, and whole cells are first boiled 2 min in this buffer before sonication. The supernatant obtained after centrifugation of the extract (48,000 *g* for 30 min) is suitable for SDS polyacrylamide gel analysis. The efficiency of extraction is generally greater for gram-negative bacteria (80–95%) than for gram-positive bacteria (60–85%). It is even more difficult to extract yeast cells in this way, and for yeast, extracts can first be prepared by using a French pressure cell.[22]

For SDS-polyacrylamide gel analysis we have routinely used the discontinuous system described by Laemmli,[23] in both tube-gel and slab-gel formats. The former is especially useful when large protein loads (0.1–0.3 mg) are to be applied, and in these cases we include 6 *M* urea in the separating and stacking gels to help prevent protein precipitation. Analysis of the radioactive biotinyl polypeptides is accomplished by slicing, digesting and counting the gels by routine procedures.[15,24] Estimation of the molecular weights of the detected biotinyl polypeptides is made by comparison to standard curves constructed with marker polypeptides.[25,26]

During the purification of biotin enzymes, biotinyl subunits from such enzymes, or biotinyl peptides, it is very convenient to use preparations

---

[21] The SDS buffer used was 2% SDS, 2% 2-mercaptoethanol in 62 m*M* Tris·HCl, pH 6.8.
[22] E. A. Logue and R. R. Fall, unpublished observations.
[23] U. K. Laemmli, *Nature (London)* **227**, 680 (1970).
[24] Several suitable methods are described by C. W. Helleiner and W. H. Wunner, *Anal. Biochem.* **39**, 333 (1971).
[25] K. Weber and M. Osborn, *J. Biol. Chem.* **244**, 4406 (1969).
[26] D. M. Neville, Jr., *J. Biol. Chem.* **246**, 6328 (1971).

containing a radioactive biotinyl prosthetic group and monitor radioactivity. In the case of *E. coli* acetyl-CoA carboxylase, the enzyme dissociates into subunits upon rupture of the cells, and for purification of the
BCCP[27] subunit we used [³H]biotin-labeled cells and measured [³H] for
the detection of [³H]BCCP.[28] [³H]Labeled BCCP preparations proved
very valuable in studies on the proteolytic degradation of BCCP[14] to
smaller biotinyl polypeptides, and for detection of [³H]biotinyl peptides in the amino acid sequencing of BCCP$_{(9100)}$.[27,29] We have also
used [³H]biotin-labeled preparations of *P. citronellolis* acetyl-CoA,
3-methylcrotonyl-CoA, and geranyl-CoA carboxylases to monitor these
preparations and to identify the biotinyl subunit of the purified enzymes
after SDS–polyacrylamide gel electrophoresis.[8,30]

It is important to verify that the radiolabel in isolated biotin proteins
resides specifically in the biotinyl prosthetic group, especially in light of reports that several microorganisms can degrade biotin added to the growth
medium.[31,32] This can be conveniently done as described by Tanabe *et
al.*,[33] by digesting the radiolabeled protein with proteinase K (Merck) and
measuring the release of radiolabeled biocytin[34] as the major labeled product. We have carried out this type of verification in the cases of *E. coli*
[³H]biotin-BCCP[35] and of *P. citronellolis* [³H]biotin-3-methylcrotonyl-
CoA carboxylase and [³H]biotin-geranyl-CoA carboxylase.[30]

Summary of Microbial Biotinyl Polypeptides

Several microbial biotin enzymes have been isolated, and the molecular weights of the biotinyl subunits were determined. These data are summarized in Table II (biotinyl polypeptides known to be produced via proteolytic degradation of larger, native biotinyl polypeptides[3,14] are not included). Two major conclusions can be derived from this information.
First, it is apparent that there is a very large size distribution in the biotinyl subunits of these enzymes, even within the same type of enzyme

[27] BCCP refers to the biotin carboxyl carrier protein subunit of *E. coli* acetyl-CoA carboxylase. BCCP$_{(9100)}$ is an 82 residue proteolytic fragment of BCCP.
[28] R. R. Fall and P. R. Vagelos, *J. Biol. Chem.* **247**, 8005 (1972).
[29] M. R. Sutton, R. R. Fall, A. M. Nervi, A. W. Alberts, P. R. Vagelos, and R. A. Brad-
shaw, *J. Biol. Chem.* **252**, 3934 (1977).
[30] R. R. Fall and M. L. Hector, *Biochemistry* **16**, 4000 (1977).
[31] K. Ogata, this series, Vol. 18A, p. 397.
[32] S. Iwahara, D. B. McCormick, and L. D. Wright, this series, Vol. 18A p. 404.
[33] T. Tanabe, K. Wada, T. Okazaki, and S. Numa, *Eur. J. Biochem.* **57**, 15 (1975).
[34] Biocytin is ε-*N*-biotinyl-L-lysine.
[35] R. R. Fall, unpublished observations; independently verified in the amino acid sequence
determination of [³H]biotin-BCCP$_{(9100)}$ (see footnote 29).

TABLE II

SUMMARY OF NATIVE BIOTINYL POLYPEPTIDES ISOLATED FROM MICROORGANISMS[a]

Enzyme	Molecular weight	Probable substructure[b]	Source	Reference[c]
Acetyl-CoA carboxylase	~22,000	1	E. coli, A. aerogenes	28, d
			Achromobacter	3
	25,000	1	P. citronellolis	8
	190,000	3	Yeast	See 3
	250,000	3	Yeast	22
Propionyl-CoA carboxylase	~23,000	1	P. citronellolis	e
	63,000	?	M. phlei	d
3-Methylcrotonyl-CoA carboxylase	73,000	2	P. citronellolis	30
	96,000	2	Achromobacter	See 3
Geranyl-CoA carboxylase	75,000	2	P. citronellolis	30
Urea carboxylase	210,000	2	Yeast	22
Pyruvate carboxylase	65,000	2	P. citronellolis	f
	~120,000	3	Yeast and Arthrobacter globiformis	See 4
	137,000	?	Yeast	22
Transcarboxylase	12,000	1	P. shermanii	10
Oxaloacetate decarboxylase	~60,000	?	A. aerogenes	d

[a] This table was adapted and expanded from that recently published by Wood and Barden[4]; data collected February, 1978. This table does not include known proteolytic fragments of biotinyl polypeptides.

[b] A probable substructure of 1 means that the biotinyl polypeptide contains only the biotin subsite; 2 means that it contains the biotin subsite plus one of the other catalytic subsites on the same polypeptide chain (the biotin carboxylase subsite in the case of Achromobacter 3-methylcrotonyl-CoA carboxylase); 3 means that the biotin, biotin donor, and biotin acceptor subsites are contained on one polypeptide chain.

[c] Numbers refer to text footnotes; letters refer to table footnotes.

[d] R. La Frinere and R. R. Fall, unpublished observations.

[e] R. R. Fall, unpublished observations.

[f] R. E. Barden, B. L. Taylor, F. Isohashi, W. H. Frey, II, G. Zander, J. C. Lee, and M. F. Utter, Proc. Natl. Acad. Sci. U.S.A. 72, 4308 (1975).

(such as the acetyl-CoA carboxylases). As discussed by Obermayer and Lynen,[3] this is due in part to different degrees of fusion of the subsites of these enzymes.

A second major conclusion is that in microorganisms that contain more than one type of biotin enzyme there is no evidence for sharing a common biotinyl subunit among the enzymes. Thus, *S. cerevisiae* and *P. citronellolis,* which contain three and five different biotin enzymes, respectively, contain the corresponding number of different, distinct biotinyl polypeptides (cf. Table II). Despite this array of different biotinyl polypeptides, recent amino acid sequence determinations of the biotin attachment site of several different biotin enzymes suggests that there may be considerable sequence homology at the biotinyl sites of all these proteins.[4,19,29]

Acknowledgments

Many of the described procedures were developed while the author was a postdoctoral fellow in the laboratory of Dr. P. Roy Vagelos. Other procedures were developed in the course of research supported in part by grants from the National Science Foundation (PCM 75-16251) and the National Institutes of Health (HL 16628).

[67] Identification of Biocytin

By K. DAKSHINAMURTI and P. M. GILLEVET

Wright *et al.*[1] isolated biocytin (ϵ-N-biotinyl-L-lysine) from yeast, and its structure was determined by Peck *et al.*[2] Kosow and Lane[3] showed that in propionyl-CoA holocarboxylase biotin is covalently linked to the ϵ-amino group of a lysine residue. The identification of biocytin in protein hydrolyzate is, thus, crucial to the elucidation of the nature of the bond involved in the interaction between the prosthetic group biotin and the apoprotein in various biotin enzymes. The following method is a modification of an earlier procedure[4] for the identification of biocytin using high-voltage electrophoresis.

In Vivo Labeling of Biotin Proteins. Biotin-deficient rats[5] are injected

[1] L. D. Wright, E. L. Cresson, H. R. Kegg, T. R. Wood, R. L. Peck, D. E. Wolf, and K. Folkers, *J. Am. Chem. Soc.* **74,** 1996 (1952).
[2] R. L. Peck, D. E. Wolf, and K. Folkers, *J. Am. Chem. Soc.* **74,** 1999 (1952).
[3] D. P. Kosow and M. D. Lane, *Biochem. Biophys. Res. Commun.* **7,** 439 (1962).
[4] S. Litvak, R. L. O. Boeckx, and K. Dakshinamurti, *Anal. Biochem.* **30,** 470 (1969).
[5] K. Dakshinamurti and C. Cheah-Tan, *Can. J. Biochem.* **46,** 75 (1968).

intraperitoneally with 5 μCi of d-[carbonyl-^{14}C]biotin (57 mCi/mmol) in saline per 100 g body weight 16 hr before sacrifice. The rats are killed by decapitation. The livers are quickly removed, rinsed in cold saline, and homogenized in 3 volumes of medium (0.24 M sucrose, 25 mM KCl, 1 mM MgCl$_2$ in 50 mM Tris·HCl, pH 7.6). The crude homogenate is centrifuged at 600 g for 10 min in a Sorvall RC-2B centrifuge, the pellet is discarded, and the supernatant is centrifuged at 105,000 g for 90 min in a Spinco Model L ultracentrifuge. The supernatant is fractionated with ammonium sulfate; the 0–40% fraction is collected and dissolved in and dialyzed against water for 4 hr. This fraction contains 93% of the supernatant radioactivity.

Proteins labeled *in vivo* are subjected to digestion with Pronase (Calbiochem). In a final volume of 20 ml, the digestion mixture contains: 200 mg of protein ($\sim$100,000 cpm); 20 mg of Pronase; 800 μmol of sodium phosphate, pH 7.4, and 500 μl of absolute ethanol. Incubation is carried out at 37° for 48 hr.

Sephadex G-10 Chromatography. A 3-ml aliquot of the hydrolyzate is clarified by centrifugation, applied to a Sephadex G-10 column (1.5 $\times$ 86 cm), and eluted with water. Three-milliliter fractions are collected at a flow rate of 1 ml/min. The elution profile of radioactivity and transmittance at 280 nm is given in Fig. 1. Fractions 46–56, which contain most of the applied radioactivity, are pooled and lyophilized.

Two-Dimensional Mapping of Labeled Biocytin. The lyophilized material is taken up in 200 μl of water. An aliquot of 50 μl is applied on Whatman 3 MM paper and subjected to electrophoresis in a Savant

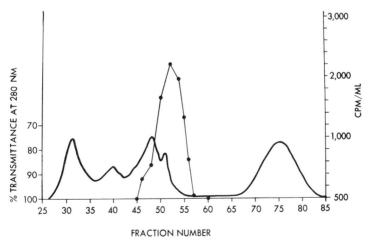

FIG. 1. Sephadex G-10 chromatography of Pronase hydrolyzate of [^{14}C]biotin proteins. —, Percent transmittance at 280 nm; ●, ^{14}C.

high-voltage apparatus at 3000 V for 1 hr at pH 1.9 (glacial acetic acid/98% formic acid/water, 87/25/888, v/v). On a separate sheet, standards of d-[carbonyl-^{14}C]biotin, biocytin, alanine, and tyrosine are applied and subjected to similar electrophoresis. After drying, the labeled biotin (on the standard sheet) is located by scanning for radioactivity with a Varian aerograph series 6000 gas-flow scanner. Both the standard and test sheets are subjected to chromatography in the second dimension, 90° to the direction of electrophoresis, for 8–9 hr using n-butanol/glacial acetic acid/water (80/20/20, v/v) as the solvent. After drying, both sheets are sprayed with ninhydrin and heated at 70° for 5 min. The location of labeled biotin in the chromatographic sheet is determined using a radioactivity scanner. The peptides and biocytin are visualized by ninhydrin. Figure 2 shows the relative positions of the applied standards. The R_f of the standards are: biotin (0.08 × 0.58), biocytin (0.37 × 0.16), alanine (0.65 × 0.17), and tyrosine (0.41 × 0.22). The ninhydrin spot on the test

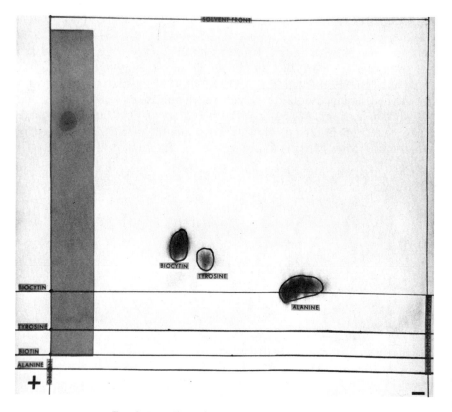

Fig. 2. Two-dimensional mapping of standards.

sheet containing the Pronase hydrolyzate of the biotin-protein, corresponding to the biocytin spot on the standard sheet was cut out, placed in a liquid scintillation vial with 15 ml of the scintillant (2.46 g of 2,5-diphenyloxazone and 9.2 mg of 1,4-bis-2-(5-phenyl oxazolyl)benzene in 1842 ml of toluene plus 1158 ml of absolute ethanol) and counted in a Beckman LS 200 liquid scintillation counter. The radioactivity applied on the paper was quantitatively recovered in the ninhydrin spot.

Alternatively, the ^{14}C-radioactivity on the test sheet can be located using autoradiography. Figure 3 shows that after the two-dimensional mapping the radioactive material is identified as a single spot on the radioautogram. Using visualization with ninhydrin, the separation of biocytin from other peptides in the hydrolyzate is evident. The Sephadex G-10 step separates the amino acids from biocytin and the dipeptides. Separate electrophoresis and paper chromatographic steps of our previous procedure[4] have been combined in this two-dimensional mapping.

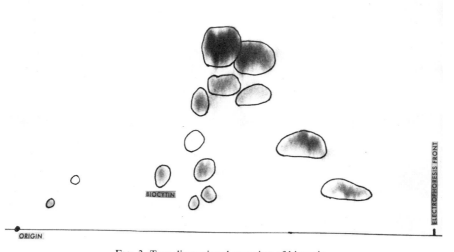

FIG. 3. Two-dimensional mapping of biocytin.

Pyridoxine, Pyridoxamine, and Pyridoxal: Analogs and Derivatives

[68] Fluorometric Assay of Pyridoxal

By M. S. Chauhan and K. Dakshinamurti

Principle

The reductive amination[1-3] of pyridoxal hydrochloride (I) with methyl anthranilate (II) and sodium cyanoborohydride at pH 4.5–5.0 and 35°–38° yields a highly fluorescent amine (III), which is used for the fluorometric assay of pyridoxal in ethanol at pH 7. This method can be applied to the determination of pyridoxal phosphate as such or after hydrolyzing it with hydrochloric acid or with acid phosphatase.[4] Pyridoxamine and pyridoxine are readily oxidized to pyridoxal.[5,6] Hence, the assay procedure described can be applied to these compounds too.

[1] R. F. Borch, M. D. Bernstein, and H. D. Durst, *J. Am. Chem. Soc.* **93**, 2897 (1971).
[2] R. F. Borch and H. D. Durst, *J. Am. Chem. Soc.* **91**, 3996 (1969).
[3] J. R. Bull, D. G. Hey, G. D. Meakins, and E. E. Richards, *J. Chem. Soc., C.* 2077 (1967).
[4] S. Takanashi and Z. Tamusa, *J. Vitaminol. (Kyoto)* **16**, 129 (1970).
[5] A. N. Wilson and S. A. Harrie, *J. Am. Chem. Soc.* **73**, 4693 (1951).
[6] D. Heyl, *J. Am. Chem. Soc.* **70**, 3434 (1948).

METHODS IN ENZYMOLOGY, VOL. 62

Reagents

> Pyridoxal hydrochloride
> Methyl anthranilate
> Sodium cyanoborohydride
> Absolute methanol
> Absolute ethanol
> Chloroform
> Methanolic HCl, 4 *N* prepared by bubbling dry HCl gas through absolute methanol

Experimental

Preparation of Methyl-N-pyridoxylanthranilate (III). Methyl anthranilate (450 mg in 5 ml of absolute methanol) is added to a solution of pyridoxal hydrochloride (100 mg in 10 ml of absolute methanol), and the pH of the solution is brought to 4.5–5.0 with methanolic HCl. The reaction mixture immediately turns yellow. It is warmed at 35°–38° in a water bath for 7 min, and a freshly prepared solution of sodium cyanoborohydride (20 mg in 5 ml of absolute methanol) is added. The solution turns colorless immediately. It is kept at 35°–38° for a further period of 15 min while the pH of the reaction mixture is maintained at 4.5–5.0 with methanolic HCl. At the end of this time, the reaction mixture is brought to pH 2–3 with methanolic HCl. The organic solvent is removed on a rotary evaporator at 35°–38°. The residue is dissolved in water (5 ml), and the pH is brought to 2–3 by adding a few drops of 5 *M* HCl. Solid sodium bicarbonate is then added until the pH of the solution is 8.0–8.5. The reaction mixture is extracted with chloroform (5 × 20 ml). The organic solvent of the extract is removed on a rotary evaporator. The residue is taken up in a small volume of chloroform–methanol (1:1) and chromatographed on a silica gel G plate (2 mm, 20 × 20 cm) using a mixture of chloroform and methanol (9:1, v/v) as solvent. The amine band is scraped off and eluted with a mixture of chloroform and methanol (1:1, v/v). The organic solvent is removed, and the residual amine is collected (95 ± 5% of theoretical yield based on the amount of pyridoxal hydrochloride used). The amine (III) is crystallized from a mixture of methanol, benzene, and petroleum ether (m.p. 168°–169°).

The above procedure is applied to 1.0, 0.5, and 0.1 mg pyridoxal hydrochloride, and the amount of the amine is determined fluorometrically in absolute ethanol at pH 7.0. Recovery determined using a standard curve (concentration × fluorescence intensity) is 95 ± 5% in each instance.

Characterization of the Amine (III). The infrared spectrum of the amine (III) shows an ester carbonyl at 1681 cm^{-1} while the UV spectrum shows λ_{max} at 254, 282, and 348 nm. The nuclear magnetic resonance spectrum in D_4-methanol shows two methyl singlets at $\delta 2.5$ and 3.84, and the methylene singlets occur at $\delta 4.67$ and 4.77. The four aromatic protons of the benzene ring occur at multiplets and are in the range of $\delta 6.5-8.2$. The hydroxy and amino group protons are not seen in the spectrum owing to their exchange with the D_4-methanol. The aromatic proton in the pyridine ring occurs as a slightly broad singlet at $\delta 8.09$. The mass spectrum of the amine (III) shows the molecular ion at $m/e = 302$, and the base peak is at $m/e = 151$ corresponding to the loss of methylanthranilate. The fluorescence spectrum in absolute ethanol has an excitation maximum at 354 nm and an emission maximum at 425 nm.

Interfering Substances. The amount of sodium cyanoborohydride is critical for the reaction as an excess may lead to the reduction of the pyridinium ring to form dihydropyridine compounds.[7,8]

Limit of Determination. This procedure can easily determine 0.15–1.5 ng of methyl-*N*-pyridoxyl anthranilate per milliliter on an Aminco–Bowman spectrophoto-fluorometer. This corresponds approximately to a range of 0.1–1.0 ng of pyridoxal per milliliter in the original sample.

Applications. This method can be applied to the assay of pyridoxal in biological material such as tissues, serum, milk, and urine.

[7] P. S. Anderson and R. E. Lyle, *Tetrahedron Lett.*, 153 (1964).

[8] P. S. Anderson, W. E. Krueger, and R. E. Lyle, *Tetrahedron Lett.*, 4011 (1965).

[69] Fluorometric Determination of Pyridoxal Phosphate in Enzymes

By Elijah Adams

Principle. This method[1] utilizes the high fluorescence yield of pyridoxal phosphate after reaction with cyanide.[2] The nature of the fluorescent product has been studied by Ohishi and Fukui.[3] Pyridoxal phosphate is released from protein on denaturation with trichloroacetic acid. The pH is then adjusted to that appropriate for reaction with KCN; after incubation with KCN, the sample is brought to the correct pH for maximal fluorescence of the pyridoxal phosphate–cyanide product. The method as

[1] E. Adams, *Anal. Biochem.* **31**, 118 (1969).

[2] V. Bonavita, *Arch. Biochem. Biophys.* **88**, 366 (1960).

[3] N. Ohishi and S. Fukui, *Arch. Biochem. Biophys.* **128**, 606 (1968).

described can easily measure 0.1 nmol of pyridoxal phosphate, corresponding to 5 μg of a 50,000 molecular weight protein containing one molar equivalent of the coenzyme. When tested with several enzymes,[1] the present method gave results in close agreement with those of an absorbance spectrum method based on formation of the phenylhydrazone,[4] but it is much more sensitive than the latter method.

Reagents

Potassium phosphate buffer, 5 mM, pH 7.4

Trichloroacetic acid, 11%

Dipotassium hydrogen phosphate, 3.3 M

Potassium cyanide, 20 mM

Phosphoric acid, 28% (85% phosphoric acid diluted 1 to 3 with water)

Potassium acetate, 2 M (2 M acetic acid adjusted with KOH to give pH 3.8)

Pyridoxal phosphate, 2 μM, 0.2 μM, as standards

Procedure. In a small acid-washed glass tube a sample of the enzyme to be measured (containing between 0.1 and 1.0 nmol of pyridoxal phosphate) is diluted to 0.2 ml with potassium phosphate buffer (pH 7.4). Trichloroacetic acid (11%, 0.2 ml) is added, and the mixture is kept at 50° for 15 min. Without necessarily cooling, 0.14 ml of 3.3 M K$_2$HPO$_4$ and 0.05 ml of 20 mM KCN are added, and the mixture is heated at 50° for 25 min. (At this point solutions can be kept at room temperature for at least

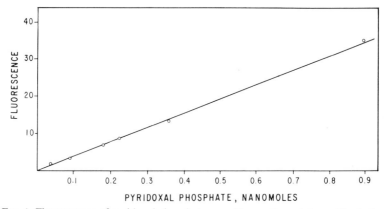

Fig. 1. Fluorescence of pyridoxal phosphate–cyanide reaction mixtures. Excitation was at 325 nm, emission at 420 nm. Fluorescence units are direct-scale readings (meter multiplication × 1) from the Aminco–Bowman spectrophotofluorometer.

[4] H. Wada and E. E. Snell, *J. Biol. Chem.* **237**, 133 (1962).

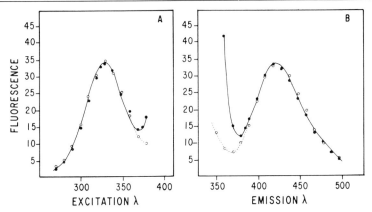

FIG. 2. Excitation spectrum (A) and emission spectrum (B) for pyridoxal phosphate ($\bigcirc$) and alanine racemase ($\bullet$). The sample of pyridoxal phosphate contained 0.23 nmol, and the sample of enzyme contained 18 μg. Ordinate units are direct-scale readings (meter multiplication $\times$ 0.3) from the Aminco–Bowman instrument. For the emission spectrum, excitation was at 325 nm; for the excitation spectrum, emission at 420 nm was measured. The enzyme was a sample described separately. [G. Rosso, K. Takashima, and E. Adams, *Biochem. Biophys. Res. Commun.* **34**, 134 (1969)].

several hours before proceeding.) H_3PO_4 (28%, 0.07 ml) and 1.0 ml of 2 M potassium acetate are added to achieve a final pH of 3.8. The final volume is 1.66 ml. Any visible protein precipitate formed in the earlier denaturation step can be removed by centrifugation, but slight turbidity has not significantly altered fluorescence.

The solution is excited at 325 nm, and the emission at 420 nm is determined. The fluorometer is standardized with pyridoxal phosphate solutions, first incubating with KCN and K_2HPO_4 as above and then following the subsequent steps described. Figure 1 shows the fluorescence of such mixtures of pyridoxal phosphate standards treated with cyanide. Figure 2 shows the similarity of excitation and emission spectra for pyridoxal phosphate and alanine racemase.

Comments. The method as outlined has been tested only with relatively pure protein samples. It will not measure pyridoxal, which has maximal fluorescence in alkaline solution.[2] Methods for the fluorometric measurement of both pyridoxal and pyridoxal phosphate in the same tissue sample have been described; they are based on chromatographic separation of the two forms.[5,6] A similar method has been extended to the measurement of additional B_6 analogs in tissue.[7] Because tissue extracts

[5] M. Yamada, A. Saito, and Z. Tamura, *Chem. Pharm. Bull.* **14**, 482 (1966).
[6] Z. Tamura and S. Takanashi, this series, Vol. 18A, p. 471.
[7] Y. H. Loo and L. Badger, *J. Neurochem.* **16**, 801 (1969).

may contain interfering substances, the fluorometric method for pyridoxal phosphate has been modified by including an internal standard.[8] A fluorometric method for pyridoxal phosphate based on its reaction with semicarbazide rather than KCN has been described[9]: semicarbazide has the possible advantage of wider reaction capability with the aldehyde group of Schiff-base bound pyridoxal phosphate. However, in several enzymes studied,[2] in which such a Schiff base is present, no limitation of the cyanide method was apparent.

[8] K. M. Grigor, D. von Redlich, and D. Glick, *Anal. Biochem.* **50,** 28 (1972).
[9] S. K. Srivastava and E. Beutler, *Biochim. Biophys. Acta* **304,** 765 (1973).

[70] High-Voltage Electrophoresis and Thin-Layer Chromatographic Separation of Vitamin B6 Compounds

By ERNEST E. MCCOY, CARLO COLOMBINI, and KEN STRYNADKA

A number of methods are available for separation of the six forms of vitamin B_6. Although column chromatography[1] permits excellent quantitation of B_6 compounds, it is time consuming. Methods which are more rapid include paper and thin-layer chromatography or high-voltage electrophoresis of B_6 compounds. Volume 18[2] outlined in detail methods for separation of B_6 compounds using thin-layer chromatography. This chapter will place emphasis of high-voltage electrophoresis and two-dimensional, thin-layer chromatographic methods for separation of vitamin B_6 compounds.

Thin-Layer Electrophoresis of Vitamin B6 Compounds

Preparation of Plates

To 30 g of cellulose powder MN 300 HR (Machery Nagel & Co.) is added 160 ml of water. The slurry is mixed with a mechanical stirrer for 30 min, then poured into a Shannon Unoplan Thin-Layer Spreader set at 500 μm thickness. The slurry is sufficient to make three cellulose-coated 46×20 cm glass plates. The plates are air-dried for 24 hr before use.

[1] H.-G. Tiselius, *Clin. Chim. Acta* **40,** 319 (1972).
[2] H. Ahrens and W. Korytnyk, this series, Vol. 18A, p. 489.

Standard Solutions

Pyridoxine, pyridoxal, pyridoxamine, pyridoxal 5-phosphate, and pyridoxamine 5′-phosphate were obtained from Calbiochem. Pyridoxine 5′-phosphate was a gift from Dr. E. Snell. Each B_6 vitamer is individually dissolved in 50 mM sodium acetate buffer, pH 4.9, at a concentration of 1.0 mg/ml. A solution is prepared containing all 6 forms at a concentration of 1.0 mg/ml for each vitamer. Solutions are kept in amber-colored bottles, aluminum-foil wrapped, and frozen at $-20°$ between use. Pyridoxal 5′-phosphate is prepared fresh weekly.

Thin-Layer Electrophoresis

From 0.1 to 5.0 μg of vitamin B_6 standards are either applied individually or as a mixture and air dried. The application is made in a darkened room. Samples are applied at the origin, which is located 3 cm from the center and toward the cathode side of the plate. Standard samples are ap-

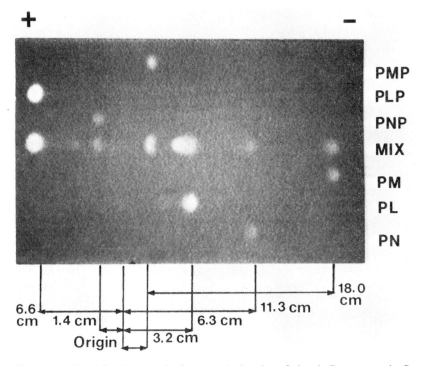

Fig. 1. An ultraviolet photograph of expected migration of vitamin B_6 compounds. See text for details.

plied at 2-cm intervals at the origin. The plate can be wetted either by pipette or sprayed with 50 mM acetate buffer, pH 4.9. The tanks are filled with approximately 300 ml of the same buffer and connected to the plate by buffer-saturated wicks of Whatman 3 M cellulose paper. To minimize evaporation, the plate is covered with Saran wrap and the aluminum block of a Savant high-voltage electrophoresis apparatus is cooled to 4°. Electrophoretic separation is achieved with a current of 800 V for 4 hr. To avoid photodecomposition, electrophoresis is done in a dark or near-dark room. On completion of electrophoresis, plates are removed and dried under a stream of warm air. The plate is placed under UV light (Chromata-veu, Ultra-Violet Products, Inc.), and spots are outlined. A UV photograph of expected migration of vitamin B$_6$ compounds is shown in Fig. 1. The spots are scraped into vials and extracted three times with 0.5 ml of 0.1 M phosphate buffer pH 7.4. The vitamers are quantitated using an Aminco–Bowman Spectrophotofluorometer according to the method of Storvich.[3]

Activation and fluorescent wavelengths are PN 325:400 nm; PM 325:405 nm; PMP 330:400 nm; and PNP 330:405 nm. Recovery is >90% of added standards by this method. PL and PLP are quantitated using the cyanohydrin method of Bonavita.[4] The activation and fluorescent wavelengths are for PL 360 and 430 nm; for PLP 315 and 415 nm. The recovery for PL and PLP is >90%.

Separation of Radioactive B$_6$ Vitamers in Tissue Extracts

Separation of radiolabeled vitamin B$_6$ compounds in tissue extracts can be achieved with this method. Tissue extracts are prepared as previously described.[5] To facilitate localization, 2 μl of a standard solution containing all six vitamers is applied at 2-cm intervals at the origin on the plate. Five-microliter aliquots of tissue extracts containing radioactive vitamin B$_6$ compounds are then added on top of the standards. An application of tissue extract without standard is applied to determine whether substances are present that would interfere with migration of standards. To one of the spaces 2 μl of standard alone is applied. The plates are placed in the horizontal Savant electrophoresis apparatus and run in the dark as described for standard solutions. Details of quantitation of the radioactive vitamers have been described.[5] Using this method the metab-

[3] C. A. Storvich, E. M. Benson, M. A. Edwards, and M. J. Wooding, in "Methods of Biochemical Analysis" (D. Glick, ed.), Vol. 12, p. 183. Wiley (Interscience), New York, 1964.
[4] V. Bonavita, Arch. Biochem. Biophys. **88**, 366 (1960).
[5] C. E. Colombini and E. E. McCoy, Anal. Biochem. **34**, 451 (1970).

olism of [^{14}C]pyridoxine in mouse brain, liver, and carcass from 1 min to 4 days has been determined.[6]

Comment

Well defined separation of all six forms of vitamin B_6 can be achieved with this method with very little trailing. Recovery of vitamers added in a spot, which is then scraped and extracted or vitamers spotted, electrophoresed, scraped and extracted, are very similar. The spot between PNP and PLP in the mixture of six vitamers has not been identified with the electrophoretic separation.

Thin-Layer Chromatography of Vitamin B_6 Compounds

Volume 18A[2] of this series reviewed thin-layer separation of B_6 compound. This remains an excellent source as well for reagents to locate B_6 compounds. Recently, Mahuren and Coburn[7] have published a two-way chromatographic method designed primarily to separate radioactive vitamin B_6 compounds derived from ^{3}HPN. We have used their method with modification to determine recovery of small amounts of nonradioactive B_6 vitamers separated by thin-layer chromatography.

Preparation of Plates

Fifty grams of silica gel type H (particle size 10–40 μm) are added to 150 ml of water, and a slurry is prepared. Glass plates 20 × 20 cm are placed on a Shannon Unoplan Thin-Layer Spreader set at 250 μm thickness. Plates are spread and allowed to air dry over night before use.

Reagents

Isoamyl alcohol, acetone, 2-butanol, ethanol, diethylamine, 1.5 N NH$_4$OH. If isoamyl alcohol and diethylamine are not redistilled, streaking of vitamers will likely occur.

Thin-Layer Chromatography

Standard mixtures of B_6 vitamers prepared as for electrophoresis are used. Two to five micrograms each of PL, PN, PM, PLP, PNP, PMP are applied to one corner of the silica-coated plate in a darkened room. The

[6] C. F. Colombini and E. E. McCoy, *Biochemistry* **9**, 533 (1970).
[7] J. D. Mahuren and S. P. Coburn, *Anal. Biochem.* **82**, 246 (1977).

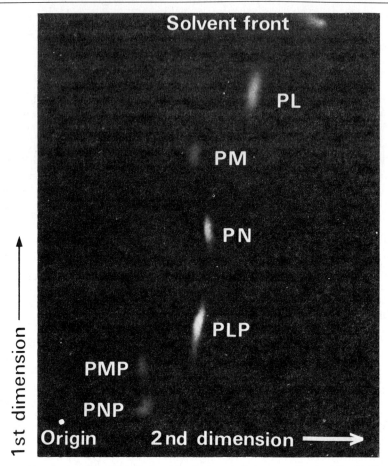

Fig. 2. An ultraviolet photograph of separated vitamers. See text for details.

plates are developed in the first direction in a solvent mixture of isoamyl alcohol : acetone : water : diethylamine (24 : 18 : 6 : 8) to a distance of 15 cm, then dried under a stream of warm air. The plates are run in the second direction of 10 cm in 2-butanol : ethanol : 1.5 N NH_4OH : isoamyl alcohol : diethylamine (20 : 6 : 7 : 2 : 2). All operations are carried out in a darkened room to avoid photodecomposition of PLP. The B_6 vitamers are located by their native fluorescence, marked, and scraped into tubes. The vitamers are extracted with three successive 0.5-ml amounts of 0.1 M phosphate buffer, pH 7.4. The vitamers are quantitated by fluorescence according to Storvich and recoveries calculated. The recovery is variable as trailing occurs with some vitamers. A UV photograph of separated vitamers is shown in Fig. 2. When B$_6$ vitamers are spotted on the plate individually then extracted, approximately 95% of the vitamer is extractable. When a mixture is applied and two-way chromatography is carried out, recovery varies from 72% for PN to 50% for PMP. Mahuren and Coburn[7] reported that, when the method is used with radioactive vitamin B$_6$, recovery of added radioactivity varies between 80 and 90%. It is important that high-grade solvents be used and isoamyl alcohol and diethylamine redistilled to minimize streaking.

Comment

Thin-layer chromatography has the advantage of requiring only simple equipment. High-voltage electrophoretic separation is rapid, and recovery of the vitamers is less variable than with the chromatographic method.

[71] High-Performance Chromatography of Vitamin B$_6$[1]

By A. K. WILLIAMS

Recent emphasis on nutritional research and labeling requirements for many consumer products has resulted in a considerable demand for simple, rapid and inexpensive methods for the analysis of many nutrients, both naturally occurring and synthetic.

Vitamin B$_6$ has historically been determined by microbiological growth response. Since the three major forms of the vitamin (pyridoxol, pyridoxal, and pyridoxamine) have been shown to elicit different growth responses of the test organism[2] it is necessary, by this technique, to assume an average growth response or to separate the vitamers prior to microbiological assay.[3]

Several assay methods have been developed utilizing the advantages of gas chromatography. These methods[4-6] all require derivativization of the vitamin to produce compounds that may then be separated and quantitated by gas–liquid chromatography. These methods, while rapid and in-

---

[1] The mention of firm names or trade products does not imply that they are endorsed or recommended by the Department of Agriculture over other firms or trade names not mentioned.

[2] W. P. Parrish, H. W. Loy, Jr., and O. L. Klive, *J. Assoc. Agric. Chem.* **38**, 506 (1955).

[3] E. W. Toepfer and J. Lehman, *J. Assoc. Agric. Chem.* **44**, 426, (1961).

[4] A. J. Sheppard and A. R. Prosser, this series, Vol. 18A, 494 (1970).

[5] W. Korytnyk, this series, Vol. 18A, 500 (1970).

[6] A. K. Williams, *J. Agric. Food Chem.* **22**, 1, 107 (1974).

METHODS IN ENZYMOLOGY, VOL. 62

expensive, suffer from two major disadvantages. Derivatization can, in some cases, present problems of incomplete reaction with the derivatizing agent. This problem can be largely overcome by the inclusion of internal standards. The second, and probably more significant, disadvantage is that derivatization of the vitamins precludes recovery of the compounds in their original form and quantity.

While pyridoxol, pyridoxal, and pyridoxamine have been successfully separated by ion-exchange techniques, the results obtained by these low-resolution resins were of a preparative nature and did not lend themselves to a rapid, quantitative analysis.

The recent developments in high-resolution, high-pressure column chromatography have proved to be extremely useful in the analysis of vitamin B_6 moieties.[7]

High-Pressure Cation-Exchange Chromatography

Principle

Aqueous solutions of pyridoxol, pyridoxal, and pyridoxamine are chromatographed on a small column of finely sized cation-exchange resin. Resins having a styrene–divinylbenzene matrix and sufonic acid exchange groups have proved satisfactory both in their resolving capacity and in their ability to withstand the relatively high pressures utilized.

Vitamins are eluted isocratically with 0.7 M ammonium formate. Samples too complex to resolve by the isocratic system may be analyzed by utilizing the increased resolving power of gradient elution.

Preparation of Samples

Because of the potentially wide use of chromatographic vitamin B_6 analysis, no universal clean-up procedure can be recommended. Our limited experience in clean-up of vitamin B_6 from foodstuffs has met with variable success because of the many interfering compounds present in the samples.

Many pharmaceutical products, such as vitamin capsules, tablets, and liquid preparations, can be analyzed by disintegrating the material (if necessary) in water, and extracting with ether or chloroform. The resulting aqueous phase is clarified by centrifugation and appropriately diluted with water, and an aliquot is introduced onto the column.

In cases where the vitamins are in relatively simple mixtures, such as enzyme reactions or other protein-containing solutions, they may be de-

---

[7] S. C. Chen, P. O. Brown, and D. M. Rosie, *J. Chromatogr. Sci.* **15**, 218, (1977).

proteinized by addition of trichloroacetic or perchloric acid followed by centrifugation and removal of the precipitant from the supernatant by extraction with 5% tri-N-octyl amine in ether.[6] The aqueous phase can usually be chromatographed without further treatment.

Column Preparation

The resin, Aminex A-5 (Bio-Rad Laboratories, Richmond, California), is washed with 2.0 M NH$_4$OH, allowed to settle, and the supernatant decanted. This procedure is repeated three times. The resin is then washed on a Büchner funnel with distilled water until the excess NH$_4$OH is removed. The resin is suspended in 0.7 M ammonium formate buffer (pH 5.60) to give a thick slurry.

A glass column, 0.3 × 150 mm (Chromatronix, Inc., Berkeley, California), in which the steel bed support is replaced with a tightly packed plug of glass wool is cleaned and filled with the resin slurry. The column is attached to a high-pressure liquid pump and the 0.7 M formate buffer passed through the column at a rate sufficient to produce about 600 psi back pressure.

It is advisable at this point not to connect the column effluent to a detector or other device since resin particles are present in the effluent stream until packing is complete.

When the resin has settled, the supernatant is removed and replaced with additional resin slurry. This procedure is repeated until the column is filled. Pumping is continued for 4–6 hr at a flow rate of 0.3 ml/min. This flow will develop a back pressure of 600–800 psi.

Instrumentation

A high-pressure chromatographic system capable of generating pressure of at least 1000 psi and a flow rate of 0.3 ml/min is required.

This system should be equipped with a suitable injector of either the loop or septum type and an ultraviolet absorption detector (254 or 280 nm). The detector should have a low volume cell (7–15 μl) and a path length of about 10 mm.

Standards

Standard solutions of pyridoxol, pyridoxal, and pyridoxamine are prepared by dissolving the authentic hydrochloride salt in water to give a final concentration of 1 mg of free base per milliliter. These solutions are stable for at least 6 weeks when frozen.

Convenient working standards are prepared by diluting 1.0 ml of standard to 100 ml with the formate buffer. This will provide solutions containing 10.0 ng free base per microliter.

Method

Isocratic Elution. The sample is introduced onto the column and elution is begun with the 0.7 M formate buffer at a flow rate of 0.3 ml/min. Figure 1 shows a typical chromatogram obtained from 100 ng of pyridoxal (peak 1) and 200 ng each of pyridoxol and pyridoxamine (peaks 2 and 3, respectively).

The dose-response plots for the three vitamins are shown in Fig. 2. The vitamers all exhibit a linear response within a range of 40–300 ng of vitamin base.

The column, when operated isocratically requires no further equilibration, and additional samples can be analyzed immediately after the last peak emerges.

Gradient Elution. It is often advantageous, when analyzing complex samples, to utilize the greater resolving power of gradient elution.

This procedure entails elution with a linear gradient of ammonium formate buffer from 10 mM, pH 6.5, to 1.0 M, pH 6.6.

Gradient-forming devices, which are provided with many commercial

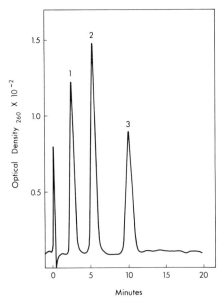

Fig. 1. Chromatography of vitamin B_6. Reprinted with permission from *J. Agric. Food Chem.* **23,** 5 (1975). Copyright by American Chemical Society.

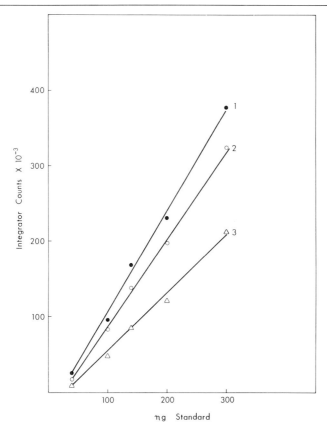

Fig. 2. Dose-response plots for pyridoxal, pyridoxol, and pyridoxamine: ●, pyridoxal; ○, pyridoxol; △, pyridoxamine. Reprinted with permission from *J. Agric. Food Chem.* **23,** 5, (1975). Copyright by American Chemical Society.

high-pressure chromatographs, produce excellent results and are highly recommended. If such a device is not available, satisfactory results may be obtained by placing 30 ml of starting buffer in a small mixing chamber; the gradient formed by adding 1.0 M buffer to the chamber at the rate of 0.15 ml/min. Elution is carried out by pumping from the mixing chamber at 0.3 ml/min.[8] This procedure produces a linear gradient and results in sharp, symmetrical peaks (Fig. 3).

Although the last peak (pyridoxamine) elutes at about 0.4 M formate, the gradient is continued to 1.0 M in order to remove extraneous materials that may be present in the sample. At the completion of the gradient, the

[8] C. A. Burtis and R. L. Stevenson, Publ. No. A-1012 Varian Aerograph, Walnut Creek, California.

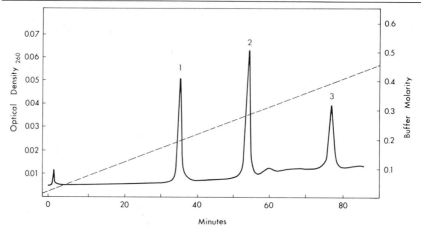

Fig. 3. Gradient-elution chromatography of vitamin B_6: —, optical density; ---, buffer concentration. Reprinted with permission from *J. Agric. Food Chem.* **23,** 5 (1975). Copyright by American Chemical Society.

column is reequilibrated with starting buffer before another analysis is performed.

Paired-Ion Chromatography

Principle

In order to utilize the speed and simplicity of reverse-phase chromatography, the vitamers are complexed with 1-heptanesulfonic acid. The resulting lipophilic ion pair is then chromatographed on a 30×0.39 cm column of μ Bondapak C18 (Waters Associates, Inc., Milford, Massachusetts).

The nonpolar nature of this column allows excellent resolution of the vitamer ion-pairs when eluted with an appropriate solvent containing the complexing counterion. This method of analysis has the distinct advantage of ease of recovery of the separated vitamin components, since the eluting solvent contains practically no buffers or other difficulty removable compounds.

Column Preparation

The column, μ Bonapak C18, obtained prepacked and tested from Waters Associates, has been found to afford adequate resolution and reproducibility.

The column is prepared by passing 25 ml of 5% acetic acid through the column at a flow rate of 1.0 ml/min. This procedure is repeated with 5% acetic acid in methanol, and finally the column is equilibrated with the eluting solvent at 0.75 ml/min. This solvent is prepared from a stock counterion solution consisting of 4.5 M sodium 1-heptane sulfonate in glacial acetate acid. The counterion, 1.8 ml, is diluted to 1 liter with 10% isopropanol. The resulting solvent is made ready for use by filtering through a 0.45 μm filter.

Instrumentation and Standards

This system requires the same instrumentation as the ion-exchange method, and standards are prepared in the same manner except that water is used for dilution instead of formate buffer.

Method

The sample may be prepared in any convenient solvent, such as water or eluting solvent. It is not necessary to have the counterion present in the sample, as complexing is instantaneous when the sample contacts the eluting solvent.

Sample (1–10 μl) is introduced onto the column and elution is continued. Figure 4 shows a typical chromatogram obtained by this method.

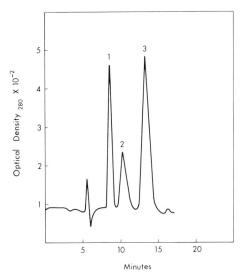

FIG. 4. Reverse-phase chromatography of vitamin B$_6$: peak 1, pyridoxal; peak 2, pyridoxol; peak 3, pyridoxamine.

Pyridoxal, pyridoxol, and pyridoxamine (peaks 1, 2, and 3, respectively) emerge at 516, 615, and 795 sec.

The column is ready for a second sample as soon as the last peak has emerged and a steady base line is established.

After repeated use it may become necessary to "clean" the column. This can be acheived by the procedure described under Column Preparation. It is important to observe the precautions and recommendations of the manufacturer when using these high-performance columns.

After repeated use the column can degenerate, with loss of resolution between peaks 1 and 2. Resolution can usually be regained by eliminating isopropanol from the eluting solvent and increasing the flow to 3 ml/min.

[72] Carbon-13 Nuclear Magnetic Resonance Spectroscopy of the Vitamin B6 Group[1]

By HENRY H. MANTSCH and IAN C. P. SMITH

Nuclear magnetic resonance (NMR) is now the most powerful method to determine the structure, conformation, and dynamics of complex molecules. During the 1960's the predominant nucleus studied was ^{1}H, mainly owing to its relative ease of detection. Much information on the vitamin B$_6$ group of compounds has been derived by ^{1}H NMR.[2,3] Early in the 1970's it became possible to obtain ^{13}C NMR spectra from reasonable concentrations of complex molecules;[4,5] therefore, a considerable number of studies have been reported on the B$_6$ vitamins, on the variations of their structures with pH, and on the interactions of the vitamins with amino acids and enzymes. In addition, ^{31}P NMR has been applied in a few cases.[6-9]

In the present chapter we shall describe the fundamentals of the ^{13}C NMR method and its operating techniques, and outline the results that have been obtained to date for the vitamin B$_6$ group of compounds.

[1] N. R. C. C. Publication No. 16544.
[2] W. Korytnyk and H. Ahrens, this series, Vol. 18A, p. 475.
[3] B. Paul and W. Korytnyk, *J. Heterocycl. Chem.* **13**, 701 (1976).
[4] J. B. Stothers, "Carbon-13 NMR Spectroscopy." Academic Press, New York, 1972.
[5] F. W. Wehrli and T. Wirthlin, "Interpretation of Carbon-13 NMR Spectra." Heyden, London, 1976.
[6] M. Martinez-Carrion, *Eur. J. Biochem.* **54**, 39 (1975).
[7] S. J. W. Busby, D. G. Gadian, G. K. Radda, R. E. Richards, and P. J. Seeley, *FEBS Lett.* **55**, 14 (1975).
[8] K. Feldmann and E. J. M. Helmreich, *Biochemistry* **15**, 2394 (1976).
[9] K. Feldmann and W. E. Hull, *Proc. Natl. Acad. Sci. U.S.A.* **74**, 856 (1977).

METHODS IN ENZYMOLOGY, VOL. 62

^{13}C Nuclear Magnetic Resonance

The properties of ^{13}C and ^{1}H are compared in Table I. Both nuclei have spin $\frac{1}{2}$, which makes many aspects of ^{13}C interpretable in terms familiar from ^{1}H NMR. Carbon-13 has a lower magnetic moment than ^{1}H, and resonates at 25 MHz in a magnetic field of 23 kilogauss. Owing to its lower magnetic moment, ^{13}C is intrinsically more difficult to observe than ^{1}H. A further difficulty lies in the low natural abundance of ^{13}C; this turns to our advantage, however, in that no coupling between ^{13}C nuclei is observed in spectra taken at natural abundance. In cases where sensitivity becomes a problem, compounds can be enriched in ^{13}C at specific positions at increasingly reasonable cost.

The chemical shift range of ^{13}C, that is, the range over which one can detect different types of carbon, is much greater than for ^{1}H. There is a similarity to ^{1}H spectra in the relative positions of the resonances. Note in Fig. 1 that aliphatic carbons have the smallest chemical shifts, that of the methyl group being considerably less than those of the hydroxymethyl groups. The carbons of the conjugated ring system have large chemical shifts whose exact positions depend upon the nature of the substituents. If spectra are taken with appropriate attention to relaxation times and digital resolution, the areas under the resonances are proportional to the number of carbon atoms of a particular type of carbon. The intensities of the resonances in Fig. 1 due to carbons 2–5 are lower than expected on the basis of the relative numbers of carbon atoms of a given type (the intensity of the C-3 resonance should be equal to that of C-6). This is because C-3 has no directly attached hydrogen, and therefore has a much longer spin-lattice relaxation time (T_1) than C-6. If free induction decays are taken repeatedly at intervals short relative to T_1, partial saturation of the resonance occurs. This phenomenon provides a first means to assign the various resonances. The value of the chemical shift, taken in view of those already known for related model compounds, provides a second means, and the multiplicity of the proton coupled resonances (see below) provides the third.

TABLE I
COMPARISON OF THE PROPERTIES OF ^{13}C AND ^{1}H

Property	^{13}C	^{1}H
Spin	1/2	1/2
Frequency (MHz)	25.2	100
Sensitivity	0.016	1.00
Abundance (%)	1.1	99.9
Chemical shift range (ppm)	220	12

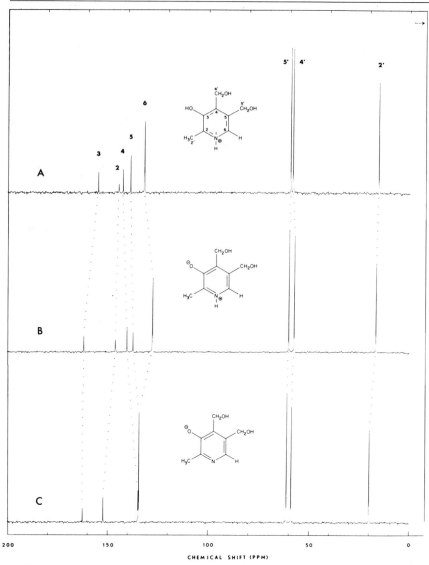

FIG. 1. Fourier-transformed ^{13}C NMR spectra (25.2 MHz, 12-mm tubes, 0.4 M in D$_2$O, 32°, 10,000 transients) of pyridoxine as a function of pH. The chemical shifts are relative to external tetramethylsilane. pH values are uncorrected meter readings in D$_2$O: (A) pH 2.3; (B) pH 6.8; (C) pH 12.5.

Coupling between ^{13}C and attached or nearby ^{1}H can lead to very complex spectra. Therefore it is customary to remove these couplings by high-power irradiation over a range of frequencies sufficient to cover the entire ^{1}H spectrum (broad band decoupling). Concomitant with spectral simplification, ^{1}H-decoupling leads to an increase in signal intensity due to multiplet collapse and to a cross-relaxation phenomenon known as the nuclear Overhauser effect (NOE). The latter can produce an intensity increase of up to 3-fold for small molecules, but the magnitude of the increase diminishes with increasing molecular size. On the other hand, the nature of the multiplet patterns is a sensitive indicator of the number of attached or neighboring hydrogens, and therefore provides a means to assign individual resonances. It is possible to obtain ^{13}C spectra with retention of ^{13}C–^{1}H couplings and the NOE, or vice versa, by means of so-called gated decoupler pulse sequences. The reader is referred to the text by Wehrli and Wirthlin[5] for details of the above.

Coupling between ^{13}C and ^{31}P can be used to determine the site of attachment of the phosphate group; when this coupling occurs across three chemical bonds, its magnitude is indicative of the dihedral angle between the planes containing the ^{13}C and the ^{31}P nuclei.[10]

The spectra of pyridoxamine 5'-phosphate in Fig. 2 illustrate the utility of couplings between ^{13}C and ^{31}P or ^{1}H in assigning resonances. Our problem is to distinguish in this region of the spectrum between the resonances of the methyl, hydroxymethyl, and aminomethyl carbons. Although this could in fact be done on the basis of chemical shifts alone, the coupling patterns provide a useful confirmation. Spectrum 2A is proton-decoupled; therefore, the small doublet observed on the resonance at ca. 63 ppm must be due to coupling with ^{31}P of the phosphate group, identifying this resonance as due to C-5'. Spectrum 2B is proton- and phosphorus-coupled. The triplet pattern centered at ca. 38 ppm is clearly due to the aminomethyl carbon, C-4', and the quartet pattern centered at 20 ppm to the methyl carbon, C-2'.

Finally, the relaxation mechanism for ^{13}C is relatively simple, and the T_1 and T_2 values can be used to study molecular dynamics.[5]

Most ^{13}C spectra are obtained by the Fourier transform method.[11,12] Nuclei are excited by a radiofrequency pulse of sufficient power to cover the frequency range of interest; the decay of the induced magnetization (free induction decay, transient) is stored in a computer and time-averaged to improve the signal-to-noise ratio. Fourier transformation of this decay yields a normal absorption versus frequency spectrum. As no time is spent scanning regions of the spectrum devoid of resonances, the time required to achieve a given signal intensity is reduced relative to that of the continuous wave approach by a factor of 50–70. Most spectrometers are under a high level of computer control; it is therefore possible

[10] R. D. Lapper, H. H. Mantsch, and I. C. P. Smith, *J. Am. Chem. Soc.* **95**, 2878 (1973).

[11] T. C. Farrar and E. D. Becker, "Pulse and Fourier Transform NMR." Academic Press, New York, 1971.

[12] D. Shaw, "Fourier Transform NMR Spectroscopy." Elsevier, Amsterdam, 1976.

to arrange a variety of automatic sequences for when the decoupler is on, what type of pulses are applied, measurement of relaxation times, and the magnitude of the NOE, etc. As the data are stored in digital form in the computer, an accurate printout of resonance positions, intensities, and chemical shifts is available automatically.

Contemporary spectrometers operate at frequencies from 20–90 MHz. The higher the frequency, the greater is the separation of resonances and the detection sensitivity. Disadvantages of high frequency observation are heating of aqueous solutions due to the high ^{1}H-decoupling power required and the earlier onset of loss of NOE with increasing molecular size.

Assignments of the ^{13}C Resonances of the Parent B_6 Vitamins

The assignments of the individual carbon resonances in the proton-decoupled spectra of the parent B_6 vitamins are summarized in Table II. Each spectrum consists essentially of eight well-separated resonances. These can be divided into two groups, the five pyridine ring carbons that constitute the aromatic pyridine ring skeleton, and the three substituent carbons. The numbering system of the individual carbon atoms, as recommended by IUPAC and IUB is shown in Fig. 1A.

TABLE II
^{13}C CHEMICAL SHIFTS OF THE PARENT VITAMINS[a]

Compound	pH	Pyridine ring carbons					Substituents		
		C-2	C-3	C-4	C-5	C-6	C-2'	C-4'	C-5'
Pyridoxine	2.3	143.8	153.8	141.7	137.9	130.9	15.5	58.1	59.2
	6.8	145.3	160.9	139.5	136.4	126.7	16.7	57.3	59.8
	12.5	151.2	161.6	133.8	133.5	133.5	20.0	58.6	60.8
Pyridoxamine	1.5	144.2	154.6	137.3	139.3	132.3	16.3	35.9	59.9
	7.0	146.1	164.0	133.5	137.0	124.6	16.5	37.9	60.3
	12.7	151.1	161.4	137.7	133.8	133.8	20.4	37.5	61.4
Pyridoxamine 5'-phosphate	1.6	144.5	154.7	138.2	136.4	132.9	16.3	36.1	63.0
	7.0	146.0	163.7	134.0	135.8	125.1	16.4	37.6	62.9
	12.5	151.1	161.2	137.5	132.2	134.1	20.3	37.6	63.8
Pyridoxal	1.8	145.1	150.2	141.1	139.3	126.7	15.4	99.6	71.0
	6.2	146.5	160.0	140.0	138.1	118.7	16.0	100.5	71.0
	12.5	153.4	162.2	131.8	135.0	127.3	19.7	134.0	67.7
Pyridoxal 5'-phosphate	2.6	b	b	b	b	131.0	15.4	88.7	62.4
	6.2	152.4	164.7	127.0	136.7	125.9	17.1	197.3	63.1
	12.4	157.5	168.4	125.2	133.8	129.4	20.1	197.7	63.6

[a] Chemical shifts are at 32°C and relative to external tetramethylsilane.
[b] These resonances were not observable at this pH.

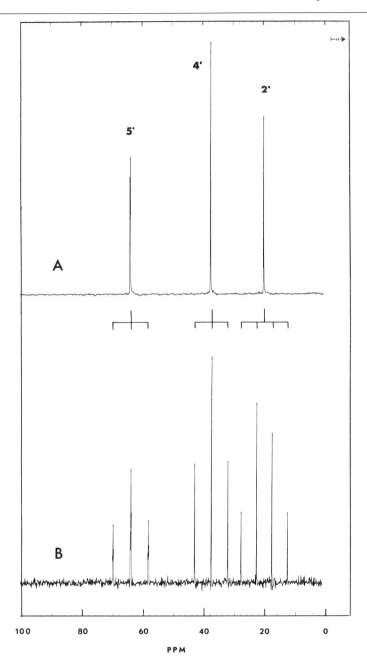

Fig. 2. ¹³C NMR spectra of pyridoxamine 5′-phosphate in the 0 to 100 ppm region with (trace A, 29,000 transients) and without hydrogen decoupling (trace B, 69,000 transients).

Carbons with directly attached hydrogens, such as C-2′, C-5′, C-4′, and C-6, are readily identified by the multiplicity of their one-bond proton couplings, as shown in Fig. 2. The assignment of the substituent carbons C-2′, C-5′, and C-4′ is straightforward. The C-2′ methyl carbon resonates between 15 and 20 ppm, the C-5′ methylene between 60 and 70 ppm, while

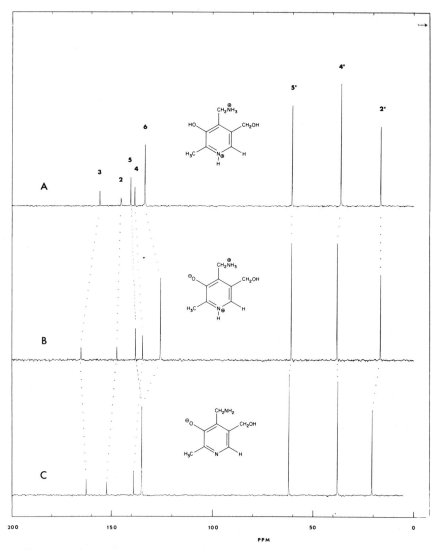

FIG. 3. ^{13}C NMR spectra (25.2 MHz, 12-mm tubes, 0.4 M in D$_2$O, 32°, 15,000 transients) of pyridoxamine as a function of pH: (A) pH 1.5; (B) pH 7.0; (C) pH 12.7.

the C-4' chemical shift can vary between 35 and 200 ppm depending upon the chemical nature or type of substitution.

The five resonances of the pyridine ring carbon atoms fall in a fairly narrow range from 118 to 168 ppm. The assignment of C-6 is simple by virtue of its large one-bond coupling to H-6 in the ^{1}H-coupled spectra. In all proton-decoupled spectra the C-6 resonance is clearly recognizable by its high intensity relative to those of the quaternary carbons.

The ^{13}C NMR spectra of pyridoxine and pyridoxamine are shown in Figs. 1 and 3, respectively. These spectra are proton-decoupled and show only one resonance for each carbon atom. In the corresponding spectrum of pyridoxamine 5'-phosphate (Fig. 4) the C-5' and C-5 resonances appear as doublets owing to the proximity of the phosphorus atom. The coupling of the vicinal and geminal carbon atoms to the phosphorus atom help in identifying the C-5' and C-5 resonances, and the three-bond vicinal coupling provides additional information with regard to the stereochemistry about the O—C-5' bond. Upon phosphorylation of pyridoxamine, the C-5' and C-5 chemical shifts increase by 2.6 and 1.2 ppm, respectively, while those of C-4 and C-6 decrease by 0.5 ppm. Assignment of C-4', C-5', and C-2' is achieved by reference to the ^{1}H-coupled spectra (Fig. 2) as described earlier.

To aid the assignment of the carbon-13 resonances in the vitamin B₆ compounds given in Table II, the ^{13}C NMR spectra of various monosubstituted pyridines and benzenes at acid, neutral, and alkaline pH values were used.[13-15] The chemical shifts of the various carbon atoms of the vitamins can be predicted fairly accurately by considering the effects of a number of substituents to be additive.[4,5,13-15]

pH Dependence of ^{13}C Chemical Shifts

The effect of pH on vitamin B₆ compounds has been studied by ^{1}H NMR[2] and by UV absorption spectroscopy.[16] However, ^{13}C NMR spectroscopy is a much more powerful tool for such investigations, since the ionization of the various heteroatoms influences profoundly the chemical shifts of the framework carbons.

Once the chemical shift assignments have been made for a vitamin at one pH, it is relatively simple to make assignments at other pH values by following the changes produced by small increments of pH.[13-15,17] As a

---

[13] R. D. Lapper, H. H. Mantsch, and I. C. P. Smith, *Can. J. Chem.* **53**, 2406 (1975).

[14] T. H. Witherup and E. H. Abbott, *J. Org. Chem.* **40**, 2229 (1975).

[15] R. C. Harruff and W. T. Jenkins, *Org. Magn. Reson.* **8**, 548 (1976).

[16] R. J. Johnson and D. E. Metzler, this series, Vol. 18A, p. 433.

[17] F. Gallais, R. Haran, J.-P. Laurent, and F. Nepveu-Juras, *C. R. Acad. Sci. Ser. C* **284**, 29 (1977).

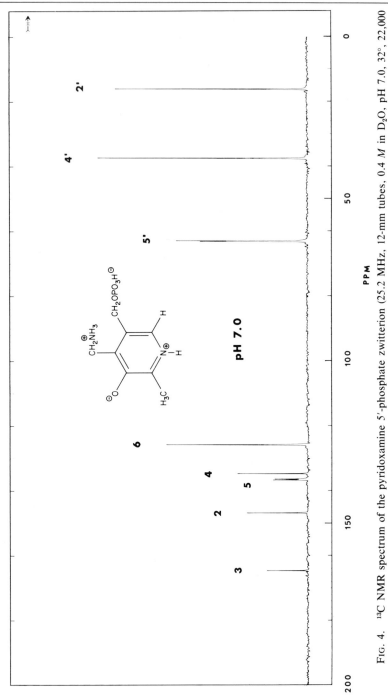

Fig. 4. ^{13}C NMR spectrum of the pyridoxamine 5'-phosphate zwitterion (25.2 MHz, 12-mm tubes, 0.4 M in D_2O, pH 7.0, 32°, 22,000 transients).

by-product, these ^{13}C NMR titration curves yield the corresponding pK values. The directions and magnitudes of the pH dependence of ^{13}C chemical shifts reflect effects such as changes in the electron density at particular atoms and in the difference between the energies of the molecular ground and excited states. It is important to understand the pH effects in the vitamin B$_6$ group since enzymes will be able to influence the reactivity of these coenzymes through selective protonation or deprotonation of the various functional groups. The ^{13}C NMR titration curves show clearly that the first deprotonation of the vitamin B$_6$ compounds occurs at the phenolic carbon and the second at the pyridinium nitrogen. The dissociation of a phosphate proton has little effect on the chemical shifts of any of the B$_6$ carbons.

As illustrated by the correlation diagrams in Figs. 1, 3, and 5, deprotonation of the phenol moiety to yield the zwitterionic form of the vitamin causes a considerable increase in the chemical shift for C-3 (trace B) relative to that of the fully protonated species (trace A), while the chemical shifts of C-4 and C-6 decrease.

In the second deprotonation step, the C-2 and C-6 resonances (trace C) undergo a large increase in chemical shift relative to those of the zwitterionic form (trace B), while the chemical shift of C-4 decreases. Deprotonation results in small but significant changes at the other aromatic pyridine ring carbons, and at the aliphatic substituent carbons (see Table II). In complex mixtures the well separated C-2' resonance provides a good indicator of the degree of protonation of the pyridine nitrogen; in the protonated vitamins it resonates around 15 to 16 ppm while in the completely deprotonated form it shifts to 20 ppm.

The most drastic pH-dependent spectral changes occur in the case of pyridoxal and pyridoxal 5'-phosphate. The situation is complicated by the fact that pyridoxal at neutral pH, and particularly pyridoxal 5'-phosphate below pH 4.0, are virtually insoluble; even protracted accumulations for the latter at pH 2.6 (trace A in Fig. 5 with 240,000 transients, requiring over 24 hr) did not yield measurable resonances for the quaternary ring carbons.

The absence of a resonance at ca. 200 ppm in the ^{13}C NMR spectra in Fig. 6 demonstrates clearly that at all pH values pyridoxal does not exist as the free aldehyde, but as the hemiacetal. However, the chemical shift of the C-4' carbon atom at pH 12.5 suggests that the hemiacetal form is in rapid equilibrium with an appreciable population of the free aldehyde. The peak at 134.0 ppm is not a "pure" resonance, but a weighted average of the chemical shift of the hemiacetal form at 100 ppm and that of the free aldehyde at 197 ppm. If the equilibrium is fast on the ^{13}C time scale (rate > the difference in chemical shifts in Hz), only one resonance is

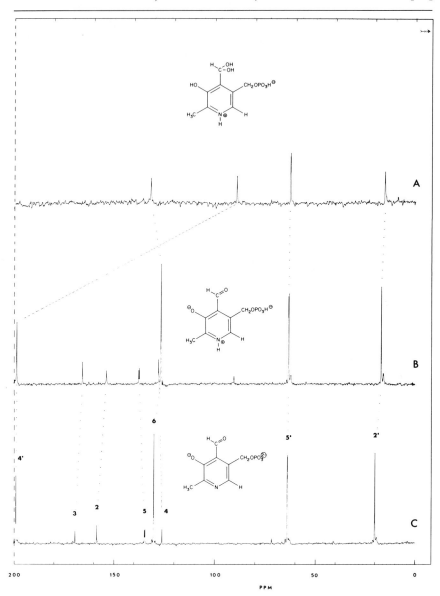

Fig. 5. ^{13}C NMR spectra of pyridoxal 5'-phosphate: 25.2 MHz, 32°, 12-mm tubes, saturated solution and 240,000 transients for trace A; 0.4 M in D_2O, 150,000 and 190,000 transients for traces B and C, respectively; (A) pH 2.6; (B) pH 6.2 (C) pH 12.4.

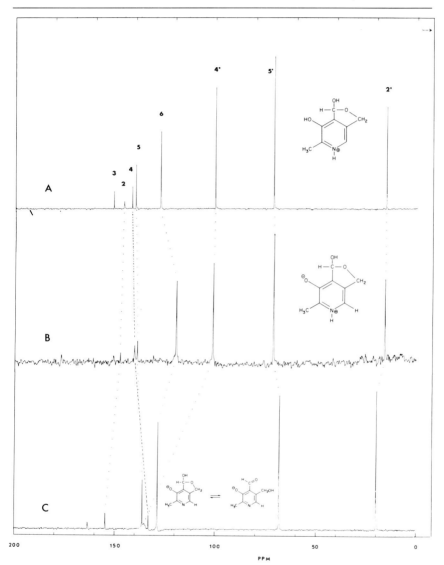

FIG. 6. pH-Dependence of pyridoxal ^{13}C NMR spectra: 25.2 MHz, 32°, 0.4 M in D$_2$O and 15,000 transients for traces A and C, saturated solution and 165,000 transients for trace B; (A) pH 1.8; (B) pH 6.2; (C) pH 12.5.

seen; however, upon slowing down the exchange rate, such as by cooling, one would expect to observe separate resonances for the two forms.

Pyridoxal 5′-phosphate clearly cannot form an intramolecular hemiacetal such as pyridoxal. This is evident from the ^{13}C chemical shifts, which show a preference for the free aldehyde at high and neutral pH, shifting to the hydrated aldehyde at lower pH. However, in this case the equilibrium between free and hydrated aldehyde is slow enough at room temperature to allow separate resonances for each form. As shown in Fig. 5B, at pH 6.2 there is a resonance at 89 ppm due to C-4' in a hydrated aldehyde as well as an aldehydic resonance at 197 ppm. At a pH value of 4.0 (not shown in Fig. 5) there are equal amounts of the two forms, as indicated by the equal intensities of the two resonances.

Vitamin B_6 Adducts with Amino Acids

Carbon-13 NMR has been used to study Schiff base and carbinolamine derivatives obtained from vitamin B_6 and various amino acids.[15,18-20] The nonenzymic formation of aldimine derivatives (I), which occurs over a wide pH range, can be easily followed by monitoring the individual chemical shifts of the corresponding vitamin. Assignments of the aldimine carbon resonances is made by comparison with the spectrum of the parent vitamin.

In aldimines of pyridoxal 5'-phosphate, the first proton titrated on increasing the pH is that on the pyridinium nitrogen, as indicated by a large change of the C-6 chemical shifts between pH 4 and 7. The titration curves of the C-2 and C-6 chemical shifts for this deprotonation are asymmetric, in contrast to the symmetry observed for the corresponding deprotonation of pyridoxamine. It has been suggested[15] that this asymmetry reflects an increased contribution of the mesomeric form (II).

(I)   (II)   (III)

Structural variations of the amino acid part of the aldimines are seen primarily in the C-3, C-4, and C-4' chemical shifts. Although geometric isomerism is possible about the carbon–nitrogen double bond,[21] only the trans isomer was found for the vitamin B_6 adducts with amino acids.[19] Iso-

[18] M. H. O'Leary and J. R. Payne, J. Biol. Chem. **251**, 2248 (1976).
[19] D. K. Dalling, D. M. Grant, W. J. Horton, and R. D. Sagers, J. Biol. Chem. **251**, 7661 (1976).
[20] B. H. Jo, V. Nair, and L. Davis, J. Am. Chem. Soc. **99**, 4467 (1977).
[21] C. Chang, T.-L. Shieh, and H. G. Floss, J. Med. Chem. **20**, 176 (1977).

topically enriched amino acids, as well as pyridoxal 5'-phosphate ^{13}C-enriched in positions 4' and 5', were used at low concentrations to detect intermediates formed in nonenzymic or in enzyme-dependent reactions.[18,19] Unlike the C-4' resonance, the chemical shift of C-5' is rather insensitive to structural changes.

The resonance of C-4' changes most dramatically on conversion of vitamin B$_6$ into various condensation products with amines or amino acids; the total span of ^{13}C chemical shifts expected for this carbon is about 100 ppm, making it a very sensitive indicator of the state of enzyme-bound pyridoxal 5'-phosphate.

The C-4' resonance occurs between 161 and 166 ppm for a variety of aldimines; resonances between 66 and 69 ppm were assigned to short-lived carbinolamine species (III).[19,20]

Besides the structural information provided by the chemical shifts of the labeled carbon atoms, the resonance widths can be used to evaluate the rigidity of binding of the vitamin coenzyme to a protein or enzyme. The rotational correlation times of the ^{13}C nuclei, which measure the rates of rotation of the nuclei relative to the magnetic field of the NMR spectrometer, can be estimated from the corresponding transverse relaxation times T_2^*. These T_2^* values are obtained from the observed half bandwidths $\Delta\nu_{1/2}$ ($\Delta\nu_{1/2} = 1/\pi T_2^*$). In the presence of D-serine dehydratase, the half bandwiths of the C-4' and C-5' resonances in ^{13}C-enriched pyridoxal 5'-phosphate, which are attributed to the Schiff base form of the bound coenzyme, are 24 ± 3 and 48 ± 3 Hz, respectively.[18] Since the linewidths of ^{13}C NMR spectra of macromolecules are governed primarily by ^{13}C–^{1}H dipolar relaxation, and since C-5' has two attached hydrogens, the resonance of C-5' for the bound coenzyme is expected to be twice as broad as that of C-4'.

The rotational correlation times of the pyridoxal 5'-phosphate–D-serine dehydratase system at pH 7.8, calculated from the above-mentioned linewidths, are 16 ± 2 nsec for both labeled carbon atoms, indicating that the coenzyme is quite rigidly bound to the enzyme. It is expected that in the future more use will be made of ^{13}C-enriched vitamin B$_6$ derivatives to study binding to enzymes and to explore the involvement of the coenzymes in the enzymic mechanism.

Acknowledgment

We are very grateful to Mrs. Adèle Martin for her expert efforts toward the figures.

[73] ^{13}C NMR Spectroscopy of the Vitamin B$_6$ Group

By ROBERT J. JAWORSKI and MARION H. O'LEARY

Proton nuclear magnetic resonance (NMR) spectroscopy has played an important role in a variety of biochemical studies.[1] Because of the greater difficulty of obtaining satisfactory ^{13}C spectra, ^{13}C NMR has only recently begun to figure importantly in such investigations. The ^{13}C nucleus gives a weaker signal than the ^{1}H nucleus both because of its smaller gyromagnetic ratio and because of its lower abundance; the two factors combined give an approximate 6000-fold difference in sensitivity at natural abundance.[2] The advent of pulsed Fourier transform spectroscopy together with proton decoupling methods has alleviated this difficulty to a considerable degree in recent years. Moderately dilute solutions ($10^{-2} M$) of small molecules can now be studied with comparative ease.[2]

In spite of the sensitivity problem, ^{13}C NMR has several potential advantages over ^{1}H NMR in biochemical applications. ^{13}C NMR spectra have a larger chemical shift range than ^{1}H NMR spectra and are potentially simpler because of the general absence of homonuclear coupling.[2]

Compounds of the vitamin B$_6$ family have been studied extensively by ^{1}H NMR spectroscopy.[3] Only recently, however, have these same compounds been examined by ^{13}C NMR[4–12] The purpose of this article is to re-

[1] R. A. Dwek, "Nuclear Magnetic Resonance in Biochemistry, Applications to Enzyme Systems," Oxford Univ. Press, London and New York, 1973; T. L. James, "Nuclear Magnetic Resonance in Biochemistry: Principles and Applications," Academic Press, New York, 1975; K. Wuthrich, "NMR in Biological Research: Peptides and Proteins," Elsevier, New York, 1976.

[2] R. A. Komoroski, I. R. Peat, and G. C. Levy, *in* "Topics in Carbon-13 NMR Spectroscopy" (G. C. Levy, ed.), Vol. 2, p. 176. (Interscience), New York, 1976; S. N. Rosenthal and J. H. Fendler, *Adv. Phys. Org. Chem.* **13**, 279 (1976); W. Egan, H. Shindo, and J. S. Cohen, *Annu. Rev. Biophys. Bioeng.* **6**, 383 (1977); F. R. N. Gurd and P. Keim, this series, Vol. 17, p. 836.

[3] W. Korytnyk and H. Ahrens, this series, Vol. 18A, p. 475.

[4] M. H. O'Leary and J. R. Payne, *J. Biol. Chem.* **251**, 2248 (1976).

[5] R. C. Harruff and W. T. Jenkins, *Org. Magn. Reson.* **8**, 548 (1976).

[6] R. D. Lapper, H. H. Mantsch, and I. C. P. Smith, *Can. J. Chem.* **53**, 2406 (1975).

[7] T. H. Witherup and E. H. Abbott, *J. Org. Chem.* **40**, 2229 (1975).

[8] D. K. Dalling, D. M. Grant, W. J. Horton, and R. D. Sagers, *J. Biol. Chem.* **251**, 7661 (1976).

[9] B. H. Jo, V. Nair, and L. Davis, *J. Am. Chem. Soc.* **99**, 4467 (1977).

[10] C. J. Chang, T. L. Shieh, and H. G. Floss, *J. Med. Chem.* **20**, 176 (1977).

[11] R. E. Hill, I. Miura, and I. D. Spenser, *J. Am. Chem. Soc.* **99**, 4179 (1977).

[12] F. Gallais, R. Haran, J.-P. Laurent, and F. Nepveu-Juras, *C. R. Acad. Sci. Ser. C* **284**, 29 (1977).

view these studies and to provide useful information regarding methodology and potential for further studies in this area.

Instrumentation and Methods for Obtaining Spectra

Sample Preparation. Ordinary aqueous solutions can be used for ^{13}C NMR studies. Although the presence of 10–20% D$_2$O in the the solvent is useful, the use of 100% D$_2$O is not necessary. As an alternative to the addition of D$_2$O to the solution, a coaxial insert tube inside the sample-containing NMR tube can be used instead. The deuterium signal from the D$_2$O is used as a heteronuclear lock.

The chemical shift standard of choice in ^{13}C NMR spectra is dioxane, which shows a singlet in the proton-decoupled ^{13}C NMR spectrum at 67.4 ppm relative to tetramethylsilane (TMS). The dioxane (1–2%) can be added either to the sample or to the D$_2$O-containing insert, if one is used. Diamagnetic susceptibility corrections for an internal versus an external standard are small enough to be neglected (<0.2 ppm).[13]

Carbon-containing solvents and buffers obviously pose potential problems, but they can often be used provided that the concentration of the sample is not too low (<10^{-2} M) and the sample ^{13}C resonances are not obscured by the solvent or buffer signal(s). Fourier-transform spectra in which the concentration of ^{13}C nuclei in the solvent or buffer is more than about 100-fold higher than that in the sample are likely to have dynamic range problems.[2]

With present state-of-the-art instrumentation (Varian XL-100-15, Varian CFT-20, Jeol PS/FFT-100, or Bruker HX90E), spectra of natural abundance samples at concentrations of 0.3–0.5 M can be obtained within an hour using 11–12 mm sample tubes under the condition of full (wide band) proton decoupling. Such decoupling simplifies the ^{13}C spectra and provides a signal enhancement potentially as high as a factor of 2.99 due to the nuclear Overhauser effect.[2] Spectra can be obtained at lower concentrations by use of ^{13}C-enriched materials or longer data accumulation times. Higher sample concentrations or longer data accumulation times are needed for experiments utilizing partial (i.e., off-resonance or single frequency) or gated decoupling because of varying degrees of reduced nuclear Overhauser enhancement.[14] Such spectra frequently provide valuable structural information.

The present lower concentration limit for obtaining single carbon resonances of a 90% enriched ^{13}C nucleus bound to a macromolecule appears

[13] G. C. Levy and J. D. Cargioli, *J. Magn. Reson.* **6**, 143 (1972).
[14] G. C. Levy and G. L. Nelson, "Carbon-13 Nuclear Magnetic Resonance for Organic Chemists." Wiley (Interscience), New York, 1972.

to be about $5 \times 10^{-4} M$. The use of 12-mm sample tubes is recommended in working with such concentrations, since the larger tube provides about a factor of 3 greater sensitivity over standard 5-mm sample tubes. Usually 1.5–2 ml of sample are required, along with a well-fitting vortex plug. Recently, Allerhand et al. reported ^{13}C NMR spectra of single carbon resonances of proteins using 20-mm sample tubes and protein concentrations of 5–20 mM.[15]

Because of the long data accumulation times required for measurements with macromolecules, traces of carbon-containing contaminants that cause no problem under less strenuous conditions may now provide numerous unwanted resonances in the NMR spectrum. Unbound contaminants can generally be distinguished from the protein resonances by the sharpness of the peaks. Extensive dialysis of the sample is recommended before performing the NMR experiment as a precaution against contamination. Turbidity, protein denaturation, and bacterial contamination are also potential problems when obtaining spectra of macromolecules, since the activity and integrity of the macromolecule over the data acquisition period can be affected by these factors.

A number of compounds of the vitamin B_6 family are excellent metal chelators, and appropriate precautions must be taken in preparing samples. Traces of paramagnetic metals can broaden the resonance lines sufficiently to impair assignment of resonances.[8] Higher concentrations of such ions may eliminate the resonances completely.

Interpreting Spectra. Biochemists and chemists have been conditioned by their experience with ^{1}H NMR spectra to expect particular kinds of structural and environmental effects in NMR spectra. However, what is true for ^{1}H NMR spectra may not be true for ^{13}C spectra, and a degree of caution is advised. For example, ionization of protonated carboxyl and amino groups leads to chemical shift changes in ^{13}C NMR which are opposite in direction to those in ^{1}H NMR.[2] The most reliable guide to interpreting ^{13}C spectra is, of course, an appropriate collection of properly assigned spectra of analogous compounds. A variety of such collections exist.[16]

[15] E. Oldfield and A. Allerhand, *J. Biol. Chem.* **250**, 6409 (1975); E. Oldfield, R. S. Norton and A. Allerhand, *J. Biol. Chem.* **250**, 6368 (1974); E. Oldfield, R. S. Norton, and A. Allerhand, *J. Biol. Chem.* **250**, 6381 (1974); K. Dill and A. Allerhand, *J. Am. Chem. Soc.* **99**, 4508 (1977); D. J. Wilbur and A. Allerhand, *FEBS Lett.* **74**, 272 (1977); K. Ugurbil, R. S. Norton, A. Allerhand, and R. Bersohn, *Biochemistry* **16**, 886 (1977); D. J. Wilbur and A. Allerhand, *J. Biol. Chem.* **252**, 4968 (1977).
[16] L. F. Johnson and W. C. Jankowski, "Carbon-13 Nuclear Magnetic Resonance Spectroscopy. A Collection of Assigned, Coded and Indexed Spectra," Wiley (Interscience), New York, 1972; E. Breitmaier and W. Voelter, "^{13}C NMR Spectroscopy, Methods and Applications", Verlag Chemie, Weinheim/Bergstr., 1974; Sadtler Standard Spectra, Sadtler Research Laboratories, Inc., Philadelphia, Pennsylvania, 1977.

The use of peak integrals in ^{1}H NMR has been a useful tool in assigning individual signals. The correspondence between area and concentration is not so clear in ^{13}C NMR because of variable contributions of nuclear Overhauser effects, relaxation effects, and other things.[14]

Spectra[17]

The system used for numbering pyridoxine and its congeners is shown in (I). The compounds to be discussed differ principally in the state of the

(I)

4' carbon and in the presence or the absence of a phosphate ester linkage to the 5' carbon.

The B$_6$ vitamins have fully decoupled ^{13}C spectra which are characterized by at least eight distinguishable resonances, one for each of the five aromatic carbons and one for each of the three side-chain carbons. Under certain conditions a single carbon may show more than a single resonance. For example, the 4' carbon of pyridoxal 5'-phosphate can show separate resonances for the aldehyde and the aldehyde hydrate.

Qualitatively, the spectra of the B$_6$ vitamins can be described as follows: the methyl carbon (C-2') is at highest field, appearing in the range 14–20 ppm downfield from tetramethylsilane. Methylene carbon C-5' appears at 58–70 ppm. Carbon C-4' appears in a variety of positions in various compounds. The five carbons of the pyridine ring resonate in the range 117 to 170 ppm. Within these, C-3 is always at lowest field because of the effect of the directly bound oxygen. C-2 is usually the next carbon upfield. This assignment can be confirmed by observing the coupling of this resonance to the three hydrogens of the C-2' methyl group in undecoupled spectra. C-6 shows a large (about 180 Hz) coupling to the attached hydrogen. In phosphorylated compounds of the vitamin B$_6$ group, C-5 is characterized by its coupling to phosphorus ($J = 8$ Hz). In unphosphorylated compounds distinction between C-5 and C-4 is often difficult.

In the following sections we describe the spectra of a number of compounds of the vitamin B$_6$ group. More detailed information concerning peak assignments, conditions, etc., can be obtained from the original references.

[17] All the chemical shifts reported in this paper are relative to tetramethylsilane (TMS).

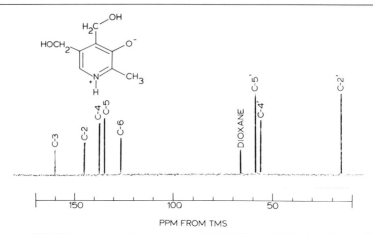

Fig. 1. ¹³C NMR spectrum of pyridoxine at 25.16 MHz in 0.1 M phosphate buffer, pH 8.2, containing 18% D_2O.

Pyridoxine. ¹³C NMR spectra of pyridoxine have been reported by Harruff and Jenkins,[5] by Lapper *et al.*[6] and by Witherup and Abbott.[7] The ¹³C NMR spectrum of pyridoxine at pH 8.2 is given in Fig. 1. All carbons except C-4, C-4′, C-5, and C-5′ can easily be identified as described above. Distinction between C-4′ (57.2 ppm) and C-5′ (59.8 ppm) is made on the basis of the fact that in the gated-decoupled spectrum (not shown) C-5′ shows an appreciable coupling ($J \cong 3$ Hz) to the proton on C-6, whereas C-4′ shows no such coupling. C-5 (135.6 ppm) and C-4 (138.3 ppm) can be distinguished by specific proton decoupling of the methylene protons. Both resonances then appear as sharp doublets due to coupling to the C-6 proton with coupling constants of about 2 Hz and 6 Hz, respectively.[18]

The chemical shifts of the aromatic carbons of pyridoxine, particularly

---

[18] The assignment of the resonances for C-4 and C-5 in pyridoxine has been based on specific proton decoupling experiments that have assumed that three-bond couplings in the vitamin B_6 series are larger than two-bond couplings. This general observation was originally derived from studies[19,20] on pyridine derivatives. In these same studies, however, particularly for 2- and 3-substituted pyridines, the $^2J_{C-5,H-6}$ value was larger than the $^3J_{C-4,H-6}$ value. The opposite trend is noted for pyridoxal 5′-phosphate, where assignment of C-4 and C-5 is unambiguous. By analogy, therefore, the assignments for C-4 and C-5 were made in pyridoxine, as well as pyridoxamine and pyridoxal, assuming the same trend in J value is followed. Recently, F. Gallais *et al.*[12] have published data on perturbations in the ¹³C chemical shifts of pyridoxine induced by chelation to Cd^{2+}. C-2, C-3, C-4 (as assigned above), and C-4′ resonances are principally affected. The C-5 resonance is virtually unperturbed.

[19] M. Hansen and Y. Jacobson, *J. Magn. Reson.* **10**, 74 (1973).

[20] Y. Takeuchi and N. Dennis, *J. Am. Chem. Soc.* **96**, 3657 (1974).

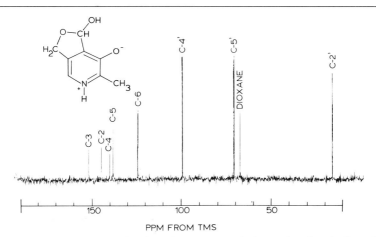

FIG. 2. ^{13}C NMR spectrum of pyridoxal at 25.16 MHz in 0.1 M phosphate buffer, pH 3.4, containing 18% D_2O.

C-3, C-2, and C-6, vary with pH. The titration curves described by these chemical shifts provide pK_a values of approximately 5.4 for the phenolic hydroxyl and 9.8 for the pyridinium nitrogen, consistent with values obtained from ultraviolet spectra.[21]

Pyridoxal. Spectra of pyridoxal have been reported by O'Leary and Payne,[4] by Harruff and Jenkins,[5] by Lapper *et al.*,[6] and by Witherup and Abbott.[7] The NMR spectrum at pH 3.4 is shown in Fig. 2. Most of the chemical shifts are similar to those in pyridoxine, except C-5′ (70.7 ppm) and C-4′ (99.6 ppm). The presence of the cyclic hemiacetal (II) is indicated by the shift of C-4′.

Harruff and Jenkins[5] and Lapper *et al.*[6] have suggested that above pH 10 the hemiacetal form of pyridoxal is destabilized and the predominant species are a mixture of free aldehyde and aldehyde hydrate. Above about pH 12.4, the anionic form of the aldehyde hydrate (III) is the predominant species.

(II)   (III)

[21] R. J. Johnson and D. E. Metzler, this series, Vol. 18A, 433 (1970).

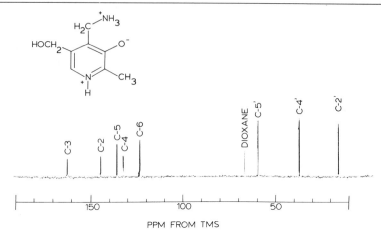

Fig. 3. ^{13}C NMR spectrum of pyridoxamine at 25.16 MHz in 0.1 M phosphate buffer, pH 6.7, containing 18% D_2O.

Pyridoxamine. The ^{13}C NMR spectrum of pyridoxamine at pH 6.7 is shown in Fig. 3. Most resonance assignments are reasonably straightforward and follow by analogy with spectra of other compounds. Results of single-frequency decoupling experiments indicate that the positions of the resonances for C-4 (133.3 ppm) and C-5 (136.7 ppm) are reversed compared to those in other compounds. C-4' (37.6 ppm) is easily distinguished from C-5' (59.9 ppm) by the upfield shift caused by the nitrogen substituent.

Pyridoxal 5'-Phosphate. Spectra of this compound have been reported by O'Leary and Payne,[4] by Harruff and Jenkins,[5] by Lapper *et al.*,[6] by Witherup and Abbott,[7] by Jo, Nair, and Davis,[9] and by Dalling *et al.*[8] The spectrum at pH 8.1 is shown in Fig. 4. The C-4' resonance occurs at 197.1 ppm, indicating that the free aldehyde is the predominant form under these conditions. The C-5' resonance (62.9 ppm) is coupled to phosphorus ($J = 3$ Hz). Distinction between C-4 (125.9 ppm) and C-5 (134.5 ppm) is easily made based on the coupling of C-5 to phosphorus ($J = 8$ Hz) and the different proton couplings to the two carbons in the undecoupled spectrum.

The values of the carbon–phosphorus coupling constants in pyridoxal 5'-phosphate and pyridoxamine 5'-phosphate and their relationship to the conformation of the phosphate side chain have been considered by Lapper *et al.*[6] The large values of the couplings in the case of these vitamin B_6 compounds suggest that the preferred rotamer about the O-C-5' bond at all pH values has the phosphorus atom trans to C-5.

The pH dependences of the ^{13}C chemical shifts in pyridoxal 5'-

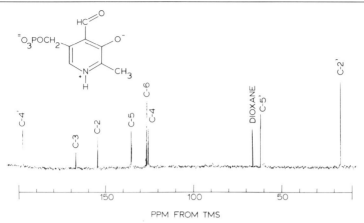

FIG. 4. ^{13}C NMR spectrum of pyridoxal 5'-phosphate at 25.16 MHz in 0.1 M phosphate buffer, pH 8.1, containing 18% D$_2$O.

phosphate are shown in Fig. 5. It is interesting to note that the chemical shift of C-5' is virtually independent of the ionization state of the attached phosphate group, the total chemical shift range for this carbon being less than 1 ppm. The chemical shift of C-4' is also virtually independent of pH. Below pH 7 a signal at 89 ppm due to the aldehyde hydrate appears. The chemical shift of C-2' changes by about 4 ppm over the pH range from 1.5 to 11.2.

The pH dependences of the chemical shifts of the aromatic carbons are much more striking than those of the aliphatic carbons. Deprotonation

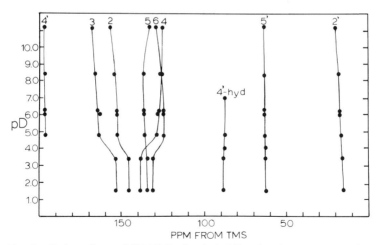

FIG. 5. pH dependence of ^{13}C NMR chemical shifts of pyridoxal 5'-phosphate.

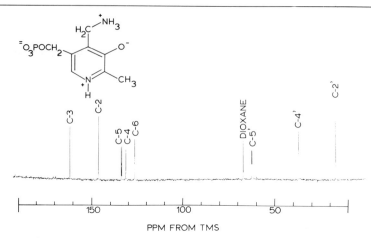

Fig. 6. ^{13}C NMR spectrum of pyridoxamine 5′-phosphate at 25.16 MHz in 0.1 M phosphate buffer, pH 8.1, containing 18% D_2O.

of the phenolic hydroxyl group causes a 15 ppm downfield shift in the resonance of C-3, giving a pK_a value of 4, consistent with the value obtained from ultraviolet spectra.[21] The other aromatic carbon resonances also reflect this ionization. Note that some resonances shift downfield and some shift upfield on ionization. Deprotonation of the pyridine nitrogen also causes significant changes in the chemical shifts of the aromatic carbons, giving an estimated pK_a value of 8.6, close to the values based on potentiometric titrations[22] and UV determinations.[21]

Pyridoxamine 5′-Phosphate. Spectra of this compound have been reported by O'Leary and Payne,[4] by Harruff and Jenkins,[5] by Lapper *et al.*,[6] by Witherup and Abbott,[7] and by Dalling *et al.*[8] The spectrum at pH 8.1 is shown in Fig. 6. The resonance of carbon C-4 (but interestingly not that of C-4′) is sensitive to the ionization state of the 4′ amino group and provides a pK_a value near 11 for this group, consistent with the value reported by Morozov *et al.*[23] The other variations in the chemical shifts of the aromatic carbons (Fig. 7) are similar to those observed with pyridoxal 5′-phosphate.

Pyridoxal 5′-Phosphate-n-butylamine Schiff Base. Because of the importance of Schiff bases in pyridoxal phosphate-dependent enzymes, a number of studies have been devoted to the determination of ^{13}C NMR spectra of such Schiff bases. The spectrum of a typical Schiff base at pH 8.3 is shown in Fig. 8. C-4′ in the Schiff base appears at 166.2 ppm, shifted

[22] F. J. Anderson and A. E. Martell, *J. Am. Chem. Soc.* **86**, 715 (1964).
[23] Y. V. Morozov, N. P. Bazhulian, M. Y. Karpeiskii, V. I. Ivanov, and A. I. Kuklin, *Biofizika* **11**, 228 (1966).

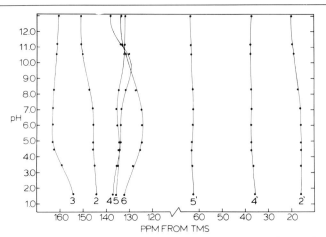

FIG. 7. pH dependence of ^{13}C NMR chemical shifts of pyridoxamine 5'-phosphate.

upfield by more than 30 ppm from its position in the free aldehyde. The resonances due to the four carbons of the butyl group in the Schiff base are also visible in Fig. 8, as are the resonances due to a trace of free *n*-butylamine.

O'Leary and Payne[4] have shown that different tautomeric forms of the Schiff base can be distinguished by ^{13}C NMR spectroscopy. In organic solvents the Schiff base exists as an enolimine tautomer (IV), λ_{max} 335 nm, as opposed to the ketoenamine form (V), λ_{max} 405 nm, which predom-

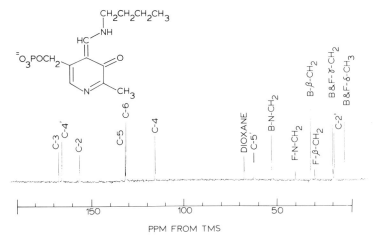

PPM FROM TMS

FIG. 8. ^{13}C NMR spectrum of Schiff base formed by reaction of pyridoxal 5'-phosphate with *n*-butylamine at 25.16 MHz in 0.1 *M* phosphate buffer, pH 8.3, containing 18% D$_2$O. F-free; B-bound.

inates in aqueous solution. C-2, C-3, C-4, and C-6 experience the greatest changes in chemical shift with change in solvent.

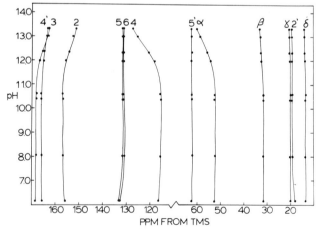

(IV)   (V)

The pH profiles for the chemical shifts of the carbons in this compound are shown in Fig. 9. Again, the greatest sensitivity to protonation and deprotonation is manifested in the chemical shifts of the aromatic carbons. Most of these shifts are similar to those observed with pyridoxal 5'-phosphate and pyridoxamine 5'-phosphate. In addition, deprotonation of the imine nitrogen affects the chemical shifts of C-4, C-3, and the α-carbon of the bound amine by 3–5 ppm. The pK_a value is 12.6. It is interesting that this ionization does not affect the chemical shift of C-4'.

Chang et al.[10] have considered the analysis of $^{13}C-{}^{1}H$ coupling constants in determining the degree of coplanarity and intramolecular hydrogen bonding in pyridoxal Schiff bases. Their preliminary studies with model compounds indicate that the smaller values for the two-bond $^{13}C-{}^{1}H$ coupling constants in these Schiff bases compared to pyridoxal 5'-phosphate itself may not be due to a change in the conformation of the C-4–C-4' bond.

FIG. 9. pH dependence of ^{13}C NMR chemical shifts of the Schiff base formed by reaction of pyridoxal 5'-phosphate with n-butylamine.

TABLE I

^{13}C Chemical Shifts of Schiff Bases Obtained by Reaction of Pyridoxal 5'-Phosphate with Various Amines

Amine	pH	Chemical shift (ppm)								Reference
		C-2	C-3	C-4	C-5	C-6	C-2'	C-4'	C-5'	
Methylamine	7.5	156.0	166.8	115.6	131.6	131.0	19.0	166.8	61.9	8
n-Butylamine	7.4	156.3	168.4	116.2	132.8	129.0	18.5	165.8	62.1	4
n-Butylamine in dioxane/CHCl$_3$	3:1	144.9	156.2	120.1	131.4	137.8	18.9	163.8	61.4	4
Isobutylamine	7.3	155.4	166.9	115.3	131.4	129.7	18.0	165.6	61.4	5
α-Methylglutamic acid	7.8	156.4	167.5	115.2	131.3	130.1	18.4	161.9	61.3	5
Valine	8.4	155.4	166.9	115.9	131.8	129.7	18.0	164.7	61.4	5,8
γ-Aminobutyric acid	7.1	155.4	166.7	115.5	131.6	129.7	18.1	165.6	61.4	5

^{13}C NMR spectra of a number of additional Schiff bases of pyridoxal 5'-phosphate are summarized in Table I. Further information, including chemical shifts of additional carbons, pH dependences, etc., can be obtained from the original references.

Other Compounds. NMR spectra of a number of additional derivatives of pyridoxal 5'-phosphate and pyridoxamine 5'-phosphate are summarized in Table II.

Enzyme Studies

Although ^{13}C NMR has obvious potential as a means of studying pyridoxal 5'-phosphate in enzymes, sensitivity considerations require that such studies be made with ^{13}C-enriched coenzyme. Pyridoxal 5'-phosphate labeled with ^{13}C to the extent of greater than 90% in both C-4' and C-5' has been synthesized and spectra of this compound bound to D-serine dehydratase have been reported.[4]

For a protein the size of D-serine dehydratase (MW 45,500) using this enriched coenzyme, satisfactory ^{13}C NMR spectra can be obtained in about 20–24 hr using a protein concentration of 1.4 mM (65 mg/ml). Use of a higher protein concentration permits obtaining a spectrum of the same quality in a shorter time. It should be noted that the time required to obtain a spectrum of a given quality varies with the square of the protein concentration (all other factors being equal); thus, there is a real premium on high protein concentrations.

In addition to the problems of protein availability and concentration, macromolecular size is an important consideration. Size is manifested in two parameters, relaxation time and linewidth. Assuming that the primary mechanism controlling ^{13}C nuclear relaxation in these spectra is direct ^{13}C–^{1}H dipolar relaxation,[2] the spin-lattice relaxation time, T_1 is given by

$$\frac{1}{T_1} = \frac{N}{10} \left[(\hbar^2 \gamma_C^2 \gamma_H^2)/r_{CH}^6 \right] f(\tau_{eff})$$

where N is the number of directly bonded hydrogens, γ_C and γ_H are the magnetogyric ratios of ^{13}C and ^{1}H, respectively, r_{CH} is the C-H bond distance (assumed 1.09 Å) and $f(\tau_{eff})$ is a complex function of the effective correlation time for rotational reorientation. The resonance linewidth, W, is given by

$$W = \frac{1}{\pi T_2} = \frac{N}{20\pi} \left[(\hbar^2 \gamma_C^2 \gamma_H^2)/r_{CH}^6 \right] f(\tau_{eff})$$

where T_2 is the transverse relaxation time. τ_{eff}, the effective relaxation time, can be considered to be composed of two factors: τ_c, the rotational

TABLE II
^{13}C CHEMICAL SHIFTS OF MISCELLANEOUS COMPOUNDS OF THE VITAMIN B_6 GROUP

Compound	pH	Chemical shift, ppm								Reference
		C-2	C-3	C-4	C-5	C-6	C-2'	C-4'	C-5'	
N-n-Butylpyridoxamine 5'-phosphate	7.8	146.0	163.6	131.6	132.0	125.4	16.4	45.5	62.5	4
Aldamine formed from pyridoxal 5'-phosphate and 1,3-diaminopropane	13.5	152.1	161.2	133.9	130.3	132.8	19.7	69.5	62.9	4
Thiazolidine formed from pyridoxal 5'-phosphate and cysteine	6.8	146.3	162.0	133.9	133.5	125.8	15.9	63.8	62.0	4

correlation time of the entire macromolecule, and τ_g, the rotational correlation time of the nucleus in question relative to the macromolecule:

$$1/\tau_{eff} = 1/\tau_c + 1/\tau_g$$

If the nucleus in question is rigidly bound to the macromolecule, then τ_g becomes very large, and $\tau_{eff} = \tau_c$. If there is some degree of motional freedom (for example, if the nucleus is attached to a flexible side chain of the protein), τ_{eff} can be substantially shorter than the rotational correlation time of the macromolecule.

The implication of the above equations for ^{13}C NMR spectra of pyridoxal 5'-phosphate bound to proteins is that as the overall molecular weight of the protein becomes higher, the ^{13}C resonance becomes harder to see because of the increase in linewidth. Excellent spectra have been obtained for D-serine dehydratase, but to date it has not been possible to obtain useful spectra of glutamate decarboxylase (MW 300,000). At the present time the useful upper limit on molecular weight may be near 100,000.

D-*Serine Dehydratase.* D-Serine dehydratase (MW 45,500) was resolved and reconstituted with ^{13}C-enriched pyridoxal 5'-phosphate.[4] The ^{13}C NMR spectrum in Fig. 10A was obtained with a protein concentration of 65 mg/ml (^{13}C concentration 1.3 mM) in a 12-mm sample tube on a Varian XL-100-15 spectrometer in about 24 hr. The spectral width in Fig. 10A was 5500 Hz, the data acquisition time (interval between pulses) was 0.1 sec, and the pulse width was 57°. The spectrum represents the result of 857,000 transients. A sensitivity enhancement factor was utilized in the data reduction, and this factor contributed 0.96 Hz to the observed linewidth.

A spectrum of D-serine dehydratase under similar conditions but without ^{13}C enrichment in the coenzyme is shown in Fig. 10B.

The difference between Fig. 10A and 10B is two resonances, the coenzyme resonance at 168.2 ppm, which must be due to the 4'-carbon in the enzyme–pyridoxal 5'-phosphate Schiff base, and the resonance at 66.4 ppm, which is that of the 5'-carbon in the enzyme-bound coenzyme. The linewidths for both resonances (43.5 Hz for C-4' and 66.4 Hz for C-5') indicate that the carbons belong to enzyme-bound coenzyme. The absence of a C-4' aldehyde resonance at 197 ppm indicates that the concentration of free pyridoxal 5'-phosphate in solution is very low. This is an important consideration, because the chemical shift of the 5'-carbon in both free and bound coenzyme is virtually identical, and the presence of a measurable amount of free coenzyme would cause an error in the estimation of the linewidth of C-5'.

The assignment of the resonance near 168 ppm to the 4'-carbon in the

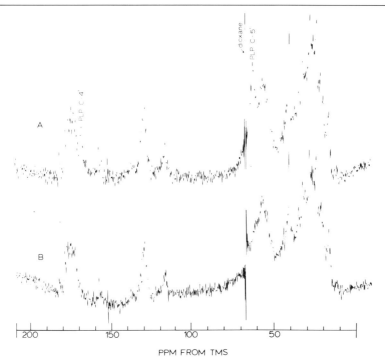

FIG. 10. (A) ¹³C NMR spectrum of D-serine dehydratase containing [4',5'-¹³C₂]pyridoxal 5'-phosphate at 25.16 MHz in 0.1 M phosphate buffer, pH 7.8, containing 18% D₂O. The protein concentration is 65 mg/ml. (B) Same as (A), but with coenzyme not enriched in ¹³C.

enzyme-bound Schiff base is strengthened by the observation that reduction of the enriched holoenzyme produces a product whose spectrum lacks the resonance near 168 ppm but has a new resonance at 41.7 ppm (Fig. 11), as expected for an enzyme-bound secondary amine. The linewidths for the C-4' resonance (74 Hz) and the C-5' resonance (74 Hz) are similar to that observed for the C-5' resonance in the native enzyme (66.4 Hz), indicating that the coenzyme is bound as rigidly in the reduced form as in the native form.

The pH dependence of the ¹³C NMR spectrum of D-serine dehydratase over the range from pH 6.2 to 10.6 reveals little change in either shift or linewidth of the two ¹³C coenzyme resonances. However, the NMR spectrum at high pH (Fig. 12) shows two resonances in the Schiff base region, one at 168.3 ppm and one at 167.4 ppm. The spectrum also shows a small peak at 198 ppm due to the presence of free pyridoxal 5'-phosphate in the solution. This resonance appears despite the fact that the enzyme was extensively dialyzed prior to determination of the NMR spectrum. Pre-

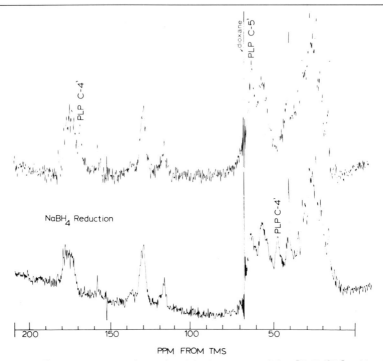

FIG. 11. ^{13}C NMR spectra of D-serine dehydratase containing [4',5'-^{13}C$_2$]pyridoxal 5'-phosphate at 25.16 MHz in 0.1 M phosphate buffer, pH 8.3, containing 18% D$_2$O, before and after reduction with NaBH$_4$.

sumably, over the several hours required to obtain the spectrum, a small amount of coenzyme dissociates from the enzyme. This enzyme has a second binding site for pyridoxal 5'-phosphate,[24] and the second resonance near 168 ppm is presumably due to binding of some coenzyme to that site.

Other Enzyme Systems. If one operates within the limits of protein size previously mentioned, ^{13}C NMR has the potential for answering several questions concerning structure and binding capability in other pyridoxal 5'-phosphate-dependent enzyme systems. The sensitivity of the C-4' resonance to structural changes indicates the potential for confirming the presence of an aldamine-type structure involving another enzyme group in addition to the lysine amino group responsible for binding the cofactor to the protein and also the possibility of identifying this second group. Also, one can examine a series of enzyme-pseudosubstrate com-

[24] Y. Z. Huang and E. E. Snell, *J. Biol. Chem.* **247**, 7358 (1972).

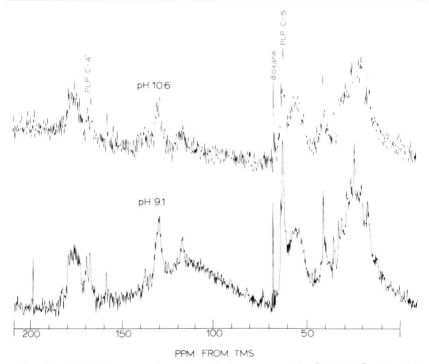

FIG. 12. ^{13}C NMR spectra of D-serine dehydratase containing [4′,5′-^{13}C$_2$]pyridoxal 5′-phosphate at 25.16 MHz in 0.1 M phosphate buffer containing 18% D$_2$O at pH 9.1 and 10.6.

plexes and obtain information about binding rigidity, structural chemistry, and stereochemistry.

Biosynthesis. ^{13}C is currently being used in many biosynthetic studies in preference to ^{14}C because the position of a ^{13}C label can usually be determined with NMR techniques without the need for chemical degradation. Hill *et al.*[11] have recently reported a study of the biosynthesis of pyridoxine in *E. coli B* grown on [1,3-^{13}C$_2$]glycerol as the sole carbon source. Carbons C-2′, C-3, C-4′, C-5′, and C-6 of pyridoxine were found to be enriched in ^{13}C, confirming earlier studies using ^{14}C.

Further Applications. Dalling, Grant, Horton, and Sagers[8] have studied nonenzymic reactions of pyridoxal 5′-phosphate with glycine, alanine, valine, serine, and several other model compounds using ^{13}C NMR and isotopically enriched amino acids. They have proposed a mechanism for the deamination of amino acids by pyridoxal 5′-phosphate.

Jo, Nair, and Davis[9] have used ^{13}C NMR techniques in an attempt to detect intermediate carbinolamine species in the formation of a Schiff base between pyridoxal 5′-phosphate and alanine.

Recently, Gallais, Haran, Laurent, and Nepveu-Juras[12] have studied the influence of Cd^{2+} cation on the ^{13}C NMR spectrum of pyridoxine. Chemical shift perturbations at pH 6.8 appear to indicate a 1:1 chelate between pyridoxine and the Cd^{2+} cation.

[74] Synthesis and Biological Activity of Vitamin B₆ Analogs[1]

By W. KORYTNYK

This article updates the earlier review in Volume 18A[2] and covers the period between 1969 and 1977 of work done in the author's laboratory and in collaboration with others.

General methods for the modification of different positions of vitamin B₆ have been developed and are described under appropriate headings. Specific examples of synthesis of enzymically and biologically interesting analogs have also been included. These descriptions have been supplemented by tables, which summarize enzymic and biological activities of the analogs and also provide references to their synthesis.

The major developments covered are as follows:

Chemistry

1. General and convenient methods for the single modification in positions 2,[3,4] 4,[5-11] 5,[12-16] and 6.[17]

[1] The work was supported in part by U.S. Public Health Service Grants CA-08793 and CA-13038.

[2] W. Korytnyk and M. Ikawa, this series, Vol. 18A, p. 524.
[3] W. Korytnyk, S. C. Srivastava, N. Angelino, P. G. G. Potti, and B. Paul, *J. Med. Chem.* **16**, 1096 (1973).
[4] W. Korytnyk and N. Angelino, *J. Med. Chem.* **20**, 745 (1977).
[5] W. Korytnyk and B. Paul, *J. Med. Chem.* **13**, 187 (1970).
[6] W. Korytnyk, H. Ahrens, and N. Angelino, *Tetrahedron* **26**, 5415 (1970).
[7] W. Korytnyk and H. Ahrens, *J. Heterocycl. Chem.* **7**, 1013 (1970).
[8] W. Korytnyk, G. B. Grindey, and B. Lachmann, *J. Med. Chem.* **16**, 865 (1973).
[9] A. C. Ghosh and W. Korytnyk, *Tetrahedron Lett.*, 4049 (1974).
[10] W. Korytnyk, A. C. Ghosh, P. G. G. Potti, and S. C. Srivastava, *J. Med. Chem.* **19**, 999 (1976).
[11] W. Korytnyk and P. G. G. Potti, *J. Med. Chem.* **20**, 567 (1977).
[12] W. Korytnyk, N. Angelino, B. Lachmann, and P. G. G. Potti, *J. Med. Chem.* **15**, 1262 (1972).
[13] F. E. Cole, B. Lachmann, and W. Korytnyk, *J. Heterocyclic. Chem.* **9**, 1129 (1972).

2. Methods for the modification of positions 2 and 4 either simultaneously or stepwise.[18]

3. Methods for systematic oxidative modification of the 4- and 5-hydroxymethyl groups resulting in the synthesis of some metabolites and their analogs[19]; 5-homologs have also been modified in this manner.[15]

Biologically Active Antagonists

1. Transition-state analogs of vitamin B$_6$ are considered to be 4-deformyl-4-vinylpyridoxal (4-VPAL, II-3) and its congeners. These analogs are effective antagonists of vitamin B$_6$ in several systems; they also compete with pyridoxal for pyridoxal phosphokinase-catalyzed phosphorylation.[20]

2. The phosphorylated analog of 4-VPAL and its derivatives and the corresponding 4-ethynyl derivative (II-14) have been found to be potent inhibitors of pyridoxine phosphate oxidase.[11,20].

3. The 4-ethynyl analog (II-14) was found to be a cofactor-site-directed irreversible inhibitor of apo-aspartate aminotransferase.[21]

4. Several analogs of pyridoxol phosphate modified in position 5 were found to be substrates of pyridoxine phosphate oxidase indicating structural features for such activity.[16]

5. Several pyridoxol analogs modified at position 5 were found to be substrates of pyridoxine dehydrogenase.[12]

6. The enzyme activities of analogs modified in position 4 have been correlated with their inhibitory activities against cells in culture.[20]

Several methodologies, such as thin-layer chromatography (TLC), nuclear magnetic resonance (NMR), mass spectrometry, and gas chromatography have been reviewed in Volume 18A of the series. NMR resonances are expressed in parts per million (ppm) units (δ) from Me$_4$Si, or from dioxane ($\delta 3.70$), as an internal standard.

In naming the compounds, preference is given to semisystematic names that relate them to vitamin B$_6$.

[14] W. Korytnyk and B. Lachmann, *J. Med. Chem.* **14,** 641 (1971).

[15] W. Korytnyk and H. Ahrens, *J. Med. Chem.* **14,** 947 (1971).

[16] W. Korytnyk, B. Lachmann, and N. Angelino, *Biochemistry* **11,** 722 (1972).

[17] W. Korytnyk and S. C. Srivastava, *J. Med. Chem.* **16,** 638 (1973).

[18] W. Korytnyk and P. G. G. Potti, *J. Med. Chem.* **20,** 1 (1977).

[19] B. Paul and W. Korytnyk, *J. Heterocycl. Chem.* **13,** 701 (1976).

[20] W. Korytnyk, M. T. Hakala, P. G. G. Potti, N. Angelino, and S. C. Chang, *Biochemistry* **15,** 5458 (1976).

[21] I. Yang, C. M. Harris, D. E. Metzler, W. Korytnyk, B. Lachmann, and P. G. G. Potti, *J. Biol. Chem.* **250,** 2917 (1975).

Compounds are designated either by Roman numerals alone or by a Roman and Arabic numeral joined by a hyphen. The latter designation refers to the listing of the compound in the table that is indicated by the Roman numeral.

Acetonides of Pyridoxol

An improved method for the synthesis of the two acetonides (II and III) of pyridoxol (I) has been developed.[3]

$\alpha^4,3\text{-}O\text{-}Isopropylidenepyridoxol$ (II). To a suspension of pyridoxol (I) hydrochloride (1.0 g, 4.86 mmol) in Me_2CO (25 ml, dried over $CaSO_4$) and 2,2-dimethoxypropane (15 ml), p-toluenesulfonic acid (3.7 g, 19.4 mmol) is added, and the mixture is stirred for 16 hr. The resulting dark-brown solution is neutralized with K_2CO_3 and then concentrated in vacuo. After extraction with $CHCl_3$, washing with H_2O, and drying (Na_2SO_4), the $CHCl_3$ is removed, and the residue is crystallized from EtOH-Et_2O: yield 0.64 g (63%). The preparation has been scaled up, starting with 41 g (0.2 mol) of pyridoxol hydrochloride, and yields 31.8 g (76%) of (II).

$\alpha^4, \alpha^5\text{-}O\text{-}Isopropylidenepyridoxol$ (III). To pyridoxol (I, free base, 100 mg, 0.59 mmol) in Me_2CO (2 ml, dried over $CaSO_4$), 0.5 ml of 2,2-dimethoxypropane and finally 112 mg (0.59 mmol) of p-toluenesulfonic acid are added, and the solution is stirred for 20 hr. The resulting precipitate is filtered, washed with a little Me_2CO, and poured into a cold saturated $NaHCO_3$ solution while it is being stirred. The resulting solution is extracted with EtOAc continuously for 6 hr, and the EtOAc is dried

(Na$_2$SO$_4$) and evaporated. TLC gives a single product: m.p. 182°; yield 120 mg (94%).

Modification of Position 2

The key intermediate 3-*O*-benzyl-2-formyl-α^4,α^5-isopropylidene-2-norpyridoxol (VIII) is obtained by the following steps.[3] The advantage of this approach is the mildness of deprotection (typically, heating with 1 N HCl on a steam bath for 1 hr), which allows the introduction of reactive groups into position 2.

3-O-Benzylpyridoxol (IV). Pyridoxol (I) hydrochloride (2.0 g, 0.97 mmol) is added to a solution of NaOEt [from Na (0.5 g) in 25 ml of absolute EtOH], and the resulting solution is stirred at room temperature for 1 hr. The latter solution is cooled in ice, and a solution of dimethylphenyl-benzylammonium chloride (3 g) in EtOH (25 ml) is added slowly, with stirring. The mixture is stirred at room temperature for 2 hr and is then evaporated *in vacuo*. Twice, dry C_6H_6 is added to the residue and evaporated to remove traces of EtOH. Dry benzene (25 ml) is added to the residue, and the reaction mixture is gently refluxed for 0.5 hr and is then evaporated completely *in vacuo*. Water (25 ml) is added, and the mixture is extracted several times with $CHCl_3$. The combined $CHCl_3$ extracts are dried over $CaSO_4$ and evaporated *in vacuo*, when crystalline material separates out. It is cooled, filtered, washed with a little Et_2O, and dried. Yield 1.43 g (57%); m.p. 113°–114°. The hydrochloride has a melting point of 198°–199°.

3-O-Benzyl-α^4, α^5-O-isopropylidenepyridoxol (V). To a solution of (IV) (9.2 g, 35.5 mmol) in Me_2CO (dried over $CaSO_4$, 100 ml), 2,2-dimethoxypropane (15 ml), and *p*-toluenesulfonic acid (6.98 g, 36.8 mmol) are added and the mixture is stirred at room temperature for 4.5 days, until only a trace of the starting material can be detected by TLC. The reaction mixture is poured into ice-cold 10% aqueous Na_2CO_3 with rapid stirring, and the alkaline solution is extracted with $CHCl_3$, washed with H_2O, dried (Na_2SO_4), and evaporated. The oily product does not crystallize and is used for preparing the N-oxide.

3-O-Benzyl-α^4, α^5-O-isopropylidenepyridoxol N-Oxide (VI). The oily product (V) from the previous experiment is dissolved in $CHCl_3$ (100 ml, dried over $CaSO_4$), and the solution is cooled to 0°. To this solution, rapidly stirring, is added *m*-chloroperbenzoic acid (7.22 g, 0.0355 mol on the basis of 85% purity of the reagent). The reaction mixture is stirred at room temperature for 2 hr until no starting material is detected by TLC, then washed with Na_2SO_4 (10%, 2 × 100 ml), $NaHCO_3$ (5%, 2 × 100 ml), and H_2O. After drying (Na_2SO_4), $CHCl_3$ is evaporated; the residual oil is used in the next experiment. Part of the oil is crystallized (Et_2O-petroleum ether) and recrystallized (EtOH-Et_2O): m.p. 109°–112°.

3-O-Benzyl-α^4, α^5-isopropylidene-α^2-hydroxypyridoxol (VII). A solution of (VI) (the total amount of the oily material from the preceding experiment) in CH_2Cl_2 (dried over $CaSO_4$, 75 ml) is cooled (ice bath). Trifluoroacetic anhydride (2.0 ml) is added while the solution is being stirred, and more (5.0 ml) is added after 5 min. The reaction mixture is stirred for 8 hr at room temperature, until no starting material can be detected by TLC. Then it is cooled to 0°, and MeOH (30 ml) is added while stirring is continued. The solvents are then evaporated. The oil is taken up in $CHCl_3$ and is washed with Na_2CO_3 (20%), followed by H_2O until neutral. It is then crystallized from EtOH–$CHCl_3$, yielding 8.24 g (74%, based on amount of IV).

3-O-Benzyl-2-formyl-α^4, α^5-isopropylidene-2-norpyridoxol (VIII). To a solution of (VII) (630 mg, 2 mmol in $CHCl_3$ (60 ml), MnO_2 (2.25 g) is added, and the suspension is stirred for 24 hr at room temperature. The reaction mixture is worked up in the usual way. The oily residue is crystallized from a mixture of Et$_2$O and petroleum ether, yielding 500 mg (80%): m.p. 52°; IR $\nu_{max}^{(KBr)}$ 1705 cm^{-1} (C=O); NMR (CDCl$_3$) 5.08 (CH$_2$), 4.94 (CH$_2$), 4.86 (CH$_2$), 10.30 (CHO), 7.50 (C$_6$H$_5$), 8.40 (C$_6$—H).

Details of the synthesis of 2-vinyl-2-norpyridoxol (I-2) are given below.[3]

(VIII) $\xrightarrow{\text{CH}_2=P\phi_3}$ (IX) $\xrightarrow{\text{H}_3\text{O}^+}$ (I-2)

3-O-Benzyl-a[4]*, α*[5]*-O-isopropylidene-2-vinyl-2-norpyridoxol (IX)*. Triphenylmethylphosphonium bromide (900 mg, 3.43 mmol) is added under N$_2$ to a stirred and cooled (0°) mixture of a 1.6 M solution of *n*-butyllithium (1.5 ml) and a solution of 0.5 g of a 1 : 1 mixture of potassium *tert*-butoxide and *tert*-butyl alcohol in 8 ml of Et$_2$O. The resulting mixture is stirred at room temperature for 2 hr. Then (VIII) (450 mg, 1.43 mmol) in Et$_2$O (3 ml) is added within 15 min, and stirring is continued for another 12 hr. The reaction mixture is filtered, and the filtrate is diluted with Et$_2$O. The dilute solution is treated with saturated NaHSO$_3$ and NH$_4$Cl solutions, and finally with H$_2$O. After drying (MgSO$_4$), the solvent is evaporated to an oil, and the vinyl compound is separated from (C$_6$H$_4$)$_3$PO by chromatography on silica gel, using EtOAc as the eluent: yield 358 mg (80%); NMR (CDCl$_3$) δ1.47 [(CH$_3$)$_2$C], 4.89 (3 × CH$_2$), 5.43–6.67 (2-CH=CH$_2$), 7.00–7.73 (2-CH=CH$_2$), 7.50 (C$_6$H$_5$), 8.23 (C$_6$—H); IR ν_{max} (neat) 1605 cm^{-1} (C=C).

2-Vinyl-2-norpyridoxol (I-2) Hydrochloride. A solution of (IX) (300 mg) in 1 N HCl (10 ml) is heated on a stream bath for 1 hr. After evaporation *in vacuo*, the residue is evaporated several times with H$_2$O to remove excess HCl and benzyl alcohol: m.p. 201°–203° dec. from EtOH; yield 170 mg (77%); NMR (D$_2$O) δ4.78 (5-CH$_2$), 5.02 (4-CH$_2$), 5.90–6.50 (2-CH=CH$_2$), 6.83–7.35 (2-CH=CH$_2$), 8.27 (C$_6$—H); IR λ_{max} (KBr) 3340, 3200, 3100, (OH), 1612 cm^{-1} (C=C); UV λ_{max} (0.1 N HCl) 324 nm (ϵ 6485); UV λ_{max} (0.1 N NaOH) 353 nm (ϵ 6360); UV λ_{max} (pH 7) 367 nm (ϵ 6800).

The 2-amino (I-5) and 2-halogens (I-7 and I-8) analogs are synthesized as shown on p. 460[4]:

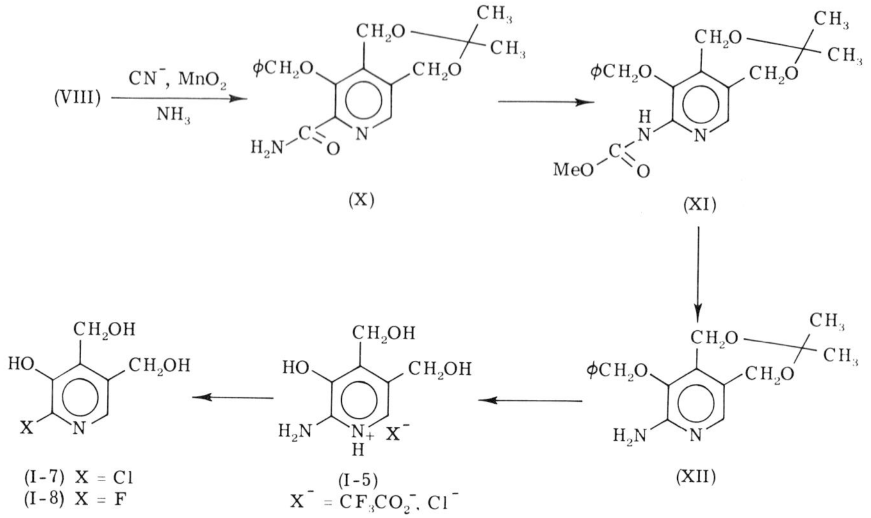

(X)   (XI)

(I-7) X = Cl
(I-8) X = F

(I-5)

$X^- = CF_3CO_2^-$. Cl^-

(XII)

As shown in the above scheme, the 2-formyl group in (VIII) is converted to a 2-carbamoyl group as in (X), by applying Gilman's cyanohydrin procedure. The carboxamide, on Hoffman reaction, gives the 2-methoxycarbonylamido derivative (XI), which on treatment with base gives the 2-amino compound (XII). Treatment of (XII) with HCl gives I-5, $X = Cl^-$, but the yield is unsatisfactory, and a better yield is obtained on treatment with trifluoroacetic acid [I-5 $X = CF_3CO_2$]. Reaction of I-5 with nitrous acid and HCl gives the 2-chloro analog I-7 (X = Cl) and with nitrous acid and HF the 2-fluoro analog I-8. As with other halogenated derivatives of vitamin B_6, the introduction of halogen abolishes the zwitterionic character of the parent compound, with the result that the analogs are soluble in organic solvents and extractable with ether.

Biological activities of these analogs are summarized in Table I, and details of syntheses are provided below.

3-O-Benzyl-α^4, α^5-O-isopropylidene-2-norpyridoxol-2-carboxamide (X). Compound (VIII) (220 mg) is suspended in 16 ml of isopropyl alcohol saturated with ammonia at 0°. Powdered NaCN (200 mg) is added, followed by MnO_2 "B" (900 mg). After stirring for 10 min, another portion of MnO_2 (900 mg) is added, and the suspension is stirred for 4 hr. After filtration, the solid is washed with chloroform (200 ml), and the chloroform layer is washed with water and dried (Na_2SO_4). The solvent is removed under vacuum, and the residue is crystallized from a mixture of Et_2O and petroleum ether: yield 180 mg (78%); m.p. 160°; IR ν_{max}^{KBr} 3425 (sharp NH stretching), 3200, 3280 (br), 1680 cm^{-1}(C=O).

3-O-Benzyl-α^4, α^5-isopropylidene-2-(methoxycarbonylamino)-2-norpy-

ridoxol (XI). A solution of (X) (80 mg, 2.44 mmol) in a mixture of MeOH (5 ml) and 0.1 N NaOH (2 ml) is added to 0.38 ml (2.68 mmol) of a commercial bleaching solution (5.25% NaOCl) and kept for 2–3 min at 50° while being stirred. The mixture is allowed to reach room temperature; after 30 min the solvent is evaporated, and the residue is extracted with ethyl acetate, washed with water, and dried (Na$_2$SO$_4$). After evaporation, the residue is dissolved in a few drops of Et$_2$O, and petroleum ether is added until turbidity develops. The product crystallizes: yield 66 mg (75%); m.p. 125°–129°; IR ν_{max}^{KBr} 3315 (NH), 1745 cm^{-1} (C=O); NMR (CDCl$_3$) 2 × CH$_3$ 1.41, OMe 3.63, 2 × CH$_2$ 4.78, CH$_2$ 4.88, Ph 7.43, C$_6$—H 7.88, NH 9.33.

2-Amino-3-O-benzyl-α^4,α^5-O-isopropylidene-2-norpyridoxol (XII). A solution of (XI) (0.23 g, 0.64 mmol) in MeOH (4 ml), to which 1 N NaOH (4 ml) is added, is refluxed for 1.5 hr on a steam bath. After the MeOH is evaporated, the compound is extracted with CHCl$_3$, and the chloroform layer is dried (Na$_2$SO$_4$). After evaporation of chloroform, anhydrous ether is added, and the compound crystallizes: yield 0.177 g (92%). The compound is recrystallized from a mixture of ether and petroleum ether: m.p. 111.5°–114°; IR ν_{max}^{KBr} 3480, 3458, 3280 (br), 1630 cm^{-1}; NMR (CDCl$_3$) 2 × CH$_3$ 1.43, 2-NH$_2$ and 3 × CH$_2$ 4.70, 4.66, 4.82, Ph 7.39, C$_6$—H, 7.59.

2-Amino-2-norpyridoxol Trifluoroacetate (2-Amino-3-hydroxy-4,5-pyridinedimethanol Trifluoroacetate) (I-5). A suspension of (XII) (0.165 g, 0.55 mmol) in trifluoroacetic acid (TFA) is stirred for 7.5 hr at room temperature and then heated on a steam bath for 0.5 hr. After evaporating the trifluoroacetic acid, the residue is dissolved in water, and the solution is extracted with chloroform. The aqueous layer is evaporated, and the residue is taken up in ethyl acetate. On the addition of ether, the compound crystallizes. Recrystallization from a mixture of ether and petroleum ether yields 66 mg (44.8%): m.p. 122°–124°.

2-Chloro-2-norpyridoxol (2-Chloro-3-hydroxy-4,5-pyridinedimethanol) (I-7). A solution of (I-5, X$^-$ = CF$_3$CO$_2^-$) (410 mg, 1.53 mmol) in concentrated HCl (4 ml) is cooled to −10° and stirred; NaNO$_2$ (130 mg, 1.89 mmol) is added in small portions over a period of 5–10 min, with stirring. Stirring is continued for another 10 min and then for 1 hr at 0°. The solution is evaporated to dryness, the residue is dissolved in water, and the solution is extracted four times with ether. After evaporation of ether, the residue is dissolved in chloroform, and then some ether is added. On addition of petroleum ether, the compound crystallizes. The yield is 72 mg (24.7%), m.p. 113°–118°, after recrystallization from a mixture of EtOH, ether, and petroleum ether.

2-Fluoro-2-norpyridoxol (2-Fluoro-3-hydroxy-4,5-pyridinedimethanol)

TABLE I
VITAMIN B_6 ANALOGS MODIFIED IN POSITION 2

No.	Compound	Structure	Synthesis[a]	Biological activity[a]
I-1	2-Formyl-2-norpyridoxol	R_2 = CHO	3	Inhibits TA3 cells[b] (ID_{50} 2×10^{-4} M)[3]
I-2	2-Vinyl-2-norpyridoxol	R_2 = CH=CH$_2$	3,[ ]	Inhibits TA3 cells[b] (ID_{50} 9×10^{-6} M; not reversed by pyridoxal)[3]
I-3	2-Ethynyl-2-norpyridoxol	R_2 = C≡CH	3	Inhibits TA3 cells[b] (ID_{50} 2×10^{-4} M)[3]
I-4	2-(β-Chloro)vinyl	R_2 = CH=CHCl	3	Inhibits TA3 cells[b] (ID_{50} 1×10^{-4} M)[3]
I-5	2-Amino-2-norpyridoxol	R_2 = NH$_2$	4,[ ]	Inhibits TA3 cells[b] (ID_{50} 1×10^{-5} M; reversed by pyridoxal)[4]

I-6	2-(N-Methylamino)-2-norpyridoxol	R_2 = NHCH$_3$	4	Inhibits TA3 cells[b] (15% at 1×10^{-4} M)[4]
I-7	2-Chloro-2-norpyridoxol	R_2 = Cl	4,[ ]	Inhibits pyridoxal phosphokinase (K_1 24 μM, competitively)[4]; inhibits TA-3 cells[b] (ID$_{50}$ 4.5 $\times$ 10^{-5} M)[4]
I-8	6-Fluoro-2-norpyridoxol	R_2 = F	4,[ ]	
I-9	α²-Aminopyridoxol	R_2 = CH$_2$NH$_2$	4,[ ]	Inhibits TA-3 cells[b] (25% inhibition at 1×10^{-5} M)[4]
I-10	2-Norpyridoxol-2-carboxylic acid	R_2 = CO$_2$H	4	—
I-11	2-Norpyridoxol-2-carboxamide	R_2 = CONH$_2$	4	Inhibits TA-3 cells[b] (40% at 1×10^{-4} M)
I-12	2-Norpyridoxol-2-carboxaldehyde hydrazone	R_2 = CH=NHNH$_2$	4	Inhibits TA-3 cells[b] (K_1 2.3 $\times$ 10^{-4} M)[4]
I-13	2-Norpyridoxol-2-carboxaldehyde azine	R_2 = (CH=N)$_2$	4	Inhibits TA-3 cells[b] (K_1 1.6 $\times$ 10^{-5}); not reversed by pyridoxal[4]

[a] Numbers refer to text footnotes: on line in column 4; superscripts in column 5. [ ] refers to the synthesis described in this chapter.
[b] Mouse mammary adenocarcinoma cells in culture; see footnote 20.

(*I-8*). To a solution of I-5 (216 mg) in concentrated aqueous HF solution (48%, 3 ml), $NaNO_2$ (73 mg, 1.05 mmol) is gradually added while the solution is being stirred in a polyethylene centrifuge tube at $-10°$ (ice–salt bath) for about 1 hr. During this time, the solution is allowed to reach room temperature gradually. After neutralization with 5 *N* NaOH, the solution is extracted several times with ether. The ether extract is chromatographed on an Amberlite CG-50 column (H^+ form, 200–400 mesh, 1.3 × 15 cm), using water as the eluent. The combined fractions are evaporated and then crystallized from a mixture of ether and petroleum ether. Yield: 55 mg (35.7%); m.p. 70°–75°.

Modification of Position 4

The most general and convenient method for the modification of position 4 starts with 3,α^5-*O*-dibenzylpyridoxal (XIII), the synthesis of which has already been described in this series.[2] The application of the Wittig reaction to this intermediate affords (XIV), and the subsequent debenzylation affords the 4-deformyl-4-vinyl-pyridoxal[8] and several of its α,β-unsaturated congeners[11,20] (XV). Experimental details for the Wittig reaction have been described in the preceding section for (IX). The debenzylation step is accomplished either by heating with dilute hydrochloric acid or, in the case of sensitive groups, with neat trifluoroacetic acid.

In general, the Wittig reaction produces both geometric isomers, which have been separated by crystallization in the case of the methyl vinyl derivatives.[20] However, other derivatives have not been separated and have been tested as mixtures of the cis and trans isomers for their inhibitory activity.[11,20]

Phosphorylation of the analog (XV) to (XVI) is carried out with polyphosphoric acid and followed by hydrolysis with 1 N HCl.[20]

The 4-ethynyl analog (II-13) and its 5′-phosphate (II-14) are obtained by essentially similar series of steps. The blocked ethynyl derivative is obtained from the β-chlorovinyl derivative (XVII), which on treatment with NaNH$_2$ in liquid NH$_3$ is dehydrochlorinated to (XVIII).

3, α^5-O-Dibenzyl-4-(β-chlorovinyl)-4-deformylpyridoxal (XVII). Compound (XIII)[2] (1.5 g, 4.3 mmol) is allowed to react with chloromethylene-triphenylphosphorane (generated from chloromethyltriphenylphosphonium chloride, 2.3 g, 6.7 mmol) according to the general procedure of the Wittig reaction described in the case of the corresponding 2-vinyl compound (IX) to give 3,α^5-O-dibenzyl-4-(β-chlorovinyl)-4-deformylpyridoxal isomers; NMR (CDCl$_3$, cis and trans isomers) 2-CH$_3$ 2.53, 3 × CH$_2$ 4.32, 4.38, 4.42, 4.63 to 5.00; CH=CH 5.38 to 7.17 (complex), 2 × C$_6$H$_5$ 7.40 to 7.45, C$_6$H 8.38; IR $\bar{\nu}_{max}^{neat}$ 1605 cm^{-1} (C=C), 700 cm^{-1} (C—Cl); ultraviolet $\lambda_{max}^{95\% \; alcohol}$ 253 nm.

3, α^5-O-Dibenzyl-4-(β-chlorovinyl)-4-deformylpyridoxal (XVII) Hydrochloride. Compound (XVII) (50 mg, 0.13 mmol) is converted into its hydrochloride by treatment with ethereal HCl. The hydrochloride is crystallized from alcohol–ether (45 mg, yield 82%, m.p. 128°–130° with decomposition); NMR (CDCl$_3$) 2-CH$_3$ 2.78, 3 × CH$_2$ 4.70, 4.73 to 5.02,

$2 \times C_6H_5$ 7.95, CH=CH$_2$ 5.87 to 7.18 (complex) C$_6$H 8.47. IR $\bar{\nu}_{max}^{KBr}$ 1610 cm^{-1} (C=C); ultraviolet $\lambda_{max}^{95\% \ alcohol}$ 253 nm.

3, α⁵-O-Dibenzyl-4-ethynyl-4-deformylpyridoxal (XVIII). Sodamide (4 mmol) in liquid ammonia is prepared and kept cooled in Dry Ice–alcohol. Compound (XVII) (450 mg, 1.18 mmol) in anhydrous ether (10 ml) is added dropwise during 30 min while the reaction mixture is stirred vigorously. Stirring is continued overnight at room temperature. The sodium acetylide formed during the reaction is carefully hydrolyzed with ammonium chloride solution. The ethynyl compound (XVIII) is extracted with ether, the ether layer is washed with water, dried over anhydrous MgSO$_4$, and filtered. The filtrate is evaporated to an oil (305 mg, yield 73%); NMR (CDCl$_3$) 2-CH$_3$ 2.47, —C≡CH, 3.68, 3 × CH$_2$ 4.67, 4.70, 5.27, 2 × C$_6$H$_5$ 7.43, 7.47, 6-H 8.50. IR $\bar{\nu}_{max}^{neat}$ 2100 cm^{-1} (C=C), 3120 cm^{-1} (≡C—H).

3, α⁵-O-Dibenzyl-4-ethynyl-4-deformylpyridoxal (XVIII) Hydrochloride. The free base (XVIII) (30 mg, 0.87 mmol) is taken into anhydrous ether (2 ml), and ethereal HCl is added. Excess HCl is removed by evaporation under vacuum. The hydrochloride is crystallized from acetone–ether (m.p. 120°–122° with decomposition, 23 mg; yield 69%); NMR (CDCl$_3$) 2-CH$_3$ 2.55, C≡CH, 3.72, 3 × CH$_2$ 4.73, 4.83, 5.18, 2 × C$_6$H$_5$ 7.45, 7.47, 6-H 8.53. IR $\bar{\nu}_{max}^{KBr}$ 2100 cm^{-1} (C=C), 3130 cm^{-1} (≡C—H).

4-Ethynyl-4-deformylpyridoxal (II-13). Compound (XVIII) (150 mg, 0.39 mmol) is taken up in trifluoroacetic acid (15 ml) and refluxed for 15 hr. The acid is removed by evaporation, and the trifluoroacetate salt is crystallized from ethyl acetate–petroleum ether (m.p. 125°–127° with decomposition, 100 mg, yield 90%); NMR (D$_2$O), 2-CH$_3$ 2.60, —C≡CH 4.65, 5-CH$_2$ 4.80, 6-H 8.20. IR $\bar{\nu}_{max}^{KBr}$ 3380 cm^{-1}, 3310 cm^{-1} (OH), 3110 cm^{-1} (C=C), 1685 cm^{-1} (C=O), m.s. molecular ion peak at 163 agrees with molecular weight —CF$_3$COOH.

4-Ethynyl-4-deformylpyridoxal phosphate (II-14). 4-Ethynyl-4-deformylpyridoxal (II-13) trifluoroacetate (50 mg, 0.18 mmol) is taken up in a phosphorylating mixture (0.2 ml) consisting of 1 part of P$_2$O$_5$ and 1.3 parts of H$_3$PO$_4$ and heated to 60°–65° for 4 hr in a oil bath. It is shaken at frequent intervals. Water (1 ml) is added and heating is continued on a boiling water bath for 30 min. HCl (1 N, 3 ml) is then added, and the solution is heated on a steam bath for another 30 min. It is cooled and diluted to 10 ml. Charcoal is added (2 g), and the solution is stirred well. After 10 min, it is filtered using filter paper, and the residue is washed with water (20 ml). The absorbed phosphate ester is eluted with 2% ammonia until the washings give no absorption at 350 nm. The filtrate is evaporated to 1 ml and applied to an Amberlite column (CG-50 H$^+$ form) and the compound is eluted with water. The fractions showing the same ultraviolet absorption spectra are combined and evaporated to dryness. Compound II-14 is crystallized from water–acetone (m.p. 164°–168° with decomposition, 26 mg, yield 55%). This material is found to contain $\frac{1}{2}$ molecule of NH₄OH (possibly as a salt) as shown by analytical data; NMR (D₂O), 2 CH₃ 2.42, —C≡C—H 4.30, 5-CH₂, 4.92, 5.04 (doublet $J_{P,H}$ = 7.5 Hz), 6-H 7.68. IR $\bar{\nu}_{max}^{KBr}$ 3260 cm⁻¹ (≡C—H); 2110 cm⁻¹ (C≡C), 1150 cm⁻¹ (P═O), 1035 cm⁻¹ (P—O—C).

4-Amino-4-norpyridoxol (II-1) is another useful intermediate for the modification of the 4-position. It has been obtained[10] from 3,5-di-O-benzylpyridoxal (XIII). Advantage has been taken of the method developed by Gilman for the conversion of aldehydes to amides using the oxidation of the initially formed cyanohydrin and subsequent reaction with ammonia. The amide thus obtained has been subjected to the Hoffman reaction, to give the blocked 4-amino-4-norpyridoxol. Treatment of the latter with HCl gives the 4-amino-4-norpyridoxol (II-1), which has also been obtained photochemically from pyridoxal oxime.[9] A modified Sandmeyer reaction on 4-amino-4-norpyridoxol gives the 4-bromo analog. Details of the synthesis are analogous to that described for position 2 (VIII → X → XI → XII → I-5 → I-7) in the preceding section.

Attempts to obtain the cyanohydrin of pyridoxal (XIX) have not been successful. Instead of the expected product, two cyclic derivatives are obtained (XX and XXI, respectively) depending on the pH of the experiment.[22]

Both products break down to pyridoxic acid or its lactone.[22] This breakdown is the basis of one of the standard fluorometric methods for the determination of pyridoxal phosphate in biological material.[23]

In Table II, the synthesis and biological activity of vitamin B₆ analogs modified on the position 4 are summarized.

Modification of Position 5

Modification of position 5 has been explored most extensively, primarily because of the ease of preparation of the key intermediate, α^4,3-O-isopropylidene pyridoxol (II).[3]

[22] W. Korytnyk, H. Ahrens, N. Angelino, and G. Kartha, *J. Org. Chem.* **38**, 3793 (1973).
[23] Z. Tamura and S. Takanishi, this series, Vol. 18A, p. 471.

TABLE II. VITAMIN B_6 ANALOGS MODIFIED IN POSITION 4

No.	Compound	Structure	Synthesis[a]	Biological activity[a]
II-1	4-Amino-4-norpyridoxol	$R_4 = NH_2$	10	Inhibits TA3 cells[b] $(K_1\ 10^{-5}\ M)$
II-2	4-Bromo-4-norpyridoxol	$R_4 = Br$	10	—
II-3	4-Vinyl-4-deformylpyridoxal	$R_4 = CH=CH_2$	8	Substrate of pyridoxal phosphokinase[20]; vitamin B_6 antagonist in mice[8] and cells in culture[20]; potentiates inhibitory activity on serine transhydroxymethylase in the spleen tissue of mice[c]
II-4	4-Vinyl-4-deformylpyridoxal 5'-phosphate		20	Inhibitor of pyridoxine phosphate oxidase[20]; binds to apoaspartate transaminase[21] and apo-L-arginine decarboxylase[d]
II-5	3-O-Methyl-4-vinyl-4-deformylpyridoxal		20	Substrate of pyridoxal phosphokinase[20]
II-6	4-Vinyl-4-deformylpyridoxal N-Oxide		20	Inhibits growth of TA3 cells[b]; some substrate activity for pyridoxal phosphokinase[20]

II-7	4-Vinyl-4-deformyl-5'-deoxypyridoxal	(structure: CH$_2$=HC, HO, CH$_3$, H$_3$C, N)	20	Inhibitor of pyridoxal phosphokinase[20]
II-8	4-Ethyl-3-hydroxy-5-(hydroxymethyl)-2-methylpyridine	R$_4$ = CH$_2$CH$_3$	20	Substrate of pyridoxal phosphokinase[20]
II-9	4-Propyl-4-deformylpyridoxal	R$_4$ = CH$_2$CH$_2$CH$_3$	20	Inhibits TA3 cells[b]
II-10	α^4-Homopyridoxol	R$_4$ = CH$_2$CH$_2$OH	11	Not tested
II-11	α^4-Methylpyridoxal	R$_4$ = C=O(CH$_3$)	11	Not tested
II-12	α^4-(1-Chlorovinyl)-4-deformylpyridoxal	R$_4$ = CCl=CH$_2$	11	Inhibits TA3 cells[b] (K_1 9 × 10^{-6} M)
II-13	4-Ethynyl-4-deformylpyridoxal	R$_4$ = C≡CH	11, 21, [ ]	Inhibits pyridoxal phosphokinase[11]; inhibitor of TA3 cells[b] (ID$_{50}$ 8 × 10^{-6} M)[11]
II-14	4-Ethynyl-4-deformylpyridoxal 5'-phosphate	(structure: CH≡C, HO, CH$_2$OPO$_3$H$_2$, H$_3$C, N)	21,[ ]	Reacts irreversibly with apoaspartate aminotransferase,[21] inhibits pyridoxine phosphate oxidase[11]
II-15	N-Acetyl-2-(α^4-pyridoxylidene)glycine	R$_4$ = CH=CHNHAcCO$_2$H	7	
II-16	2-(α^4-Pyridoxyl)-2-oxacetic acid	R$_4$ = CH$_2$COCO$_2$H	7	

[a] Numbers refer to text footnotes: on line in column 4; subscripts in column 5. [ ] refers to the syntheses described in this chapter.

[b] Mouse mammary adenocarcinoma cells in culture, see footnote 20.

[c] Y. V. Bukin, V. A. Draudin-Krylenko, and W. Korytnyk. *Biochem. Pharmacol.* (1979). In press.

[d] E. E. Snell, personal communication, see footnote 20.

TABLE III

VITAMIN B_6 ANALOGS MODIFIED IN POSITION 5

Common structure:

No.	Compound	Structure	Synthesis[a]	Biological activity[a]
III-1	3-Hydroxy-4-(hydroxymethyl)-2-methyl-5-vinylpyridine	$R_5 = CH{=}CH_2$	12	Substrate of pyridoxine dehydrogenase[12]
III-2	α^5-Deoxy-α^5-methylidyne-pyridoxol	$R_5 = C{\equiv}CH$	12	Substrate of pyridoxine dehydrogenase[12]
III-3	3-Chloro-1-(5-pyridoxyl)-2-propanone	$R_5 = CH_2CH_2COCH_2Cl$	14	Inhibits apotryptophase[14]
III-4	1-C5'-Homopyridoxol phosphate	$R_5 = CH_2CH_2O{-}\overset{\overset{\displaystyle O}{\|}}{P}{-}OH$, O^-	16	Substrate of pyridoxine phosphate oxidase[16]
III-5	DL-5'-Methylpyridoxol phosphate	$R_5 = \overset{CH_3}{\underset{H}{\overset{\|}{C}}}{-}O{-}\overset{\overset{\displaystyle O}{\|}}{P}{-}OH$, O^-	16	Substrate of pyridoxine phosphate oxidase
III-6	Pyridoxol-5'-methyl phosphonate	$R_5 = CH_2O{-}\overset{\|}{P}{-}CH$, O^-	13, 16	

[a] Numbers refer to text footnotes: on line in column 4; subscripts in column 5.

The phosphorylation of (II) using 2-cyanoethyl phosphate has been described in this series[2] and is applicable for the synthesis of analogs III-4, III-5, and III-6.[16]

The synthesis and biological activity of these analogs are summarized in Table III.

Substitutions in Position 6

Most of the 6-halogenated derivatives have been prepared.[17] Two methods of synthesis have been employed.

The direct method takes advantage of the activation of position 6 by the 3-OH group. An appropriate derivative, such as the α^4,α^5-O-isopropylidenepyridoxol (III) or the ethyl hemiacetal of pyridoxal is treated with a halogenating agent (e.g., *tert*-butyl hypochlorite).[17]

In the indirect method 6-aminopyridoxol serves as the starting material. The 6-fluoro derivative is obtained by a modified Schiemann reaction, and other halogenated analogs by a Sandmayer synthesis.[17]

Either of the methods is far simpler than the multistep synthesis of 6-chloropyridoxol reported by Blackwood *et al.*[24]

6-Chloropyridoxol (Table IV). An Example of the Direct Method

(III)   (XXII)   (IV-2)

To α^4,α^5-O-isopropylidenepyridoxol (III, 500 mg, 2.39 mmol) in *tert*-butyl alcohol (20 ml) is added *tert*-butyl hypochlorite (0.25 ml). The reaction mixture is stirred for 30 min, direct light being excluded. The solvent is distilled off *in vacuo,* and the yellowish residue is shaken with benzene, when some starting material (80 mg) separates out. The C$_6$H$_6$ solution is evaporated, and ether and petroleum ether are added to the residue when α^4,α^5-O-isopropylidene-6-chloropyridoxol (XXII) separates out; the yield is 400 mg (82%), m.p. 157°–159°.

The product is hydrolyzed by stirring it in 1 N HCl (25 ml) overnight. After removal of the solvent, the residue is crystallized from MeOH-Et$_2$O: yield of IV-2 is 350 mg; m.p. 188°.

[24] R. K. Blackwood, G. B. Hess, C. E. Larrabee, and F. J. Pilgrim, *J. Am. Chem. Soc.* **80,** 6244 (1958).

TABLE IV

VITAMIN B_6 ANALOGS MODIFIED IN POSITION 6

No.	Compound	Structure	Synthesis[a]	Biological activity
IV-1	6-Fluoropyridoxol	X = F	17, [ ]	Inhibits pyridoxal phosphokinase (K_i = 50 μM; competitively) Inhibits TA-3 cells[b]
IV-2	6-Chloropyridoxol	X = Cl	17, 24 [ ]	
IV-3	6-Bromopyridoxol	X = Br	17	
IV-4	6-Aminopyridoxol	X = NH_2	17, 25 [ ]	Inhibits TA-3 cells[b] in culture (ID_{50} 6 × 10^{-5} M)
IV-5	6-Fluoropyridoxal	R_4 = CHO, R_5 = H	17, [ ]	Inhibits TA-3[b] (ID_{50} 1 × 10^{-4} M) and Sarcoma 180 (ID_{50} 4 × 10^{-6} M) cells in culture; activates apophosphorylase b in the presence of inorganic phosphite[c]
IV-6	6-Fluoropyridoxamine	R_4 = CH_2NH_2, R_5 = H	17, [ ]	Inhibits TA3 cells[b] (ID_{50} 8 × 10^{-5} M)
IV-7	6-Fluoropyridoxamine 5′-phosphate	R_4 = CH_2NH_2, R_5 = PO_3H_2	17, [ ]	Inhibits TA3 cells[b] (ID_{50} 1 × 10^{-4} M)
IV-8	6-Fluoropyridoxal oxime	R_4 = CH=NOH, R_5 = H	17, [ ]	Inhibits TA3 cells[b] (ID_{50} 4 × 10^{-6} M)

[a] Numbers refer to text footnotes. [ ] refers to the synthesis described in this chapter.

[b] Mouse mammary adenocarcinoma cells in culture; see footnote 20.

[c] D. J. Graves, R. F. Parrish, R. J. Uhling, and W. Korytnyk, in "Regulatory Mechanisms of Carbohydrate Metabolism" (V. Esman, ed.),

Indirect Methods: 6-Fluoro Analogs.

The synthesis of 6-fluoro analogs of vitamin B$_6$, an example of an indirect method of synthesis, is shown in the reaction sequence below:

(XXIII)   →   (IV-4)   →   (IV-1)

(IV-5) (hemiacetal)   ⇌   (IV-5) (open)

(IV-7)   ←   (IV-6)   ←   (IV-8)

6-Aminopyridoxol (IV-4). This analog and intermediate is prepared from 6-phenylazopyridoxol (XXIII) by reduction as described by Katritzky *et al.*[25] In our hands, catalytic reduction (Pd/C) gives an improved yield of the pure product, but sodium dithionite reduction of the azo compound gives a product that cannot be readily purified from the residual inorganic salt.

To 6-phenylazopyridoxol (XXIII) (4.0 g, 14.6 mmol) in ethanol (400 ml), 10% Pd/C (500 mg) is added, and the compound is hydrogenated in a Parr hydrogenator for 20 hr at 25 psi.[2] The catalyst is filtered, the solvent is evaporated, and the residue is steam-distilled to remove aniline. The aqueous solution is concentrated and is acidified with 12 *N* HCl. The solid is crystallized from cold EtOH, giving a yield of 2.44 g (84%), m.p. 180°–185° dec.

6-Fluoropyridoxol (IV-1). Compound (IV-4) (722 mg, 3.49 mmol) is dissolved in 40% HBF$_4$ (16 ml), and the solution is cooled to −5°; solid

[25] A. R. Katritzky, H. Z. Kucharska, M. J. Tucker, and H. M. Wuest, *J. Med. Chem.* **9,** 620 (1966).

NaNO$_2$ (300 mg) is added slowly with stirring, and stirring is continued at 10° for 2 hr. After addition of 5 N NaOH at 10° until pH 3–4, the solution is extracted with Et$_2$O (six times), washed with H$_2$O, dried (Na$_2$SO$_4$), and concentrated. The crystalline material thus obtained is recrystallized from Et$_2$O or Me$_2$CO–Et$_2$O, yielding 225 mg (34%) of IV-1, m.p. 155°–156°.

6-Fluoropyridoxal (IV-5). To a solution of IV-1 (56.7 mg, 0.30 mmol) in H$_2$O (20 ml), through which N$_2$ is bubbled to prevent overoxidation, MnO$_2$ (220 mg) is added, and the mixture is stirred for 2.5 hr. After filtration the solution is concentrated and worked up by either of the following two methods. (A) The solution is streaked on a preparative TLC plate, the streak is developed with ethyl acetate, and the yellow zone is eluted with a large quantity of ether. On concentration of the ether solution, (IV-5) (41.2 mg) crystallizes. (B) Trituration of the yellow residue with water filtration and recrystallization of the residual crystals from warm ether gives 32 mg of (IV-5), m.p. 140°–141°.

6-Fluoropyridoxal Oxime (IV-8). The reaction product from oxidation of IV-1 (15 mg) with MnO$_2$ (50 mg) is heated with NH$_2$OH·Cl (25 mg) and NaOAc (50 mg); the mixture is stirred for 0.5 hr, and the precipitate is filtered off. After washing with water, the oxime is crystallized from ethyl alcohol: yield 11 mg; m.p. 235°–236° dec.

6-Fluoropyridoxamine Hydrochloride (IV-6). A solution of the oxime (IV-8) (35 mg, 0.17 mmol) in acetic acid (15 ml) is hydrogenated in the presence of 10% Pd/C (38 mg) for 20 hr at atmospheric pressure. The hydrogenation product is converted to the HCl salt by treatment with 6 N HCl, repeated evaporation with ethyl alcohol, and recrystallization from Et$_2$O-MeOH, giving 19 mg (48%) of IV-6, m.p. 190°–195° dec.

6-Fluoropyridoxamine Phosphate (IV-7). 6-Fluoropyridoxamine hydrochloride (IV-6) (60 mg, 0.27 mmol) and polyphosphoric acid (2 ml, prepared from 5.2 g of 85% H$_3$PO$_4$ and 4.0 g of P$_2$O$_5$) are mixed, and the mixture is heated on a steam bath at 60° for 2 hr. Then 1 N HCl (2.5 ml) is added, and the mixture is heated on a steam bath for 15 min. The solution is concentrated *in vacuo,* and concentrated NH$_4$OH is added to neutralize the acids. The product is chromatographed on an Amberlite CG 50(H$^+$) column with H$_2$O. Fractions with a UV absorption of 290 nm are combined, evaporated, and crystallized from EtOH. The last traces of NH$_4$Cl are removed by sublimation (at 5×10^{-5} mm), and the compound is crystallized from EtOH–Et$_2$O, yielding 42 mg (46%) of IV-7, m.p. 220°–225° dec.

Vitamin B$_6$ Analogs Modified in Positions 4 and 5.

In order to convert 5-modified pyridoxol analogs to analogs of the other forms of the vitamin (e.g., pyridoxal and its 5'-phosphate) or its metabolites, the 4-hydroxymethyl groups are oxidized to the aldehyde and carboxylic acid.[15] Manganese dioxide is used as the oxidizing agent of choice. Conditions for this oxidation have to be varied for each compound. The length of the side chain in 5-homologs has a profound effect on the oxidizability of the 4-CH$_2$OH group, probably because of the ring-chain tautomerism of these compounds.[15] Thus conditions[2,26] that have been worked out earlier for the oxidation of pyridoxol to pyridoxal and 4-pyridoxic acid cannot be applied.

Representative examples of oxidation to the 4-aldehydes and 4-carboxylic acids are provided.[15] (Table V).

α^5-*Homopyridoxal (V-1).* α^5-Pyridoxylmethanol·HCl (XXIV) (60 mg)[27] is dissolved in H$_2$O (20 ml), to which concentrated H$_2$SO$_4$ (0.02 ml) is added. After addition of MnO$_2$ (500 mg),[29] the reaction mixture is shaken vigorously for 70 sec, then the MnO$_2$ is filtered off and washed with H$_2$O (25 ml). (The time for the reaction and washing should not exceed 6 min.) The filtrate is extracted with CHCl$_3$ (3 × 50 ml), which removes α^5-homo-4-pyridoxic acid lactone (XXV), 9.8 mg (20%), m.p. 112°–114°.

The aqueous layer is evaporated *in vacuo* at 30°, dissolved in H$_2$O (1 ml), and carefully neutralized by the addition of solid NaHCO$_3$, when crystallization occurs. The yield is 36.5 mg (74%) of V-1 melting around 135°, which is recrystallized from H$_2$O –Me$_2$CO: TLC, 1:1 MeOH-CHCl$_3$, R_f 0.77, retarded by boric acid[28] to R_f 0.46; NMR (in 1 N NaOD), (2-CH$_3$) 2.29, (α^5-CH$_2$) 2.73 (tr $J = 6$ Hz), (β^5-CH$_2$) 3.20 (tr $J = 6$ Hz), (α^4-H) 6.35 (board), (C$_6$—H) 7.03; at a neutral pH peaks appear

(XXV)   (XXIV)   (V-1) (hemiacetal)

(V-3)

[26] H. Ahrens and W. Korytnyk, *J. Heterocylc. Chem.* **4**, 625 (1967).

[27] W. Korytnyk, B. Paul, A. Bloch, and C. A. Nichol, *J. Med. Chem.* **10**, 345 (1967).

[28] H. Ahrens and W. Korytnyk, *Anal. Biochem.* **30**, 413 (1969); this series, Vol. 18A, p. 489.

[29] O. Mancera, G. Rosenkranz, and F. Sondheimer, *J. Chem. Soc. (London)*, 2189 (1953).

TABLE V
VITAMIN B$_6$ ANALOGS MODIFIED IN POSITIONS 4 AND 5

No.	Compound	Structure	Synthesis[a]	Biological activities[a]
V-1	α^5-Homopyridoxal		15, [ ]	Growth inhibition of *S. carlsbergensis* (ID$_{50}$ 5×10^{-5} M)[15]
	Oxime of V-1		15	Inhibits pyridoxal phosphokinase[15]
V-2	α^4-Deoxy-α^5-pyridoxylmethanol		15	Inhibitor of pyridoxal phosphokinase (K_1 5×10^{-4} M)[15]
V-3	3-Hydroxy-5-(2-hydroxyethyl)-2-methylpyridine-4-carboxylic acid		15, [ ]	—
V-4	1-C$^{5'}$-Homopyridoxal phosphate		16	Binds to apoglutamic acid decarboxylase, 5% cofactor activity; binds to apo-L-arginine decarboxylase, 10% cofactor activity; binds to apo-D-serine dehydratase, 8% cofactor activity[16,b]
V-5	3-Hydroxy-5-(3-hydroxypropyl)-2-methylpyridine-4-carboxaldehyde Oxime of V-5		15	Inhibitor of pyridoxal phosphokinase
V-6	5-(3-Hydroxypropyl)-2,4-dimethyl-3-pyridinol		15	—

		Structure	Ref.	
V-7	3-Hydroxy-5-(3-hydroxypropyl)-2-methylpyridine-4-carboxylic acid		15	—
V-8	α^5-Thiobenzoylpyridoxal		15	Growth inhibitor of *S. carlsbergensis* (ID_{50} 5×10^{-6} M)
V-9	α^5-Thiopyridoxal		16	—
V-10	DL-5'-Methylpyridoxal phosphate		16, [ ]	Binds to apo-D-serine dehydratase: 24% cofactor activity; binds to apoaspartate amino transferase and apoglutamic acid decarboxylase, 3% cofactor activity[16,b]
V-11	Pyridoxal 5'-Methylphosphonate		16	—
V-12	2-Methyl-3-hydroxy-5-formylpyridine-4-carboxylic acid		19	Metabolite of vitamin B₆[c]
V-13	2-Methyl-3-hydroxy-4-formylpyridine-5-carboxylic acid		19	—

(continued)

TABLE V (*Continued*)

No.	Compound	Structure	Synthesis[a]	Biological activities[a]
V-14	2-Methyl-3-hydroxypyridine-4,5-dicarboxylic acid		19	Metabolite of vitamin B_6[c]
V-15	5-(β-Chlorovinyl)-5-norpyridoxal		d	—

[a] Numbers refer to text footnotes: on line in column 4; subscripts in column 5. [   ] refers to the syntheses described in this chapter.
[b] E. E. Snell, *Vitam. Horm. (New York)* **28**, 265 (1970).
[c] R. W. Burg, V. W. Rodwell, and E. E. Snell, *J. Biol. Chem.* **235**, 6164 (1960).
[d] W. Korytnyk and P. G. G. Potti, unpublished results, 1978.

broadened, and α^4-H is at 6.07 ppm; UV, $\lambda_{max}^{0.1\,N\,NaOH}$ 218.5 (ϵ 6700); 247 (ϵ 5800), 310 nm (ϵ 5750); $\lambda_{max}^{0.1\,N\,HCl}$ 230 (ϵ 2500) shoulder, 292 nm (ϵ 6750); λ_{max}^{pH7} 255 (ϵ 4400), 312 (ϵ 7450); IR no C=O group (KBr pellet).

3-Hydroxy-5-(2-hydroxyethyl)-2-methylpyridine-4-carboxylic   Acid (4 → 5) Lactone (XXV). A solution of (XXIV) (200 mg) in H$_2$O (5 ml) is added to a mixture of MnO$_2$ (1.65 g, prepared according to the method of Mancera *et al.*[29]), H$_2$O (15 ml), and concentrated H$_2$SO$_4$ (0.1 ml) and is stirred for 1 hr at room temperature. MnO$_2$ is filtered off and washed with H$_2$O (30 ml). The filtrate is shaken with CHCl$_3$ (3 × 30 ml). The CHCl$_3$ extract is washed with H$_2$O (25 ml), dried (CaSO$_4$), and evaporated, giving 125 mg (76%) of lactone XXV, m.p. 119°. Recrystallization from Et$_2$O – petroleum ether gives the analysis sample: m.p. 120°; NMR (CDCl$_3$); (2-CH$_3$) 2.52, (α^5-CH$_2$) 3.05 (tr $J = 6.0$ Hz broad), (β^5-CH$_2$) 4.65 (tr $J = 6.0$ Hz, sharp), (C$_6$—H) 7.98 (broad), (3-OH) 10.57 (disappears on addition of D$_2$O); UV, $\lambda_{max}^{0.1\,N\,HCl}$ 247 (ϵ 2200, sh) 329 (ϵ 6550); IR, λ_{max}^{Nujol} 1700 cm^{-1} (C=O).

3-Hydroxy-5-(2-hydroxyethyl)-2-methylpyridine-5-carboxylic   Acid (V-3). 3-Hydroxy-5-(2-hydroxyethyl)-2-methylpyridine-4-carboxylic acid (4 → 5) lactone (XXV) (25.0 mg) is dissolved by gentle heating in 1 N NaOH (3 ml). After adding 0.1 N HCl to pH 8, the solution is evaporated to 1.0 ml. More 0.1 N HCl is added until, at pH 6, the acid crystallizes. After cooling in ice, the acid is filtered, yielding 16.5 mg (60%), m.p. 226°. Recrystallization from EtOH does not raise the melting point: NMR (in 1 N NaOD); (2-CH$_3$) 2.27, (α^5-CH$_2$) 2.67 (tr $J = 7.0$ Hz), (β^5-CH$_2$) (tr $J = 7.0$ Hz), (C$_6$—H) 7.33; $\lambda_{max}^{0.1\,N\,NaOH}$ 220 nm (ϵ 10,500), 347 (ϵ 5500), 308 nm (ϵ 6400).

(III-5)   (V-10)

DL-*5'-Methylpyridoxal Phosphate (V-10)*. A solution of (III-5) (90 mg, 0.35 mmol) is dissolved in H$_2$O (10 ml) and stirred for 100 min with MnO$_2$ (800 mg, prepared according to the method of Harnfeist *et al.*[30] The pH of the reaction mixture is adjusted to 7.3 with Amberlite CG-50 (Na$^+$) resin, and is filtered (Celite filter aid). The filtrate is concentrated, then chromatographed with an Amberlite CG-50 column (wrapped in black paper), as described earlier. The combined fractions are filtered and evaporated.

[30] M. Harnfeist, A. Baverley, and W. A. Lazier, *J. Org. Chem.* **19**, 1608 (1954).

The oily residue is heated with acetonitrile containing a few drops of H_2O. The oil changes to a yellow solid material (60 mg, 63%).

Details of the conversion of the 4-hydroxymethyl group to the methyl group, to give 4-deoxypyrodoxol analogs (V-2) and (V-6) using anhydrous hydrazine, have been provided.[2] Quinone methide intermediates (XXVIa–c) have been postulated based on the substitution of deuterium into the 2- and 4-methyl groups as well as the 6-position, following the reduction with fully deuterated hydrazine.[15]

(XXVIa)   (XXVIb)   (XXVIc)

Vitamin B$_6$ Analogs Modified in Positions 2 and 4

Since it has been shown that single modification of either position 2 or position 4 of the vitamin gives biologically active compounds, it could be anticipated that introduction of appropriate groups into positions 2 and 4 simultaneously would likewise result in potent antagonists of vitamin B$_6$.[18] In the case of two or more reactive groups there is the possibility of cross-reaction within the active site. Two vinyl groups have been introduced into positions 2 and 4 (Table VI) as shown on p. 481.

3-α⁵-O-Dibenzyl-4-deformyl-4-vinylpyridoxal N-Oxide (XXVII). A solution of 172 mg (0.5 mmol) of (XIV, R = H)[8] in $CHCl_3$ (5 ml) is added to a chloroform solution (5 ml) of m-chloroperbenzoic acid during 5 min. The reaction mixture is left stirring for another 30 min and then diluted with $CHCl_3$ and shaken with 20% Na_2SO_3 solution followed by 5% $NaHCO_3$ solution. The chloroform solution is washed with water, dried ($MgSO_4$), filtered, and evaporated to an oil; yield 174 mg (96%). NMR ($CDCl_3$) 2-CH_3 2.47, 3 × CH_2 4.53, 4.63, 4.85, 4-$CH=CH_2$ 5.56–6.17 (m), 4-$CH=CH_3$ 6.53–7.08 (m), 2 × C_6H_5 7.43, 7.47, C_6H 8.33, IR ν_{max}^{neat} 1490 ($C=C$), 1205 cm^{-5} ($N \rightarrow O$).

3, α⁵-O-Dibenzyl-α²-hydroxy-4-vinyl-4-de(hydroxymethyl)pyridoxol (XXVIII). The blocked N-oxide (XXVII) (890 mg, 2.46 mmol) is taken up in dry CH_2Cl_2 (3 ml), and the solution is cooled in an ice bath. To this is added $(CF_3CO)_2O$ (1 ml), and the mixture is left stirring for 45 min at room temperature. The reaction vessel is cooled in ice, and MeOH (5 ml) is added. The solvents are evaporated, and the residue is taken up in ethyl acetate, treated with 10% $NaHCO_3$ solution, and washed with water. It is dried over anhydrous $MgSO_4$, filtered, and evaporated to an oil, giving

(XXVIII) (820 mg, yield 95%); NMR (CDCl$_3$) 4.57, 4.58, 4.68, 4.85 (4CH$_2$), 5.53–6.17 (4-CH=CH$_2$), 6.63–7.33 (4-CH=CH$_2$), 7.40 (2C$_6$H$_5$), 8.38 (C$_6$H); IR ν_{max}^{KBr} 1655 cm^{-1}.

3, α⁵-O-Dibenzyl-2,4-divinyl-4-de(hydroxymethyl)-2-norpyridoxol (*XXIX*). The alcohol (XXVIII) (500 mg, 1.38 mmol) is taken up in dry CHCl$_3$ (27 ml), and MnO$_2$ "B" is added. The reaction mixture is stirred for 14 hr and then filtered with Celite filter aid. The residue is washed with fresh warm CHCl$_3$, and the combined filtrate is evaporated to the aldehyde (XXIX) (448 mg, yield 91%): NMR (CDCl$_3$) 4.63 (5-OCH$_2$), 4.98 (3-OCH$_2$), 5.55–6.13 (4-CH=CH$_2$), 6.57–7.17 (4-CH=CH$_2$), 7.38 (2C$_6$H$_5$), 8.35 (C$_6$H); IR ν_{max}^{KBr} 1710 (C=O), 1575 cm^{-1} (C=C).

3, α⁵-O-Dibenzyl-2,4-divinyl-4-de(hydroxymethyl)-2-norpyridoxol (*XXX*). The 4-vinyl-2-aldehyde (XXIX) (100 mg, 0.25 mmol) is allowed to react with the ylide generated from methyltriphenylphosphonium bromide (200 mg, 0.56 mmol) as described earlier for the synthesis of the 2-vinyl compound (IX), giving the divinyl derivative (XXX) (69 mg, yield 70%). NMR (CDCl$_3$) 4.65, 4.67, 4.90 (3 × CH$_2$), 5.42–6.50 (CH$_2$=CH), 7.33–7.40 (complex, CH=CH$_2$), 7.50–7.55 (2 × C$_6$H$_5$), 8.43 (C$_6$H); IR ν_{max}^{neat} 1580 cm^{-1} (C=C).

TABLE VI

VITAMIN B_6 ANALOGS MODIFIED IN POSITIONS 2 AND 4

No.	Compound	Structure	Synthesis[a]	Biological activity[a]
VI-1	2,4-Divinyl-4-de(hydroxy-methyl)-2-norpyridoxol	$R_2 = R_4 = CH=CH_2$; $R_5 = H$	18.[ ]	Inhibitor of TA3 cells[b] (ID_{50} 7×10^{-6} M, partially reversible)[18]
VI-2	2,4-Divinyl-4-deformyl-2-norpyridoxal 5'-phosphate	$R_2 = R_4 = CH=CH_2$; $R_5 = PO_3H_2$	18.[ ]	
VI-3	2-Vinyl-4-deoxy-2-nor-pyridoxol	$R_2 = CH=CH$; $R_4 = CH_3$; $R_5 = H$	18	Substrate of pyridoxal phosphokinase; as inhibitor K_i 68.1 μM (noncompetitive); inhibitor of TA3 cells[b] (ID_{50} 3×10^{-6} M; partially reversible)[18]
VI-4	2-Vinyl-4-deoxy-2-nor-pyridoxol 5'-phosphate	$R_2 = CH=CH_2$; $R_4 = CH_3$; $R_5 = PO_3H_2$	18	Inhibitor of TA3 cells[b] (ID_{50} 2×10^{-4} M; partially reversible)[18]
VI-5	2-Formyl-4-deoxy-2-norpyridoxol	$R_2 = CHO$; $R_4 = CH_3$; $R_5 = H$	18	Inhibitor of TA3 cells[b] (ID_{50} 6×10^{-5} M; partially reversible)[18]
	Thiosemicarbazone of VI-5	—	18	Inhibitor of TA3 cells[b] (ID_{50} 3×10^{-6} M; not reversible)[18]

[a] Numbers refer to text footnotes: on line is column 4; subscripts in column 5; [ ] refers to the synthesis described in this chapter.
[b] Mouse mammary adenocarcinoma cells in culture, see footnote 20.

2,4-Divinyl-4-de(hydroxymethyl)-2-norpyridoxol (VI-1). The blocked vinyl intermediate (XXX) (100 mg, 0.28 mmol) is refluxed with CF$_3$COOH (10 ml) for 15 hr. The acid is removed by evaporation under vacuum. Methanol is added, and the solution is evaporated again to dryness. The divinyl compound is purified by column chromatography (silica/CHCl$_3$–EtOAc) and the material is converted into its hydrochloride by treatment with alcoholic HCl. The hydrochloride is crystallized from alcohol–ether (m.p. 156°–158° dec, 37 mg, yield 64%): NMR (D$_2$O) 4.75 (5-CH$_2$), 5.67–7.33 (2CH$_2$), 5.67–7.33 (2CH$_2$, highly complex), 8.23 (C$_6$H).

2,4-Divinyl-4-deformyl-2-norpyridoxal 5'-Phosphate (VI-2). The divinyl derivative (VI-1) is phosphorylated with polyphosphoric acid as described earlier, for IV-6, giving the 5'-phosphate (m.p. 234°–242°, 23 mg, yield 36%); IR ν_{max}^{KBr} 3350 (OH), 1590 cm^{-1} (C=C).

[75] Vitamin B$_6$ Antagonists of Natural Origin[1]

By HAROLD J. KLOSTERMAN

Vitamin B$_6$ antagonists are generally one of two types: (a) structural analogs of a B$_6$ vitamer; or (b) carbonyl reagents—reactive hydrazines or hydroxylamines that are capable of forming hydrazones or oximes with pyridoxal or pyridoxal phosphate. A limited number of examples of substituted hydrazines and hydroxylamines have been reported to occur in plants or fungal secretions.[2]

Of the eight compounds listed in Table I, only canaline (VI) and cycloserine (VII) have been examined extensively as inhibitors of pyridoxal enzymes. A few studies have included 1-amino-D-proline (I), methylhydrazine (II), and O-amino-D-serine (VIII). The other carbonyl reagents listed in Table I (compounds III, IV, and V) should form hydrazones with pyridoxal and pyridoxal phosphate and should be antagonists for enzymes requiring pyridoxal phosphate, or for pyridoxal phosphokinase.

There are no known examples of naturally occurring vitamin B$_6$ antagonists which are structural analogs of the B$_6$ vitamers. The reaction product of a carbonyl reagent with pyridoxal or pyridoxal phosphate may be considered to be an analog of vitamin B$_6$. These substituted oximes and hydrazones of pyridoxal phosphate probably function by binding on the apoenzyme at the binding site for pyridoxal phosphate. In addition, the

---

[1] Financial support was provided by the North Dakota Agricultural Experiment Station.
[2] H. J. Klosterman, *J. Agric. Food Chem.* **22**, 13 (1974).

TABLE I
VITAMIN B_6 ANTAGONISTS OF NATURAL ORIGIN

Reactive agent	Precursor	Source	Reference[c]
A. Substituted hydrazines			
1-Amino-D-proline (I) (structure: pyrroline ring with COOH, N, NH₂)	Linatine (Ia) (structure: pyrroline ring with COOH, N, HN–γ-glutamyl)	*Linum usitatissimun*	1
Methylhydrazine (II) H_2NNHCH_3	Gyromitrin (IIa) $CH_3CH{=}N{-}N(CH_3)CHO$ (and homologs)	*Gyromitra esculenta*	2,3
α-(1-Methylhydrazino)acetic acid $H_2NN(CH_3)CH_2COOH$	Negamycin (IIIa) $H_2NCH_2CH(OH)CH_2CH_2CH(NH_2)CH_2CONHN(CH_3)COOH$	*Streptomyces purpeofuscus*	4
3-(1, 3, 5-Hexatrienyl)phenyl-hydrazine (IV)[a,b] (structure: benzene ring with H₂NNH and (CH=CH)₂CH=CH₂)	Spinamycin (IVa) (structure: benzene ring with HOOCCH₂CH₂CONHN and (CH=CH)₂CH=CH₂)	*Streptomyces albinospinus*	5
4-Hydroxymethylphenyl-hydrazine (V)[a] (structure: benzene ring with HOH₂C and NHNH₂)	Agaritine (Va) (structure: benzene ring with HOH₂C and NHNH–γ-glutamyl)	*Agaricus* spp.	6

B. Substituted hydroxylamines

L-Canaline (VI) $H_2NOCH_2CH_2CH(NH_2)COOH$	Canavanine (VIa) $H_2NC(=NH)NHOCH_2CH_2CH(NH_2)COOH$	Lotoidae spp.	7
D-Cycloserine (VII) 	—	*Actinomyces* spp.	8
O-Amino-D-serine (VIII) $H_2NOCH_2CH(NH_2)COOH$	D-Cycloserine (VII)	Rat urine	9

a Unstable compound; has not been isolated or synthesized.

b The hexahydroderivative has been prepared.[5]

c References: (1) H. J. Klosterman, G. L. Lamoureux, and J. L. Parsons, *Biochemistry* **6**, 170 (1967).
(2) P. H. List and P. Luft, *Arch. Pharmazie (Weinheim)* **302**, 143 (1969).
(3) H. Pyssalo, *Naturwissenschaften* **62**, 395 (1975).
(4) S. Shibahara, S. Kondo, K. Maeda, and H. Umezawa, *J. Am. Chem. Soc.* **94**, 4353 (1972).
(5) H. Naganawa, T. Takita, K. Maeda and H. Umezawa, *J. Antibiot. (Tokyo)* **21**, 241 (1968).
(6) B. Levenberg, *J. Biol. Chem.* **239**, 2267 (1964).
(7) J. D. Williamson and L. C. Orchard, *Life Sci.* **14**, 2481 (1974).
(8) F. A. Kuehl *et al.*, *J. Am. Chem. Soc.* **77**, 2344 (1955).
(9) T. Yasumitsu, T. Takao, and Y. Kakimoto, *Biochem. Pharmacol.* **25**, 253 (1976).

nature of the carbonyl reagent may influence the effectiveness of the hydrazone or oxime to function as an inhibitor. L-Cycloserine is a more effective inhibitor than D-cycloserine (VII) for the α-dialkylamino acid transaminases of *Pseudomonas fluorescens*[3] and *P. cepacia*.[4] The D- and L-forms of cycloserine react equally with enzyme-bound coenzyme, but the chirality of the cycloserine–coenzyme complex results in a more effective inhibitor from the L-form than from the D-form.

It may be expected that the enantiomers of aminoproline (I) and canaline (VI) would display differential effectiveness as inhibitors of pyridoxal enzymes, as was observed with D- and L-cycloserine. It may also be expected that the other carbonyl reagents in Table I would each demonstrate unique inhibitory properties when exposed to a variety of pyridoxal enzymes. Harik and Snyder[5] found that ornithine decarboxylase is inhibited more effectively by α-hydrazinoornithine than by other α-hydrazinoacids or alkylhydrazines.

The hydrazones and oximes of pyridoxal are generally not inhibitors of enzymes requiring pyridoxal phosphate as a coenzyme. These derivatives of pyridoxal effectively inhibit pyridoxal phosphokinase.[6]

Carbonyl reagents have been used to aid in the resolution of pyridoxal enzymes. With those enzymes that bind pyridoxal phosphate firmly, the apoenzyme may often be obtained by first allowing the holoenzyme to react with a carbonyl reagent followed by dialysis to remove excess reagent and the reagent–pyridoxal phosphate complex. Cycloserine has been most commonly used for this purpose. By adjusting the molar ratio of holoenzyme to carbonyl reagent (cycloserine) it is possible to distinguish between catalytically active coenzyme and coenzyme bound at sites that are not catalytically active.[7]

The reaction products between carbonyl reagents and pyridoxal enzymes have been established in some instances. With substituted hydrazines the products are the respective hydrazones of pyridoxal phosphate. It would appear that the free amino group of D-cycloserine (VII) would form a Schiff base with pyridoxal phosphate. However, Beeler and Churchich found that L-canaline (VI), D-cycloserine (VII), and aminoxyacetic acid form compounds with pyridoxal phosphate, all of which have identical fluorescence emission spectra, and they concluded that in each case the oxime derivative is formed.[8] With D-cycloserine, a rearrangement of the initial Schiff base product leads to the formation of the oxime of *O*-

[3] G. B. Bailey, O. Chotamangsa, and K. Vuttivej, *Biochemistry* **9**, 3243 (1970).
[4] C. A. Lamartiniere, H. Itoh, and W. B. Dempsey, *Biochemistry* **10**, 4783 (1971).
[5] S. I. Harik and S. H. Snyder, *Biochim. Biophys. Acta* **327**, 501 (1973).
[6] D. B. McCormick and E. E. Snell, *J. Biol. Chem.* **236**, 2085 (1961).
[7] J. E. Churchich and J. Bieler, *Biochim. Biophys. Acta* **229**, 813 (1971).
[8] T. Beeler and J. E. Churchich, *J. Biol. Chem.* **251**, 5267 (1976).

amino-D-serine (VIII).[9] The reaction product of pyridoxal and D-cycloserine, however, is the Schiff base, not the oxime.[10] O-Amino-D-serine has been found in rat urine as a metabolite of rats injected with D-cycloserine.[11]

Methods of synthesis have been described for 1-amino-D-proline (I) and the enantiomer,[12] α-(1-methylhydrazino)acetic acid (III),[13] L-canaline (VI) (from canavanine),[14] and O-amino-D-serine [from D-cycloserine (VII)].[15] D- and DL-cycloserine are available from commercial sources at moderate cost.

4-Hydroxymethylphenylhydrazine (V) is obtained by enzymic degradation of agaritine (Va), but is unstable and has been characterized as the glyoxylate derivative.[16] Presumably the derivative of (V) and pyridoxal phosphate could be prepared for enzyme studies. The structure of 3-(1,3,5-hexatrienyl)phenylhydrazine (IV) has been deduced from structural studies, but is extremely unstable and has not been prepared. The hexahydroderivative of (IV) is stable and has been isolated by the acid hydrolysis of hydrogenated spinamycin.[17]

Methods for the preparation of some of the naturally occurring carbonyl reagents listed in Table I and their phosphopyridoxylidene derivatives are presented along with examples of the use of the carbonyl reagents in the study of enzymes.

1-Amino-D-Proline (I)

The procedure of Klosterman et al.[12] is suitable for the preparation of gram quantities of either antipode of 1-aminoproline. 1-Amino-D-proline may also be obtained by hydrolysis of linatine, but the relatively low abundance of linatine makes this a poor alternative.

1-Nitrosoproline. Both enantiomorphs of 1-nitrosoproline are prepared from the corresponding isomers of proline by a preparative modification of the method of Hamilton and Ortiz.[18] To 1 g of D- and L-proline and 6 g of sodium nitrite in 10 ml of water, 16 ml of 3 M sulfuric acid are

[9] R. M. Khomuton, M. Y. Karpeisky, and E. S. Severin, *in* "Chemical and Biological Aspects of Pyridoxal Catalysis" (E. E. Snell, P. M. Fasella, A. Braunstein, and A. R. Fanelli, eds.), p. 313. Pergamon, Oxford, 1963.

[10] E. H. Abbott and A. E. Martell, *J. Am. Chem. Soc.* **92,** 1754 (1970).

[11] T. Yasumitsu, T. Takao, and Y. Kakimoto, *Biochem. Pharmacol.* **25,** 253 (1976).

[12] H. J. Klosterman, G. L. Lamoureux, and J. L. Parsons, *Biochemistry* **6,** 170 (1967).

[13] A. Carmi, G. Pollack, and H. Yellin, *J. Org. Chem.* **25,** 44 (1960).

[14] J. D. Williamson and L. C. Orchard, *Life Sci.* **14,** 2481 (1974).

[15] C. H. Stammer, *J. Org. Chem.* **27,** 2957 (1962).

[16] B. Levenberg, *J. Biol. Chem.* **239,** 2267 (1964).

[17] H. Naganawa, T. Takita, K. Maeda, and H. Umezawa, *J. Antibiot.* **21,** 241 (1968).

[18] P. B. Hamilton and P. J. Ortiz, *J. Biol. Chem.* **184,** 607 (1950).

added over a period of 10 min. The reaction mixture is extracted four times with 25 ml portions of ether, the ether extract is dried over calcium chloride and evaporated to dryness at 0° under reduced pressure. The crystals that separate are recrystallized from 1 : 1 ether–benzene to give 0.9 g of 1-nitroso-D- or L-proline, m.p. 109°–110° dec, $[\alpha]_D^{24} - 173°$ (for 1 nitroso-L-proline) and $+175°$ (for 1-nitroso-D-proline) (c 2, water). The ultraviolet absorption spectrum shows λ_{max} 347 nm (ϵ 13,800) in water.

1-Aminoproline. 1-Nitroso-D-proline or 1-nitroso-L-proline (1 g) is dissolved in 50 ml of 50% acetic acid, and the solution is cooled in an ice bath. Zinc dust (4 g) is gradually added over a 15-min period while the reaction mixture is stirred vigorously and maintained below 10° by an ice bath. The course of the reaction is followed by measuring the rate of disappearance of the nitrosoproline absorption at 347 nm and is complete after 15 min. The unreacted zinc dust is removed by filtration, and the zinc is precipitated from solution by hydrogen sulfide. The precipitated zinc sulfide is removed by filtration, and the filtrate is evaporated to dryness. The residual syrup is dissolved in 3 ml of absolute ethanol and 1-amino-D-proline (I) or 1-amino-L-proline separates as slightly yellow crystals. A yield of 0.6 g is obtained, m.p. 155°–156°. For the D-isomer (I) $[\alpha]_D^{24} +113°$ (c 2, 0.5 M HCl); for the L-isomer $[\alpha]_D^{24} -111°$ (c 1.6, 0.5 M HCl).

The derivative prepared by allowing 1-amino-D-proline to react with 3-nitrobenzaldehyde melts at 123°–124°, $[\alpha]_D^{24} +112°$ (c 0.6, ethanol). The 2-hydroxy-5-nitrobenzylidene derivative of 1-amino-D-proline melts at 136°–137° and shows $[\alpha]_D^{24} + 154°$ (c 0.35, ethanol). The corresponding derivatives prepared from 1-amino-L-proline show the same melting points as those obtained from the D-isomer, and the specific rotations are identical in magnitude, but of opposite sign.

1-Nitrosoproline and 1-aminoproline are stable compounds and may be stored dry at $-20°$ for extended periods. Solutions of 1-aminoproline decompose rapidly and should be prepared just before use.

The above procedure for the preparation of 1-aminoproline from 1-nitrosoproline works well when small quantities (1 g or less) of 1-nitrosoproline are used. When larger quantities are used, the yields are always low, and the product is difficult to purify.

The physiological activity of the aminoprolines has not been thoroughly investigated. In acute toxicity tests, the L-isomer is about twice as toxic as the D-isomer.[19]

Phosphopyridoxylidene Derivative of 1-Aminoproline. Pyridoxal phosphate (1 mmol) is dissolved in 10 ml of water at 50°. 1-Aminoproline (D

---

[19] K. Sasaoka, T. Ogawa, K. Moritoko, and M. Kimoto, *Biochim. Biophys. Acta* **428**, 396 (1976).

or L) (1 mmol) is added. The phosphopyridoxylidene derivative separates after a few minutes as pale yellow crystals that are collected by filtration and washed with ice water; yield 70%. The product is of high purity and suitable for use in enzyme studies, λ_{max} 374 nm (pH 6.0).

Preparation of Apoaspartate-Aminotransferase from Escherichia coli B by use of 1-amino-D-proline.[20] Aspartate aminotransferase (AAT) (EC 2.6.1.1) is isolated from *E. coli* B by the procedure of Chesne and Pelmont.[21] The AAT activity is determined by means of a coupled reaction with malate dehydrogenase (EC 1.1.1.37) according to the method of Karmen.[22] The enzyme preparation shows absorption maxima at 320 and 385 nm, characteristic of pyridoxal holoenzymes, and has a specific activity of 48 U per milligram of protein. Aspartate-aminotransferase activity is retained during prolonged dialysis in 20 mM phosphate buffer, pH 8.0, indicating firm coenzyme binding.

For resolution of the holoenzyme, 4.5 units of AAT (100 μg of protein) in 3.0 ml of 20 mM phosphate buffer (pH 8.0) containing 1 mM dithiothreitol and 1-amino-D-proline (I) (2.3 mM) are allowed to stand at 25° for 30 min. The absorption spectrum is altered during the reaction and a new absorption maximum develops at 345 nm. The enzyme is completely inactivated. To remove excess aminoproline, the sample is applied to a column of BioGel P4 that has been equilibrated with 20 mM phosphate buffer (pH 8.0) containing 1 mM dithiothreitol, and the protein is eluted with the same buffer. The absorption spectrum of the protein fraction is unchanged from that found before chromatography, indicating that the coenzyme–aminoproline complex is firmly bound to the protein.

Dialysis against 1.0 M phosphate buffer (pH 6.5) removes the inhibitor complex and produces apo-ATT. The absorption spectrum of apo-ATT shows no absorption in the region of 300–450 nm. The enzyme activity is fully restored by the addition of pyridoxal phosphate to apo-ATT.

Addition of phosphopyridoxylidene–aminoproline (see above) to apo-ATT results in the same firm association of enzyme and inhibitor, as is observed when the inhibitor complex is formed *in situ* with the holoenzyme.

Methylhydrazine (II)

Methylhydrazine has been shown to be an inhibitor of ornithine decarboxylase[5] and pyridoxal phosphokinase,[6] and probably inhibits a wide variety of pyridoxal enzymes.

[20] This section was contributed by B. P. Sleeper, North Dakota State University, Fargo, North Dakota.
[21] S. Chesne and J. Pelmont, *Biochimie* **55**, 237 (1973).
[22] A. Karmen, *J. Clin. Invest.* **34**, 131 (1955).

Methylhydrazone of Pyridoxal and Pyridoxal Phosphate. The procedure is that of Furst and Gustafson.[23] The calculated amount of pyridoxal or pyridoxal phosphate is dissolved in a small volume of water and heated to 50°. An equimolar amount of methylhydrazine (II) dissolved in 5 ml of water is added, and the temperature is kept at 50° for an additional 1–2 hr. Crystals of the hydrazone separate on cooling and are recrystallized from benzene–petroleum ether.

α-(1-Methylhydrazino)acetic Acid (III)

Preparation of α-(1-Methylhydrazino)acetic Acid (III). The synthesis procedure is that of Carmi *et al.*[13] To 36 g (0.76 mol) of methylhydrazine in 200 ml water is added gradually 13.5 g (0.143 mol) of monochloroacetic acid and left at room temperature for 4–5 days. Excess methylhydrazine is distilled *in vacuo*. The residue is dissolved in 600 ml of water, and 100 ml acetone is added, lowering the pH from 8 to 4. The solution is passed through a Duolite A7 column and concentrated *in vacuo* to dryness. The residue is dissolved in 20 ml of water and added dropwise to 500 ml of absolute ethanol. The yield is 8 g (54.5%); m.p. 153°–154°. The *m*-nitrobenzylidene derivative melts at 128°–130°.

The α-(1-methylhydrazino)acetic acid is obtained in poor yield by mild acid hydrolysis of negamycin (IIIa).[24]

L-Canaline (VI)

L-Canaline (VI) is obtained by the enzymic hydrolysis of canavanine (VIa) and has been found to be a very effective inhibitor of pyridoxal phosphokinase[6] and a variety of enzymes requiring pyridoxal phosphate.[25] The concentration of L-canaline required to achieve 50% inhibition varies with enzyme systems. The effect of L-canaline on several enzymic reactions is shown in Table II. The inhibitory effects of L-canaline on pyridoxal phosphate-containing enzymes are probably the result of its nonenzymic, irreversible binding with pyridoxal phosphate.

Preparation of L-Canaline (VI) from Canavanine.[14] Arginase (60–70 units, Sigma) is dissolved to a concentration of 3 mg/ml in 50 mM Tris·HCl buffer, pH 7.6, containing 1 mM $MnCl_2$, and dialyzed for 48 hr at 4° against two changes of a 100-fold greater volume of the same buffer. Insoluble material is sedimented at 4000 g for 15 min, and chromatographi-

[23] A. Furst and W. R. Gustafson, *Proc. Soc. Exp. Biol. Med.* **124,** 172 (1967).
[24] S. Shibahara, S. Kondo, K. Maeda, and H. Umezawa, *J. Am. Chem. Soc.* **94,** 4353 (1972).
[25] E. L. Rahiala, M. Kekomaki, J. Janne, A. Raina, and N. C. R. Raiha, *Biochim. Biophys. Acta* **227,** 337 (1971).

TABLE II

SUMMARY OF THE STUDY OF THE EFFECTS OF L-CANALINE
ON DIFFERENT ENZYMIC REACTIONS[a]

Enzyme	EC number	Pyridoxal phosphate (M)	L-Canaline causing 50% inhibition (M)	Incubation system[b]
1. Ornithine transcarbamylase	2.1.3.3	None	$>2.5 \times 10^{-2}$	II
2. Tyrosine aminotransferase	2.6.1.5	4×10^{-5}	4×10^{-5}	I
3. Ornithine-ketoacid amino-transferase	2.6.1.13	None	3×10^{-6}	I,P
4. Ornithine decarboxylase	4.1.1.17	2×10^{-4}	1.0×10^{-4} to 1.5×10^{-4}	I,II,P
5. 5-Hydroxytryptophanase	4.1.1.28	5×10^{-4}	1×10^{-4}	II,P
6. Diamine oxidase	1.4.3.6	None	5×10^{-5}	II

[a] Source: E. L. Rahiala, M. Kekomaki, J. Janne, A. Raina, and N. C. R. Raiha, *Biochim. Biophys. Acta* **227**, 337 (1971).

[b] I, constant L-canaline concentration, the concentration of pyridoxal phosphate varied; II, constant pyridoxal phosphate concentration, the concentration of L-canaline varied; P, additional analyses were carried out with different preincubation systems, namely canaline + enzyme, canaline + pyridoxal phosphate, and enzyme + pyridoxal phosphate.

cally homogeneous L-canavanine (Sigma) is dissolved in the recovered supernatant to 12 mg per milliliter final concentration. This reaction mixture is incubated at 37° for 24 hr. After incubation, 2 ml of the enzyme digest are applied to a Sephadex G-10 column (30 cm × 1.5 cm) equilibrated previously with 0.14 M NaCl in 10 mM phosphate buffer, pH 7.3, and eluted with the same buffer. Fractions of the eluent are examined for the presence of protein by absorption at 280 nm, for canaline by the Jaffe reaction,[26] and for urea by a quantitative diacetyl monoxime reaction.[27]

Protein is eluted at the void volume followed by Jaffe-reacting material (L-canaline). Urea is found in the third peak. The solution of L-canaline is suitable for enzyme studies and may be stored at −50°. The concentration of L-canaline may be determined by a quantitative Jaffe reaction.[26]

L-Canaline has also been prepared by an enzyme preparation from jack bean (*Canavalia ensiformis*).[27] DL-Canaline has been prepared from γ-butyrolactone in a 5-step process.[28]

Resolution of Cystathionase by Use of L-Canaline.[8] Rat liver cystathionase is prepared according to the method of Matsuo and Greenberg[29]

[26] D. Hunninghake and S. Grisolia, *Anal. Biochem.* **16**, 200 (1966).

[27] G. A. Rosenthal, *Anal. Biochem.* **51**, 354 (1973).

[28] D. D. Nyberg and B. E. Christensen, *J. Am. Chem. Soc.* **79**, 1222 (1957).

[29] Y. Matsuo and D. M. Greenberg, *J. Biol. Chem.* **230**, 545 (1958).

omitting the crystallization step. The last step of the purification of the enzyme is a stepwise elution from a hydroxyapatite column with phosphate buffer (pH 7.5) as described by Mushawar and Koeppe.[30] Enzymic activity is assayed by measuring α-ketobutyric acid production with L-homoserine as substrate.[31] Cystathionase shows a strong absorbance at 420 nm (pH 7.4).

Cystathionase at a concentration of 6 μM is allowed to react with canaline (50 μM) in 0.1 M phosphate buffer (pH 7.4). The absorption spectrum of the reaction mixture changes over a period of 10 min. The absorbance band found at 420 nm for cystathionase disappears and a new absorption band appears at 320 nm. The absorption band at 320 nm is due to the formation of a phosphopyridoxal–canaline complex whereas the absorption band at 420 nm is due to the covalent linkage of coenzyme to the enzyme. Kinetic studies show that the reaction of L-canaline with cystathionase proceeds more rapidly than does the reaction of L-canaline with pyridoxal phosphate. The results indicate that the enzyme-bound pyridoxal phosphate groups differ in their reactivity toward L-canaline.

D-Cycloserine (VII) and O-Amino-D-Serine (VIII)

D-Cycloserine (VII) is cleaved by acid hydrolysis to O-amino-D-serine (VIII).[15] O-Amino-D-serine has been found in rat urine after injection of D-cycloserine.[11] O-Amino-D-serine is the next lower homolog of L-canaline and has the same chemical properties as L-canaline. The inhibitory effects of O-amino-D-serine on two enzyme systems are shown in Table III. D-Cycloserine is a less effective inhibitor for these two enzymes by two to three orders of magnitude.[11]

Preparation of O-Amino-D-serine (VIII) by Acid Hydrolysis of D-cycloserine (VII).[15] (i) O-AMINO-D-SERINE MONOHYDROCHLORIDE. A solution of 20 g of D-cycloserine in 100 ml of 6 N hydrochloric acid is heated at 60° in an oil bath for 3 hr. The solution is evaporated to dryness *in vacuo*. The residue is twice dissolved in absolute ethanol and evaporated to dryness to remove residual hydrochloric acid. After 3 hr *in vacuo*, the residue is dissolved in 150 ml of boiling absolute ethanol and the solution is allowed to cool to room temperature. Twelve milliliters of pyridine are added dropwise to the stirred solution. The pink precipitate is collected on a filter, washed with ethanol, and dried. The product is dissolved in 60 ml of water and filtered; the filtrate is slowly (2 hr) diluted with 150 ml of absolute ethanol. After standing overnight at 5°, the O-

[30] I. K. Mushawar and R. Koeppe, *J. Biol. Chem.* **248**, 7407 (1973).
[31] F. W. Sayre and D. M. Greenberg, *J. Biol. Chem.* **220**, 787 (1956).

TABLE III

EFFECT OF O-AMINO-D-SERINE ON D-β-AMINOISOBUTYRATE:PYRUVATE AND
β-ALANINE:α-KETOGLUTARATE AMINOTRANSFERASES[a]

Concentration of O-amino-D-serine (M)	% Inhibition of D-β-aminoisobutyrate:pyruvate aminotransferase[b]
1.66×10^{-7}	13
1.66×10^{-6}	66
1.66×10^{-5}	94
1.66×10^{-4}	98
	% Inhibition of β-alanine:α-ketoglutarate aminotransferase
1.4×10^{-6}	63
1.4×10^{-5}	93
1.4×10^{-2}	99

[a] Source: T. Yasumitsu, T. Takao, and Y. Kakimoto, *Biochem. Pharmacol.* **25**, 253 (1976).

[b] An aqueous solution of O-amino-D-serine was added to the enzyme reaction to obtain the final concentrations given in the table.

amino-D-serine·HCl is collected by filtration and dried. Yield 12.5 g (41%), m.p. 143°–145°, $[\alpha]_D^{25}$ − 18.3 (c 2.1, 1 N HCl).

(ii) O-AMINO-D-SERINE (VIII). A solution of 500 mg of O-amino-D-serine monohydrochloride (above) in 3.5 ml of 3 N ammonium hydroxide is cooled in an ice bath and 10 ml of a 1:1 2-propanol–ethanol mixture is added. The dropwise addition of acetic acid to the solution causes crystallization of the zwitterion form of O-amino-D-serine (VIII). Another 10 ml of the 2-propanol–ethanol solution are added, the mixture is left at 5° for 1 hr, and the crystals are collected by filtration. The yield of O-amino-D-serine is 296 mg (77%), m.p. 166°–168° dec., $[\alpha]_D^{25}$ − 24.4 (c 2.13, 1 N HCl).

Identification of Catalytically Active Pyridoxal Phosphate Groups by Use of DL-*Cycloserine.*[7] Rat liver cystathionase is prepared according to the method of Matsuo and Greenberg,[29] but not crystallized. The enzyme shows an absorption maximum at 410 nm and contains 3.1 pyridoxal phosphate residues per mole of protein (150,000 molecular weight). Cystathionase (10 ml, 1 mg protein/ml) is incubated with 20 mM DL-cycloserine for 30 min at 25° in 0.1 M phosphate buffer (pH 7.4) and dialyzed against two changes of the same buffer for 14 hr at 4°. In order to ensure complete removal of cycloserine, the dialyzed sample is filtered through a Sephadex G-25 column (1 × 25 cm) previously equilibrated with 0.1 M phosphate buffer, pH 7.4, at 4°. The protein eluted from the column shows no cystathionase activity, but still shows an absorption maximum at 410 nm characteristic of pyridoxal phosphate bound to protein. Analysis for pyridoxal phosphate shows that only one pyridoxal phosphate group had been released by treatment with cycloserine.

For reconstitution experiments cycloserine-treated cystathionase (0.8 mg protein per milliliter) is allowed to react with excess pyridoxal phosphate (10:1 molar ratio) at 25°. Approximately 85% of the cystathionase activity is restored after 60 min incubation.

When the molar ratio of cofactor to protein is 2:1, 80% of the activity is restored after 3 hr of incubation at 25°.

The role of the second and third pyridoxal phosphate residues has not been established.

Complete Resolution of Rat Liver Cystathionase.[32] Cystathionase at a concentration of 5 mg/ml in 0.1 M phosphate buffer, pH 7.4, is allowed to react with 200-fold molar excess of D-cycloserine at 4° for 1 hr. The incubation mixture is then dialyzed against two changes of 0.1 M phosphate buffer, pH 7.4, for 14 hr at 4°. The absorption spectrum shows negligible absorption in the spectral region 300–500 nm indicating complete removal of all pyridoxal phosphate groups.

Catalytic activity is restored by incubation of apocystathionase prepared as above with 10 μM pyridoxal phosphate.

Preparation of Apoglutamate–Aspartate Transaminase.[33] Glutamate–aspartate transaminase (Boehringer) is purified by hydroxyapatite chromatography according to the method of Jenkins et al.[34] This preparation has a specific activity of 30 units per milligram of protein and exhibits an absorption maximum at 362 nm (pH 8.5), indicating the presence of pyridoxal phosphate. For resolution, the enzyme is dissolved in 20 mM sodium phosphate buffer, pH 6.8, and dialyzed against several changes of 0.1 M NaH$_2$PO$_4$ at 4° for 6 hr to bring the enzyme solution to pH 4.3. The enzyme at a concentration of 0.4 mg of protein per milliliter is allowed to react with excess cycloserine (10 mM) for 1 hr at 30°. After dialysis at 4° against 0.1 M Tris-acetate buffer, pH 8.2, the enzyme shows 10% of original activity.

For reconstitution, the dialyzed enzyme solution, pH 8.2, which contains 0.3 mg of protein per milliliter is incubated with 10 mM pyridoxal phosphate at 30° for 1 hr. Approximately 90% of the original activity is restored by the treatment with pyridoxal phosphate.

[32] K. Oh and J. E. Churchich, *J. Biol. Chem.* **248,** 7370 (1973).
[33] J. E. Churchich, *J. Biol. Chem.* **242,** 4414 (1967).
[34] W. T. Jenkins, D. A. Yphantis, and J. W. Sizer, *J. Biol. Chem.* **234,** 51 (1959).

Enzymic activity is determined in 0.1 M Tris-acetate buffer, pH 8.2, at 32° by measuring the increase in optical density at 280 nm due to formation of oxaloacetate in the transaminase reaction.[35]

Attempts to obtain resolution of the enzyme at a pH closer to neutral have been unsuccessful. It is concluded that to obtain resolution the experimental conditions must be sufficient to ensure cleavage of the covalent and electrostatic bonds that link the cofactor to the enzyme.

[35] I. W. Sizer and W. T. Jenkins, this series, Vol. 5 [94a], p. 677.

[76] Spin-Labeled Vitamin B$_6$ Derivatives: Synthesis and Interaction with Aspartate Aminotransferase

By A. Yu. Misharin, O. L. Polyanovsky, and V. P. Timofeev

Spin-labeled analogs of coenzymes,[1-4] substrates[5-7] and prosthetic groups[8] have proved to be a useful tool for elucidating structure and conformational changes in proteins. The present paper is concerned with the preparation of spin-labeled vitamin B$_6$ derivatives and their interaction with the apoenzyme and holoenzyme of L-aspartate:2-oxoglutarate aminotransferase (EC 2.6.1.1; AAT). The iminoxyl group of spin-labeled derivatives has been used not only as a "reporter" group, but also as a specific oxidant of the cysteine residue. This paper covers our work[9-13] in which paramagnetic analogs of pyridoxal 5'-phosphate (PLP) and pyridoxamine 5'-phosphate (PMP) are applied for studying AAT.

[1] H. Weiner, *Biochemistry* **8**, 526 (1969).

[2] A. S. Mildvan and H. Weiner, *Biochemistry* **8**, 552 (1969).

[3] A. S. Mildvan and H. Weiner, *J. Biol. Chem.* **244**, 2465 (1969).

[4] S. W. Weidman, G. R. Drisdale, and A. S. Mildvan, *Biochemistry* **12**, 1874 (1973).

[5] G. C. K. Roberts, J. Hannan, and O. Jardetzky, *Science* **165**, 504 (1969).

[6] R. Cooke and J. Duke, *J. Biol. Chem.* **246**, 6360 (1971).

[7] C. F. Chignell and R. H. Erlich, *in Int. Biophys. Congr. 4th* Abstr. I, 89 (1972).

[8] T. Asakura, J. S. Leigh, M. R. Drott, T. Yonetani, and B. Chance, *Proc. Natl. Acad. Sci. U.S.A.* **68**, 861 (1971).

[9] A. Yu. Misharin, A. V. Azhayev, and O. L. Polyanovsky, *Izv. Acad. Nauk SSSR, Ser. Khim.* **1975**, 1185 (1975).

[10] A. Yu. Misharin, V. P. Timofeev, and O. L. Polyanovsky, *Mol. Biol. (SSSR)* **9**, 113 (1975).

[11] A. Yu. Misharin, O. L. Polyanovsky, and V. P. Timofeev, *FEBS Lett.* **41**, 131 (1974).

[12] A. Yu. Misharin, A. V. Azhayev, V. P. Timofeev, and O. L. Polyanovsky, *Bioorg. Khim. (SSSR)*, **1**, 257 (1975).

[13] A. Yu. Misharin, Thesis, Univ. of Moscow, 1975.

Synthesis of Spin-Labeled Vitamin B₆ Derivatives

The synthesis of vitamin B_6 derivatives is presented in Scheme 1. The compounds thus obtained can be regarded as paramagnetic analogs of PLP (I), PMP (II), and N-pyridoxylamino acids (III–V), which inhibit coenzyme binding[14–15]

The 5'-phosphate group of PLP is esterified with the radical 2,2,6,6-

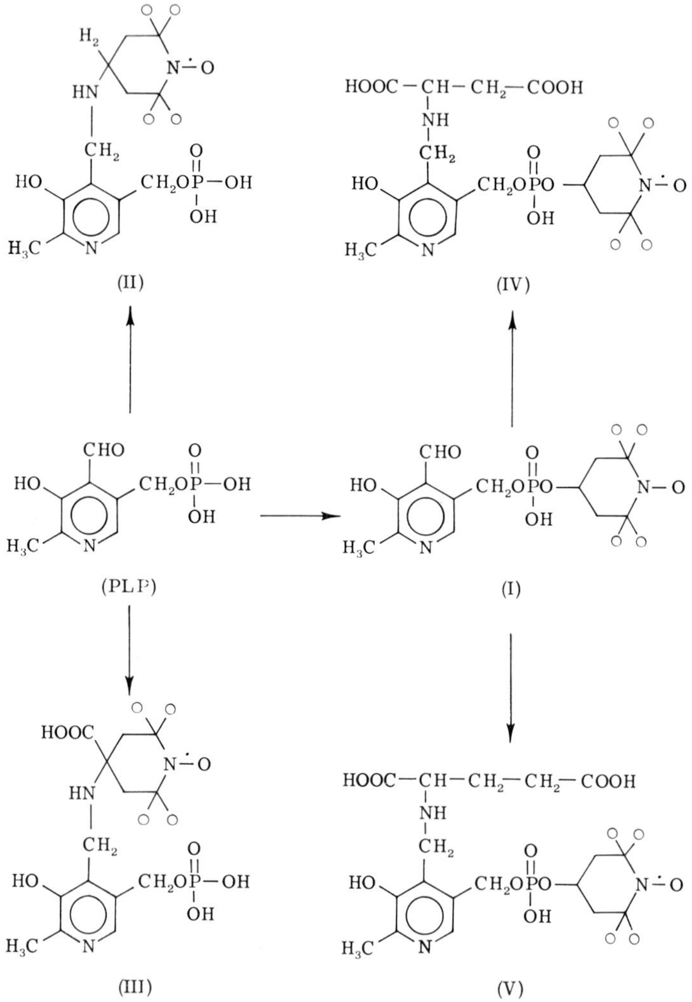

Scheme 1

[14] A. E. Braunstein, in "The Enzymes" (P. Boyer, ed.), 3rd ed., Vol. 9, p. 379. Academic Press, New York, 1973.

[15] E. S. Severin, G. K. Kovaleva, and L. P. Sastchenko, Biokhimiya (SSSR) 30, 469 (1972).

tetramethyl-1-oxyl-4-hydroxypiperidine (VI) using the carbodiimide method. A typical reaction of orthohydroxyaldehydes, i.e., formation of the stable Schiff bases with amines and amino acids, is used to prepare (II)–(V). The radicals 2,2,6,6-tetramethyl-1-oxyl-4-aminopiperidine (VII) and 2,2,6,6-tetramethyl-1-oxyl-4-amino-4-carboxypiperidine (VIII) as well as L-aspartic and L-glutamic acids serve as amino-containing components. The Schiff bases obtained are reduced, without being isolated, by NaBH$_4$ in methanol (NaBH$_4$ does not react with the iminoxyl group[16]).

All the compounds have been isolated by chromatography on an anion-exchange cellulose column. The electron spin resonance (ESR) spectrum for all the compounds is a triplet, the constant of hyperfine structure in 0.1 N sodium phosphate buffer (pH 7.2) is 16.2 ± 0.2 G.

Interaction of Spin-Labeled Vitamin B$_6$ Derivatives with the Apoenzyme of AAT

The apoenzyme of AAT (apo-AAT) is stable enough and capable of binding an analog of the coenzyme.[14] This results in a typical maximum in the circular dichroism (CD) spectrum, the appearance of which should be attributed to the asymmetrical surroundings of the chromophore. The interaction of the apo-AAT with (I)–(V) is also accompanied by a characteristic Cotton effect in the absorption band for analogs (Fig. 1).

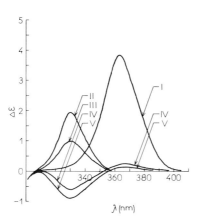

FIG. 1. Circular dichroism spectra of the apo-AAT complexes with (I)–(V). The concentration of the protein is 0.2 mM; that of B$_6$-derivatives (I)–(V), 0.5 mM.

[16] E. G. Rozantzev "Stable Iminoxyl Radicals." Khimia, Moscow, 1970.

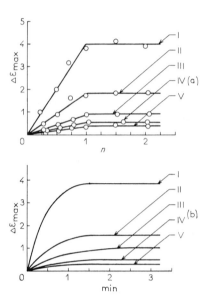

FIG. 2. Circular dichroism titration of apo-AAT with (I)–(V). (a) Dependence of $\Delta\epsilon_{max}$ (see Fig. 1) on the amount of an analog [(I)–(V)] being added (the spectra are recorded 10 min after the addition of each sample). (b) Dependence of $\Delta\epsilon_{max}$ on the time during which AAT is incubated with (I)–(V) (a 2-fold excess of the analog). The concentration of the protein is 0.2 mM, pH 5.2; n is the ratio between the concentration of (I)–(V) and apo-AAT.

The (apo-AAT)-I complex does not have pH-indicator properties. The typical maximum in the CD spectrum at 430 nm is not observed upon decreasing the pH to 4.8. As has been shown earlier,[17,18] alkyl phosphodiesters of PLP and 5′-alkylphosphonic derivatives of pyridoxal do not possess the pH-indicator property either. Apparently the spectral properties of the coenzyme in the active site are influenced by the ionization state of the analog 5′-phosphate group.

Complete binding of (I)–(V) with the apoenzyme occurs at an apo-AAT : B$_6$-derivative ratio close to the equimolar one (Fig. 2). Equilibrium is established for all the analogs within 2–3 min, which corresponds to k_2 (the rate constant of a bimolecular reaction between the apoenzyme and an analog of the coenzyme) = 10^4 mol^{-1} min^{-1}. The presence of an optically active maximum at 430 nm (pH 5.2) in the spectrum of the native holoenzyme makes it possible to follow the analog displacement from the ac-

[17] M. L. Fonda and R. I. Johnson, *J. Biol. Chem.* **245**, 2709 (1970).
[18] M. L. Fonda, *J. Biol. Chem.* **246**, 2230 (1971).

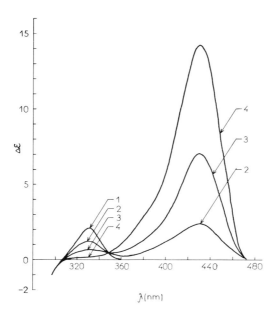

FIG. 3. Interaction of pyridoxal 5'-phosphate (PLP) with the AAT-II complex at pH 5.2 (the circular dichroism spectra). The concentration of apo-AAT is 50 μM; that of (II) is 50 μM. 1, apo-AAT-II complex; 2–4, the same in the presence of PLP. The concentrations of PLP are 12.5 μM (2), 25 μM (3), and 50 μM (4).

tive site (Fig. 3). The family of curves displays an isosbestic point; this proves that the vitamin B$_6$ analogs (I)–(V) are specifically bound at the active site.[10,11]

Spin-Labeled Derivatives (I)–(V) as Reagents Modifying the Cysteine Residue in the Coenzyme Binding Site

Interaction of (I)–(V) with apo-AAT results in the quantitative and irreversible disappearance of the ESR signal. The ESR spectrum of a free radical in solution is not observed once a spin-labeled analog has been displaced by the coenzyme, or upon the action of denaturing agents.

Figure 4 presents the results of the apo-AAT titration with the spin-label (I) (formation of the complex is registered by the CD method, whereas disappearance of the free radical is followed by the ESR spectrum).

Mercaptans and cysteine residues in peptides and proteins are known to react with iminoxyl radicals, reducing them to corresponding hydroxy-

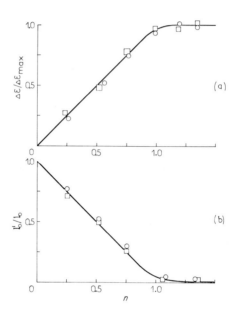

Fig. 4. (a) A change in the circular dichroism (CD) amplitude upon formation of the apo-AAT:(I) complex. $\Delta\epsilon/\Delta\epsilon_{max}$, relative value of CD amplitude at 362 nm. (b) A change in the amplitude of the central component in the ESR spectrum of the radical upon interaction with apo-AAT. I_0 is the amplitude of the central component in the ESR spectrum of the free label; I_0' is the amplitude of the central component in the ESR spectrum in the presence of the apoenzyme; n is as specified for Fig. 2. The apoenzyme concentration is 0.2 mM. Similar dependences have been obtained for interaction of apo-AAT with (II)–(V).

lamines.[19–21] The binding of (I)–(V) at the active site of AAT is accompanied by the saturation of the free valency; this suggests a specific reaction of the spin-labeled derivatives (I)–(V) with the thiol group of the cysteine residue.

Further studies were intended (a) to prove that the thiol group is modified upon specific binding of (I)–(V) at the AAT active site and (b) to elucidate to which of the five cysteine residues of AAT this thiol group belongs.

Determination of the primary structure of AAT[22] and investigation of

[19] J. D. Morrissett and H. R. Drott, *J. Biol. Chem.* **244,** 5083 (1969).

[20] G. A. Corker, M. P. Klein, and M. Calvin, *Proc. Natl. Acad. Sci. U.S.A.* **56,** 1356 (1965).

[21] G. J. Giotta and H. H. Wang, *Biochem. Biophys. Res. Commun.* **46,** 1576 (1972).

[22] Yu. A. Ovchinnikov, A. E. Braunstein, C. A. Egorov, O. L. Polyanovsky, N. A. Aldanova, M. Yu. Feigina, V. M. Lipkin, N. G. Abdulaev, E. V. Grishin, A. P. Kiselev, N. N. Modyanov, and V. V. Nosikov, *Dokl. Akad. Nauk SSSR* **207,** 728 (1972).

the reactivity of thiol groups[23-26] makes it possible to classify the cysteine residues of AAT.

As was found by determining free sulfhydryl groups with the method of Boyer,[27] once (I)–(V) are bound to apo-AAT and carboxymethylated apo-AAT (the apoenzyme preparation obtained from AAT preliminarily carboxymethylated at Cys residues 45 and 82), one thiol group is modified. Consequently, one sulfhydryl group belonging either to Cys-390 or to Cys-252 reacts with paramagnetic analogs, because the SH group of Cys-191 is not modified in native and carboxymethylated AAT with Boyer's method.

To distinguish between these two residues (Cys-390 and Cys-252), comparisons were made on the effect of syncatalytic modification[24] on the native AAT and AAT, which was preliminarily treated with spin-labeled vitamin B_6-derivatives (B_6-R·) according to the scheme:

Holo-AAT $\xrightarrow{\text{carboxymethylation}}$ carboxymethylated holo-AAT $\xrightarrow{-\text{PLP}}$

carboxymethylated apo-ATT $\xrightarrow{B_6\text{-R·}}$ (B_6-R·)-AAT $\xrightarrow{+\text{PLP}}$ reconstituted holo-AAT

The activity of the reconstituted AAT is 65–75% of the original one; the enzyme contains one thiol group. Treatment with N-ethylmaleimide (NEM) in the presence of substrates does not cause alkylation of the thiol group and only slightly decreases enzyme activity. Similar modification of the native AAT results in almost complete inactivation of the enzyme and disappearance of Cys-390 as was demonstrated by Birchmeier et al.[24] and confirmed by our experiments.

These results suggest that it is Cys-390 that is being modified once the spin-labeled derivatives (I)–(V) are bound at the AAT active site. Here, mild one-electron oxidation of the SH group of Cys-390 causes only a slight decrease (25–35%) in enzyme activity. Having compared data on the modification of Cys-390 by various reagents, one may conclude that inactivation upon this modification depends on the structure of a modifying agent. The V_{max} for the reconstituted AAT (see next section) indicates that oxidation of Cys-390 has no effect on the catalytic properties of the

[23] O. L. Polyanovsky, V. V. Nosikov, S. M. Deev, A. E. Braunstein, E. V. Grishin, and Yu. A. Ovchinnikov, FEBS Lett. 35, 322 (1973).

[24] W. Birchmeier, K. J. Wilson, and P. Christen, J. Biol. Chem. 248, 1751 (1975).

[25] O. L. Polyanovsky, V. P. Timofeev, M. J. Shaduri, A. Yu. Misharin, and M. V. Volkenstein, Biochim. Biophys. Acta 220, 357 (1973).

[26] V. V. Nosikov, E. V. Grishin, S. M. Deev, O. L. Polyanovsky, A. E. Braunstein, and Yu. A. Ovchinnikov. Mol. Biol. (SSSR) 8, 406 (1974).

[27] P. D. Boyer, J. Am. Chem. Soc. 76, 4331 (1954).

enzyme. Partial inactivation upon the modification of this residue should be attributed only to changes in substrate binding.

Therefore, paramagnetic vitamin B_6 derivatives (I)–(V) turn out to be useful as oxidizing reagents with specific affinity, and make it possible to locate Cys-390 in close proximity to the region of the coenzyme binding. Recent studies using the techniques of peptide chemistry and bifunctional reagents have confirmed independently the localization of Cys-390 in the region adjacent to the AAT active site.[28]

Formation of Enzyme–Inhibitor Complexes of AAT with Spin-Labeled Vitamin B_6 Derivatives

The paramagnetic vitamin B_6 derivatives (III)–(V) can form complexes with the holoenzyme of AAT. Titration of the native AAT with (III)–(V) induces typical changes in the CD spectrum. Since, at pH 5.2, PLP absorbs at 378 nm and does not absorb at 430 nm, the absorption spectra suggest the complete specific binding of PLP at the active site. A decrease in the amplitude of the Cotton effect at 430 nm characterizes therefore a change in the asymmetry of the coenzyme surroundings in the AAT active site rather than competition for the active site between the coenzyme and an analog. Moreover, a negative extremum is observed in the CD spectrum at 328 nm, i.e., within the absorption region of the analogs (III)–(V). This finding as well as low values of $\Delta\epsilon/\epsilon_{430}$ indicate that (III)–(V) produce complexes with the holoenzyme (Fig. 5). Similar absorption and CD spectra are obtained when PLP is added to the AAT-B_6-R· complex in 10-fold excess to the analog.

Since no complexes were known so far to be formed between the derivatives of vitamin B_6 and the holoenzyme, it was of interest to find why such complexes are produced.

We have employed inhibitor analysis to establish whether or not (III)–(V) are fixed by interaction between the carboxy groups and the cation centers of the substrate binding site, or due to nonspecific binding with a protein.

It is well known that β-erythrohydroxyaspartic acid interacts with AAT to yield a semiquinoid structure, with maximum absorption at 492 nm, which is slowly transaminated.[29,30] Since the rate of transamination of β-erythrohydroxyaspartic acid is far less than the rate of formation of the semiquinoid intermediate, the spectrum of AAT hardly changes in the

---

[28] S. M. Deyev, G. A. Afanasenko, and O. L. Polyanovsky, *Bioorg. Khim. (SSSR)* **3**, 816 (1977).

[29] W. T. Jenkins, *J. Biol. Chem.* **236**, 1121 (1961).

[30] W. T. Jenkins, *J. Biol. Chem.* **239**, 1742 (1964).

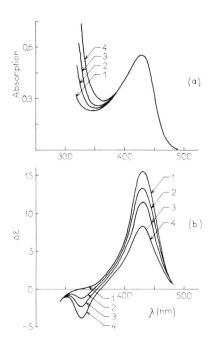

FIG. 5. Interaction of holo-AAT with (III) at pH 5.2. (a) Absorption spectra; (b) circular dichroism spectra; (1) AAT; complex AAT-III at different concentrations of (III): (2) 50 μM; (3) 0.125 mM; (4) 0.5 mM. The concentration of AAT is 50 μM.

presence of β-erythrohydroxyaspartic acid within 2–3 min.[30] Figure 6 presents the spectra of AAT in the presence of β-erythrohydroxyaspartic acid and the paramagnetic analogs (III)–(V).

A decrease in the typical maximum at 492 nm indicates that β-erythrohydroxyaspartic acid competes with the analogs (III)–(V).

The thiol group of Cys-390 is modified when (III)–(V) form complexes with the holoenzyme, as also happens in the case of their interaction with the apoenzyme; this has been supported by experiments similar to those described in a previous section.[12,13] The spectral characteristics, enzyme activity, and number of titratable SH groups in AAT and carboxymethylated AAT treated with (III)–(V) are identical, after separation from the excess of B_6-radicals, to those of reconstituted AAT.[12]

A further decrease in enzyme activity is found with an increase in the concentration of the analogs. In other words, compounds (III)–(V) are effective inhibitors of transamination of the substrate amino acids. The type of AAT inhibition by the vitamin B_6 derivatives (III)–(V) is determined by

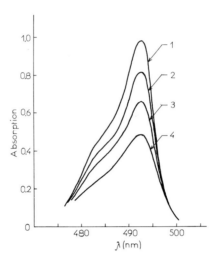

FIG. 6. Interaction of AAT with β-erythrohydroxyaspartic acid in the presence of (IV). All the spectra are recorded 1 min after mixing 30 μM; β-erythrohydroxyaspartic acid is 0.8 mM. Concentrations of (IV) are: (I) 0; (2) 0.1 mM; (3) 0.2 mM; (4) 0.5 mM.

the method of Lineweaver and Burk[31] from the plot of the L-aspartic acid transamination rate in the presence of the compounds (III)–(V) (Fig. 7). Lines 1 and 2 intersect the ordinate ($1/V$) at one point; therefore, the values of V_{max} for transamination of L-aspartic acid are the same in the presence of either the native enzyme or the reconstituted AAT. The value of V_{max} (200 sec^{-1}) obtained graphically coincides with the data of kinetic studies.[32] The graph (Fig. 7) indicates that the inhibition is competitive, which makes it possible to determine the effective K_i by the Dixon method.[31] The effective values of K_i determined graphically are within the range of 1–3 μM.

The paramagnetic compounds (III)–(V) are effective competitive inhibitors of AAT. The ESR spectra of (III)–(V) in the presence of AAT characterize immobilization of the radical fixed within the substrate-binding site. At an excess of any of the analogs, there appears in the ESR spectrum an immobilized component corresponding to the complex of the reconstituted AAT and a spin-label. The value of τ_c of the enzyme–inhibitor complexes is 10^{-9} sec for all three compounds. The τ_c for the (apo-AAT)–(III) complex, the conformational mobility of which is more

[31] M. Dixon and H. Webb, "Enzymes." Nauka, Moscow, 1966.
[32] B. E. S. Banks, M. P. Bell, A. J. Lawrens and C. A. Vernon, *in* "Chemistry and Biology of Pyridoxal Catalysis" (A. E. Braunstein, E. S. Severin, E. E. Snell, and Yu. M. Tortchinsky, eds.). Nauka, Moscow, 1968.

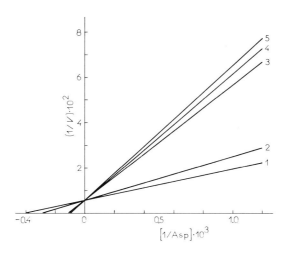

FIG. 7. Competitive inhibition of AAT with spin-labeled vitamin B$_6$ derivatives (III)–(V). 1, Holo-AAT; 2, reconsituted AAT; 3–5, the same as 2 in the presence of spin-labeled derivatives (III), (IV), and (V), respectively. Concentrations are: AAT, 6.7 nM; 2-oxoglutaric acid, 6.7 mM; (III)–(V) 2 μM, pH 8.3.

restricted, hardly differs from the apo-AAT complexes with more "flexible" spin-labels (IV) and (V). Consequently, the structure of a label has but a slight effect on the ESR spectrum.

Spatial Correlations within the Active Site of AAT

The above data make it possible to speculate on certain spatial and functional correlations within the active site of AAT. The oxidative – reductive reaction between the iminoxyl fragment of the vitamin B$_6$ analogs (I)–(V) and the thiol group of Cys-390 proceeds with the formation of specific complexes of (I)–(V) with the apoenzyme and the holoenzyme; this is confirmed by the coincidence of the rates for the complex formation and the iminoxyl group reduction (within experimental error).[12,13] It can be assumed therefore that, once the substituted pyridine analog is located within the active site, the iminoxyl fragment can approach the sulfhydryl group by the distance of the van der Waals radii. Structure (III) differs from (II) only by the presence of the carboxy group. Their interaction with the apoenzyme is similar, but (III) forms the enzyme–inhibitor complex with the holoenzyme, and (II) does not. Apparently, the presence of the α-carboxy group and the total charge of the molecule (-2) are important for the formation of the enzyme–inhibitor complexes of the vitamin B$_6$ spin-labeled derivatives (III)–(V).

According to published data,[14,15] the conformation of N-substituted residues of aspartic acid (in IV) and glutamic acid (in V) in the active site is similar to that of the natural substrate. It would be plausible therefore to explain the similarity of the AAT complexes with the analogs (III)–(V) by assuming that the carboxy group of (III) is coordinated with the cation center of the substrate-binding site. The distance between the carboxy group and the N-O· fragment in the molecule is determined by the size of the piperidine ring of the label. This makes it possible to estimate the distance between the thiol group reacting with the iminoxyl fragment (Cys-390) and the cation center in the protein involved in immobilization of the carboxy group.

It has not been possible hitherto to evaluate the conformation of the analogs (I)–(V) bound within the active site. Therefore, molecular models have been used to provide approximate evidence on the distance between Cys-390, which reacts with the iminoxyl group, and the center of the coenzyme pyridine ring, as well as the cation center involved in the binding of the substrate α-carboxy group. These distances are $\simeq 8$ Å and 5 Å, respectively.

Experimental Procedure

General

UV spectra were recorded with a Specord UV-Vis (DDR). The ESR spectra were recorded with a modified EPR-II instrument (USSR) and a Varian E-4 model (USA). The unpaired electron concentration was assayed by comparing it with the reference sample.

The CD spectra were recorded with a Dichrographe II model (Jouan-Russel, France) in cuvettes having optical paths of 0.5, 1, 2, and 5 cm. Thin-layer chromatography was conducted on FND cellulose plates as well as on Silufol and Silufol-UV-254 (Cavalier, Czechoslovakia) plates in the systems n-butanol–96% ethanol–5% NH$_4$OH–CH$_3$COOH, 10:10:10:1 (A), and n-propanol–water, 7:3 (B).

Analytical electrophoresis was run on Whatman 3 MM paper at pH 8.0 and 30 V/cm. Reference substances: PLP and PMP. Radicals: 2,2,6,6-tetramethyl-1-oxyl-4-hydroxypiperidine (VI), 2,2,6,6-tetramethyl-1-oxyl-4 aminopiperidine (VII), 2,2,6,6-tetramethyl-1-oxyl-4 amino-4-carboxypiperidine (VIII) were synthesized by known methods.[9,16, 33–35]

[33] A. Yu. Misharin, L. M. Vinokurov, and O. L. Polyanovsky, *Izv. Acad. Nauk SSSR, Ser. Khim. 1974*, 1897 (1974).

[34] E. G. Rozantzev and Yu. V. Kochanov, *Izv. Akad. Nauk SSSR Ser. Khim. 1966*, 1477 (1966).

[35] A. Rassat and P. Rey, *Bull. Soc. Chim. Fr. 1967*, 815 (1967).

The residues Cys-45 and Cys-82 in AAT preparations are carboxymethylated according to Polyanovsky *et al.*[23]

Carboxymethylated AAT is alkylated with NEM in the presence of L-glutamic acid as described by Birchmeier *et al.*[24]

Carboxymethylated AAT was prepared by the method of Bocharov *et al.*[36]

Preparation of Spin-Labeled Vitamin B$_6$ Derivatives [2,2,6,6-
tetramethyl-1-oxylpiperidine-4-(pyridoxal-5′)-phosphate (I)]

Trioctylamine (4 mmol) is added to 1.0 g (4 mmol) of PLP and evaporated 5 times to dryness with 10 ml of absolute pyridine. The salt thus obtained is dissolved in 50 ml of absolute pyridine, and 6.9 g (4 mmol) of freshly recrystallized (VI) and 4.1 g (20 mmol) of *N,N'*-dicyclohexylcarbodiimide are added to the solution. The mixture is stirred in a closed, dark flask for 4 days at 20°. Then 150 ml of water are added, and the mixture is allowed to stand for 24 hr. The precipitate of *N,N'*-dicyclohexyl urea is filtered off and washed with water; the washes are pooled with the filtrate. The total solution is evaporated, dissolved in 20 ml of water, extracted with an ether–hexane mixture (1:1), diluted with water to a volume of 2 liters, separated from the precipitate being formed, and, after adjusting the pH to 7.8, isolated from the reaction mixture by column chromatography on DE-32 Whatman cellulose equilibrated with a 2 mM NH$_4$HCO$_3$ solution (pH 7.8). A solution containing 2 mmol of the substance in 2 liters of water is loaded on a 3.5 × 30 cm column and eluted with a linear NH$_4$HCO$_3$ gradient (2 mM to 0.1 M). Fractions having typical absorption and ESR spectra are pooled and evaporated *in vacuo* with added water. This procedure is repeated 5 times. The residue is dissolved in 20 ml of 96% ethanol. The suspension that does not dissolve is separated by centrifugation, the ethanol solution is evaporated and the residue is dehydrated over P$_2$O$_5$. Yield of (I) is 0.4 g (42%).

Preparation and Reduction of the Schiff Bases (Synthesis of
spin-labels (II–V)

A saturated aqueous solution containing 10 mmol of L-aspartic acid, L-glutamic acid or (VII) is mixed with an equal volume of triethylamine, evaporated to dryness, then repeatedly evaporated with 10 ml of triethylamine. The residue is dissolved in 25 ml of absolute methanol. The methanol solutions of the amino acids or (III) are added to a solution con-

[36] A. L. Bocharov, T. V. Demidkina, O. L. Polyanovsky, and M. Ya. Karpejsky, *Mol. Biol. (SSSR)* **7**, 620 (1973).

CHARACTERISTICS OF SPIN-LABELED VITAMIN B_6 DERIVATIVES (I–V)

Compound	Yield (%)	$R_f{}^a$ A	$R_f{}^a$ B	$\lambda_{max}(\epsilon)^b$ pH 1.0	$\lambda_{max}(\epsilon)^b$ pH 7.0	$\lambda_{max}(\epsilon)^b$ pH 14.0
(I)	42	0.64	0.47	292 (12800)	287 (3600)	308 (4900)
					328 (7600)	389 (6300)
					381 (3100)	
(II)	78	0.17	0.15	293 (9000)	253 (6700)	245 (5800)
					325 (7600)	308 (6900)
(III)	23	0.13	0.05	291 (8700)	252 (6400)	247 (6000)
					327 (8000)	309 (7300)
(IV)	32	0.38	0.11	294 (8900)	253 (7000)	247 (6000)
					328 (7300)	308 (7300)
(V)	45	0.48	0.20	294 (8600)	252 (7500)	246 (5200)
					327 (7600)	309 (7400)

^{a} For systems A and B, see Experimental Procedure.
^{b} The values of molar extinction coefficients (ϵ) are calculated from the absorption spectra of (I)–(V) assuming the unpaired electron concentration to be 6.0×10^{23} spins per mole.

taining 10 mmol of trioctylammonium salts of (I) or (VIII). The solutions are stirred for 24 hr at 20° in the presence of 1 ml of triethylamine and evaporated to dryness. Then 50 ml of absolute ethanol followed with portions of 0.42 g (10 mmol) of dry $NaBH_4$ are added to the residue, stirred for 6 hr (the reaction is controlled by the disappearance of a long-wavelength maximum in the absorption spectrum), and mixed with 10 ml of acetone; the solvent is evaporated after 10 min. The residue is dissolved in the minimum volume of water and loaded on a Sephadex G-10 column (1.5 × 120 cm). Fractions absorbing at 325 nm are diluted with water to 2 liters, and, after the pH is adjusted to 7.8, chromatographed as described above. The solvent is evaporated, and then the residue is evaporated 5 times with 10 ml of water and freeze-dried. The resultant yellow powder is dehydrated over P_2O_5.

All the compounds are chromatographically and electrophoretically homogeneous. The electrophoretic mobility of (II) is the same as that of pyridoxamine 5'-phosphate whereas the mobilities of (III)–(V) are twice as high. The yields and spectral and chromatographic characteristics of the compounds are given in Table I.

Determining the Number of Free SH Groups in AAT Preparations[25]

All measurements were carried out with a double-beam Cary 16 (USA) spectrophotometer. One cuvette is filled with 1.8 ml of a solution containing 0.3–0.5 mg of protein per milliliter of sodium acetate buffer, pH 5.5; the second cuvette, with the same volume of the buffer (control).

Then 0.1-ml aliquots of a 10% SDS solution are added to either cuvette, followed with 0.2-ml aliquots of a titrated 0.1 mM PCMB solution. The absorbance at 250 nm is plotted vs. the number of moles of PCMB being added (with a correction for dilution). The number of SH groups is expressed by the equation

$$n = m^*/VC$$

where n = number of SH groups, m^* = number of PCMB moles corresponding to the bending point in the graph for the increment of optical density at 250 nm plotted against the amount of added PCMB (with a correction for dilution); V = sample volume in the cuvette (ml); C = protein concentration (mol/ml).

Determining the Rate of Reduction of the Radicals (III)–(V)

Measurements were done with an EPR-II model. The amplitude of the central component in the ESR spectrum is registered every 60 sec. The concentrations of apo-AAT and carboxymethylated apo-AAT are 0.1 mM ÷ 2(0.1) mM. The concentration of the analog varies from 0.2 to 0.4 mM. The concentration of the radical in solution is measured by comparing the amplitude of the ESR spectrum with the reference sample.

The CD measurements of the complex formation are conducted in parallel, using the same solutions of the protein and spin-labels, in a 2-cm cuvette at the maximum of the CD spectrum.

Determining the Rate of Transamination of L-Aspartic Acid in the Presence of Inhibitors[12]

Three milliliters of a mixture containing 2-oxoglutaric acid (6.7 mM), L-aspartic acid (the concentration varied within the range of 0.67 mM to 13 mM), and the spin-labeled inhibitor (III) (IV, V) (2 μM) are put into the cuvette of a SP-4A (USSR) spectrophotometer, the cuvette being thermostatted at 37°. The rate of enzymic transamination of L-aspartic acid is determined by the formation of oxaloacetic acid. Thus, the number of moles of aspartic acid having reacted with 1 mol of the enzyme per second is expressed by the equation

$$\text{Moles aspartic acid} = d/(\epsilon \cdot C_{AAT} \cdot 60)$$

where d = increment of optical density per minute; ϵ = 600 (molar extinction coefficient of oxaloacetic acid at 280 nm) C_{AAT} = concentration of the enzyme in the cuvette (moles).

The effective values of K_i for the compounds (III)–(V) are determined in a similar way. The concentration of the inhibitors varies within the range of 67 mM to 3.4 μM.

Acknowledgments

The authors express grateful thanks to Academician A. E. Braunstein for his encouragement and continued support in this work and to Dr. M. Verkhovtseva for the translation.

[77] The Use of 4'-N-(2,4-Dinitro-5-fluorophenyl)-pyridoxamine 5'-Phosphate

By F. Riva, A. Giartosio, C. Borri Voltattorni,
and C. Turano

4'-N-(2,4-Dinitro-5-fluorophenyl)pyridoxamine 5'-phosphate (FDNP-

PMP) has been designed to behave as an affinity label for vitamin B_6-dependent apoenzymes, being characterized by a phosphopyridoxyl moiety, which endows it with high affinity for pyridoxal 5'-phosphate (pyridoxal-P) binding sites, and by a fluorinated dinitrophenyl moiety, which is the chemically reactive group.

Its synthesis and properties have been reported elsewhere.[1]

All the pyridoxal-P-dependent enzymes, so far examined, reversibly bind FDNP-PMP with a satisfactory affinity, but not all are labeled by the reagent. The labeling occurs when the FDNP moiety reacts with a nucleophile of the protein located within a definite distance from the coenzyme binding site, giving a stable covalent derivative. The resulting enzyme is irreversibly inactivated, because the coenzyme binding site is blocked.

The reactivity of the enzymes that are labeled[2] varies appreciably. Consequently the methods used to follow the reaction must be adapted to this variability.

As an example of the different situations encountered in the use of the reagent and of the kind of information that can be acquired, we describe here the reaction of FDNP-PMP with two representative enzymes, namely cytosolic and mitochondrial apo-aspartate transaminases from pig heart.

[1] F. Riva, A. Giartosio, and C. Turano, this series, Vol. 46, p. 441.
[2] F. Riva, A. Giartosio, C. Borri Voltattorni, A. Orlacchio, and C. Turano, *Biochem. Biophys. Res. Commun.* **66**, 863 (1975).

Cytosolic Apo-aspartate Aminotransferase

The reaction of cytosolic apo-aspartate aminotransferase (apo-AATc) (10 μM) with FDNP-PMP (12–15 μM) is performed at pH 8 in 50 mM Tris·HCl buffer, in the dark at room temperature, and is followed monitoring (1) the catalytic activity, (2) the spectral variations, and (3) the resulting chemical modifications.

Measurements of Catalytic Activity

At various times aliquots from the reaction mixture are diluted in 1 M phosphate buffer, pH 5.3, to a protein concentration of 10 μg/ml and incubated for 1 hr in the presence of 0.5 mM pyridoxal-P. This treatment is required to displace the reversibly bound inhibitor so that the decrease of specific activity measured is due only to the irreversible covalent binding of the inhibitor.[3]

Similar procedures should be used whenever one deals with an apoenzyme displaying high affinity for the phosphopyridoxyl moiety of the inhibiting reagent.

When the reaction is followed by this method, assaying the activity according to Karmen,[4] it is found that a maximum inhibition of about 80% is reached after 20 hr. Activity measurements carried out directly on aliquots of the reaction mixture, without any preincubation with pyridoxal-P, show that a slow activation of the apoenzyme occurs, reaching a maximum of about 20% within 20 hr.

These observations can be explained assuming the reactions shown in Scheme 1.

Two competing reactions take place after the formation of the apoenzyme–reagent complex: irreversible labeling of the protein and slow

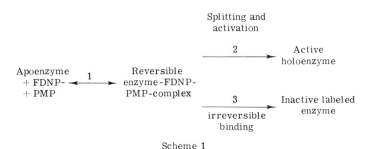

Scheme 1

[3] V. Raso and B. D. Stollar, *Biochemistry* **14**, 591 (1975).
[4] A. Karmen, *J. Clin. Invest.* **34**, 131 (1955).

formation of active coenzyme. The extent of the labeling depends on the ratio between the rates of these two reactions.

In the case of apo-AATc, this ratio is 4 at pH 8 and decreases at lower pH.

Indirect evidence supporting Scheme 1 is provided by the observation that the nonfluorinated analog of FDNP-PMP,[1] which is a competitive inhibitor toward pyridoxal-P and is unable to bind irreversibly to the protein, produces a slow activation of the apoenzyme, presumably according to reactions 1 and 2 of the scheme.

Measurement of Spectral Variations

The absorption spectrum of the reaction mixture in the visible range at zero time is the sum of the spectra of the apoenzyme and of free

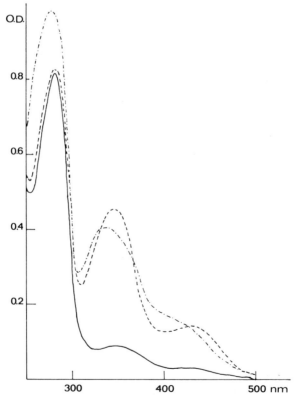

Fig. 1. Absorption spectra of apo-AATc (———), carbamylated apo-AATc (—·—·—), and apo-AATm (———) after 24 hr of incubation with FDNP-PMP, followed by filtration on Sephadex G-25 in 50 mM Tris·HCl buffer, pH 8.

FDNP-PMP at the chosen pH, with a maximum at 350 nm and a shoulder at 410 nm.[1]

During the reaction, spectral variations occur that are correlated to the irreversible loss of catalytic activity. These consist in an increase in absorption at 350 nm and in the appearance of a new peak at 430 nm.

The contribution to this spectrum of nonreacted FDNP-PMP can be eliminated by gel filtration on Sephadex G-25 in 50 mM Tris·HCl buffer, pH 8. The spectrum obtained after 24 hr of incubation and removal of the excess reagent is shown in Fig. 1. From the experiments with model compounds[1], it may be attributed to the reaction of a protein amino group with the reagent.

To demonstrate the irreversible binding of the chromophore, the protein is denatured in 5% trichloroacetic acid (a condition in which any non-

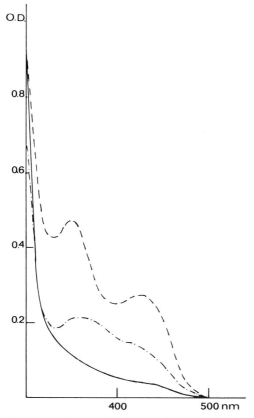

FIG. 2. Absorption spectra of apo-AATc (— — —), carbamylated apo-ATTc (—·—·—), and apo-AATm (———) after 24 hr of incubation with FDNP-PMP, precipitation by 5% trichloroacetic acid, and redissolution in M sodium hydroxide.

covalently bound material is removed from the protein) and redissolved in 1 *M* sodium hydroxide. The spectrum of this solution (Fig. 2) shows an absorption in the visible range that can be attributed to the label.

Chemical Analysis of the Modified Protein

To identify the residues involved in the reaction, the protein is subjected to enzymic digestion and the resulting peptides are separated by the usual techniques. The labeled peptides can be easily followed taking advantage of their yellow color, and the presence of the phosphate group allows their separation according to Strausbauch and Fischer.[5]

The results indicate the labeling of Lys-258, which is the lysyl residue known to bind pyridoxal-P in aldiminic linkage in the native holoenzyme.[6-8] Traces of a peptide containing Cys-45 were also found.

These results suggest that Scheme 1 might be modified as shown in Scheme 2.

If reaction 3 of Scheme 1 is indeed subdivided in two different pathways, a selective modification of Lys-258 should abolish reaction along one of the pathways, favoring the second reaction.

Modified AATc

The lysine that binds pyridoxal-P in aldiminic linkage in apo-AATc can be selectively modified by cyanate with a consequent loss of catalytic activity. However, because cyanate is a small molecule, the other groups present at the active site are still exposed after the lysine residue has been

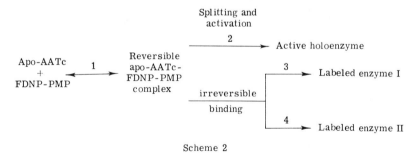

Scheme 2

[5] P. H. Strausbauch and E. H. Fischer, *Biochemistry* **9,** 233 (1970).

[6] Y. Morino and T. Watanabe, *Biochemistry* **8,** 3412 (1969).

[7] A. E. Braunstein, *in* "The Enzymes" (P. Boyer, ed.,), 3rd ed., Vol. 9, p. 379. Academic Press, New York, 1973.

[8] S. Doonan, H. J. Doonan, R. Hanford, C. A. Vernon, J. M. Walker, L. P. Airoldi, F. Bossa, D. Barra, M. Carloni, P. Fasella, and F. Riva, *Biochem. J.* **149,** 497 (1975).

carbamylated, and the resulting enzyme is still able to bind the coenzyme.[9]

On the assumption that also FDNP-PMP will bind to the modified protein, cyanate-inactivated apo-AATc is treated with the reagent in order to see whether other group(s) of the protein might be labeled.

The reaction is performed under the same conditions used for native apo-AATc and followed as previously described, except that activity cannot be assayed. Spectral variations occur, and the spectrum of the enzyme after removal of the excess reagent, shows a shoulder in the 420-nm region and an asymmetric maximum at 350 nm (Fig. 1).

The spectrum of trichloroacetic acid-denatured protein shows the presence of bound material absorbing in the visible range (Fig. 2).

As shown by the reaction with model compounds,[1] such features suggest that the label is bound to a sulfhydryl group of the protein.

A peptide analysis according to the methods described above, indicates that Cys-45, which is known to be "exposed"[7,8] on the enzyme surface, is labeled.

Mitochondrial Aspartate Aminotransferase

The mitochondrial isozyme of aspartate aminotransferase (AATm) has a primary structure closely resembling that of AATc, with about 47% homology.[10]

This homology, the fact that the two enzymes catalyze the same reaction and the high affinity for the coenzyme, make it reasonable to assume that the active site is similar in the two isozymes. But when apo-AATm is allowed to react with FDNP-PMP, in the same experimental conditions used for AATc, a different result is obtained. The initial inhibition is reversed just by incubation with pyridoxal-P; thus, after 8 hr of reaction, the specific activity reaches 89% of the value for the untreated apoenzyme. A slow spontaneous activation occurs, which can be evaluated, as in the case of AATc, by direct activity measurements without addition of pyridoxal-P (38% of activation after 8 hr of reaction).

These data can be interpreted according to Scheme 1, assuming that apo-AATm reacts with FDNP-PMP through reactions 1 and 2, but that reaction 3, corresponding to the covalent binding, does not occur appreciably in this system.

Moreover, the spectral variations occurring in the reaction mixture differ from those observed with apo-AATc, in that there is a slow disap-

[9] J. C. Slebe and M. Martinez-Carrion, *J. Biol. Chem.* **251**, 563 (1976).
[10] D. Barra, F. Bossa, S. Doonan, H. M. A. Fahmi, G. J. Hughes, K. Y. Kakoz, F. Martini, and M. Petruzzelli, *FEBS Lett.* **83**, 241 (1977).

pearance of the shoulder at 410 nm, while a slight increase in absorption is noticed at 340 nm. These variations may be attributed to the splitting of the bond between the nitrogen and the dinitrophenyl group in the reagent molecule.

After 24 hr of incubation, the spectrum of the enzyme freed from excess reagent is very similar to that of native aldehydic enzyme (Fig. 1), and after trichloroacetic acid denaturation a spectrum similar to that of native apoenzyme is obtained (Fig. 2).

It is not surprising that a cysteinyl adduct is not formed in this case, since AATm contains no thiol corresponding to Cys-45 of AATc.[10]

However, the lysyl residue at the active site (Lys-258), known to bind the aldehydic coenzyme, should be available.

Comments

FDNP-PMP produces irreversible labeling and allows one to identify fragments near the active site.

The behavior of pyridoxal-P-dependent apoenzymes toward this reagent is always characterized by a high affinity; however, the labeling reaction may be different, and the examples reported here show that it can be greatly affected by small differences in the active-site architecture of very similar enzymes.

Light-stimulated cleavage of reduced pyridoxal-P–enzyme complexes has been recently described.[11] In the present case, the activation reaction takes place also in the dark when the reagent is bound to the active site of the enzyme.

The mechanism of this reaction is not yet explained, but it could be assumed that in FDNP-PMP the dinitrophenyl moiety acts as a strong electron-withdrawing group, contributing to the labilization of the compound.

An activation by FDNP-PMP has been noticed in other pyridoxal-P-dependent apoenzymes and is particularly rapid in dopa decarboxylase from pig kidney.

Acknowledgment

This work has been supported by the Center of Molecular Biology of the C.N.R.

[11] J. M. Ritchey, I. Gibbons, and H. K. Schachman, *Biochemistry* **16,** 4584 (1977).

[78] Immobilization of Pyridoxal 5'-Phosphate and Pyridoxal 5'-Phosphate-Dependent Enzymes on Sepharose

By SEI-ICHIRO IKEDA and SABURO FUKUI

Immobilized derivatives of PLP (pyridoxal 5'-phosphate) may be useful for the purification by affinity chromatography of enzymes requiring this coenzyme or for the studies of coenzyme–apoenzyme interaction. Furthermore, considerable attention is being given to the application of soluble or insoluble macromolecularly bound PLP to bioreactors for analytical, biochemical, or technological purpose.

This article deals with (1) the preparative procedures for three types of Sepharose-bound PLP (SP-A, SP-B, and SP-C in Fig. 1) and their properties, (2) the application of these immobilized PLP for the affinity chromatography of apo-tryptophanase from *Escherichia coli* and for the immobilization forms of some PLP enzymes consisting of multisubunits, e.g., apo-tryptophanase, through biospecific binding to the active center of the one subunit.

Preparation of Sepharose-Bound Derivatives of PLP

Three types of Sepharose-bound PLP retaining all or most of the functional groups (Fig. 1) are prepared according to the following procedures.[1,2] 6-Immobilized PLP (SP-A) is obtained by coupling diazotized *p*-aminobenzamidohexyl-Sepharose to the 6-position of PLP. *p*-Aminobenzamidohexyl-Sepharose, which is prepared from Sepharose 4B as described by Cuatrecasas,[3] is diazotized by treatment with sodium nitrite (0.1 M) in 0.5 N HCl for 7 min at 4°. The resulting diazonium salt of the Sepharose derivative (10 ml) is added to 10 ml of 0.2 M borate buffer (pH 8.0) containing 40 mg of PLP. The coupling reaction is allowed to proceed for about 8 hr at 4°. The PLP-bound Sepharose thus obtained is thoroughly washed with water and then heated up to *ca.* 60° for 30 min to destroy unreacted diazonium derivative.

N-Immobilized and 3-O-immobilized PLP derivatives (SP-B and SP-C) can be obtained by coupling of PLP to bromoacetamidohexyl-Sepharose under suitable conditions. The preparation of SP-B is carried out in dimethylformamide in which the pyridine nitrogen of PLP is mostly

[1] S. Ikeda and S. Fukui, *Biochem. Biophys. Res. Commun.* **52,** 482 (1973).

[2] S. Ikeda, H. Hara, and S. Fukui, *Biochim. Biophys. Acta* **372,** 400 (1974).

[3] P. Cuatrecasas, *J. Biol. Chem.* **245,** 3059 (1970).

METHODS IN ENZYMOLOGY, VOL. 62

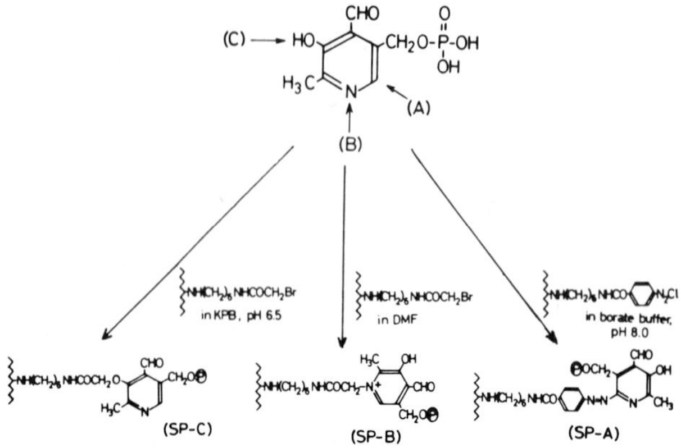

FIG. 1. Structures of three kinds of Sepharose-bound pyridoxal 5'-phosphate and their preparation procedures.

nonprotonated and the 3-OH group is nonionized. In contrast, the preparation of SP-C is carried out in 0.1 M potassium phosphate buffer (pH 6.0) in which the pyridine nitrogen of PLP is mostly protonated and the 3-OH group is ionized.[4,5] Bromoacetamidohexyl-Sepharose is prepared from Sepharose 4B according to the procedure described by Cuatrecasas.[3] This Sepharose derivative (5 ml) is added to 5 ml of dimethylformamide (50% v/v) containing 10 mM PLP. The coupling reaction is allowed to proceed for about 72 hr at room temperature in the dark. The reaction product is thoroughly washed with 0.1 M potassium phosphate buffer (pH 9.0 and pH 5.5, alternatively) and then treated with excess ethanolamine (50 mM, pH 7.0) for 6 hr to block the remaining reactive groups of the Sepharose derivative. The PLP-bound Sepharose thus obtained is stored at 4° in 0.1 M borate buffer (pH 5.3).

For the preparation of SP-C, bromoacetamidohexyl-Sepharose (5 ml) is added to 5 ml of 0.1 M potassium phosphate buffer (pH 6.0) containing 10 mM PLP and 0.1 M DL-valine. The mixture is then treated as in the preparation of SP-B.

Properties of Sepharose-Bound PLP Derivatives

Spectral Characterization

Treatment of SP-A (brown) (250 mg) with 0.1 M sodium dithionite (2 ml) in 0.2 M sodium borate (pH 9.0) results in liberation of a

[4] D. E. Metzler and E. E. Snell, J. Am. Chem. Soc. 77, 2431 (1955).
[5] D. Heinert and A. E. Martell, J. Am. Chem. Soc. 85, 183, 188 (1963).

pyridoxine-like substance having an absorption maximum at about 316 nm at pH 9.0 and at 282 nm at pH 3.0, respectively. Katritzky *et al.*[6] synthesized 6-aminopyridoxine by reduction with sodium dithionite of 6-phenylazopyridoxine, which was prepared by coupling diazotized aniline to pyridoxine. These facts strongly suggest that 6-aminopyridoxine 5′-phosphate (6-amino-PINP) or a closely related compound is released from SP-A by the dithionite treatment. From the optical density of this pyridoxine derivative liberated, it is estimated that the SP-A preparation tested contains about 1.5 mol of PLP per gram of Sepharose.

Absorption spectra of SP-B and SP-C (both are yellow) are recorded with suspensions of these immobilized PLP derivative (200 mg, wet gel) in 2 ml of glycerol (75% v/v) containing 0.1 N NaOH or 0.1 N HCl. A corresponding suspension of bromoacetamidohexyl-Sepharose is used as a blank.

Both SP-B and SP-C show absorption maxima at 295 nm in 0.1 N HCl. On the other hand, in 0.1 N NaOH, SP-B shows an absorption maximum at 388 nm as observed in free PLP or N-methyl PLP[7] and SP-C shows an absorption maximum at 315–335 nm as observed in 3-O-methyl PLP.[5,7] These results are consistent with the expected structures of SP-B and SP-C. The PLP contents in SP-B and SP-C can be calculated from the absorbance at 295 nm in 0.1 N HCl using the molar absorption coefficient of the free coenzyme, 6700 mol^{-1} cm^{-1}.[8]

More clear evidence for the structures of SP-B and SP-C can be obtained by the absorption spectra of the substances liberated from these gels by acid treatment. The unstable 4-formyl group of SP-B or SP-C is reduced chemically by adding 1 mg of NaBH$_4$ to 200 mg (wet) of the gel suspended in 2 ml of 0.1 M potassium phosphate buffer (pH 8.5). When these reduced PLP analogs are treated with 2 ml of 6 N HCl for 12 hr at room temperature, pyridoxine-like substances are liberated from the gels. The eluate from the reduced SP-B shows an absorption maximum at 290 nm in 6 N HCl and an absorption peak at 325 nm at both pH 7 and 10. These spectra are analogous to that of N-methylpyridoxine.[4] The eluate from the reduced SP-C shows an absorption maximum at 290 nm in 6 N HCl and at 280 nm at pH 10. These spectra resemble that of 3-O-methylpyridoxine.[4] The treatment of these Sepharose-bound PLP with 6 N HCl is considered to hydrolyze the peptide bond in the arm linking PLP to Sepharose matrix.

[6] A. R. Katritzky, H. Z. Kucharska, M. J. Tucker, and H. M. Wuest, *J. Med. Chem.* **9**, 620 (1966).
[7] A. Pocker and E. H. Fischer, *Biochemistry* **8**, 5181 (1969).
[8] E. A. Peterson and H. A. Sober, *J. Am. Chem. Soc.* **62**, 3198 (1940).

TABLE I
CHARACTERIZATION OF THREE KINDS OF IMMOBILIZED PLP DERIVATIVES

Immobilized PLP derivatives	Immobilized position of PLP	PLP content (μmol PLP/g Sepharose)	Nonenzymic catalytic activity[c] (μmol PLP/g Sepharose)	Affinity for vitamin B_6 enzymes
SP-A	6	1.5[a]	1.5–3.0	+
SP-B	1-N	1.2[b]	0.8–1.1	+
SP-C	3-O	0.84[b]	0–trace	+

[a] Calculated from the spectral data of the substance liberated from SP-A gel by treatment with sodium dithionite (see S. Ikeda and S. Fukui, Biochem. Biophys. Res. Commun. **52**, 482 (1973).

[b] Calculated from the direct spectral measurement of SP-B and SP-C (see footnote 2).

[c] Calculated from the catalytic activity of free PLP.

Catalytic Activity for Nonenzymic Cleavage of Tryptophan

Pyridoxal or its appropriate analogs are well known to catalyze the nonenzymic cleavage of tryptophan in the presence of Cu^{2+} [9]; both Sp-A and SP-B show catalytic activity. On the other hand, SP-C does not exhibit any appreciable catalytic activity in the model system, since it lacks the essential 3-hydroxyl group.[10]

Characterization of the three kinds of immobilized analogs of PLP prepared as mentioned above is summarized in Table I.

Applications

Application of Sepharose-Bound PLP Derivatives to Affinity Chromatography[11]

Since all the immobilized derivatives of PLP possess a free 4-formyl group and a 5'-phosphate group, which are known to be essential for binding with apo-PLP enzymes, it is expected that these compounds have specific affinities for such apoproteins.

For example, apo-tryptophanase in the crude extracts from *E. coli* B/lt 7-A[12] is specifically adsorbed on a SP-C column and eluted from the column with 0.5 M potassium phosphate buffer (pH 7.0) containing 0.5 mM PLP, as shown in Fig. 2A.[11] The specific tryptophanase activity in the fraction increased 6 to 8 times as compared with that of the initial crude extracts, although some inactivation of the enzyme is observed during the course of affinity chromatography owing to appreciable instability of apo-tryptophanase under the experimental conditions employed.

On the other hand, a Sepharose-bound pyridoxine 5'-phosphate derivative of the same type does not serve as biospecific adsorbent. Like apo-tryptophanase, tyrosine phenol-lyase from *E. intermedia*[13] is efficiently adsorbed on a column of Sepharose-bound PLP and eluted with 0.5 M potassium phosphate buffer (pH 7.0) containing 0.5 mM PLP. Thus, it is demonstrated that a suitable immobilized derivative of PLP, such as SP-C, serves as an excellent affinity adsorbent for different kinds of vitamin B_6 enzymes having an appropriate affinity for the coenzyme. Although pyridoxamine 5'-phosphate (PMP) bound to Sepharose through its 4-aminomethyl group has been reported to be successfully used for the

[9] E. McEvoy-Bowe, *Arch. Biochem. Biophys.* **113**, 167 (1966).

[10] E. E. Snell, *Vitam. Horm.* (*New York*) **16**, 77 (1958).

[11] S. Ikeda, H. Hara, S. Sugimoto, and S. Fukui, *FEBS Lett.* **56**, 307 (1975).

[12] W. A. Newton, Y. Morino, and E. E. Snell, *J. Biol. Chem.* **240**, 1211 (1965).

[13] H. Yamada, H. Kumagai, N. Kashima, H. Torii, H. Enei, and S. Okumura, *Biochem. Biophys. Res. Commun.* **46**, 370 (1972).

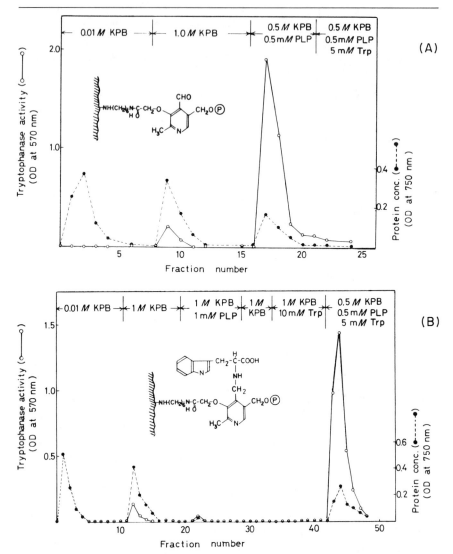

Fig. 2. Elution patterns of apo-tryptophanase from Sepharose-bound pyridoxal 5'-phosphate column (A) and from a Sepharose-bound tryptophan-pyridoxal 5'-phosphate complex. Crude extracts from *Escherichia coli* B/lt 7-A (3.0 mg; specific tryptophanase activity, 0.4 unit per milligram of protein) was applied to a column of the indicated adsorbent (0.5 × 5 cm). Elution was carried out with the indicated eluting agent at 25° at a flow rate of space velocity (SV) = 5 hr^{-1}. Fractions (0.5 ml) were collected and assayed for both tryptophanase activity and protein concentration. For the details, see S. Ikeda, H. Hara, S. Sugimoto, and S. Fukui, *FEBS Lett.* **56,** 307 (1975).

purification of tyrosine aminotransferase[14] and aspartate aminotransferase,[15] this compound lacks a free 4-formyl group essential for the biospecific binding to most of vitamin B_6 enzymes. Hence, this immobilized derivative of PMP has only limited usefulness in the affinity chromatography of vitamin B_6 enzymes.

As shown in Fig. 2B, apo-tryptophanase shows a peculiar behavior toward a Sepharose-bound PLP–tryptophan complex (SP-Trp) prepared by the reduction of the Schiff base formed between SP-C and tryptophan.[2] The enzyme is desorbed only when the column is treated with a mixture of potassium phosphate buffer, PLP, and tryptophan. Thus, SP-Trp serves as a more specific adsorbent than SP-C. Various phosphopyridoxylamino acids have structures analogous to those proposed for intermediate coenzyme–substrate complexes during the course of PLP-dependent enzyme reactions. Such phosphopyridoxylamino acids have been shown to bind efficiently to apoproteins of various PLP enzymes.

Application of Sepharose-Bound PLP Derivatives for Immobilization of Vitamin B_6 Enzymes

Most of vitamin B_6 enzymes are known to consist of subunits and have multiple binding sites. For instance, *E. coli* tryptophanase consists of four subunits, each of which has one PLP-binding site.[12,16] This enzyme catalyzes α,β-elimination and β-replacement reactions of tryptophan and its analogs.[12,16] The α,β-elimination reactions are found to be reversible.[17] All the above-mentioned Sepharose-bound PLP derivatives are useful for the immobilization of tryptophanase and other vitamin B_6 enzymes of multisubunit structures through the biospecific binding at the active site of the one subunit.[18] Immobilization of tryptophanase by this method is compared with two well-known methods in which the enzyme is immobilized directly or through a spacer on Sepharose as insoluble carrier (Fig. 3). METHOD 1: The enzyme is immobilized directly on CNBr-activated Sepharose [(A) in Fig. 3]. For example, CNBr-activated Sepharose 4B (0.5 g) is mixed with 0.20 mg of apo- or holo-tryptophanase dissolved in 1.0 ml of 0.1 M potassium phosphate buffer (pH 7.0). In a glass funnel

[14] E. S. Severin, N. N. Gulyaev, E. N. Khurs, and R. M. Khomutov, *Biochem. Biophys. Res. Commun.* **35**, 318 (1969).

[15] C. B. Voltattorni, A. Orlacchio, A. Giartosio, F. Conti, and C. Turano, *Eur. J. Biochem.* **53**, 151 (1975).

[16] Y. Morino and E. E. Snell, *J. Biol. Chem.* **242**, 2800 (1967).

[17] T. Watanabe and E. E. Snell, *Proc. Natl. Acad. Sci. U.S.A.* **69**, 1086 (1972).

[18] S. Fukui, S. Ikeda, M. Fujimura, H. Yamada, and H. Kumagai, *Eur. J. Biochem.* **51**, 155 (1975).

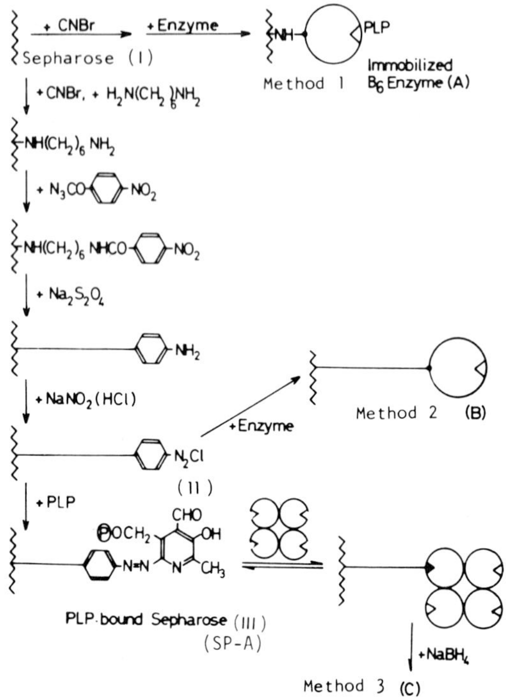

FIG. 3. Principles for immobilization of vitamin B_6 enzymes (e.g., tryptophanase) by different covalent binding methods.

equipped with a sintered-glass filter, the mixture is gently agitated for 24 hr at 4°. The resulting insolubilized tryptophanase is thoroughly washed with 0.1 M potassium phosphate buffer (pH 7.0). METHOD 2: The enzyme is immobilized on Sepharose through a reactive side arm. A diazonium derivative of Sepharose and a bromoacetyl derivative of Sepharose are used for immobilization. For example, diazotized p-aminobenzamidohexyl-Sepharose (0.5 g) is mixed with 1.0 ml of 0.1 M potassium phosphate buffer (pH 8.0) containing 1.0 mg of apo- or holo-tryptophanase. The coupling reaction is allowed to proceed for 15 min at 4°. The immobilized tryptophanase thus obtained [(B) in Fig. 3] is thoroughly washed with 0.1 M potassium phosphate buffer (pH 8.0). METHOD 3: The enzyme is immobilized on Sepharose through PLP previously bound to Sepharose. Three kinds of Sepharose-bound PLP (SP-A, SP-B, and SP-C in Fig. 1) are used. For example, immobilization of apo-tryptophanase on SP-A is carried out as follows. Wet SP-A (0.5 g) is mixed with 1.0 ml of 0.1 M potassium phosphate buffer (pH 7.0) containing 0.50 mg of apo-tryptophanase. The mixture is incubated for 20 min at 37°. The resulting SP-A-apo-tryptophanase complex is reduced with NaBH$_4$ and then washed thoroughly with a mixture of 0.1 M potassium phosphate buffer (pH 6.0), 5 mM PLP, and 11.4% ammonium sulfate [(C) in Fig. 3].

The third immobilization method, employed first by the present authors,[1,18] is different from the above-mentioned two well-known methods in that this new immobilization method is based on the specific affinity between the coenzyme and the active center of apoenzyme. Since the Schiff base linkage between the 4-formyl group and the ε-amino group of the lysine residue at the active center of tryptophanase is not sufficiently strong to obtain a stable immobilized enzyme, the linkage is fixed by reduction with NaBH$_4$. The conditions for immobilization and the enzyme activities of immobilized enzyme preparations obtained by these methods are summarized in Table II.

In the case of tryptophanase, the third immobilization method, especially the use of SP-A, is most effective with respect to both immobilization efficiency (ca. 60%) and activity of the resulting preparation (about 80% relative activity of the free counterpart). As shown in Fig. 3, the binding of apo-tryptophanase to Sepharose-bound PLP occurs at the active site of one subunit of tetrameric apo-tryptophanase under our experimental conditions. Although this subunit coupled to Sepharose-bound PLP becomes inactive by the above-mentioned reduction treatment of Schiff base linkage, the remaining three subunits seem to retain most of the original activities. In fact, the fairly high activity of the immobilized enzyme by the biospecific coupling to Sepharose-bound PLP derivatives indicates that the extent of conformational distortion of enzyme molecule caused by the immobilization method would be considerably smaller than the cases of the other immobilization methods.

Of the three Sepharose-bound PLP derivatives tested, SP-A is the best for immobilization of apo-tryptophanase. These results may be explained as follows: SP-A has all the functional groups of PLP necessary for the appearance of its coenzyme activity and binding to apoenzyme. Moreover, the PLP content of SP-A used was higher than that of SP-B and of SP-C (See Table I). In the case of SP-B, the alkylation of the pyridine-nitrogen of PLP moiety would result in a decreased affinity for apoprotein and affect the reactivity of the 4-formyl group as well as ionization of the 3-hydroxy group. In the case of SP-C, the 3-hydroxy group of PLP moiety is blocked. This may cause a decrease in the affinity for apoprotein by steric and/or electronic hindrance. As mentioned before, SP-A and SP-B serve as catalysts in the nonenzymic α,β-elimination reaction of tryptophan, whereas SP-C does not exhibit any catalytic activity. However, SP-C, as in the cases of SP-A and SP-B, maintains the functional groups

TABLE II

COMPARISON OF IMMOBILIZATION EFFICIENCY AND ACTIVITY OF RESULTING IMMOBILIZED PREPARATION OF TRYPTOPHANASE AND
TYROSINE PHENOL-LYASE BY DIFFERENT IMMOBILIZATION METHODS

Expt. No.	Method[a]	Sepharose derivative used[b] (g)	Type and amount of enzyme used (mg)	Efficiency of immobilization[c] (%)	Relative activity[d]
1	1	I	0.50	Trp-ase(apo) 0.20	73	34
2	1	I	0.50	(holo) 0.20	68	42
3	2-A	II-A	0.50	(holo) 1.00	42	10
4	2-B	II-B	0.50	(holo) 0.50	65	35
5	3-A	SP-A	0.50	(apo) 0.50	81	21
6	3-B	SP-B	0.50	(apo) 0.50	21	51
7	3-C	SP-C	0.50	(apo) 0.50	33	48
1'	1	I	1.09	Tyr-ase(apo) 2.00	59	41
2'	1	I	1.36	(apo) 5.00	98	40
3'	1	I	1.06	(holo) 2.00	60	30
4'	2-B	II-B	0.44	(apo) 0.50	94	41
5'	2-B	II-B	0.49	(holo) 0.50	94	33
6'	3-A	SP-A	0.52	(apo) 0.50	58	29

[a] Method 1: Enzyme was bound directly on CNBr-activated Sepharose 4B. Method 2: Enzyme was bound on Sepharose 4B through a suitable spacer. Method 3: Enzyme was bound to Sepharose through PLP.

[b] Sepharose derivative used: I, CNBr-activated; II-A, diazotized p-aminobenzamidohexyl-Sepharose; II-B, bromoacetamidohexyl-Sepharose.

[c] Expressed by the ratio of the amount of protein immobilized on each Sepharose derivative to the amount of protein originally used for the immobilization.

[d] The ratio of the specific activities of immobilized enzymes to those of free counterparts using the same amount of enzyme protein. The details of the assay methods are mentioned by S. Fukui, S. Ikeda, M. Fujimura, H. Yamada, and H. Kumagai, $Eur. J. Biochem.$ **51**, 155 (1975).

necessary for the binding to apoproteins of various vitamin B_6 enzymes. Hence, all these three Sepharose-bound PLP derivatives can be applicable to both the affinity chromatography and the immobilization of vitamin B_6 enzymes consisting of multisubunits.

The usefulness of Sepharose-bound PLP derivatives for immobilization of vitamin B_6 enzymes change depending on the structure of enzyme molecule to be immobilized.

For example, in the case of tyrosine phenol-lyase from *E. intermedia* having two PLP-binding sites per molecule,[19] the catalytic activity of the immobilized preparations obtained with Sepharose-bound PLP derivatives were inferior to that of the counterparts immobilized on CNBr-activated Sepharose directly or through a suitable spacer (Table II). The reason may be explained as follows. Binding of the one catalytic center of the enzyme with Sepharose-bound PLP blocks this site and renders it inactive. The resulting distortion of the enzyme molecule should have more serious effects on the activity of the remaining catalytic center than the case of tetrameric tryptophanase.

Immobilization of tryprophanase and tyrosine phenol-lyase changes, to some extent, the properties of these enzymes. Upon immobilization, the pH optima of both enzymes shift 0.5–1.0 pH unit to the alkaline side. Both immobilized enzymes show higher thermal stability and resistance to a denaturing agent, such as guanidine-HCl, than their free counterparts.[18]

Immobilized tryptophanase and tyrosine phenol-lyase are useful not only for the analyses of tryptophan, tyrosine, and their analogs, but also for the production of these amino acids. L-Tryptophan, L-tyrosine, and their analogs, such as 5-hydroxy-L-tryptophan and L-DOPA, can be efficiently produced in batch systems or continous-flow systems using these immobilized enzymes as biocatalysts.[20]

[19] H. Kumagai, H. Yamada, H. Matsui, H. Ohgishi, and K. Ogata, *J. Biol. Chem.* **245**, 1767, 1773 (1970).

[20] S. Fukui, S. Ikeda, M. Fujimura, H. Yamada, and H. Kumagai, *Eur. J. Appl. Microbiol.* **1**, 25 (1975).

[79] Pyridoxal 5'-Phosphate and Analogs as Probes of Coenzyme–Protein Interaction[1]

By Bob In-yu Yang *and* David E. Metzler

Pyridoxal phosphate (pyridoxal-P) provides a built-in indicator at the active sites of a substantial number of enzymes. Its absorption spectrum, which is often radically altered by substrates, quasisubstrates, or inhibitors, provides a means of directly monitoring intermediates in the catalytic process. A variety of analogs of the coenzyme are available and can often be substituted for pyridoxal-P to good advantage. The reaction sequence may be slowed. Intermediates may be stabilized, and insight into the mode of binding of the coenzyme may be obtained. Some analogs cause covalent modification of enzymes. Pyridoxal phosphate and related compounds have been widely used as modifying reagents for nonpyridoxal-P-dependent enzymes and other proteins. These aldehydes form Schiff bases with lysine side chains, often at specific locations in a protein. Reduction with borohydrides provides a convenient means of irreversibly attaching the aldehyde.

In this chapter we consider the following: sources of vitamin B_6-dependent enzymes; reversible removal of pyridoxal-P; reconstitution with analogs; analysis of absorption spectra, circular dichroism, and kinetics of reconstituted enzymes. The use of analogs in crystallographic studies is described in this volume [80].

Preparation of Pyridoxal Phosphate-Dependent Enzymes

Classical methods, such as salt fractionation, ion-exchange chromatography, and crystallization, are usually employed. Several techniques that seem to be particularly suitable for these enzymes are: (a) heating in the presence of a competitive inhibitor to inactivate other proteins, (b) chromatography on hydroxyapatite, and (c) chromatography on carboxymethyl cellulose or carboxymethyl Sephadex.

A number of enzymes of this class that have been prepared in homogeneous or near-homogeneous form are listed in the table. Methods of preparation are given in the sources indicated. Procedures for isolation of some other enzymes of this class are given in this series, Volumes 17A and 17B.

The two enzymes aspartate aminotransferase and glutamate decarboxylase have provided much of the authors' experience and the illustrative

[1] Supported in part by NIH Grant AM-01549. We wish to express our appreciation to Carol Harris Metzler for assistance.

material for this chapter. We, therefore, summarize in the following paragraphs suitable methods for their preparation.[2]

Aspartate Aminotransferase, Cytosolic Isoenzyme. The enzyme may be isolated from pig hearts or chicken hearts. The procedure of Jenkins and Sizer[3] should be modified as suggested by Martinez-Carrion *et al.*[4] to avoid the use of maleate buffer, which modifies sulfhydryl groups during the heat treatment, and to separate the subforms. The quantities mentioned are appropriate for 16 kg of pig hearts. Grind the fresh heart muscle, freed of excess fat and fibers, with a meat grinder. Do not use a blender in preparing the cytosolic enzyme. Mix the ground tissue with 0.5 volume of 50 mM glutarate buffer, pH 6, containing 5 mM EDTA. With continuous stirring heat to 60° and add 2.1 g of α-ketoglutaric acid. Then heat at 75° for 20 min. Strain through cheesecloth, cool rapidly to 5°, and fractionate with ammonium sulfate between 52 and 68% saturation (328 g/liter, then an additional 106 g/liter). A broader cut (47% saturation; 290 g/liter to 72% saturation; an additional 170 g/liter) will increase the yield significantly but may require more column volume in subsequent steps. Take up in a small volume (about 250 ml for a preparation from 16 kg of hearts) of 20 mM potassium phosphate (no sodium) buffer, pH 6.8. Add 2 ml of 0.2 M neutral α-ketoglutarate. Dialyze against the same buffer to completely remove ammonium sulfate. Prepare a 2.6 × 40 cm column of hydroxyapatite, preferably in the granular form described by Mazin *et al.*[5] (Note: the latter may be obtained in higher yield by using K_2HPO_4 rather than Na_2HPO_4. See also Siegelman *et al.*[6] for another preparation that avoids the heating steps.)

Again add 2 ml of 0.2 M α-ketoglutarate; apply to the hydroxyapatite column and elute with 80 mM potassium phosphate buffer, pH 6.8. Record the absorption spectra of the golden yellow fractions. There is often a small amount of a heme protein with a sharp absorption maximum at 410 nm. It is totally removed in the final step of chromatography on CM-Sephadex. However, we discard fractions in which the absorbance at 410 nm exceeds that at 364 nm (the peak for bound pyridoxal-P).

Dialyze the pooled yellow fractions from the hydroxyapatite column (after concentration by ultrafiltration, if desired) against 40 mM sodium acetate buffer, pH 5.1. Apply the enzyme to a 2.6 × 35 cm column of

---

[2] We are indebted to W. T. Jenkins, M. Fonda, and H. Ueno for help in the preparation of this section.

[3] I. W. Sizer and W. T. Jenkins, this series, Vol. 5 [94a].

[4] M. Martinez-Carrion, C. Turano, E. Chiancone, F. Bossa, A. Giartosio, F. Riva, and P. Fasella, *J. Biol. Chem.* **242**, 2397 (1967).

[5] A. L. Mazin, G. E. Sulimova, and B. F. Vanyushin, *Anal. Biochem.* **61**, 67 (1974).

[6] H. W. Siegelman, G. A. Wieczorek, and B. C. Turner, *Anal. Biochem.* **13**, 402 (1965).

HIGHLY PURIFIED PYRIDOXAL PHOSPHATE AND PYRIDOXAMINE PHOSPHATE-DEPENDENT ENZYMES

Enzyme	EC No.	Source	Literature[n]		
			Holo	Crystallization	Apo
Aminotransferases (transaminases)					
Alanine (glycine)	2.6.1.2	Pig heart	1	—	a
		Rat liver,	2, 3[b,c]	2	a
D-Alanine-D-glutamate	2.6.1.21	*Bacillus subtilis*	4	—	4
γ-Aminobutyrate	2.6.1.19	Rat brain	5,[d] 6[c]	—	—
		Mouse brain	7[a]	—	—
		Pig brain	8[b]	—	—
		Rabbit brain	9[c]	—	
Aspartate (glutamic–oxaloacetic transaminase, glutamic aspartic transaminase)	2.6.1.1	Pig heart	10,[b] 11,[d] 12[c]	13, 14	12, 15, 16
		Beef liver	17	—	18
		Chicken heart	19,[d] 20,[d] 21[b]	19, 20, 21	19, 21
		Chicken heart and liver	22[b,d]	—	—
		Beef kidney	23[d]	—	—
		Escherichia coli	24[f]	—	25
Cysteine	2.6.1.3	Rat liver	26	—	—
Glutamine	2.6.1.15	Rat liver	27	—	a
Imidazolylacetol phosphate	2.6.1.9	*Salmonella typhimurium*	28	—	—
Kynurenine	2.6.1.7	Rat liver	29	—	—
Leucine	2.6.1.6	Rat liver	30[e]	—	
Lysine	2.6.1.36	*Flavobacterium lutescens*	31	31	31
Ornithine	2.6.1.13	Rat kidney	32, 33, 34	32, 33, 34	37
		Rat liver	35, 36	35, 36	
Phosphoserine	2.6.1.52	Sheep brain	38	—	38
Tyrosine	2.6.1.5	Rat liver	39, 40, 41[b]	—	42
		E. coli	24[f]	—	25
Branched chain	2.6.1.42	Pig heart	43, 44	—	43
		Pigeon brain	45	—	—

Enzyme	EC no.	Source			
Serine pyruvate (Phe[His]-pyruvate)[g]	2.6.1.51	S. typhimurium	46	46	46
		Mouse, dog, cat liver	47, 48	—	—
Pyridoxamine pyruvate	2.6.1.30	Rat liver	49, 50	—	—
		Pseudomonas	51[h]	51	51
Racemases					
Alanine	5.1.1.1	Pseudomonas	52, 53	—	—
Arginine	5.1.1.9	P. graveoleus	54	54	54
Nonspecific	5.1.1.10	P. striata	55	—	55
Enzymes catalyzing β elimination					
Alliin lyase (allinase, S-alkyl-L-cysteine lyase)	4.4.1.4	Acacia farnesiana seedlings	56	—	—
L-Serine dehydratase[i]	4.2.1.13	Rat liver	57, 58	57	57
D-Serine dehydratase	4.2.1.14	E. coli	59, 60	59, 60	59, 60
Threonine dehydratase[i]					
Biodegradative	4.2.1.16	Clostridium tetanomorphum[j]	61	61	—
		E. coli[k]	62	62	63
		B. subtilis	64	64	—
Biosynthetic		S. typhimurium	65	—	65
Tryptophanase	4.1.99.1	E. coli	66[h]	66	66[h]
		B. alvei	67	—	67
Tyrosine phenol-lyase	4.1.99.2	E. intermedia	68[h]	68	68[h]
Enzymes catalyzing β replacement					
β-Cyanoalanine synthetase	4.4.1.9	Blue lupine seedlings	69	—	—
Cystathionine β-synthetase (serine sulfhydrylase, β-thionase, methylcysteine synthetase)	4.2.1.22	Chicken liver	70	—	—
O-acetylserine sulfhydrylase (cysteine synthetase)	4.2.99.8	Rat liver	71	—	71
		S. typhimurium	72	—	—
		Rape leaves (Brassica chinensis)	73, 74[l]	—	—
Tryptophan synthetase, β₂ subunit	4.2.1.20	E. coli	75[h]	75	75[h]

(continued)

TABLE (*Continued*)

Enzyme	EC No.	Source	Literature[a]		
			Holo	Crystallization	Apo
Enzymes catalyzing γ-elimination or replacement					
Cystathionine γ-lyase (homoserine dehydratase, γ-cystationase)	4.4.1.1	Rat liver	76, 77	77	76, 78
L-Methionine γ-lyase (methioninase)	4.4.1.11	*Pseudomonas oxalis*	79	—	80
Cystathionine γ-synthetase	4.2.99.9	*S. typhimurium*	81	—	81
Decarboxylases					
Arginine	4.1.1.19	*Lathyrus sativus* seedling	82	—	—
		E. coli	83	83	83
(Biosynthetic)		*E. coli*	84	84	84
Aromatic amino acid	4.1.1.28	Pig kidney	85	—	*a*
Glutamate	4.1.1.15	Mouse brain	86	—	—
		E. coli	87	—	88
Lysine	4.1.1.18	*E. coli*	89	—	89
Ornithine	4.1.1.17	*Bacterium cadaveris*	90	90	*a*
		Rat liver	91	—	—
		Mouse fibroblast (SV-40 transformed)	92	—	—
Dialkyl amino acid	4.1.1.64	*P. cepacia*	93	93	—
		P. fluorescens	94	—	—
Aspartate β-decarboxylase	4.1.1.12	*Alcaligenes faecalis*	95	—	95
		Achromobacter	96	96	—
		Pseudomonas dacunhae	97	97	—
Aldol-type cleavage					
Serine *trans*-hydroxymethylase	2.1.2.1	Rabbit liver	98	—	99

Glycogen phosphorylase	2.4.1.1	Rat liver	100[e]	100	—
		Lamb liver	101		
		Rabbit muscle	102	102	103, 104
		Rat muscle	105	105	103
CDP-4-keto-6-deoxy-D-glucose-3-dehydrogenase (E1)		Pasteurella pseudotuberculosis	106[h,m]	—	106[h,m]

[a] Resolution is difficult.

[b] Cytosolic isoenzyme.

[c] Shown to be identical to glycine aminotransferase.

[d] Mitochondrial isoenzyme.

[e] Both cytosolic and mitochondrial isoenzymes.

[f] Aspartate and tyrosine (aromatic) aminotransferases purified simultaneously.

[g] Identical to the inducible phenylalanine (histidine)-pyruvate aminotransferase isoenzyme 1.

[h] Isolated as apoenzyme.

[i] L-Serine dehydratases are also active on L-threonine and vice versa.

[j] Activated by ADP.

[k] Activated by AMP.

[l] Has O-acetylhomoserine sulfhydrylase activity as well.

[m] Pyridoxamine 5'-phosphate rather than pyridoxal-P is the coenzyme.

[n] Literature:

1. M. H. Saier and W. T. Jenkins, J. Biol. Chem. **242**, 91 (1967). See also this series, Vol. 17A [180].
2. T. Matsuzawa and H. Segal, J. Biol. Chem. **243**, 5929 (1968). See also this series, Vol. 17A [6].
3. T. Noguchi, Y. Takada, and R. Kido, Hoppe-Seyler's Z. Physiol. Chem. **358**, 1533 (1977).
4. M. Martinez-Carrion and W. T. Jenkins, J. Biol. Chem. **240**, 3538 (1965). See also this series, Vol. 17A [9].
5. V. Yu. Vasil'ev and V. P. Eremin, Biokhimiya **33**, 1143 (1968).
6. M. Maître, L. Ciesielski, C. Cash, and P. Mandel, Eur. J. Biochem. **52**, 157 (1975).
7. A. Schousboe, J.-Y. Wu, and E. Roberts, Biochemistry **12**, 2868 (1973).
8. M. Bloch-Tarday, B. Rolland, and P. Connard, Biochimie **56**, 823 (1974).
9. R. A. John and L. J. Fowler, Biochem. J. **155**, 645 (1976).
10. M. Martinez-Carrion, C. Turano, E. Chiancone, F. Bossa, A. Giartosio, R. Riva, and P. Fasella, J. Biol. Chem. **242**, 2397 (1967).
11. C. M. Michuda and M. Martinez-Carrion, Biochemistry **8**, 1095 (1969).
12. Y. Morino, S. Tanase, T. Watanabe, H. Kagamiyama, and H. Wada, J. Biochem. (Tokyo) **82**, 847 (1977).

(continued)

13. A. Arnone, P. H. Rogers, J. Schmidt, C.-N. Han, C. M. Harris, and D. E. Metzler, *J. Mol. Biol.* **112**, 509 (1977).

14. V. V. Borisov, S. N. Borisova, G. S. Kachalova, N. I. Sosfenov, B. K. Vainshtein, Yu. M. Torchinsky, and A. E. Braunstein, *J. Mol. Biol.*, in press. See also V. V. Borisov *et al.*, *Dokl. Akad. Nauk SSSR Biol. Sci. Sect.* (Engl. transl.) **235**, 203 (1977).

15. V. Scardi, P. Scotto, M. Iaccarino, and E. Scarano, *Biochem. J.* **88**, 172 (1963).

16. W. T. Jenkins and L. D'Ari, *J. Biol. Chem.* **241**, 2845 (1966).

17. Y. Morino, H. Itoh, and H. Wada, *Biochem. Biophys. Res. Commun.* **13**, 348 (1963).

18. H. Wada and E. E. Snell, *J. Biol. Chem.* **237**, 127 (1962).

19. L. H. Bertland and N. O. Kaplan, *Biochemistry* **9**, 2653 (1970).

20. H. Gehring, P. Christen, G. Eichele, M. Glor, J. N. Jasonius, A.-S. Reimer, J. D. G. Smit, and C. Thaller, *J. Mol. Biol.* **115**, 97, (1977).

21. L. H. Bertland and N. O. Kaplan, *Biochemistry* **7**, 134 (1968).

22. E. J. Shrawder and M. Martinez-Carrion, *J. Biol. Chem.* **248**, 2140 (1973).

23. R. Scandurr and C. Cannella, *Eur. J. Biochem.* **26**, 196 (1972).

24. C. Mavrides and W. Orr, *Biochim. Biophys. Acta* **336**, 70 (1974).

25. C. Mavrides and W. Orr, *J. Biol. Chem.* **250**, 4128 (1975).

26. M. P. C. Ip, R. J. Thibert, and D. E. J. Schmidt, *Can. J. Biochem.* **55**, 958 (1977).

27. A. J. Cooper and A. Meister, *Biochemistry* **11**, 661 (1972).

28. R. G. Martin and R. F. Goldberger, *J. Biol. Chem.* **242**, 1168 (1967). See also this series, Vol. 17B [147].

29. T. Noguchi, Y. Minatogawa, E. Okuno, M. Nakatani, M. Morimoto, and R. Kido, *Biochem. J.* **151**, 399 (1975).

30. T. Ikeda, Y. Konishi, and A. Ichihara, *Biochim. Biophys. Acta* **445**, 622 (1976).

31. K. Soda and H. Misono, *Biochemistry* **7**, 4110 (1968). See also this series, Vol. 17B [169].

32. C. C. Kalita, J. D. Kerman, and H. J. Strecker, *Biochim. Biophys. Acta* **429**, 780 (1976).

33. Y. Sonada, I. Suemori, and N. Katunuma, *Biochim. Biophys. Acta* **220**, 42 (1970).

34. W. T. Jenkins and H. Tsai, this series, Vol. 17A [31].

35. T. Matsuzawa, T. Katsunuma, and N. Katunuma, *Biochem. Biophys. Res. Commun.* **32**, 161 (1968).

36. C. Peraino, L. G. Bunville, and T. M. Tahmisian, *J. Biol. Chem.* **244**, 2241 (1969).

37. E. Kominami, Y. Banno, K. Chichibu, T. Shiotani, Y. Hamaguchi, and N. Katunuma, *Eur. J. Biochem.* **51**, 51 (1975).

38. H. Hirsch and D. M. Greenberg, *J. Biol. Chem.* **242**, 2283 (1967). See also this series, Vol. 17B [179].

39. F. A. Valeriote, F. Auricchio, G. M. Tomkins, and D. Riley, *J. Biol. Chem.* **244**, 3618 (1969). See also this series, Vol. 17A [80].

40. A. Belarbi, C. Bollack, N. Befort, J. P. Beck, and G. Beck, *FEBS Lett.* **75**, 221 (1977).

41. W. Roewekamp, C. E. Sekeris, and J. Staerk, *FEBS Lett.* **73**, 225 (1977).

42. S. Hayashi, D. K. Granner, and G. M. Tomkins, *J. Biol. Chem.* **242**, 3993 (1967).

43. R. T. Taylor and W. T. Jenkins, *J. Biol. Chem.* **241**, 4396 (1966). See also this series, Vol. 17A [110].

44. K. Aki, K. Ogawa, A. Shirai, and A. Ichihara, *J. Biochem.* (Tokyo) **62**, 610 (1967). See also this series, Vol. 17A [111].

45. K. Aki, A. Yokojima, and A. Ichihara, *J. Biochem.* (Tokyo) **65**, 539 (1969). See also this series, Vol. 17A [112].

46. M. S. Coleman and F. B. Armstrong, *Biochim. Biophys. Acta* **227**, 56 (1971).

47. Y. Minatogawa, T. Noguchi, and R. Kido, *Hoppe-Seyler's Z. Physiol. Chem.* **358**, 59 (1977).

48. T. Noguchi, Y. Takada, and R. Kido, *Biochem. J.* **161**, 609 (1977).

49. T. Noguchi, E. Okuno, Y. Minatogawa, and R. Kido, *Biochem. J.* **155**, 107 (1976).

50. T. Noguchi, E. Okuno and R. Kido, *Biochem. J.* **159**, 607 (1976).

51. J. E. Ayling and E. E. Snell, *Biochemistry* **7**, 1616 (1968).

52. M. Julius, C. A. Free, and G. T. Barry, this series, Vol. 17A [10].

53. G. Ross, K. Takashima, and E. Adams, *Biochem. Biophys. Res. Commun.* **34**, 134 (1969).

54. T. Yorifuji, T. and K. Ogata, *Biochem. Biophys. Res. Commun.* **34**, 760 (1969). See also this series, Vol. 17A [41].

55. K. Soda and T. Osumi, *Biochem. Biophys. Res. Commun.* **35**, 363 (1969). See also this series, Vol. 17B [222].

56. M. Mazelis and R. K. Greveling, *Biochem. J.* **147**, 485 (1975).

57. H. Nakagawa, H. Kimura, and S. Miura, *Biochem. Biophys. Res. Commun.* **28**, 359 (1967). See also this series, Vol. 17B [182].

58. A. Nagabhushanam and D. M. Greenberg, *J. Biol. Chem.* **240**, 3002 (1965).

59. W. Dowhan, Jr. and E. E. Snell, *J. Biol. Chem.* **245**, 4618 (1970).

60. R. Labow and W. G. Robinson, *J. Biol. Chem.* **241**, 1239 (1966).

61. A. Vanquickenborne and A. T. Phillips, *J. Biol. Chem.* **243**, 1312 (1968).

62. Y. Shizuta, A. Nakazawa, M. Tokushige, and O. Hayashi, *J. Biol. Chem.* **244**, 1883 (1969). See also this series, Vol. 17B [215].

63. P. D. Whanger, A. T. Phillips, K. W. Rabinowitz, J. R. Piperno, J. D. Shada, and W. A. Wood, *J. Biol. Chem.* **243**, 167 (1968).

64. G. W. Hatfield and H. E. Umbarger, *J. Biol. Chem.* **245**, 1736 (1970).

65. C. J. Decedue, J. G. Hofler, and R. O. Burns, *J. Biol. Chem.* **250**, 1563 (1975).

66. W. A. Newton, Y. Morino, and E. E. Snell, *J. Biol. Chem.* **240**, 1211 (1965).

67. J. A. Hoch, F. J. Simpson, and R. D. DeMoss, *Biochemistry* **5**, 2229 (1966).

68. H. Kumagai, H. Yamada, H. Matsui, H. Ohgishi, and K. Ogata, *J. Biol. Chem.* **245**, 1767 (1970).

69. T. N. Akopyan, A. E. Braunstein, and E. V. Goryachenkova, *Proc. Natl. Acad. Sci. U.S.A.* **72**, 1617 (1975).

70. N. D. Lac, *Biokhimiya* **34**, 861 (1969).

71. S. Kashiwamata and D. M. Greenberg, *Biochim. Biophys. Acta* **212**, 488 (1970).

72. M. A. Becker, N. Kreidich, and G. M. Tomkins, *J. Biol. Chem.* **244**, 2418 (1969).

73. M. Masada, K. Fukushima, and G. Tamura, *J. Biochem.* (Tokyo) **77**, 1107 (1975).

74. S. Yamagata and K. Takashima, *J. Biochem.* (Tokyo) **80**, 777 (1976).

75. O. Adachi and E. W. Miles, *J. Biol. Chem.* **249**, 5430 (1974). See also: E. W. Miles and M. Moriguchi, *J. Biol. Chem.* **252**, 6594 (1977).

(continued)

76. D. Deme, O. Durieu-Trautmann, and F. Chatagner, *Eur. J. Biochem.* **20**, 269 (1971).
77. Y. Matsuo and D. M. Greenberg, *J. Biol. Chem.* **234**, 545 (1958). See also this series, Vol. 5 [126].
78. Y. Matsuo and D. M. Greenberg, *J. Biol. Chem.* **234**, 507 (1958).
79. H. Tanaka, N. Esaki, T. Yamamoto, and K. Soda, *FEBS Lett.* **66**, 307 (1976).
80. H. Tanaka, N. Esaki, and K. Soda, *Biochemistry* **16**, 100 (1977).
81. S. Guggenheim and M. Flavin, *J. Biol. Chem.* **244**, 3722 (1969).
82. S. Ramakrishna and P. R. Adiga, *Eur. J. Biochem.* **59**, 377 (1975).
83. S. L. Blethen, E. A. Boeker, and E. E. Snell, *J. Biol. Chem.* **243**, 1671 (1968).
84. W. H. Wu and D. R. Morris, *J. Biol. Chem.* **248**, 1687 (1973).
85. J. G. Christenson, W. Dairman, and S. Udenfriend, *Arch. Biochem. Biophys.* **141**, 356 (1970).
86. J. -Y. Wu, T. Matsuda, and E. Roberts, *J. Biol. Chem.* **248**, 3029 (1973).
87. R. Shukuya and G. W. Schwert, *J. Biol. Chem.* **235**, 1649 (1960).
88. T. E. Huntley and D. E. Metzler, *in* "Symposium on Pyridoxal Enzymes," (K. Yamada, N. Katunuma, and H. Wada, eds.), p. 81. Maruzen, Tokyo, 1968.
89. D. L. Sabo, E. A. Boeker, B. Byers, H. Waron, and E. H. Fischer, *Biochemistry* **13**, 662 (1974).
90. K. Soda and M. Moriguchi, *Biochem. Biophys. Res. Commun.* **34**, 34 (1969). See also this series, Vol. 17B [228].
91. M. Ono, H. Inoue, F. Suzuki, and Y. Takeda, *Biochim. Biophys. Acta* **284**, 285 (1972).
92. R. J. Boucek and K. J. Lembach, *Arch. Biochem. Biophys.* **184**, 408 (1977).
93. C. A. Lamartiniere and W. B. Dempsey, *Biochemistry* **10**, 4783 (1971).
94. G. B. Bailey and W. B. Dempsey, *Biochemistry* **6**, 1526 (1967). See also this series, Vol. 17A [116].
95. E. M. Wilson and A. Meister, *Biochemistry* **5**, 1166 (1966). See also this series, Vol. 17A [90a].
96. E. W. Wilson, *Biochim. Biophys. Acta* **67**, 345 (1963). See also this series, Vol. 17A [90].
97. I. Chibata, T. Kakimoto, J. Kato, T. Shibatani, and N. Nishimura, *Biochem. Biophys. Res. Commun.* **26**, 662 (1967).
98. L. Schirch and T. Gross, *J. Biol. Chem.* **243**, 5651 (1968). See also this series, Vol. 17B [180].
99. L. G. Schirch and M. Mason, *J. Biol. Chem.* **238**, 1032 (1963).
100. M. Fujioka, *Biochim. Biophys. Acta* **185**, 338 (1969).
101. R. J. Ulevitch and R. G. Kallen, *Biochemistry* **16**, 5342 (1977).
102. E. H. Fischer and E. G. Krebs, Vol. 5 [49a].
103. S. Shaltiel, J. L. Hedrick, and E. H. Fischer, *Biochemistry* **5**, 2108 (1966). See also J. L. Hedrick, S. Shaltiel, and E. H. Fischer, *Biochemistry* **5**, 2117 (1966).
104. S. Shaltiel, J. L. Hedrick, A. Pocker, and E. H. Fischer, *Biochemistry* **8**, 5189 (1969).
105. L. L. Sevilla and E. H. Fischer, *Biochemistry* **8**, 2161 (1969).
106. P. Gonzales-Porque and J. L. Stromminger, *J. Biol. Chem.* **247**, 6748 (1972).

CM-Sephadex. Wash with 100–200 ml of the 40 mM, pH 5.1 acetate buffer, and elute with about 0.5–1.0 liter of 60 mM acetate, pH 5.1. This may be conveniently done overnight. The relatively inactive γ-subform is removed in this step. Then elute the half-active β-subform with about 1 liter of 60 mM sodium acetate buffer of pH exactly 5.35. The pH is critical; if too low more buffer will be required. Finally wash off the fully active and most abundant α-subform with 60 mM acetate buffer of pH 5.5. The fractions selected should have an absorbance ratio $A_{430\ nm}/A_{340\ nm}$ = 2.3–3.5 or higher at pH 5.5.

Concentrate by ultrafiltration to about 20 mg/ml, filter (Millipore) into sterilized vials and store in a refrigerator. Do *not* lyophilize. The enzyme retains its full activity but appears to lose crystallizability when stored in the frozen state. When diluted 4-fold with 0.2 M acetate buffer, pH 4.7, the absorbance ratios should fall in the following ranges: $A_{430\ nm}/A_{340\ nm}$ = 3.9–4.7 or perhaps a little higher for totally homogeneous enzyme; $A_{280\ nm}/A_{430\ nm}$ = 8.0–9.0, the lower value being observed for homogeneous enzyme. The yield from 16 kg of hearts is about 350 mg or 7.6 μmol (assuming a molar absorptivity of $6.55 \times 10^4\ M^{-1}\ cm^{-1}$ at 280 nm[7] and a molecular weight of 4.63×10^4 per subunit[8]). The specific activity is $250 \pm 20\ \mu$mol/min/mg at 25°.

Because of the possibility of damage by microbial action on columns, we pretreat both hydroxyapatite and CM-Sephadex columns with 1 liter of a solution of chloretone (0.1 g/liter stock; 5 ml/liter added to buffer).

The corresponding enzyme from chicken heart has been isolated as have the mitochondrial isoenzymes from pig and chicken (Table).

Glutamate Decarboxylase from Escherichia coli

The procedure is modified[9,10] from that originally reported by Shukuya and Schwert.[11] It is simpler than that of Hager previously published in this series.[12] Grow the bacteria (*E. coli*, ATCC 11246) as described by Shukuya and Schwert but with inclusion of 1% L-glutamic acid in the growth medium and an increase in the amount of yeast extract to 0.6%. After 17–20 hr of growth at 37°, harvest the cells. Fifty liters of medium should yield 150–170 g of cell paste. Suspend the cells in water in a volume

[7] F. S. Furbish, M. L. Fonda, and D. E. Metzler, *Biochemistry* **8**, 5169 (1969).
[8] Yu. A. Ovchinnikov, C. A. Egorov, N. A. Aldanova, M. Yu. Feigina, V. M. Lipkin, N. G. Abdulaev, E. V. Grishin, A. P. Kiselev, N. N. Modyanov, A. E. Braunstein, O. L. Polyanovsky, and V. V. Nosikov, *FEBS Lett.* **29**, 31 (1973).
[9] T. E. Huntley and D. E. Metzler, *Biochem. Biophys. Res. Commun.* **26**, 109 (1967).
[10] M. L. Fonda and R. F. DeGrella, *Biochem. Biophys. Res. Commun.* **56**, 451 (1974).
[11] R. Shukuya and G. W. Schwert, *J. Biol. Chem.* **235**, 1649 (1960).
[12] L. P. Hager, this series, Vol. 17A [120].

of 400 ml and pass them once or twice through a French press at 18,000–20,000 psi. Dilute the effluent to 2 liters and centrifuge. Slowly add to the supernatant 10% (w/v) streptomycin sulfate in water until no further precipitation occurs or to a final concentration of 1% streptomycin sulfate. Stir for 1 hr and centrifuge. The supernatant may then be tested with more streptomycin to ensure that removal of nucleic acids is complete. While this and subsequent steps may be carried out at room temperature, it is preferable to work at 4°.

Adjust the pH to 4.6 and concentrate the enzyme by addition of solid ammonium sulfate to 70% saturation (436 g/liter at 0°). Stir for 30 min and centrifuge. Dissolve the precipitate in 0.2 M pyridine-HCl buffer, pH 4.6, containing 0.1 mM pyridoxal-P and 0.1 mM dithiothreitol. Bring the volume to 500 ml and add 0.66 g of solid glutaric acid to give a 10 mM concentration of this protective inhibitor. Adjust the pH to 4.6 with either glutaric acid or pyridine. Warm the solution at 37° for 1 hr, cool quickly and centrifuge.

Carry out a second ammonium sulfate fractionation between 30% and 50% saturation (164 g/liter, then an additional 117 g/liter, calculated from the volume after the first centrifugation). Stir the suspension for 30 min after each addition before centrifugation. Dissolve the precipitated enzyme in 20–40 ml of 0.1 M phosphate buffer, pH 6.0, containing 0.1 mM pyridoxal-P and 0.1 mM dithiothreitol and dialyze against the same buffer to remove ammonium sulfate. Use the same buffer for chromatography on DEAE-A50 Sephadex. Apply the dialyzed enzyme to a 2.5 × 60 cm column. Elute at a flow rate of 0.5 ml/min until a band of opalescent material is removed. Then elute with a linear gradient of 400 ml of the 0.1 M potassium phosphate buffer, pH 6.0, and 400 ml of the same buffer at a concentration of 0.4 M. Monitor the absorbance at 280 nm and the activity. Pool the selected fractions and precipitate by addition of ammonium sulfate to 70% saturation. Dissolve the precipitate in 0.1 M pyridine-HCl buffer, pH 4.6, and dialyze against the same buffer. The preceding steps should all be carried out without delay because the enzyme is relatively unstable at pH 6. The enzyme at this stage has a specific activity of about 54 μmol/min per milligram of protein at 25° (about 2.7 times this activity at 37°) when assayed in 20 mM L-glutamate, 0.1 M pyridine-HCl buffer.

The specific activity may be increased substantially by crystallization. Take up the precipitate from 70% ammonium sulfate (preceding step) in phosphate buffer, pH 6.5, and crystallize with ammonium sulfate as described by Hager.[12] Alternatively, the enzyme in pyridine buffer may be dialyzed against 0.1 M acetate buffer, pH 4.6. Crystallization is spontaneous. A simpler step for further purification is chromatography on

Sephacryl S-200 (Superfine). Use a 2.5×60 cm column of the gel with 0.1 M pyridine buffer, pH 4.6, containing 0.1 mM dithiothreitol at a flow rate of 0.3 ml/min. Concentrate the protein by ultrafiltration to about 30 mg/ml before applying to the column. Monitor at 417 nm. The ratio $A_{280 \text{ nm}} : A_{417 \text{ nm}}$ for the final product should be 9.2 or less and the specific activity about 70 μmol/min per milligram of protein at 25°. The molar absorptivity, ϵ, of the coenzyme band at 420 nm is 10,000 M^{-1} cm^{-1} at pH 4.6.[13] Based on this value of ϵ, a yield of about 150 mg (3.0 μmol) of active sites is obtained from 160 g of cell paste.

Reversible Removal of Pyridoxal Phosphate (Resolution)

For some enzymes, resolution occurs readily in the isolation process. In fact, tryptophanase and tryptophan synthetase β_2 protein of *Escherichia coli* are routinely crystallized in the apoenzyme form. For a few other enzymes, e.g., alanine aminotransferase of pig heart and rat liver, glutamine aminotransferase of rat liver, and lysine decarboxylase of *Bacterium cadaveris,* the coenzyme is so tightly bound that resolution is accomplished only with great difficulty. When possible, it is desirable to convert the bound pyridoxal-P to pyridoxamine-P, which is usually easier to remove. For aminotransferases this may be accomplished by transamination with a normal substrate. However, the transamination usually does not go to completion. Therefore, cysteine sulfinate, whose transamination product loses sulfur dioxide, thereby driving the reaction to completion, is often used. For other enzymes, quasisubstrates may induce a transamination to give pyridoxamine-P, which spontaneously dissociates from the protein. Cysteine, which forms a stable cyclic thiazolidine compound with the coenzyme, is frequently used to trap the coenzyme and aid in its removal. Hydroxylamine or other carbonyl reagents may be used in a similar way. In other instances, the presence of imidazole or other "deforming agents" may facilitate the coenzyme removal by means that are not understood. A valuable discussion of resolution and binding of analogs has been written by Snell.[14]

Apoenzymes are often not as stable as the holoenzymes and may have to be used quickly. Inclusion of 10% glycerol in the medium stabilizes the apo form of glutamate decarboxylase of *E. coli* remarkably.[15] It may be useful in other cases as well.

The table lists references to methods for reversible resolution of many

---

[13] M. L. Fonda, *J. Biol. Chem.* **246**, 2230 (1971).

[14] E. E. Snell, *Vitam. Horm.* (*New York*) **28**, 265 (1970).

[15] B. S. Sukhareva and A. S. Tikhonenko, *Mol. Biol.* **6**, 851 (1972).

pyridoxal phosphate-dependent enzymes. In addition, the following paragraphs describe a few typical and reliable procedures.

Aspartate Aminotransferase, Cytosolic Isoenzyme, α Subform. The following procedure is essentially that of Scardi *et al.*[16] It is highly reliable. Mix about 30 mg of holoenzyme in a volume of approximately 1 ml with 24 ml of 0.2 M L-glutamate (or with a smaller quantity of cysteine sulfinate[17]), pH 8.3. Then add 25 ml of 1.0 M phosphate buffer, pH 4.8, and warm the solution at 30° for 20 min. Precipitate the protein by addition of 150 ml of saturated (at 4°) ammonium sulfate solution. Collect the precipitate by centrifugation, and take it up in 1–2 ml of 0.1 M Tris·HCl or triethanolamine·HCl buffer, pH 8.3. Repeat the entire procedure. Dialyze the apoenzyme in the pH 8.3 buffer or desalt by gel filtration. Buffers of pH 4.8 to 9.5 may be used. The ratio $A_{280 \text{ nm}} : A_{330 \text{ nm}}$ at pH 8.3 should be greater than 40, and the enzymic activity should be about 0.2% of the original. Over 90% of the original activity can be restored by addition of an excess of pyridoxal-P. The apoenzyme may be stored in a refrigerator for up to a week. However, a gradual rise in absorbance at 320 nm is observed. We recommend that the apoenzyme be used quickly for critical experiments.

Glutamate Decarboxylase. The apoenzyme is prepared by allowing the enzyme to act on α-methylglutamate, which undergoes a decarboxylative transamination, converting pyridoxal-P to pyridoxamine-P, which dissociates from the enzyme.[10,18] Add solid α-methylglutamate to the holoenzyme in 0.2 M pyridine HCl, pH 4.6, to a concentration of 0.1 M. Maintain the pH at 4.6 by addition of 1 M potassium phosphate, pH 7. Remove any precipitate by centrifugation. Then dialyze the apoenzyme overnight against 0.1 M piperazine HCl buffer, pH 4.6, containing 10% (v/v) glycerol and 0.1 mM dithiothreitol. The apoenzyme retains less than 1% of its original activity but can be reactivated to 80–90% of its original activity with pyridoxal-P. The molar absorptivity of the apoenzyme subunits is 86,100 M^{-1} cm^{-1} at 280 nm.[13] Old enzyme should be warmed briefly (e.g., 15 min at 37°) and centrifuged to remove denatured protein before resolution.

Reconstitution of Apoenzymes with Pyridoxal-P, Pyridoxamine-P, and Analogs

A variety of derivatives and analogs of the coenzymes are available. The coenzymes may be methylated on the ring nitrogen, on phenolic ox-

[16] V. Scardi, P. Scotto, M. Saccarino, and E. Scarano, *Biochem. J.* **88,** 172 (1963).

[17] W. T. Jenkins and L. D'Ari, *Biochem. Biophys. Res. Commun.* **22,** 376 (1966).

[18] T. E. Huntley and D. E. Metzler, *in* "Symposium on Pyridoxal Enzymes" (K. Yamada, N. Katunuma, and H. Wada, eds.), p. 81. Maruzen, Tokyo, 1968.

ygen, or on the phosphate group. The methyl group in the 2-position may be substituted with a variety of groups. The 4-formyl group may be replaced to give analogs that are bound but cannot react normally. A large variety of modifications in the 5'-position are possible without necessarily preventing binding to apoenzymes. The 6-hydrogen can be replaced to give yet another series of analogs. Also of interest are relatively simple analogs, such as 3-pyridinemethanol phosphate,[19] and unphosphorylated pyridoxal and pyridoxamine.[20]

Methods of synthesis of analogs are summarized in this volume[21] and in Vol. 18A.[22]

Measuring Coenzymic Activity. An excess of analog is usually added to apoenzyme at relatively high concentration (e.g., 0.1 mM). After a suitable period of time, the solution is diluted and assayed. The binding of coenzyme to apoenzyme may be extremely rapid or it may take many minutes or even hours. A problem that may arise is the presence of inorganic phosphate or sulfate in the binding site for the phosphate group of the coenzyme.[23,23a] Thus, the rate of binding may be dependent on the presence or absence of multivalent anions in the apoenzyme preparation. For some enzymes, e.g., aspartate aminotransferase, a 10-fold excess (1 mM) of pyridoxal-P may be added, but for others (e.g., serine dehydratase) excess pyridoxal-P is inhibitory. For tightly bound analogs it may be better to mix with apoenzyme in a 1 : 1 ratio or to have an excess of apoenzyme present. The measured activity can then be expressed relative to the concentration of analog rather than to that of the apoenzyme. On the other hand, weakly bound analogs may partially dissociate upon dilution. In such cases the activity should be measured as a function of the concentration of coenzyme present in the assay mixture itself. For transaminases, a problem lies in the possible loss of the amine forms of coenzyme analogs from the enzyme. Pyridoxamine phosphate is less tightly bound than is pyridoxal phosphate and for some analogs the amine forms are very weakly bound. In such instances, low enzymic activity will be observed unless the assay is conducted in the presence of high concentration of the amine form.

A serious problem in measuring the catalytic activity of analogs arises from the difficulty of totally removing coenzyme during preparation of apoenzyme. For some enzymes, such as aspartate aminotransferase, the

[19] S. W. Koontz, Ph.D. Dissertation, Univ. of Wisconsin (1977).

[20] H. Wada and E. E. Snell, *J. Biol. Chem.* **237**, 127 (1962).

[21] W. Korytnyk, this volume [74].

[22] W. Korytnyk and M. Ikawa, this series, Vol. 18A [88]; V. L. Florentiev, V. I. Ivanov, and M. Ya. Karpeisky, this series, Vol. 18A [89]; S. Fukui, N. Ohishi, and S. Shimizu, this series, Vol. 18A [90].

[23] M. L. Fonda, *Arch. Biochem. Biophys.* **170**, 690 (1975).

[23a] M. L. Fonda and S. B. Auerbach, *Biochim. Biophys. Acta* **422**, 38 (1976).

background activity may be reduced to about 0.2% of the original. However, it is still difficult to detect with certainty the activity of an analog that is only 1% that of the true coenzyme.

Binding Constants. Besides determining the coenzymic activity of an analog relative to that of pyridoxal-P, it is of interest to evaluate the equilibrium constant for binding to the active site. Because many of the apoenzymes have a very high affinity for coenzyme, this is often hard to do. It is easier for analogs, which are usually less tightly bound. One obvious method is to observe the degree of saturation of the active sites by measurement of enzymic activity as a function of analog concentration. This will be successful only if the rates of binding and dissociation of the analog are high enough. A further complication for transaminases arises from the "Ping-Pong" mechanism and the need to measure binding constants separately for the two coenzyme forms.

Useful information may be obtained by observing whether or not a bound analog can be displaced by added pyridoxal-P. If so, full activity may be restored to the enzyme.

Spectrophotometric Methods. Direct observation of bound analogs by spectrophotometry and measurement of circular dichroism (CD) is often useful. The apoenzyme solution should be of relatively high concentration. For example, 0.7–1.0 ml of 0.05–0.1 mM apoenzyme (concentration based on number of active sites) is placed in a semimicro spectrophotometer cuvette. The spectrum is recorded. A small volume (e.g., 20 μl) of a solution of coenzyme or analog is added. For study of spectra of tightly bound analogs, it is desirable to add only 50–90% as much coenzyme as the calculated concentration of active sites. This seems to give cleaner spectra with less "nonspecific" binding than if an exact 1:1 equivalence is used. After mixing, the spectrum should be recorded at intervals of time until it becomes constant. Other properties, such as circular dichroism or fluorescence, may also be measured at this stage. It may then be desirable to add a portion of lower or higher pH buffer to change the pH and to add one or more portions of substrate, quasisubstrate or inhibitor with recording of spectra and CD after each addition. It is possible to do this with little or no loss of solution, hence with precisely known concentration.

A problem arises from the fact that pyridoxal-P is a general amino group reagent and may bind at many sites on the surface of a protein. Whereas most of these binding sites have low affinity constants, their presence may interfere with the observation of binding at the active site. In our experience with aspartate aminotransferase at pH 8, there is almost always a small amount of absorption at 430 nm which is not characteristic of coenzyme bound in the active site. With pyridoxal-P the amount is very small but with some analogs it is larger. This absorption band is devoid of circular dichroism and tends to diminish with time. This leads us to suspect that it represents nonspecific binding. The problem probably exists with other enzymes as well and investigators should be aware of it.

Spectrophotometers, dichrographs, and other instruments with equipment for digital acquisition of data are now readily available. It is most convenient to record spectra in digital form for subsequent analysis.[24-26] We find it very satisfactory to record 50–100 or more points on a spectrum at regular intervals of 1–2 nm or better of 100–200 cm^{-1}. For mathematical analysis it is most appropriate to plot absorbance or molar absorptivity against wave number. When data are collected at regular intervals of wavelength, a computer can be used to replot against wave number. The use of computer-assisted treatment of the data is highly recommended, for it permits easy averaging of repetitive scans and subtraction of absorbance contributed by the apoenzyme from that of the apoenzyme plus coenzyme, when desired. It is also convenient to divide the absorbances by the total concentration of added coenzyme to give apparent molar absorptivity. Plots of the latter can be obtained easily by use of the computer, regardless of the number of portions of coenzyme or substrates added to the cuvette.

The binding of coenzymes can usually be assessed directly from spectra, in particular from that of the circular dichroism (CD). If the pH is chosen correctly, many bound coenzymes will absorb at about 410–431 nm, a region in which the coenzyme itself is transparent. For higher energy bands, such as that of transaminases in the 360-nm region, it is still possible to measure binding from CD or from absorption spectra by analysis with log-normal distribution curves. This is possible even though the band of the bound analog overlaps the 390-nm band of the free aldehyde. A positive circular dichroism (CD) is observed for most bands of bound pyridoxal-P, whether in the 410–430 nm, 360, or 330 nm regions, and for bound pyridoxamine-P at about 330 nm. Thus, the extent of coenzyme binding may be estimated quite reliably from the CD spectrum. One way to assess the reactivity of very slightly active analogs is to observe directly the turnover of the enzyme in the cuvette. For transaminases this is easily done by addition of glutamate or aspartate. However, even for an analog showing less than 1% of the activity of pyridoxal-P, the turnover observed in this way may be "instantaneous." The approach can be

[24] R. J. Johnson and D. E. Metzler, this series, Vol. 18A [77].

[25] D. E. Metzler, C. M. Harris, R. J. Johnson, D. B. Siano, and J. A. Thomson, *Biochemistry* **12**, 5377 (1973).

[26] D. E. Metzler, C. M. Harris, R. W. Reeves, W. H. Lawton, and M. S. Maggio, *Anal. Chem.* **49**, 846A (1977).

modified by use of fast reaction techniques or by looking at slowly reacting quasisubstrates.

Reactions with Quasisubstrates and Inhibitors. The value of spectrophotometric studies is enhanced by the availability of slow-reacting substrates and of inhibitors that are able to participate in only part of a catalytic sequence. A good example of the latter is α-methylaspartate which appears to undergo transimination (see Scheme 2) at the active site of aspartate aminotransferase but cannot complete the transamination sequence.[27,28] α-Methylated substrates sometimes (e.g., with decarboxylases[18,29,30]) undergo side reactions. Amino acids with polar substitutents in the β-position often undergo elimination with loss of the β-substituent and modification of the enzyme or coenzyme. However, elimination does not always occur and *erythro-β*-hydroxyaspartate is an especially interesting slowly reacting substrate (about 1 sec^{-1}) with aspartate aminotransferases. It binds tightly to the enzymes and a major fraction of the enzyme may be converted to a quinonoid intermediate with characteristic spectral properties.[28,31] A slow substrate of this type may also be useful for comparison of reaction rates of very slow analogs. By making the enzymic reaction slow enough to observe directly in the spectrophotometer, relative rates for slowly reacting analogs may be measured. Formation of transient intermediates may be detected.

Interpretation of Results

While it is relatively easy to measure apparent coenzymic activity, and to record absorption spectra, interpretations may be complex. The two questions most often asked are: Does an analog bind? Does it have coenzymic activity? Several additional questions are: Is there more than one detectable species of bound coenzyme? Are there pK_a values observable for bound coenzyme? What ionic form of pyridoxal-P is bound? How many steps are detectable in the binding process? What is the environment of the bound coenzyme? Are spectral properties of any amino acid residues altered upon binding of coenzyme? What are the relative stabilities of different intermediate enzyme–substrate complexes with analogs as compared to those observed with pyridoxal-P?

[27] P. Fasella, A. Giartosio, and G. C. Hammes, *Biochemistry* **5**, 197 (1966).
[28] A. E. Braunstein, *in* "The Enzymes" (P. Boyer, ed.), 3rd ed., Vol. 9, p. 379. Academic Press, New York, 1973.
[29] M. H. O'Leary and R. L. Baughn, *J. Biol. Chem.* **252**, 7168 (1977).
[30] M. H. O'Leary and R. M. Herreid, *Biochemistry* **17**, 1010 (1978).
[31] W. T. Jenkins, *J. Biol. Chem.* **236**, 1121 (1961).

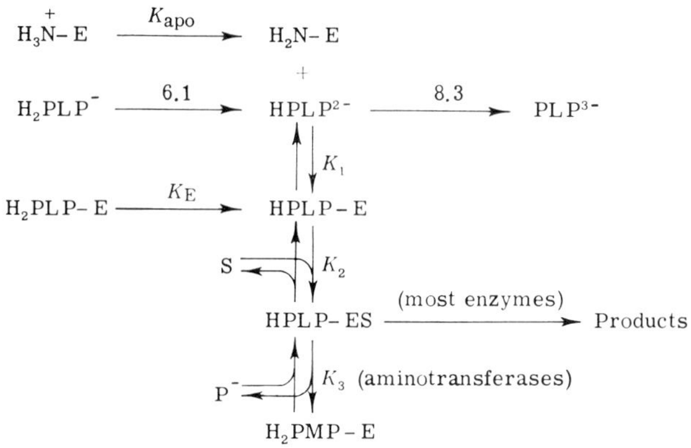

$$\overset{+}{H_3N}-E \xrightarrow{K_{apo}} H_2N-E$$

$$H_2PLP^- \xrightarrow{6.1} \overset{+}{HPLP^{2-}} \xrightarrow{8.3} PLP^{3-}$$

$$\Big\updownarrow K_1$$

$$H_2PLP-E \xrightarrow{K_E} HPLP-E$$

$$S \rightleftharpoons \Big\updownarrow K_2$$

$$HPLP-ES \xrightarrow{\text{(most enzymes)}} \text{Products}$$

$$\Big\updownarrow K_3 \text{ (aminotransferases)}$$

$$P^- \rightleftharpoons$$

$$H_2PMP-E$$

Scheme 1

Little information about the forms of pyridoxal-P bound is available for most enzymes. However, in most instances a lysine side chain, with pK_{apo} (Scheme 1) reacts in the unprotonated form with some ionic form of pyridoxal-P to form a Schiff base at the active site. The formation constant (K_1 in Scheme 1) should be defined for a pH-independent equilibrium. It may not be clear initially which ionic form of pyridoxal-P reacts, but it is reasonable to select the one that predominates near pH 7. This is the monoprotonated form with a dipolar ionic ring and doubly charged phosphate which is designated $HPLP^{2-}$ in Scheme 1. The two pK values of 6.1 (phosphate) and 8.3 (ring nitrogen) will affect the overall equilibrium in the binding process. The bound coenzyme–enzyme complex also displays pK values, and, in the case of a number of enzymes, one of these pK's is that of protonation of the imine nitrogen. For cytosolic aspartate aminotransferase, at moderate ionic strength, this pK is about 6.3. There may also be other pK values either in the free enzyme or in the enzyme–coenzyme complex that have an effect upon the binding of coenzyme at a given pH. By examining the apparent binding constant as a function of pH, it is possible not only to define K_1, but also to evaluate these pK values including pK_{apo} of Scheme 1. Such measurements appear to have been made but rarely. From the spectrum of cytosolic aspartate aminotransferase, it may be deduced that the dipolar ionic ring form of pyridoxamine-P is bound; the same is presumably true for pyridoxal-P. The state of ionization of the phosphate group is less certain. However, measurements of Fonda and Auerbach[23a] at two pH values suggest that the diionized form ($HPLP^{2-}$ of Scheme 1) is bound. The same conclusion is supported by the fact that an analog containing the trans vinylphosphonate group $-CH=CH-PO_3^{2-}$ in the 5'-position is bound less tightly to the apo form of aspartate aminotransferase at pH 9.5 than at pH 8.3.[32]

Koontz and O'Leary[19] have used a competitive equilibrium method to determine the binding constants of a large number of inhibitors to the apo form of glutamate decarboxylase. The apparent binding constant of 4-deoxypyridoxine 5'-phosphate is determined fluorimetrically in the presence of various concentrations of inhibitor. The fact that the binding constants for inhibitors such as 3-pyridinemethanol phosphate and ethyl phosphate vary linearly with pH with a slope of one in the pH range 4.6–5.5, whereas those of SO_4^{2-} and 3-pyridinemethanol sulfate do not suggest that these inhibitors bind as the phosphate dianions. Another approach is to deduce the state of protonation of the phosphate group from the chemical shift in the ^{32}P NMR spectrum. Thus, it was concluded by Martinez-Carrion[33] that for cytosolic aspartate aminotransferase the phosphate is diionized. However, for glycogen phosphorylase, Feldman and Helmreich[34] observed the monoprotonated state for the free enzyme. Chemical modification together with ^{19}F NMR has been used to establish a pK of 8.4 for the active site lysine-258 of a modified aspartate aminotransferase.[35]

Some pK_a values for enzyme–coenzyme or enzyme–analog complexes can be determined spectrophotometrically. An example is the pK of 6.3 (Scheme 1) for cytosolic aspartate aminotransferase. Large shifts of this pK for analogs with monoanionic side chains have been observed.[32,36] The corresponding constants for enzyme–inhibitor complexes may be measured by methods described by Jenkins et al.[37,38] These constants may also be determined using computer-assisted methods,[39] which provide a convenient way of calculating directly the complete absorption spectra of the enzyme–inhibitor complexes.

Examination of the absorption spectrum and circular dichroism of a bound analog may reveal additional information. It has been established by fitting with log-normal distribution curves that absorption bands have accurately reproducible widths.[32] Furthermore, band shapes are similar in absorption and in the induced circular dichroism that is ordinarily present with bound analogs.

The relative constancy of the width of absorption bands for a variety

[32] R. Miura, J. J. Likos, C. M. Metzler, and D. E. Metzler, in preparation.

[33] M. Martinez-Carrion, *Eur. J. Biochem.* **54**, 39 (1975).

[34] K. Feldmann and E. J. M. Helmreich, *Biochemistry* **15**, 2394 (1976).

[35] J. C. Slebe and M. Martinez-Carrion, *J. Biol. Chem.* **253**, 2093 (1978).

[36] F. S. Furbish, M. L. Fonda, and D. E. Metzler, *Biochemistry* **8**, 5169 (1969).

[37] W. T. Jenkins and L. D'Ari, *J. Biol. Chem.* **241**, 2845 (1966).

[38] W. T. Jenkins and L. D'Ari, *J. Biol. Chem.* **241**, 5667 (1966).

[39] M. L. Fonda and R. J. Johnson, *J. Biol. Chem.* **245**, 2709 (1970).

of pyridoxal-P-dependent enzymes has been noted previously.[40] We suspect that the widths of the aldimine bands at 410–430 nm may nearly all fall within the range 3.6×10^3 cm^{-1} (glutamate decarboxylase)[32] to about 4.0×10^3 cm^{-1}. The skewness of the bands is typically 1.4 to 1.5. Absorption bands in the 360- to 370-nm region, which presumably represent the dipolar-ionic ring form of the Schiff bases, have very similar width and skewness parameters. Bands at 330 to 340 nm representing either pyridoxamine phosphate or adducts of Schiff bases of pyridoxal-P range in width from 3.3 to 3.6×10^3 cm^{-1}. Bands in the 330-nm region that represent nondipolarionic tautomers of Schiff bases may be a little wider, e.g., 4.0×10^3 cm^{-1}. However, precise measurements do not appear to have been made.

Peaks in absorption and circular dichroism spectra have nearly identical positions. However, the bandwidth of the circular dichroism peaks is about 8% narrower than the width of the corresponding absorption peaks. With this information it is possible to use the circular dichroism to estimate the amount of an actively bound form of the coenzyme even when its absorption band is heavily overlapped by other bands which lack circular dichroism. Using lognormal distribution curves such spectra can be resolved satisfactorily.

A Schiff base of pyridoxal-P with net charge of zero on the chromophoric system can exist as three tautomers (I–III, Scheme 2) absorbing maximally at 410–430, 360–370, and about 330 nm, respectively. The tautomeric form may vary from one enzyme to another. Thus in glycogen phosphorylase the nondipolarionic tautomer (III) is bound,[41,42] but in aspartate aminotransferase it is clearly the dipolar ion (II). Some other enzymes with absorption bands at 410–430 nm may contain (I). Further-

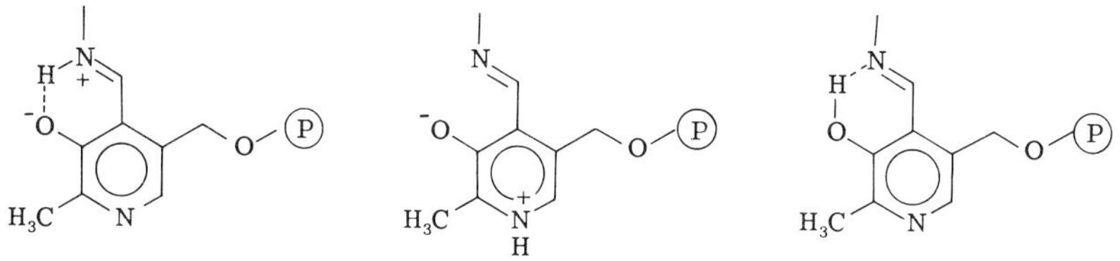

430 nm	360–370 nm	330 nm
(I)	(II)	(III)

Scheme 2

[40] L. Davis and D. E. Metzler, in "The Enzymes" (P. Boyer, ed.), 3rd ed., Vol. 7, p. 62. Academic Press, New York (1972).

[41] G. F. Johnson, J.-I. Tu, M. L. S. Bartlett, and D. J. Graves, J. Biol. Chem. 245, 5560 (1970).

[42] V. Gani, A. Kupfer, and S. Shaltiel, Biochemistry 17, 1294 (1978).

more, the N-protonated form of (I) probably exists in enzymes, such as aspartate aminotransferases and possibly glutamate decarboxylase, at low pH. For many enzymes one tautomer may predominate but smaller amounts of other tautomers may also be present. Resolution with log-normal curves should be useful in describing equilibria among tautomers, as it has been in the case of pyridoxal phosphate itself.[43]

The differences among enzymes in the preferred tautomer presumably arise in part because of differences in polarity of the environment. Thus, glycogen phosphorylase is assumed to have a hydrophobic environment that stabilizes tautomer (III). The 4-vinyl analog of pyridoxal-P is an interesting probe of the polarity of binding sites.[44] Like pyridoxamine-P, it is bound to the apo form of cytosolic aspartate aminotransferase almost entirely as the dipolar ion (Fig. 1). The analog readily tautomerizes to the nondipolar ionic form in water–methanol or water–dioxane mixtures with the small amount of remaining dipolar ionic peak shifting markedly to lower energies. The same bathochromic shift (from 339 to 360 nm) is observed when the analog binds to aspartate aminotransferase, but there is no tautomerization. Apparently some groups in the protein interact with the ring, presumably by hydrogen bonding, to stabilize the dipolar ionic ring form despite the surrounding hydrophobic environment. Aside from the differences in hydrogen bonding, the environments in the aminotransferase and phosphorylase may not be as different as is sometimes assumed. The binding of the 4-vinyl analog to the apo form of glutamate decarboxylase leads to an even greater shift to 364 nm with appearance of some vibrational fine structure.[13] The binding site appears even more hydrophobic than that of aspartate aminotransferase, but it still accommodates the dipolar ionic ring.

When pyridoxal-P binds to apoenzymes there is sometimes a narrowing of the absorption band from the usual width of about 4.0×10^3 cm^{-1} for model Schiff bases. This narrowing is much more pronounced for glutamate decarboxylase (to 3.6×10^3 cm^{-1}) than for aspartate aminotransferase (3.9×10^3 cm^{-1}). These changes could also reflect the hydrophobic nature of the environment but may also depend upon increased rigidity of the coenzyme in the active site.

Difference spectroscopy of apoenzyme with and without added coenzyme or analog may be useful in detecting perturbations of amino acid residues in the protein by the binding of coenzyme. For example, Fig. 1 is a difference spectrum for binding of the 4-vinyl analog of pyridoxal-P. In addition to the absorption band of the bound analog with a peak at 360

[43] C. M. Harris, R. J. Johnson, and D. E. Metzler, *Biochem. Biophys. Acta* **421**, 181 (1976).
[44] I.-Y. Yang, C. M. Harris, D. E. Metzler, W. Korytnyk. B. Lachmann, and P. G. G. Potti, *J. Biol. Chem.* **250**, 2947 (1975).

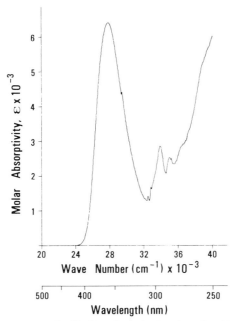

Fig. 1. Difference spectrum for binding of a 4-vinyl analog of pyridoxal-P to the apo form of aspartate aminotransferase (α-subform of the cytosolic enzyme of pig heart). The spectrum was determined by recording in digital form the spectrum of the apoenzyme. Then a small volume of a solution of the analog was added and the spectrum was recorded again. After correction for dilution, the first spectrum was subtracted from the second and the difference was plotted. See I.-Y. Yang, C. M. Harris, D. E. Metzler, W. Korytnyk, B. Lachmann, and P. G. G. Potti, *J. Biol. Chem.* **250**, 2947 (1975).

nm, there are two small peaks separated by a sharp valley at 287, 291, and 296 nm. The spacing and sharpness suggest vibrational fine structure in an absorption band. Since the binding can presumably cause perturbation in either the analog or the protein moiety, the apparent fine structure observed may theoretically arise from perturbation of either species. However, a similar perturbation is seen upon binding of pyridoxal-P itself as well as other analogs.[45] This and other evidence has led us to propose perturbation of a tryptophan residue in the apoenzyme.[44,45] Changes in the circular dichroism in the tryptophan region support this proposal.

Of special interest is analysis of absorption spectra and circular dichroism of native enzyme or of enzyme reconstituted with coenzyme analogs when combined with substrates, inhibitors, and quasisubstrates. Making use of both absorption and circular dichroism, it is possible to resolve the spectra into a series of bands and to calculate the relative stabil-

[45] B. I. Yang and D. E. Metzler, in preparation.

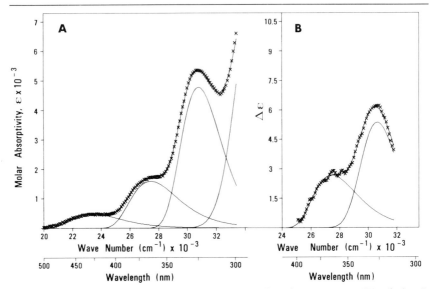

FIG. 2. Absorption spectrum (A) and circular dichroism spectrum (B) of the 5-ethylphosphonate analog of pyridoxal-P bound to the apo form of aspartate aminotransferase after addition of L-glutamate. The spectra have been resolved with log-normal distribution curves as explained in the text. The final concentrations are: apoenzyme, 60 μM; analog, 52 μM; and L-glutamate, 2.9 mM. Figures by courtesy of R. Miura and C. M. Metzler.

ity of the different enzyme-bound forms. An example is shown in Fig. 2. The spectrum shown is that of the 5-ethylphosphonate analog of pyridoxal-P in the active site of aspartate aminotransferase after addition of L-glutamate. Three absorption bands have been resolved with log-normal curves in the absorption spectrum. However, the 430-nm band is devoid of circular dichroism, indicating that it is an E–S complex, the substrate aldimine. From the circular dichroism at 333 nm, it is clear that the amino form of the analog (the transamination product) is bound to the enzyme. From the areas under the three curves and known molar areas, the molar percentages of 430, 362, and 333 nm components are 9, 26, and 65, respectively. It should be of interest to observe how these relative stabilities differ with different coenzyme analogs.

For aminotransferases it would be preferable to use a computer to evaluate spectra of the mixture of enzyme–substrate complexes as well as K_2 and K_3 of Scheme 1. Then a resolution such as that in Fig. 2 would allow calculation of equilibrium constants for interconversion of various spectral forms in the ES or EI complex. This, in turn, should provide a quantitative measure of the strength of interaction of various intermediates with the protein. For example, from Fig. 2 we conclude that the 430-nm substrate aldimine is more strongly stabilized with the 5-ethylphosphonate analog than with pyridoxal-P. However, the corresponding experiment with *erythro*-β-hydroxyaspartate shows that the key quinonoid intermediate is relatively less stable with the analog.[32]

Pitfalls

The user of coenzyme analogs should be keenly aware of several potential difficulties. Among these are the instability of many apoenzymes, the difficulty of complete resolution and the inhibition of analog binding by divalent anions. Instability or impurity of analogs may be a problem. Several analogs are reconverted to pyridoxal-P upon binding. This is true for oximes and hydrazones[46,47] for a cyanoethyl ester of pyridoxal-P[36] and in some instances for the *N*-oxide of pyridoxal-P. The weak binding of some analogs poses problems as well as advantages. The participation in secondary side reactions complicates the picture in a number of cases.

[46] M. Fujioka and E. E. Snell, *J. Biol. Chem.* **240**, 3044 (1964).
[47] Yu. M. Torchinsky, *Biokhimiya* **28**, 731 (1963).

[80] Investigation of Crystalline Enzyme–Substrate Complexes of Pyridoxal Phosphate-Dependent Enzymes[1]

By CAROL M. METZLER, PAUL H. ROGERS, A. ARNONE, DON S. MARTIN, and DAVID E. METZLER

Many pyridoxal-P dependent enzymes have been crystallized,[2] but few of the crystals have been studied either by X-ray crystallography or by other physical techniques. Recently, three groups have initiated crystallographic studies on aspartate aminotransferases. The cytosolic enzyme from chicken hearts[3] and from pig hearts[4] has been prepared in orthorhombic forms, while the mitochondrial isoenzyme of chicken heart has been crystallized in a triclinic form.[5] These are highly soluble enzymes. In

[1] Supported by NIH Grant AM-01549 and NSF Grant CHE-83665.
[2] Yang and Metzler, this volume [79], Table I.
[3] V. V. Borisov, S. N. Borisova, G. S. Kachalova, N. I. Sosfenov, A. A. Voronova, B. K. Vainshtein, Yu. M. Torchinsky, G. A. Volkova, and A. E. Braunstein, *Dokl. Acad. Nauk SSSR Biol. Sci. Sect.* (Engl. transl.) **235**, 203 (1977).
[4] A. Arnone, P. H. Rogers, J. Schmidt, C. Han, C. M. Harris, and D. E. Metzler, *J. Mol. Biol.* **112**, 509 (1977).
[5] H. Gehring, P. Christen, G. Eichele, M. Glor, J. N. Jasonius, A.-S. Reimer, J. D. G. Smit, and C. Thaller, *J. Mol. Biol.* **115**, 97 (1977).

two cases, the crystals have been prepared by the method of McPherson using polyethyleneglycol.[6] Many pyridoxal-P enzymes form very small and relatively insoluble crystals. However, if the solubility of these enzymes is increased by changing the pH or ionic composition of the buffer it may be possible to use the polyethyleneglycol method to obtain larger crystals. It will be desirable to carry out X-ray crystallographic studies on a variety of pyridoxal-P containing enzymes. Furthermore, crystals of these enzymes can be studied by spectroscopic means.[7-9] Substrates can be diffused into the crystals. Enzyme–inhibitor complexes, and in some cases enzyme–substrate complexes, can be obtained in crystalline form.[7]

In this chapter we recount our relatively recent experiences in the preparation of crystals of the cytosolic isoenzyme of aspartate aminotransferase from pig heart and of enzyme–substrate and enzyme–inhibitor complexes and their use in spectrophotometric studies. In addition, the coenzyme analog N-methylpyridoxal phosphate has been incorporated in place of the natural coenzyme and the resulting enzyme has been crystallized.[10] It should be possible to incorporate a variety of other analogs into crystals. The apoenzyme of aspartate aminotransferase can also be obtained in a form isomorphous with the native and pyridoxal-P or analogs can be diffused into the coenzyme-binding sites.[11]

Preparation of Crystals of Native Enzyme

Isolate the enzyme as described in this volume [79].[2] Separate the subforms by chromatography on CM-Sephadex. Dialyze against 40 mM sodium acetate buffer, pH 5.4, and concentrate by ultrafiltration (Schleicher and Schuell, Inc., protein concentrator) to 35–45 mg of protein per milliliter. Prepare the crystals using polyethyleneglycol (PEG) of molecular weight 6000 or 4000 by the depression-plate and sandwich-box technique of McPherson.[6] Place 20 μl of enzyme plus 20 μl of 8% PEG in each well of the plate. In the reservoir below put 25 ml of 8% PEG containing 40 mM acetate buffer. Stir each well immediately after mixing. It may be desirable to transfer the mixed solution with a pipette to a clean well to avoid growth of crystals around the edge. Crystals will appear in from 1

[6] A. McPherson, Jr., J. Biol. Chem. 251, 6300 (1976).

[7] C. M. Metzler, D. E. Metzler, D. S. Martin, R. Newman, A. Arnone, and P. H. Rogers, J. Biol. Chem. 253, 5251 (1978).

[8] Yu. V. Morozov, N. P. Bazhulina, L. V. Karklit, V. M. Kochkina, and Yu. M. Torchinsky, private communication.

[9] G. Eichele, D. Karabelnik, R. Halonbrenner, J. N. Jansonius, and P. Christen, J. Biol. Chem. 253, 5239 (1978).

[10] V. Chen and D. E. Metzler, Abstr. 176th Am. Chem. Soc. Natl. Meeting, Biol. 62 (1978).

[11] J. Schmidt, C. M. Metzler, P. H. Rogers, and A. Arnone, unpublished results.

day to 1 week. The morphology of the crystals is somewhat unpredictable. In our experience "siliconized" wells tend to produce chunky crystals suitable for X-ray diffraction, whereas unsiliconized wells may produce more flat plates, which often grow on edge and may appear as needles at first glance. In dispensing the enzyme into the wells, we have used pipettes with plastic tips. These tips should be washed thoroughly. Some of the tips, unless carefully washed, led to rapid nucleation and production of large numbers of very small crystals.

Large single crystals are required for crystallographic studies. Although these do occasionally grow spontaneously without seeding, we prefer to use the following seeding technique to routinely grow very large crystals, which can then be cut to optimal size. Choose a small, well-formed "seed" crystal from a drop of enzyme previously crystallized. Allow it to drift slowly down a glass tube (75 mm long with a 2.5-mm bore) filled with a solution that is 1% in PEG-6000 and 40 mM in sodium acetate at pH 5.4. As the seed crystal slowly sediments through the dilute PEG solution, it may partially dissolve. However, any accompanying microcrystals, which would also serve as centers for crystal growth, apparently dissolve completely or sediment so slowly that they are physically separated from the major seed crystal by the time it reaches the bottom of the tube. At this point use the glass tube as a pipette to transfer the seed crystal, along with a small amount ($\sim$5 μl) of the 1% PEG solution, to a well of a clean, siliconized glass depression tray. To this 5-μl drop add 50–100 μl of a solution which is 4% in PEG 6000, 40 mM in sodium acetate, and contains about 20 mg/ml of enzyme. Equilibrate the seed crystal in its new mother liquor against a reservoir of 8% PEG 6000 and 40 mM acetate buffer. The seeding should be done at 4°, and the growing seeds should not be disturbed for at least 5 days. The crystals are fully grown in 1–2 weeks, and they routinely attain dimensions up to to 3.5 $\times$ 2.5 $\times$ 0.6 mm without twinning.

Platelike crystals suitable for spectroscopy can be prepared by transferring very small seeds with a dissecting needle or disposable pipette into a well of low PEG concentration to wash away microcrystals and expose a new crystallizing surface. As few as possible of these washed seeds can then be transferred into a well that has been prepared for crystallization.

The triclinic form of the mitochondrial isoenzyme from chicken heart[5] is prepared in a similar manner using PEG 4000 and a glycine–NaOH buffer at a pH of 7.5–9. The crystals give an excellent diffraction pattern and well-defined spectra.[9]

A different orthorhombic form of the chicken heart cytosolic aspartate aminotransferase can be prepared as the α-methylaspartate complex.[3] A portion (100 μl) of enzyme (23 mg/ml) in 0.2 M potassium phosphate buffer (pH about 7.5) and also containing 0.1 M α-methyl-DL-aspartate, 1 mM EDTA, 15% 2-methylpentane-2,4-diol, and ammonium sulfate to 0.2 saturation is placed in a test-tube 6 mm in diameter at 4°. It is overlaid with an equal volume of 4 M CsCl, 0.4 saturated with ammonium sulfate. Large crystals are obtained in 7–10 days. These may be manipulated in a stabilizing solution of the following composition: 30 mM α-methyl-DL-aspartate, 0.15 M potassium phosphate buffer, pH 7.5, 1 mM EDTA, 0.6 saturated with ammonium sulfate.

Crystals of Enzyme–Substrate or Enzyme–Inhibitor Complexes

The enzyme may first be dialyzed rapidly against water (3 successive portions of water over a 4-hr period) and then concentrated by ultrafiltration. Mix together in a well, 10 μl of 16% PEG plus 10 μl of 4–50 mM amino acid or inhibitor. Then stir in 20 μl of the dialyzed enzyme and allow crystals to form. It may be necessary to modify conditions slightly, e.g., by increasing the PEG concentration to a final (reservoir) concentration of 10–12% or to decrease or increase the pH or the concentration of enzyme or amino acid. Crystal formation is sensitive to pH. For example, in our experience very large crystals of the native enzyme formed easily at pH 5.0, but we have been able to obtain only microcrystals at pH 4.7. Unless acetate buffer is present in the reservoir of PEG solution, the pH tends to drift upward with time, and the crystals will dissolve. The causes of this change in pH are not clear but can include bacterial contamination.

Crystals of enzyme containing analogs of pyridoxal-P can be prepared by reconstituting the apoenzyme as described in the preceding chapter and then treating it as above for the native enzyme.

Spectrophotometry

Select single crystals with dimensions approximately 1 × 1 mm and 0.1 mm thickness. In some instances pieces can be cut from larger crystals and smaller attached crystals can be cut away. Transfer may be accomplished with a dissecting needle. Suitable spectrophotometer cells may be prepared from 10 × 20 mm silica plates onto which have been laid three layers of vinyl plastic electrical tape. The latter should first be stuck together and creased in the center while a hole is cut with scissors (in the same manner as is commonly used to make a fancy valentine). Then press the tape onto a silica plate with the tapered hole opening upward. The hole should be just large enough to prevent excessive movement of the crystal. Transfer the crystal to the cell so that it lies flat on the silica plate in the center of the hole, which has been filled with mother liquor. Draw off excess mother liquor, place the second silica plate on top with a pet-rolatum seal, and tape the edges. The cell can be opened easily to permit addition of reagents.

It is necessary to know the Miller indices of the face of the crystal to interpret the spectra. Examine the crystal under a polarizing microscope. With the crystal between crossed polarizers observe the extinctions that occur for each 90° rotation of the microscope stage. For general faces of uniaxial crystals, for faces of orthorhombic crystals which contain a crys-tallographic axis or for faces of a monoclinic crystal which contain the unique axis, the extinction angles are fixed by the crystallographic sym-metry. With such faces measure and record the extinction angles on a sketch of the crystal. For a general face of a biaxial crystal (orthorhombic, monoclinic, or triclinic) the extinction angles can be functions of the wavelength. Frequently, extinction angles do not change more than $\pm 1°-2°$ over the wavelength range and satisfactory spectra can be re-corded for polarization along visually observed extinctions. Additional complexities will be involved in attempting measurements with crystals for which there are large variations in extinction directions.

Place a small pinhole at least 0.5 mm in diameter in a metal plate over the crystal. Check under a microscope for light leaks, and position the crystal in the light beam of a high sensitivity spectrophotometer equipped with a matched pair of suitable polarizers in the sample and reference beam. Calcite polarizers of the Glan–Thompson type have been found satisfactory.[12] Record the spectra with polarizers set along the two polar-ization directions. Maximum and minimum transmission (or absorbance) will be observed along the two extinction directions for any wavelength. The user should read the review by Hofrichter and Eaton[13] for interpreta-tions. (References to microspectrophotometric techniques are also given.)

Crystals may also be observed by microspectrophotometry. For ex-ample, Eichele *et al.* have studied triclinic plates of average dimensions of $0.3 \times 0.2 \times 0.01–0.03$ mm using a Zeiss UMSP I recording double-beam microspectrophotometer.[9] The vertical light beam of this instrument allows convenient horizontal mounting of the crystal, which is placed on the object table between two quartz microscope slides, the upper one being supported by two cover slips. The position of the crystal with respect to the light beam is observed through the built-in microscope. A polarizing filter, e.g., 105 UV from Polacoat (Cincinnati, Ohio), may be inserted in front of the sample position. The portion of the light beam en-

[12] D. S. Martin, Jr., *Inorg. Chim. Acta Rev.* **5**, 107 (1971).
[13] J. Hofrichter and W. A. Eaton, *Annu. Rev. Biophys. Bioeng.* **5**, 511, (1976).

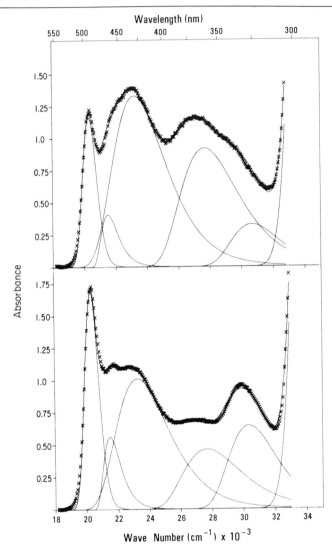

Fig. 1. Absorption spectra of an orthorhombic crystal of cytosolic aspartate aminotransferase of pig heart into which was diffused 5 μl of 4 mM DL-*erythro*-β-hydroxyaspartate at pH 6.3. The spectra were measured using plane-polarized light with the light beam along the *b* axis of the unit cell. Upper curve: Plane of polarization parallel to *c* axis of unit cell. Lower curve: Plane of polarization parallel to *a* axis (90° to *c* axis). Spectra have been resolved into components with log-normal distribution curves of peak positions 329 nm (pyridoxamine-P-product complex), 361, 431, 465, and 492 nm. The latter two narrow bands together represent the quinonoid intermediate in the transamination reaction.

tering the photomultiplier has a diameter of only 32 μm when it passes through the crystal. The transmission spectra obtained are converted to absorption spectra using the transmission of the medium surrounding the crystal as a reference value.

After measurement of spectra on a crystal of native enzyme, it is possible to soak the crystal in buffers of varying pH or substrate or inhibitor solutions and to observe changes.[7-9] Changes may be visible within a few minutes, but complete saturation of large crystals may require several hours. The concentration of active sites in the crystals is nearly 15 mM. Therefore, care should be taken to ensure that the compound being diffused into the crystal is present in adequate amount. The soaking solutions should contain PEG and should be saturated with enzyme. If the crystals crack or dissolve, it may be necessary to raise the PEG concentration or to lower the concentration of substrate or inhibitor added.

Figure 1 shows a pair of spectra obtained by diffusing DL-*erythro*-β-hydroxyaspartate into a native crystal of aspartate aminotransferase. The resulting spectrum has been resolved with log-normal distribution curves and the areas of the different bands have been recorded. These include a band at 492 nm with a shoulder at 465 nm representing the quinonoid intermediate. The latter has been represented by two narrow log-normal curves in Fig. 1. Next there is a band at 430 nm, possibly representing the enzyme–substrate complex, and a band at 362 nm that may be a mixture of enzyme–substrate complex and possibly of free enzyme. The third band at 330 nm represents principally the product complex of pyridoxamine phosphate in the enzyme.

Experiments of this type are especially easy to do with aminotransferases because of their Ping-Pong mechanism. This permits complete conversion of the enzyme into enzyme–substrate complexes[14] by addition of a single substrate product pair or even by a single substrate as in the case of Fig. 1. The approach may be less satisfactory for enzymes with unstable enzyme–substrate complexes. However, enzymic reactions in crystals may be slower than in solution and with certain coenzyme or substrate analogs may be slow enough to allow observation of intermediates. Furthermore, it is possible to study transient intermediates by means of crystallographic studies at low temperature.

The pH indicator properties of certain pyridoxal-P-dependent enzymes can also be studied by spectrophotometry. Thus Fig. 2 shows the pH titration of a crystal of mitochondrial aspartate aminotransferase obtained by Eichele *et al.*[9] using nonpolarized light. The apparent pK_a of the crystalline enzyme turned out to be 6.4 $\pm$ 0.1, which is 0.3 pH unit higher

[14] W. T. Jenkins and R. T. Taylor, *J. Biol. Chem.* **240**, 2907 (1975).

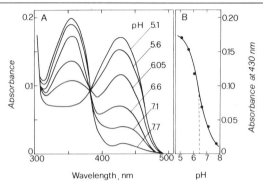

Fig. 2. Spectrophotometric pH titration of a triclinic crystal of the pyridoxal form of mitochondrial aspartate aminotransferase in 20% (w/v) polyethylene glycol/50 mM sodium phosphate/50 mM sodium acetate. (A) Starting from pH 7.7, the crystal was successively transferred into solutions of lower pH values. Equilibration time at each pH value was 5–10 min. (B) A theoretical dissociation curve was fitted to the experimental values (●) by a nonlinear least-squares computer program. Courtesy of G. Eichele, D. Karabelnik, R. Halonbrenner, J. N. Jansonius, and P. Christen, *J. Biol. Chem.* **253**, 5239 (1978).

than the value found in solution. The difference could, however, be attributed to polyethylene glycol, which was shown to increase the measured pH value of a phosphate buffer solution.

As pointed out by Eichele *et al.*,[9] a feature of polyethylene glycol as a crystallization and storage medium, that deserves special emphasis is that this polymer does not penetrate protein crystals.[6] Thus, the functional groups of the crystalline protein can be studied under solvent conditions essentially identical to those of the dissolved protein.

[81] Dithionite-Induced Changes in the Spectra of Free and Enzyme-Bound Pyridoxal Phosphate

By Elijah Adams

Treatment of free pyridoxal phosphate at neutral pH with excess dithionite produces a rapid change in the pyridoxal phosphate spectrum (Fig. 1). The reaction is not due to decomposition of dithionite to bisulfite, since the pyridoxal phosphate–bisulfite addition product has a quite distinct spectrum from that seen after dithionite treatment[1] and the apparent dithionite reaction, unlike that with bisulfite, is not abolished by the presence of excess acetone, added as a bisulfite binder (Fig. 1). The special

[1] E. Adams, *Anal. Biochem.* **31**, 484 (1969).

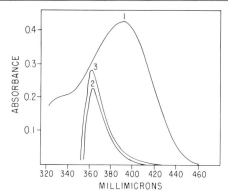

FIG. 1. Effect of sodium dithionite in large excess on the spectrum of pyridoxal phosphate. Curve 1, pyridoxal phosphate (88 μM) in potassium phosphate, 50 mM, pH 7.4, or in the same buffer containing 10% acetone; 2, immediately after addition of sodium dithionite (5.8 mM) to pyridoxal phosphate in buffer alone; 3, like 2, except that the pyridoxal phosphate solution contained 10% acetone. All spectra were obtained at room temperature with a Cary 14 recording spectrophotometer, in 1-ml volumes using matched cuvettes of 1 cm light path. Addition of reagents was always made identically to both a blank cuvette and a cuvette containing the same buffer together with pyridoxal compounds or enzymes. Dilution effects were minimized by keeping the volume of added reagents to 0.05 ml or less. Dithionite solutions were freshly made; maintenance under aerobic or anaerobic conditions did not influence spectra.

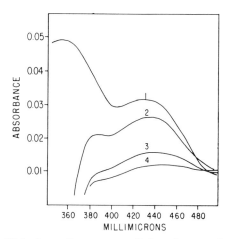

FIG. 2. Effect of dithionite on the spectrum of glutamic–aspartic transaminase. 1, enzyme, 0.5 mg/ml, in 50 mM potassium phosphate, pH 7.4; 2, immediately after adding 2.9 mM sodium dithionite; 3, 10 min after 2; 4, 20 min after 2. Other conditions as noted under Fig. 1.

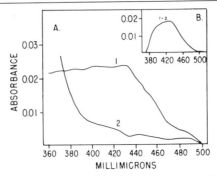

FIG. 3. Spectrum of glutamic alanine transaminase before and after dithionite. (A) 1, enzyme (1.5 mg/ml) in 0.2 N potassium acetate, pH 5.8; 2, immediately after adding 2.9 mM sodium dithionite. (B) Inset showing difference spectrum (1 minus 2, from A). Other conditions as noted under Fig. 1.

change with dithionite is slow at lower dithionite concentrations: a 7-fold molar excess of dithionite over pyridoxal phosphate produces the typical spectral change seen with a large excess of dithionite (Fig. 1), but the entire reaction is not yet complete after 15 min.[1] Similar spectral changes were seen with pyridoxal, but not with pyridoxine.[1]

Two considerations make the dithionite reaction of enzymologic interest. The spectral change might be distinctive enough to identify a pyridoxal phosphate enzyme. Additionally, the spectral shift of enzyme-bound pyridoxal phosphate on dithionite treatment should be kept in mind in interpreting presumptive flavin difference spectra. While the existence of single enzymes containing both a flavin and pyridoxal phosphate coenzyme has not been reliably documented, enzyme mixtures, or multienzyme aggregates of physiological significance, might create an occasion for this confusion.

Figures 2 and 3 show the effect of excess dithionite on the respective spectra of two pyridoxal phosphate enzymes, glutamic aspartic transaminase (Boehringer-Mannheim Corp., Catalog No. 15335 EGAK) and glutamic alanine transaminase (same source, Catalog No. 15344 EGAL). Figure 4 shows the spectrum of alanine racemase[2] before and after dithionite, and contrasts this difference spectrum with that for D-amino acid oxidase (Worthington, "DAOFF"). It is of interest that dithionite selectively reduces the absorption peak in the low 400-nm range, characteristic of several pyridoxal phosphate enzymes. In the case of glutamic alanine transaminase and alanine racemase (Figs. 3 and 4), the direct spectra of these preparations would not have provided unambiguous

[2] G. Rosso, K. Takashima, and E. Adams, Biochem. Biophys. Res. Commun. 34, 134 (1969).

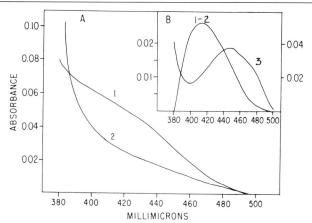

F<small>IG</small>. 4. Spectrum of alanine racemase before and after dithionite. (A) 1, Alanine racemase, 0.017 mg in 0.1 ml of 0.05 M potassium phosphate, pH 7.4, in a self-masking 0.05-ml cuvette (Beckman, No. 97250); 2, immediately after adding sodium dithionite, 0.05 mg in 0.005 ml. (B) Inset showing difference spectrum; 1-2, alanine racemase (*before* minus *after* dithionite, from data of A); 3, difference spectrum (*before* minus *after* dithionite for D-amino acid oxidase (Worthington, "DAOFF"). D-Amino acid oxidase was present at 0.5 mg/ml; otherwise conditions were exactly as for alanine racemase. Absorbance scale on left of inset refers to curve 1-2; that on right to curve 3. Other conditions as for Fig. 1.

indication of this peak, attributed to the pyridoxal phosphate–lysine azomethine complex. In the difference spectra these peaks are unequivocal.

Alanine racemase is of special interest in connection with a dithionite difference spectrum because of an earlier suggestion[3] that the enzyme contains a flavin prosthetic group together with pyridoxal phosphate, a conclusion disputed on other grounds[2] as well as the difference spectra shown here.

[3] W. F. Diven, J. J. Scholz, and R. B. Johnston, *Biochim. Biophys. Acta* **85**, 322 (1964).

[82] Assay of Pyridoxal Phosphate and Pyridoxamine Phosphate, Employing S-*o*-Nitrophenyl-L-cysteine, a Chromogenic Substrate of Tryptophanase

By C<small>LARENCE</small> H. S<small>UELTER</small> and E<small>SMOND</small> E. S<small>NELL</small>

H. Wada and his co-workers[1] were the first to describe the determination of pyridoxal-P based on the use of apotryptophanase. This method,

[1] H. Wada, T. Morisue, Y. Sakamoto, and K. Ichahara, *J. Vitaminol. (Kyoto)* **3**, 183 (1957).

later modified by Okuda *et al.*[2] and Haskell and Snell,[3] is based on the fact that the enzyme reaction is dependent on the content of the coenzyme when excess apotryptophanase is present. The assay described in this paper is built on the same principle, but the apoenzyme–coenzyme complex is determined with the use of a direct spectrophotometric assay using a chromogenic substrate, S-*o*-nitrophenyl-L-cysteine (SOPC).[4] Pyridoxamine-P is determined by the same method after conversion to pyridoxal-P by nonenzymic methods.

Assay Method

Principle. The reaction catalyzed by tryptophanase that is employed for the present assay is

S-*o*-Nitrophenyl-L-cysteine + H_2O → pyruvate + NH_3 + *o*-nitrothiophenol

This reaction provides a direct and convenient spectrophotometric assay of enzyme activity. In the presence of excess apoenzyme required to complex all pyridoxal-P, a linear relationship between the reaction rate measured at 370 nm and pyridoxal-P is observed. A plot of $\Delta A_{370\,nm}$ min^{-1} vs pyridoxal-P is used as a calibration curve to estimate the amount of coenzyme in a sample.

Pyridoxamine-P, which does not activate apotryptophanase, is first converted to pyridoxal-P by nonenzymic transmination with glyoxylate. The active coenzyme can then be estimated as described above.

Reagents. S-*o*-Nitrophenyl-L-cysteine (SOPC) is synthesized by a procedure adapted from Boyland *et al.*[5] Zinc powder (5 g) is heated with L-cysteine-free base (8.5 g) in 1 N sulfuric acid (200 ml) at 60° for 20 min, cooled, and filtered. Diazotized *o*-nitroaniline from 8.3 g in 1 N H_2SO_4 (400 ml) is added over 30–45 min to the cysteine solution containing a small dab of cuprous chloride with stirring at 0–2°; the mixture is kept at 0° for 2 hr, then left at room temperature overnight and filtered.

o-Nitroaniline, which is not completely soluble in 400 ml of 1 N H_2SO_4, is first suspended in the 1 N H_2SO_4 and heated with stirring in a hood to effect as much solution as possible. The resulting solution with some liquid *o*-nitroaniline (melted) is cooled rapidly on ice with stirring. The fine precipitate and solidified melt are then removed from the mother liquor and ground with a mortar and pestle before being resuspended in the same 1 N H_2SO_4 mother liquor for diazotization. Diazotization is

[2] K. Okuda, S. Fujii, and M. Wada, this series, Vol. 18A, p. 505.
[3] B. E. Haskell and E. E. Snell, *Anal. Biochem.* **45**, 567 (1972).
[4] C. H. Suelter, J. Wang, and E. E. Snell, *Anal. Biochem.* **76**, 221 (1976).
[5] E. Boyland, D. Manson, and R. Nery, *J. Chem. Soc. (London)*, p. 606 (1962).

completed at $0°-2°$ by adding an excess of solid sodium nitrite (5.2 g) over a 1 to 1.5-hr period and is allowed to stir until all o-nitroaniline is in solution. An additional hour should be allowed. Two grams of urea are then added and allowed to react with the excess $NaNO_2$ (45 min to 1 hr) as determined by potassium iodide paper.

The filtrate from the coupling reaction is stirred with 20 g of acid-washed charcoal, filtered off under vacuum, and washed with H_2O until free from sulfate. SOPC is eluted with a cold 5% solution of ammonium hydroxide (d 0.88) in methanol (500 ml). The eluate is adjusted to pH 7 with 6 N HCl and flash evaporated to dryness. In our experience, the 20 g of charcoal is not sufficient to adsorb all SOPC. A higher yield is obtained if the original charcoal filtrate and wash solutions are treated with the same charcoal after the methanolic ammonium hydroxide elution as often as is necessary to remove SOPC, usually a total of three treatments. SOPC is crystallized from H_2O, m.p. $173°-175°$.

Potassium phosphate, 0.5 M, pH 8.0
Glyoxylate, 40 mM
$CuSO_4$, 10 mM
Sodium borohydride, 0.25 M in 50 mM NaOH
Pyridoxal-P, 0–1600 pmol ml^{-1}
Pyridoxamine-P, 0–20 μmol ml^{-1}

Procedure. The following procedure for determination of pyridoxal-P is adopted. Buffer A (0.2 M potassium phosphate, pH 8.0, 0.4 mM dithioerythritol, 2 mM EDTA, and 20% glycerol) containing 100 μg of apotryptophanase in 0.25 ml[6] and pyridoxal-P (0–400 pmol in 0.25 ml) are incubated in a 1-ml plastic disposable cuvette at $37°$ for 40 min in the dark, or for 15 min at $50°$ in the dark. After cooling the mixture to $25°$, the newly formed holoenzyme is assayed in the same cuvette after addition of 0.5 ml of 1.2 mM SOPC. A linear relationship between the reaction rate at 370 nm and pyridoxal-P is observed when the protein concentration is in excess.

Pyridoxamine-P is converted to pyridoxal-P by nonenzymic transamination with glyoxylate as originally devised by Metzler *et al.*[7] Glyoxylate (0.5 ml), $CuSO_4$ (0.1 ml), potassium phosphate, pH 8.0 (0.2 ml), and pyridoxamine-P (0.2 ml) are incubated at $50°$ for 5 min. Then 0.1 ml of the reaction mixture is added to 0.4 ml of buffer A containing 100 μg apotryptophanase and is incubated at $50°$ for 15 min. The newly formed

---

[6] Excess apoenzyme must be present in each assay to assure complete complexation of pyridoxal phosphate. With 100 μg of apotryptophanase, the range is 0.5–105 μg of pyridoxal phosphate.

[7] D. E. Metzler, J. Olivard, and E. E. Snell, *J. Am. Chem. Soc.* **76**, 644 (1954).

holotryptophanase is assayed as outlined above by addition of 0.5 ml of 1.2 mM SOPC to the reaction mixture after cooling to 25°.

Mixture of Pyridoxal-P and Pyridoxamine-P. When both pyridoxal-P and pyridoxamine-P are to be determined in the same sample, pyridoxal-P is assayed in one portion, while a separate portion of the same sample is treated with an excess of 0.25 M NaBH$_4$ in 50 mM NaOH to reduce pyridoxal-P to the coenzymically inactive pyridoxine-P. After 15 min, the sample is treated with glyoxylate as described in the section describing the determination of pyridoxamine-P. Excess borohydride is destroyed by part of the glyoxylate and does not interfere with the pyridoxal-P determination.

This chapter does not deal with the merits or demerits of various procedures for the extraction of pyridoxamine-P or pyridoxal-P from tissues preparatory to assay. We have used the extraction and deproteinization with 1 N perchloric acid, as described by Lyon *et al.*,[8] for this purpose; the extracts are then titrated to pH 7–8 with KOH, and potassium perchlorate is removed in the cold. Portions of extracts prepared in this or other ways are then used in the assay as a source of coenzyme in the manner described earlier for the preparation of standard curves of pyridoxal-P or pyridoxamine-P.

Preparation of Apotryptophanase

Growth of Bacteria. Apotryptophanase is isolated from a constitutive mutant strain of a tryptophan auxotroph, *Escherichia coli* B/lt 7-A. Since this mutant was first isolated, two different approaches have been described for preparing large cultures of this organism. Both approaches will be noted here, since different laboratories have differing facilities, and may therefore find one procedure more convenient than the other.

The preparation of apotryptophanase described below was developed with cells grown under growth conditions previously described by Kagamiyama *et al.*[9] Slant cultures of *E. coli* B/lt7-A are maintained on agar prepared as described in Table I (Medium A). The inoculum from one or more slants is transferred to a 3-liter Fernbach flask containing 1 liter of medium B the table. After incubation for 12 hr with shaking at 37°, the entire contents are transferred to a 20-liter jar fermentor containing 12 liters of medium B. This culture is incubated for 12 hr at 37° on a New Brunswick fermentor unit with stirring at 200 rpm and aeration set at 2000 ml/min. Cells from one or several jars can be harvested, or 1 or 2 such

---

[8] J. B. Lyon, Jr., J. A. Bain, and H. L. Williams, *J. Biol. Chem.* **237**, 1989 (1962).
[9] H. Kagamiyama, H. Wada, H. Matsubara, and E. E. Snell, *J. Biol. Chem.* **247**, 1571 (1972).

COMPOSITION OF MEDIA USED FOR GROWTH OF *Escherichia coli* B/1t7-A[a]

Component	Medium A (g/liter)	Medium B (g/liter)
Major salts		
KH_2PO_4	3.0	7.5
$K_2HPO_4 \cdot 3H_2O$	9.5	21.8
$(NH_4)_2SO_4$	1.0	1.0
Minor salts[b]		
$CaCl_2 \cdot 2H_2O$	0.01	0.003
$FeSO_4 \cdot 7H_2O$	0.001	0.003
$MgSO_4 \cdot 7H_2O$	0.1	0.3
$MnSO_4 \cdot H_2O$	—	0.001
$ZnSO_4 \cdot 7H_2O$	—	0.001
$CuSO_4 \cdot 5H_2O$	—	0.001
NaCl	—	0.25
Organic components		
Glucose	1.0	—
Indole[c]	0.01	0.25
Tryptophan	—	0.33[d]
Yeast extract	—	0.1
Tryptone (Bacto)[e]	—	20.0
Casein, enzymically hydrolyzed	—	10.0
Agar	20.0	—
Tap water	100.0	100.0

[a] Reprinted from reference 9, page 1572, by permission of the American Society of Biological Chemists and modified by addition of enzymically hydrolyzed casein.

[b] Concentrated stock solutions of the mixed minor salts are prepared for later use as needed; precipitation can be avoided by addition of a few drops of HCl. Sterilization is best accomplished by passage through a 0.2 μm-pore membrane.

[c] *Escherichia coli* B/1t7-A is a deletion mutant for tryptophan synthetase, and requires indole (or tryptophan) for growth. When indole is supplied, it is converted to tryptophan through the action of tryptophanase, thus permitting growth. To avoid mutational loss of tryptophanase, the stock cultures are carried with indole rather than with tryptophan.

[d] Tryptophan solution is autoclaved separately or sterilized by passage through a 0.2-μm membrane.

[e] Peptone (Bacto) or some commercial enzymic digests of casein are equally satisfactory.

cultures can be used to inoculate a 100- or 200-liter batch fermentor.[10] The yield of wet cells is usually 100–150 g from 100 liters of culture.

A second approach was developed originally by Newton and Snell[11,12] and is described in this series.[13] A preparation of enzyme using cells

[10] Airflow through the fermentor should not exceed 2 CFM. Airflow beyond this removes indole, which then allows excessive growth of bacteria, with diminished yield of enzyme.

[11] W. A. Newton and E. E. Snell, *Proc. Natl. Acad. Sci. U.S.A.* **51**, 382 (1964).

[12] W. A. Newton, Y. Morino, and E. E. Snell, *J. Biol. Chem.* **240**, 1211 (1965).

[13] Y. Morino and E. E. Snell, this series, Vol. 17A, p. 439.

grown under these conditions is also described in the above reference. This preparation works well only when tryptophanase constitutes 10% of the soluble protein.

Other preparations of apotryptophanase from *E. coli* K12[14] and from several other microorganisms have been described.[15-17]

Purification Procedure

Step 1. Crude Extract. Thirty grams of wet cells of *E. coli* B/lt7-A are suspended in 5 volumes of 0.1 M potassium phosphate, pH 8.0, containing 0.2 mM dithioerythritol and 2 mM EDTA,[18] and homogenized in a Waring blender at high speed for 30 sec. The cells are then broken in 100-ml portions with a Branson J-17 oscillator for 4 min at 60% maximal amplitude. Alternatively, the 30 g of cells are suspended in 2.5 volumes of 0.1 M potassium phosphate, pH 8.0, containing 0.2 mM dithioerythritol and 2 mM EDTA, broken by a brief sonication and then passed through a precooled power laboratory press of the American Instrument Co. The suspension is then centrifuged at 39,000 g for 20 min.

Step 2. Protamine Sulfate Treatment. After determination of protein in the supernatant solution from step 1 by the Lowry procedure,[19] and adjusting the concentration to 10 mg ml^{-1}, the protein solution is adjusted to pH 6 with 10% acetic acid. One-fifth volume of a 20% solution of protamine sulfate is then added gradually with stirring at room temperature; stirring is continued for an additional 20 min after this addition. The precipitate is removed by centrifugation and discarded.

Step 3. Ammonium Sulfate Fractionation. The supernatant solution from step 2 is adjusted to pH 7 with 10% NH$_4$OH, before solid (NH$_4$)$_2$SO$_4$ (24.8 g/100 ml) is added gradually with stirring over a 30-min period at 4°. A pH of 7 is maintained by occasional addition of 10% NH$_4$OH. The precipitate is then removed by centrifugation and an additional 15.6 g of solid (NH$_4$)$_2$SO$_4$ is added per 100 ml of supernatant solution in the same fashion. The precipitate is collected by centrifugation and dissolved in 0.1 M potassium phosphte, pH 7.8, 2 mM EDTA, 1 mM dithioerythritol, 0.2 mM pyridoxal-P, and 11 g of (NH$_4$)$_2$SO$_4$ per 100 ml to give a final volume equivalent to one-third the volume of the supernatant solution from step

[14] J. London and M. E. Goldberg, *J. Biol. Chem.* **247**, 1566 (1972).
[15] J. D. Fenske and R. D. DeMoss, *J. Biol. Chem.* **250**, 7554 (1975).
[16] J. L. Cowell, K. Moser, and R. D. DeMoss, *Biochim. Biophys. Acta* **315**, 449 (1973).
[17] H. Yoshida, H. Kumagai, H. Yamada, and H. Matsubara, *Biochim. Biophys. Acta* **391**, 494 (1975).
[18] Potassium or, if available, ammonium phosphate buffers are preferred because, in contrast to the sodium salt, they enhance the stability of the preparation.
[19] O. H. Lowry, N. J. Rosebrough, A. L. Farr, and R. J. Randall, *J. Biol. Chem.* **193**, 265 (1951).

2. Portions of this solution, normally 100 ml, are heated to 65° in a 72° water bath for 5 min, cooled rapidly and centrifuged to remove denatured protein.

Step 4. Ammonium Sulfate Precipitation. Ammonium sulfate (25 g/100 ml) is added gradually to supernatant 3 with stirring over a 30-min period. The precipitate obtained by centrifuging is dissolved in a minimum volume of buffer A (25 mM potassium phosphate, pH 7.0, containing 1 mM EDTA, 0.2 mM dithioerythritol and 10% glycerol) and dialyzed overnight at room temperature against 100 volumes of the same buffer containing 10 mM β-mercaptoethanol. If the dialysis is completed at 4° in buffer A minus glycerol, a precipitate is occasionally obtained that is soluble at room temperature.

Step 5. Chromatography. The dialyzed solution from step 4 is applied at room temperature to a Seph-C$_7$-NH$_2$ column (2.5 × 40 cm) equilibrated with buffer A. This column should have a capacity of 300 mg of protein. Seph-C$_7$-NH$_2$ is prepared by coupling Sepharose 4B to 1,7-heptanediamine.[20] Nonabsorbed protein is washed from the column with equilibration buffer before the enzyme is eluted with a 1200-ml gradient of NH$_4$Cl (600 ml of buffer A and 600 ml of buffer A containing 0.2 M NH$_4$Cl). Tryptophanase elutes at 0.13–0.15 M NH$_4$Cl.

Step 6. Resolution of Enzyme. Fractions from the Sepharose column that contains enzyme are combined and made 10 mM in DL-penicillamine. The enzyme is then precipitated by dialysis of the combined fractions overnight in the cold against 0.1 M potassium phosphate (pH 7.0) containing 1 mM EDTA, 10 mM β-mercaptoethanol, and sufficient (NH$_4$)$_2$SO$_4$ to give 90% saturation. The precipitate is collected by centrifugation, dissolved in 0.1 M potassium phosphate, pH 7.0, containing 1 mM EDTA, 0.2 mM dithioerythritol, 10 mM penicillamine, and 0.8 M (NH$_4$)$_2$SO$_4$, diluted to 10–20 mg ml^{-1} and dialyzed overnight at 4° against 100 volumes of the same solution. This treatment with 10 mM penicillamine is repeated until the specific activity of the apoenzyme is less than 1% of that observed with an excess of added pyridoxal phosphate. Concentrations of enzyme are determined spectrophotometrically using ε = 0.795 mg ml^{-1} cm^{-1}.[21] The protein may be stored as a suspension under nitrogen in 0.1 M potassium phosphate, pH 8.0, 0.2 mM dithioerythritol, 1 mM EDTA, and 2 M (NH$_4$)$_2$SO$_4$.

Comments on Assay

The main advantages of this procedure are its simplicity, sensitivity, and the unusually wide range of coenzyme concentration (1–400

[20] S. Shaltiel and F. Er-el, *Proc. Natl. Acad. Sci. U.S.A.* **70**, 778 (1973).
[21] Y. Morino and E. E. Snell, *J. Biol. Chem.* **242**, 2800 (1967).

pmol/sample). A principal disadvantage arises from the fact that excess apoenzyme must be present in each assay to assure complete complexation of pyridoxal phosphate. Thus, relatively large amounts of a nearly homogeneous apoenzyme, which is not commercially available, are required.

The activity of the enzyme is inhibited nearly 50% by 0.3 M NH$_4^+$. Sulfate ion up to 1 M did not interfere with the assay. Recoveries of added pyridoxal-P or pyridoxamine-P should always be checked when this assay is applied. The possibility that certain samples, such as that found with yeast extract,[4] contain materials that inhibit reconstitution of holotryptophanase or its catalytic action has not been investigated in detail.

[83] Pyridoxamine (Pyridoxine) 5'-Phosphate Oxidase from Rabbit Liver

By Alfred H. Merrill, Michael N. Kazarinoff,
Haruhito Tsuge, Kihachiro Horiike,
and Donald B. McCormick

Pyridoxamine 5'-phosphate + O$_2$ + H$_2$O → pyridoxal 5'-phosphate + NH$_3$ + H$_2$O$_2$
Pyridoxine 5'-phosphate + O$_2$ → pyridoxal 5'-phosphate + H$_2$O$_2$

Pyridoxamine (pyridoxine) 5'-phosphate oxidase (EC 1.4.3.5), in conjunction with pyridoxal kinase, is responsible for the formation of the coenzyme, pyridoxal 5'-phosphate, from the B$_6$ vitamers pyridoxamine and pyridoxine. The oxidase has been detected in a variety of organisms and purified to apparent homogeneity from rabbit liver[1] and bacteria;[2] considerable purification of the yeast enzyme has also been achieved.[3]

Assay Methods

The standard assay[4] involves the formation of the phenylhydrazone of pyridoxal 5'-phosphate, which absorbs light at 410 nm (ϵ 23,000). Similar methods have been reported[5] using the 4-nitrophenylhydrazone [λ_{max} 435 nm (ϵ 39,000) in 15 N H$_2$SO$_4$, λ_{max} 550 nm (ϵ 32,000) in 3% KOH and 40%

[1] M. N. Kazarinoff and D. B. McCormick, *J. Biol. Chem.* **250**, 3436 (1975).
[2] S. Yamamoto, T. Tochikura, and K. Ogata, *Agric. Biol. Chem.* **29**, 315 (1965).
[3] H. Tsuge, A. Hirose, and K. Ohashi, *Res. Bull., Gifu Univ.* **40**, 71 (1977).
[4] H. Wada, this series, Vol. 18A [96].
[5] M. I. Turkov and K. M. Ermolaev, *Appl. Biochem. Microbiol.* **12**, 489 (1976).

ethanol] and the 2,4-dinitrophenylhydrazone [λ_{max} 415 nm (ϵ 33,000) in 1 N H$_2$SO$_4$, λ_{max} 485 nm (ϵ 31,000) in 0.5% KOH and 60% ethanol].

Continuous monitoring of the reaction is possible by either polarographically measuring O$_2$ consumption or observing the appearance of pyridoxal 5'-phosphate absorption (usually as the Schiff base with Tris buffer). Both methods are at least 10-times less sensitive than the standard assay and are complicated by product inhibition by pyridoxal 5'-phosphate.

Since N-(5'-phosphopyridoxyl)amines are good substrates for the oxidase,[6] greater sensitivity can be obtained by preparing substrates containing a fluorescent or radioactive group that is released by the oxidase reaction.[7]

One enzyme unit catalyzes the formation of 1 nmol of pyridoxal 5'-phosphate per hour under standard assay conditions.

Purification Procedure

All steps are conducted at 0–4° in dim light. Phenylmethylsulfonyl-fluoride (50 μM, Sigma) is included in all buffers through step 4 to minimize proteolysis.

Step 1. Extraction and Acid Treatment. Rabbit liver[8] (500 g) is broken into small pieces while frozen and then homogenized for 5 min at low speed in a Waring blender, together with 2 liters of 20 mM potassium phosphate buffer (pH 7) containing 0.1 mM β-mercaptoethanol. The resulting suspension is centrifuged at 18,000 g for 30 min, and the supernatant is filtered through glass wool. The pH of the extract is adjusted to 5.0 by the dropwise addition of cold 2 N acetic acid, and the resulting suspension is stirred for 10 min. The precipitate is removed by centrifugation (18,000 g for 15 min). Potassium chloride (18 g) is added to the supernatant, then water to bring the volume to 2 liters. The pH is adjusted to 5.0 with 2 N acetic acid or 10% potassium hydroxide as needed.

[6] M. N. Kazarinoff and D. B. McCormick, *Biochem. Biophys. Res. Commun.* **52**, 440 (1973).

[7] The analog, N-(5'-phospho-4'-pyridoxyl)tryptamine, has been prepared using conventional methods and is a good substrate for the oxidase (under standard assay conditions, V_{max} = 36,000 units/mg protein, K_m = 60 μM). The enzymic reaction is stopped by adding base, and tryptamine is extracted by an organic solvent. Tryptamine may be quantitated by fluorescence (λ_{ex} 285 nm, λ_{em} 365 nm in water) or counting, if radioactively labeled tryptamine (Amersham/Searle) is used to prepare the substrate.

[8] Rabbit liver (type II, young, full-fed, liquid nitrogen quick-frozen) is purchased from Pel-Freez Biologicals, Rogers, AR, and stored frozen ($-20°$) until use. Rapid freezing of the liver and addition of a protease inhibitor to subsequent tissue extracts is necessary, or proteolysis occurs.

Step 2. Ethanol Precipitation. The supernatant[9] is placed in a $-20°$ freezer and stirred for 10 min. Prechilled absolute ethanol (400 ml) is added dropwise from a separatory funnel over a period of about 25 min. The suspension is stirred for an additional 30 min and centrifuged (10,000 g for 10 min) in prechilled centrifuge bottles at $-17°$. The precipitate is gently homogenized with 100 ml of 0.1 M potassium phosphate (pH 8) using a hand-held glass TenBroeck apparatus. Insoluble material is removed by centrifugation at 24,000 g for 20 min.[10]

Step 3. DEAE-Sephadex A-50 Chromatography. The redissolved ethanol precipitate is applied to a column of DEAE-Sephadex A-50 (5 × 60 cm, equilibrated with 0.1 M potassium phosphate, pH 8). The column is washed with 250 ml of 0.1 M buffer, then the oxidase is eluted by a linear gradient between 0.1 and 0.2 M potassium phosphate (pH 8) totaling 3 liters. Fractions of high oxidase activity are pooled (the greatest activity elutes at approximately 0.16 M).

Step 4. Sephadex G-100 Chromatography. The pooled DEAE fractions are slowly brought to 80% saturation with solid ammonium sulfate (61 g/100 ml) and stirred for 1 hr. The protein precipitate is collected by centrifugation (18,000 g for 15 min), the pellet is suspended in ∼20 ml of 20 mM potassium phosphate (pH 7) and dialyzed overnight against 1 liter of this buffer. The clear, dark-yellow solution is applied to a Sephadex G-100 column (5 × 150 cm) equilibrated with 20 mM potassium phosphate (pH 7) and eluted with the same buffer. Fractions of high activity are pooled (the greatest activity elutes at ∼1.5 times the column void volume, as expected for a protein of molecular weight ∼50,000).

Step 5. Calcium Phosphate Gel Chromatography. The G-100 fractions are concentrated to ∼5 ml by ultrafiltration using an Amicon PM-10 membrane[11] and applied to a column of calcium phosphate on cellulose[12] (1.2 × 20 cm) equilibrated with 20 mM potassium phosphate (pH 7). The column is washed with 25 ml of this buffer, then the activity is eluted by a linear gradient from 0.02 to 0.1 M potassium phosphate buffer (pH 7)

[9] Temperature control is critical for the success of this step.

[10] The enzymic activity and percent yield should be determined after this step. If the yield is less than 40%, the preparation should be aborted, since the success of the subsequent steps is usually affected.

[11] After completion of the concentration, the nitrogen pressure should be released gradually to avoid frothing.

[12] Calcium phosphate gel is prepared as follows [from S. M. Swingle and A. Tiselius, *Biochem. J.* **48**, 171 (1951) and V. Massey, *Biochim. Biophys. Acta* **37**, 310 (1960)]: Add 75 g of CaO to 450 g of sucrose in 2 liters of water and agitate for several hours. Clarify the suspension by filtration, cool the filtrate to 5°, and add dropwise, with stirring over 1 hr, 45 ml of H_3PO_4 (85%). The pH will reach 9.5. Stir 4 hr at 5°, then wash exhaustively with water. Mix one part of calcium phosphate (suspended to yield 30 mg/ml) with two parts of a 10% w/v suspension of cellulose powder (CF-11, Whatman) in 20 mM potassium phosphate (pH 7). Degas and pour the column.

TABLE I

PURIFICATION OF PYRIDOXAMINE (PYRIDOXINE) 5′-PHOSPHATE OXIDASE FROM
RABBIT LIVER

Step	Volume (ml)	Total protein (mg)	Total activity (units)	Specific activity (units/mg)	Yield (%)
Crude extract	1830	124,000	970,000	7.8	100
1	2000	58,100	895,100	15.4	92
2	150	12,600	722,000	57.3	74
3	480	657	456,000	694	47
4	85	104	310,000	2,980	32
5	45	25.7	235,500	9,160	24
6	8	5.0	115,000	23,000	12
Affi-Gel 501	35	7.0	175,000	25,000	18

totaling 250 ml. Fractions with more than 6000 units of activity/A_{280} are pooled (maximal activity elutes at ~60 mM buffer).

Step 6. Preparation of Apoenzyme. The volume of the calcium phosphate gel fractions is reduced to 10 ml by Amicon PM-10 ultrafiltration. The yellow solution is dialyzed against four changes (250 ml) of 2 M potassium bromide and 0.1 mM EDTA in 0.1 M potassium acetate (pH 4) for 24 hr. The salt is then removed by exhaustive dialysis against 20 mM potassium phosphate (pH 7). Insoluble material is removed by centrifugation (24,000 g for 15 min).

The purified enzyme is shell-frozen in small aliquots and stored at −20°. The enzyme may be converted to the holo form before storage by adding excess FMN and separating the holooxidase from free FMN by Sephadex G-10 chromatography. The purification is summarized in Table I.

An Affi-Gel 501 column[13] can be substituted for steps 5 and 6; however, for reasons not yet fully understood, use of this procedure gives variable results. The enzyme from the Sephadex G-100 column is applied directly to a prewashed and equilibrated Affi-Gel 501 column (1.2 × 40 cm). Unwanted proteins are eluted with 75 ml of 20 mM potassium phosphate (pH 7), followed by 75 ml of this buffer containing 200 μM dithiothreitol. The oxidase is then eluted by buffer plus 1 mM dithiothreitol. The yellow fractions (specific activity of ~8000–9000 units/$A_{280\,nm}$) are frozen overnight at −20°, thawed, centrifuged to remove precipitated protein, and exhaustively dialyzed against 20 mM potassium phosphate

[13] Affi-Gel 501 (Bio-Rad) is regenerated by washing with five bed volumes of 50 mM sodium acetate (pH 4.8), three bed volumes of 10 mM HgCl$_2$ and 20 mM EDTA in 50 mM sodium acetate (pH 4.8), ten bed volumes of 0.2 M NaCl and 1 mM EDTA in 100 mM sodium phosphate (pH 7), and equilibrated with 20 mM potassium phosphate (pH 7).

(pH 7). Enzyme thus obtained is identical to that from step 6 but is in the holo form.

Properties

Purity. The purified enzyme electrophoreses as a single protein and activity band on analytical disc gels containing 4, 7.5, or 9% acrylamide and as a single protein band on sodium dodecyl sulfate acrylamide gels.

Stability. No significant loss of activity has been observed after four years of storage at $-20°$ of shell-frozen apo- and holoenzyme. Both are, however, rapidly denatured by repeated freeze-thawing. Dilute solutions, such as those used for activity assays, are more stable when bovine serum albumin ($\sim$10 μg/ml) is added.

Physical Properties and Amino Acid Composition. The enzyme is composed of two noncovalently attached subunits of the same apparent molecular weight (27,000). The amino acid composition is shown in Table II.[14] The amino-terminus of both subunits is blocked.

TABLE II

AMINO ACID COMPOSITION OF PYRIDOXAMINE 5'-PHOSPHATE OXIDASE

Amino acid	Residues/ 54,000 g of protein	Nearest integer of residues/54,000 g of protein
Lys	24.7	25
His	7.0	7
Arg	39.7	40
Asx	39.7	40
Thr	12.9	13
Ser	26.0	26
Glx	75.0	75
Pro	36.4	36
Gly	40.4	40
Ala	36.9	37
Val	24.1	24
Met	12.3	12
Ile	8.6	9
Leu	41.3	41
Tyr	15.6	16
Phe	25.1	25
Trp	6.4	6
Cys	5.8	6

[14] H. Tsuge and D. B. McCormick, *in* "Flavins and Flavoproteins" (K. Yagi and T. Yamano, eds.). Japan Scientific Societies Press, Tokyo, in press.

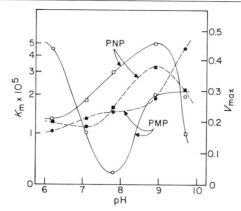

FIG. 1. Variation of K_m (solid lines) and V_{max} (broken lines) for pyridoxamine 5'-phosphate (PMP, $\bigcirc$, $\bullet$) and pyridoxine 5'-phosphate (PNP, $\square$, $\blacksquare$) with pH.

One mole of FMN is bound ($K_d = 20$ nM) per 54,000 daltons. The absorption spectra contain maxima at 448 nm (ϵ 12,000), 380 nm (ϵ 10,000), and 275 nm (ϵ 64,500). Both protein and flavin fluorescence are quenched. The CD spectrum of holoenzyme contains strong negative ellipticities in both the visible and UV regions.[15]

Coenzyme and Substrate Specificity. FMN analogs modified at positions 2, 3, 7, 8, 8α, and at the ribityl side chain are all bound by the apoenzyme but less tightly than FMN.[16,17] Riboflavin, riboflavin 5'-sulfate, and FAD are bound poorly. FMN binding is pH dependent, with tightest binding at pH 9.

The K_m and V_{max} (from Lineweaver–Burk plots) for pyridoxamine 5'-phosphate and pyridoxine 5'-phosphate are also pH dependent, as is shown in Fig. 1. Under standard assay conditions (pH 8), pyridoxamine 5'-phosphate is bound more tightly than pyridoxine 5'-phosphate. Many *N*-(5'-phosphopyridoxyl)amines are also good substrates for the enzyme.[6]

Activators and Inhibitors. Buffers, neutral and ionic compounds, and detergents activate and/or inhibit the enzyme, as is illustrated in Fig. 2.[18] Both natural substrates produce strong substrate inhibition, even at low substrate concentrations (below K_m). The product, pyridoxal 5'-phosphate, is a competitive inhibitor ($K_i = 3$ μM). Neither NH_3 nor H_2O_2 inhibits appreciably. The substrate and product analogs, 4'-deoxypyridoxine

[15] A. H. Merrill and H. Tsuge, *Fed. Proc., Fed. Am. Soc. Exp. Biol.* **37**, 1347 (Abstr. 431) (1978).

[16] M. N. Kazarinoff and D. B. McCormick, *Biochim. Biophys. Acta* **359**, 282 (1974).

[17] A. H. Merrill and D. B. McCormick, *Am. Chem. Soc. 172nd Meet.,* #148 (1976).

[18] K. Horiike, A. H. Merrill, and D. B. McCormick. *Arch. Biochem. Biophys.* (in press, 1979).

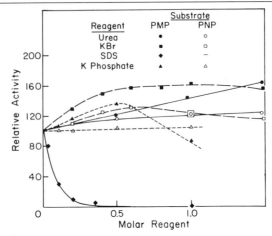

FIG. 2. Effect of various reagents on enzymic activity. Assays were conducted in the presence of excess FMN and, except for the potassium phosphate experiment, 0.5 M Tris buffer (pH 8).

5'-phosphate[19] (K_i = 1.8 μM) and pyridoxal 5'-phosphate oxime (K_i = 2 μM), are strong competitive inhibitors.

Sulfhydryl-modifying reagents, such as p-chloromercuribenzoate and S-mercuri-N-dansylcysteine, react with two thiols per mole of holoenzyme with no loss of activity. Treatment of apoenzyme with these reagents, however, results in modification of an additional cysteine and inactivation of the subsequently prepared holoenzyme.[14]

[19] W. Korytnyk, M. T. Hakala, P. G. G. Potti, N. Angelino, and S. C. Chang, *Biochemistry* **15**, 5458 (1976).

[84] Pyridox(al, amine) 5'-Phosphate Hydrolase from Rat Liver[1]

By LAWRENCE LUMENG and TING-KAI LI

Pyridox(al, amine) 5'-phosphate + H_2O → pyridox(al, amine) + P_i

The tissue content of pyridoxal 5'-phosphate may be controlled by the protein binding of this coenzyme and its hydrolysis by a cellular phos-

[1] This work was supported in part by a Veterans Administration Grant (MRS 583-5246). The authors wish to express their thanks to Mrs. Marilyn McKee for her skillful technical assistance.

phohydrolase.[2] In human erythrocytes[3] and rat liver,[4] this phosphohydrolase activity resides in plasma membranes. Kinetic and other studies strongly suggest that this coenzyme, pyridoxamine 5′-phosphate, and other phosphomonoesters, such as p-nitrophenylphosphate and β-glycerophosphate, are hydrolyzed by the same enzyme, the nonspecific alkaline phosphatase (EC 3.1.3.1). The hydrolysis of pyridoxal phosphate by purified rat liver alkaline phosphatase has not been studied. The methods described here apply mainly to pyridoxal phosphate hydrolase activity in isolated plasma membranes from rat liver and to the butanol-solubilized enzyme from the same source.

Pyridoxal phosphate hydrolase activity may be assayed by measuring either the rate of disappearance of pyridoxal phosphate or the rate of release of inorganic phosphate (P_i) or pyridoxal. The specific method to be employed depends upon the pH, substrate concentration and other conditions of the experiment. Pyridoxamine phosphate hydrolase activity is assayed by measuring the rate of formation of P_i. The assays are performed under yellow light, since pyridoxal phosphate is light-sensitive. In addition to these assay methods, the fluorometric determination of pyridoxal and pyridoxal phosphate after reaction with cyanide[5] may also be employed.

Assay Methods

Inorganic Phosphate Assay

Principle. The assay is based on the spectrophotometric determination of P_i.[6] It is suitable for the assay of activity in crude liver homogenate and subcellular fractions. One of the characteristic features of alkaline phosphatase-catalyzed hydrolysis of pyridoxal phosphate is the change in pH-rate profile with substrate concentration (Fig. 1). When the specific enzymic activity of plasma membranes is 10–18 mU per milligram of protein and the substrate concentration is ≥ 150 μM, pyridoxal phosphate hydrolase activity may be measured by this method.

   Reagents
   Sodium 5,5-diethylbarbiturate·HCl buffer, 20 mM, pH 7.0–9.8
   Magnesium chloride, 0.2 M
   Pyridoxal phosphate, 1.5 mM, is prepared immediately before use.

[2] T.-K. Li, L. Lumeng, and R. L. Veitch, *Biochem. Biophys. Res. Commun.* **61,** 677 (1974).
[3] L. Lumeng and T.-K. Li, *J. Clin. Invest.* **53,** 693 (1974).
[4] L. Lumeng and T.-K. Li, *J. Biol. Chem.* **250,** 8126 (1975).
[5] See this series, Vol. 18A [78].
[6] E. S. Baginski, P.-P. Foà, and B. Zak, *Clin. Chem.* **13,** 326 (1967).

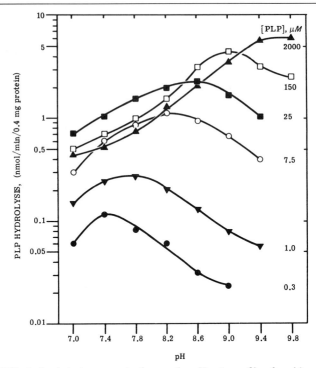

FIG. 1. Effect of substrate concentration on the pH-rate profile of pyridoxal phosphate hydrolysis by rat liver plasma membranes. P_i assay was employed to measure the activities when the substrate concentrations were 150 and 2000 μM. The tyrosine apodecarboxylase assay was used for the other pyridoxal phosphate concentrations.

Its concentration is determined spectrophotometrically at 388 nm in 0.1 M NaOH. The absorption coefficient is 6600 mol^{-1} cm^{-1}.[7]

Isolated rat liver plasma membranes, ca. 10 mg of protein per milliliter of 1 mM sodium bicarbonate, pH 7.5

Ascorbic acid in 10% trichloroacetic acid (A-TCA), 2%

Procedure. To each Erlenmeyer flask are added 1.5 ml of the barbitone sodium buffer, 0.05 ml of MgCl$_2$, 0.2 ml of pyridoxal phosphate, 0.04 ml of the plasma membrane suspension, and water to a final volume of 2.0 ml. The reaction is started by the addition of pyridoxal phosphate, and incubation is performed in a shaking water bath at 25°. Pyridoxal phosphate is omitted from a control flask. Zero-time samples are obtained by adding A-TCA before the addition of pyridoxal phosphate. At 15- and 30-min intervals, 0.5 ml of the reaction mixture is transferred to 1.5 ml of A-TCA.

[7] E. A. Peterson and H. A. Sober, *J. Am. Chem. Soc.* **76,** 169 (1954).

Precipitated protein is then removed by centrifugation, and an aliquot of the supernatant is used for P_i assay.[5]

Definition of Unit. One unit of enzyme is defined as that amount which catalyzes the formation of 1 μmol of product per minute under the above conditions and at pH 9.0. Protein concentrations are measured by the procedure of Lowry *et al.*[8] with bovine serum albumin as standard.

Pyridoxal Phosphate Assay

Principle. The assay is based on the enzymic determination of pyridoxal phosphate with tyrosine apodecarboxylase.[3] This assay is used when pyridoxal phosphate concentrations are ≤ 25 μM (cf. Fig. 1). The method is a modification of those described by Sundaresan and Coursin[9] and Chabner and Livingston.[10] The major features include: further purification of a commercially available apoenzyme isolated from *Streptococcus faecalis* to eliminate contaminating pyridoxal kinase, deproteination of reaction mixtures with TCA followed by the extraction of TCA with ether, the use of citrate as buffer, and the use of a higher concentration of L-tyrosine. At pH 7.4, the disappearance of pyridoxal phosphate can also be measured spectrophotometrically with an Aminco–Chance or DW-2 dual-wavelength, split-beam spectrophotometer (A_{392} nm versus A_{470} nm).

Reagents. In addition to those listed above:

L-[1-^{14}C]Tyrosine solution[11]: 10 mM in 0.1 M sodium citrate buffer, pH 6.0 (specific radioactivity, 25 μCi/mmol). This solution is warmed to 58°–62° before use to solubilize the tyrosine.

Pyridoxal phosphate standard solution, 20 ng/ml, in 0.1 M sodium citrate buffer, pH 6.0

Methyl benzethonium hydroxide, 1 M in methanol

Fluoralloy TLA,[12] 32.2 g in 2.5 liters of toluene and 1.25 liters of methoxyethanol

[8] O. H. Lowry, N. J. Rosebrough, A. L. Farr, and R. J. Randall, *J. Biol. Chem.* **193**, 265 (1951).

[9] See this series, Vol. 18A [85].

[10] B. Chabner and D. Livingston, *Anal. Biochem.* **34**, 413 (1970).

[11] L-Tyrosine, 1.2 g, is dissolved in 1.2 liter of 0.1 M sodium citrate buffer, pH 6.0, by heating to 80°. This solution is filtered through filter paper under vacuum. To 1 liter of this warm solution, add 1.13 g of L-tyrosine in 125 ml of 0.15 M HCl and refilter. Then add 250 μCi of L-[1-^{14}C]tyrosine, 40–60 mCi/mmol (New England Nuclear Corp., Boston, Massachusetts).

[12] Fluoralloy TLA (Beckman Instrument, Inc.) is a mixture of 30.3 g of 2-(4'-*tert*-butylphenyl)-5-(4''-biphenylyl)-1,3,4-oxadiazole (butyl-PBD) and 1.9 g of 2-(4'-biphenylyl)-6-phenylbenzoxazole (PBBO).

Tyrosine apodecarboxylase,[13] ca. 100 U/ml; diluted 1:10 with
   sodium citrate buffer, 0.1 M, pH 6.0, immediately before use
Sodium citrate buffer, 0.1 M, pH 6.0
Hydrochloric acid, 5 M
TCA, 75%
Peroxide-free diethyl ether, water-saturated

Procedure. The assay condition for pyridoxal phosphate hydrolase
activity is the same as above except that the pyridoxal phosphate concentrations are ≤25 μM. Aliquots (0.5 ml) of the reaction mixture for pyridoxal phosphate hydrolase activity assay are taken at appropriate time intervals and added to 0.05 ml of 75% TCA. After standing at room temperature for 30 min, the precipitated protein is centrifuged and discarded.
The supernatant solution is extracted four times with at least 4 volumes of
water-saturated ether. The final pH of the supernatant solution should be
≥4.5. The supernatant solution from the unknown samples are diluted as
needed with 0.1 M sodium citrate in order to fit into the range of the pyridoxal phosphate standard curve.

Pyridoxal phosphate assay is performed with 0.1 ml of tyrosine apodecarboxylase (about 1.0 U), 0–0.2 ml (0–4 ng) of the pyridoxal phosphate
standard solution or unknown, 1.0 ml of L-[1-^{14}C]tyrosine solution and 0.1
M sodium citrate buffer in a final volume of 1.4 ml. Before the addition of
L-[1-^{14}C]tyrosine, the flasks are preincubated for 30 min at room temperature to permit reconstitution of the holoenzyme. The reaction is started by
addition of L-[1-^{14}C]tyrosine, and the flasks are immediately capped with
rubber stoppers fitted with center wells. The center wells contain a filter

---

[13] Add 1 g of commercially available tyrosine apodecarboxylase (Sigma Chemical Co., St.
Louis, Missouri; 1 mg will release 300–400 μl of CO_2 in 30 min at pH 5.5 at 37° in the presence of added excess pyridoxal phosphate) to 10 ml of 10 mM sodium citrate buffer, pH
6.0. This suspension is then homogenized in a Potter–Elvehjem Teflon–glass homogenizer
and centrifuged at 25,000 g for 10 min. The supernatant solution is collected and the pellet
is resuspended in another 10 ml of 10 mM sodium citrate buffer. The resultant suspension
is then sonicated for 6 min (6 pulses of 1 min) with the use of a Savant model 500 Insonator
(0.5-inch standard sonohorn) at an output meter setting of 75. These centrifugation, resuspension, and sonication steps are repeated 5 times, and 60 ml of supernatant solution are
collected. The supernatant solution is brought to 60% saturation by addition of solid
$(NH_4)_2SO_4$ with mechanical stirring. The suspension is centrifuged at 25,000 g for 20 min
and the precipitate is discarded. The supernatant solution is brought to 85% saturation with
$(NH_4)_2SO_4$, and the precipitate is suspended in 5 ml of a buffer containing 0.3 M sodium
citrate, pH 6.0, 24% (v/v) glycerol, and 2 mM mercaptoethanol. The enzyme suspension is
then dialyzed overnight against two changes of 1 liter of the same buffer. This partially
purified apoenzyme, usually 100 U/ml (1 unit is defined as the amount of enzyme that
catalyzes the decarboxylation of 1 μmol of tyrosine per minute in the presence of excess
pyridoxal phosphate and under the conditions of the assay), is free of pyridoxal kinase, and
is stable at −20° for several months. All the above steps are carried out at near 0°.

paper and 0.2 ml of methyl benzethonium hydroxide in methanol. The flasks are then incubated at 32° in a Dubnoff shaking water bath for 20 min, and the reaction is terminated by injection of 1 ml of 5 M HCl. To ensure complete trapping of $^{14}CO_2$ after acidification, the flasks are incubated for an additional 60 min at 32°. The stopper and center well are then removed, and the well is put in a counting vial containing 5 ml of the Fluoralloy scintillation fluid. $^{14}CO_2$ is counted in a liquid scintillation spectrometer. The efficiency of this $^{14}CO_2$ trapping system is close to 100%. The standard curve with enzyme activity (cpm) plotted as function of pyridoxal phosphate concentration (0–4 ng) is linear, and the background count is low. The slope of the standard curve is usually 20,000–30,000 cpm per nanogram of pyridoxal phosphate. All the assays are done in duplicate. The reproducibility of the method is ±4.7% (±SD).

Pyridoxal Assay

Principle. In experiments at pH 9.0, the rate of formation of pyridoxal can be assayed spectrophotometrically. The absorption coefficient for pyridoxal at 308 nm is 5.05 mmol^{-1} cm^{-1}.

Reagents. Same as listed above.

Preparation of the Enzyme

Plasma membranes are isolated essentially as described by Ray.[14] This method includes two major modifications of the original method developed by Neville[15] and later modified by Emmelot et al.[16] These modifications increase the yield; however, the use of calcium ions may lead to some contamination with mitochondria and with cytoplasmic RNA in the final preparations.[16]

Solubilization of the pyridox(al, amine) phosphate hydrolase or alkaline phosphatase from isolated plasma membranes is accomplished by butanol extraction.[17] The solubilized enzyme with specific activity, 60–90 mU per milligram of protein at pH 9.0 for pyridoxal phosphate hydrolysis, is routinely obtained in this manner. This represents at least a 300-fold purification. The final yield is between 10 and 15%.

Reagents

Sodium bicarbonate, 1 mM, pH 7.5, freshly prepared, with calcium chloride 0.5 mM

[14] T. K. Ray, *Biochim. Biophys. Acta* **196**, 1 (1970).

[15] D. M. Neville, Jr., *J. Biophys. Biochem. Cytol.* **8**, 413 (1960).

[16] See this series, Vol. 31 [5].

[17] R. K. Morton, *Biochem. J.* **57**, 595 (1954).

Sodium bicarbonate, 1 mM, pH 7.5, with magnesium chloride, 5 mM
Sucrose solutions of 63.7, 48, 45, and 41% (w/w)
n-Butanol, water-saturated
Barbitone sodium buffer, 20 mM, pH 8.0
Triethanolamine buffer, 10 mM, pH 7.4

Procedure. All steps are carried out at 0 to 4°. Ten to 20 g of liver are
minced by a chilled hand press fitted with an 80-mesh (180 μm opening)
sieve. The minced tissue is suspended in 10 volumes of the NaHCO₃–
CaCl₂ buffer and homogenized in a glass, loose-fitting Dounce homogenizer using 25 gentle strokes. The homogenate is diluted further with 10
volumes of the NaHCO₃-CaCl₂ buffer and allowed to stand for 5 min with
periodic stirring. The diluted homogenate is then filtered through eight
layers of cheesecloth (20 × 24 mesh). The filtrate is transferred into six
315-ml capacity plastic bottles and centrifuged at 3000 rpm for 30 min in a
GSA rotor (Sorvall RC2-B ultracentrifuge). The supernatant solution is
discarded and the pellet (pellet 1) is resuspended in the same buffer by
gentle (5 strokes) homogenization in the Dounce homogenizer. The suspension is diluted to half the previous volume of buffer and centrifuged at
2500 rpm for 15 min. The supernatant solution is discarded and the pellet
(pellet 2) is resuspended by homogenization in the same buffer. The suspension is brought to half the previous volume and centrifuged again at
2500 rpm for 15 min. The pellet (pellet 3) is resuspended in less than 4 ml
of the NaHCO₃–CaCl₂ buffer and mixed with a 63.7% (w/w) sucrose solution to achieve a final sucrose concentration of 48%. Pellet 3, suspended
in 48% sucrose, is divided equally among 3 centrifuge tubes. Eight milliliters of 45% sucrose are then layered over this suspension, followed by 9
ml of 41% sucrose and 3.5 ml of 37% sucrose. The tubes are centrifuged in
a Spinco SW 25.1 rotor at 25,000 rpm for 2 hr. The plasma membranes
form a compact sheet between the 37% and 41% sucrose layers, and this
fraction is collected by a Pasteur pipette. After washing away the sucrose,
the plasma membranes are finally resuspended in 1 mM NaHCO₃, pH 7.5,
containing 5 mM MgCl₂. The yield of isolated plasma membrane based on
p-nitrophenyl phosphatase and pyridoxal phosphate hydrolase activities
is usually between 20 and 30%. The specific activity increases 50- to
90-fold (Table I).

Plasma membranes from two preparations are pooled and centrifuged
at 12,000 g for 30 min. The membranes are resuspended in 7 ml of 20 mM
barbitone sodium buffer, pH 8.0. Water-saturated n-butanol, 2.3 ml, is
added to the membrane suspension dropwise with agitation in an ice bath
and then sonicated for 2 min. After mixing for 30 min, the butanol and
aqueous solution are separated by centrifugation in an International PR-2
centrifuge, 3000 rpm for 45 min, using a 269 rotor. The butanol fraction is

TABLE I
PURIFICATION OF PYRIDOXAL PHOSPHATE HYDROLASE FROM RAT LIVER[a]

Fraction	Total activity[b] (mU)	Yield (%)	Specific activity (mU per mg protein)
Homogenate	1002	100	0.2
Pellet 1	552	55	0.6
Pellet 2	365	36	1.0
Pellet 3	386	38	1.4
Plasma membranes	222	22	17.6
Butanol-solubilized enzyme	125	12	82.5

[a] Assays are performed at pH 9.0.
[b] Total activities from 20 g of rat liver.

collected and the aqueous phase is reextracted with butanol. The two butanol fractions are pooled and dialyzed for 12 hr against 1 liter of 10 mM triethanolamine buffer, pH 7.4 with three changes. The yield of the solubilized enzyme is 10–15% (Table I).

Properties

The ratio of specific activities for pyridoxal phosphate and p-nitrophenylphosphate (a typical substrate for alkaline phosphatase) hydrolysis at pH 9.0 remains between 2 and 3 throughout the purification procedure. These activities also behave concordantly with respect to

TABLE II
EFFECT OF pH ON THE K_m VALUES FOR THE HYDROLYSIS OF PYRIDOXAL PHOSPHATE, PYRIDOXAMINE PHOSPHATE, AND p-NITROPHENYLPHOSPHATE BY ISOLATED PLASMA MEMBRANES OF RAT LIVER AND BUTANOL-SOLUBILIZED PHOSPHOHYDROLASE

Substrate	K_m (μM)		
	pH 7.4	pH 8.2	pH 9.0
Pyridoxal-P	2[a]	12[b]	55[c]
Pyridoxamine-P	—	25[c]	95[c]
p-Nitrophenyl-P	2[d]	6[d]	20[d]

[a] Rate of pyridoxal phosphate hydrolysis was measured spectrophotometrically with an Aminco–Chance dual-wavelength/split-beam spectrophotometer (λ_{392} versus λ_{470} nm).
[b] Rate of pyridoxal phosphate hydrolysis was determined by the tyrosine apodecarboxylase method.
[c] Rates of pyridox(al, amine) phosphate hydrolysis were measured by P_i formation.
[d] Rate of hydrolysis of p-nitrophenylphosphate was measured by the formation of p-nitrophenol.

pH-rate profile, pH $-K_m$ relationship (Table II) and response to chelating agents (the order of effectiveness being EDTA $>$ 1,10-phenan-throline $>$ 2,2′-dipyridyl), Zn^{2+}, Mg^{2+}, and inhibitors (P_i is a competitive inhibitor whereas L-tryptophan and L-homoarginine are uncompetitive inhibitors). Kinetic studies indicate that pyridoxal phosphate binds to the same enzyme site as β-glycerophosphate and phosphorylcholine. These data favor the conclusion that pyridoxal phosphate hydrolase activity in liver is mediated by the alkaline phosphatase.

The solubilized enzyme does not differ from the membrane-associated phosphatase in K_m values for p-nitrophenylphosphate and pyridoxal phosphate. However, these two preparations differ with regard to the effect of Mg^{2+}, which is required to stabilize the solubilized phosphatase.

Ohkubo et al.[18] have recently purified a solubilized alkaline phosphatase 42,000-fold from the livers of rats previously subjected to bile duct ligation. Their method may be applicable to the further purification of the pyridox(al, amine) phosphate hydrolase, although bile duct ligation may introduce an uncertain variable.

[18] A. Ohkubo, N. Langerman, and M. M. Kaplan, *J. Biol. Chem.* **249**, 7174 (1974).

[85] Plasma Binding of B6 Compounds

By RALPH GREEN and BARBARA B. ANDERSON

Vitamin B6 compounds are bound by plasma albumin.[1] The various forms of B6 are bound with different affinities to plasma protein.[2] Pyridoxal 5′-phosphate (pyridoxal-P) is the most avidly bound form, pyridoxal too is bound, but less avidly, and pyridoxine appears not to be bound at all. Binding of B6 compounds added to plasma or incubated with whole blood can be demonstrated by protein fractionation of the plasma using gel filtration column chromatography. This method also allows separation of B6 forms into "bound" or "free" fractions without their modification or loss of identity. Detection of binding is based on the measurement of identifiable B6 in association with a protein. The B6 may be identified either by microbiologic assay[3] or by radioactivity measurements if radio-labeled B6 is used.

For demonstration of binding in plasma specimens, pyridoxal-P is

[1] W. B. Dempsey and H. N. Christensen, *J. Biol. Chem.* **237**, 1113 (1962).
[2] B. B. Anderson, P. A. Newmark, M. Rawlins, and R. Green, *Nature (London)* **250**, 502 (1974).
[3] B. B. Anderson, M. B. Peart, and C. E. Fulford-Jones, *J. Clin. Pathol.* **23**, 232 (1970).

METHODS IN ENZYMOLOGY, VOL. 62

added to plasma and B_6 measured by microbiologic assay after conversion to pyridoxal by acid hydrolysis.[3] The assay organism *Lactobacillus casei* shows selective growth response to pyridoxal only. For studies with whole blood, pyridoxine is added to blood. The pyridoxine is rapidly taken up by red cells, and enzymically converted to pyridoxal-P and pyridoxal. Pyridoxal is then released into the plasma.[4] Measurement of the plasma binding of the released aldehyde form can again be carried out using the microbiologic assay. Alternatively, radiolabeled [³H]pyridoxine may be used, and plasma binding determined by radioactivity measurements following red cell conversion and release.

The methods for demonstrating plasma binding of B_6 compounds can be used for studying the relative proportions of forms of B_6 that are bound in plasma. Applied to whole blood, the technique is also useful for measuring the degree of plasma binding that occurs after red cell conversion of B_6 *in vitro*. Red cell uptake and release of B_6 compounds may be related to their plasma binding.[2]

Method

Reagents

Sephadex G-200 (Pharmacia Fine Chemicals)
Phosphate buffer, 50 mM, pH 7.4, containing 0.154 M sodium chloride
Pyridoxine and pyridoxal 5'-phosphate (Sigma Chemical Co.)
³H-labeled pyridoxine (Radiochemical Centre)
Lactobacillus casei NCIB 8010 (ATCC 7469)

Special Apparatus

Fraction collector
Column, 2.2 × 90 cm
UV spectrophotometer
Liquid scintillation spectrometer (for [³H]pyridoxine method)

Procedure

All procedures are carried out in subdued light. Incubation flasks and columns are covered with aluminum foil.

Sample Preparation. Blood is anticoagulated with heparin (125 IU of lithium heparin per 10 ml of blood). Either plasma or whole blood is then used as follows: For *plasma*, 1000 ng of pyridoxal-P in a volume of 20 μl is

[4] B. B. Anderson, C. E. Fulford-Jones, J. A. Child, M. E. F. Beard, and C. J. T. Bateman, *J. Clin. Invest.* **50**, 1901 (1971).

added to 2 ml of plasma and incubated at 37° for 30 min. For *whole blood,* 2500 ng of ^{3}H-pyridoxine (specific activity adjusted to approximately 60 mCi/mmol with unlabeled pyridoxine) in a volume of 50 μl is added to 5 ml of heparinized whole blood. The mixture is incubated at 37° for 2 hr in a shaking water bath. Blood is then centrifuged, and the plasma is removed.

Column Chromatography. Two milliliters of the plasma-pyridoxal-P mixture or the plasma obtained after incubation of whole blood with [^{3}H]pyridoxine are then subjected to protein fractionation through a Sephadex G-200 column previously prepared as follows: G-200 beads are swollen in the phosphate buffer for 3 days at room temperature. The Sephadex slurry is then cooled to 4°, degassed and poured to a height of 90 cm in a column of 2.2 cm diameter. For handling Sephadex and pouring and operation of the column, precautions described by the manufacturer are followed.

The plasma (2 ml) is applied to the top of the column and eluted with the phosphate buffer at a flow rate of 12 ml/hr. Four-milliliter fractions are selected for protein measurement and B_6 determination by microbiologic assay or measurement of ^{3}H radioactivity. After calibration of a column with respect to the elution volumes of bound and unbound B_6, only those fractions of interest need be selected for B_6 measurement by microbiologic assay or ^{3}H counting (Figs. 1 and 2). Once calibrated the column

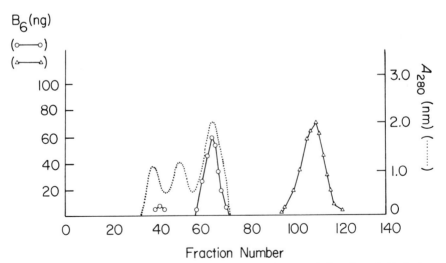

Fig. 1. Binding of pyridoxal-P added *in vitro* to plasma. Elution through Sephadex G-200. B_6 was detected by microbiologic assay (O———O). The elution position of "free" pyridoxal-P in the absence of plasma is shown (△———△). Protein was determined by absorbence at 280 nm (· · · ·).

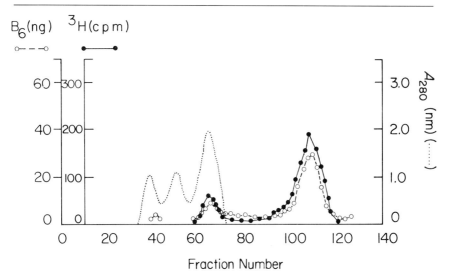

FIG. 2. Plasma protein binding of pyridoxal originating from red cell conversion of [³H]pyridoxine added to whole blood *in vitro*. Elution through Sephadex G-200. B₆ detected either by microbiologic assay (O——O) or ³H radioactivity (●——●) shows good agreement. Protein was determined by absorbence at 280 nm (· · · ·).

may be used for at least 50 determinations. In our hands, this method has given excellent reproducibility.

Elution positions of the three major protein peaks in plasma are established by measurement of absorbency at 280 nm. For measurement of B₆ by microbiologic assay, *Lactobacillus casei* NCIB 8010 (ATCC 7469) is used as described in detail previously.[3] The organism responds only to pyridoxal at concentrations down to 0.01 ng/ml. To measure pyridoxal-P, it must first be hydrolyzed to pyridoxal. For measurement of radiolabeled [³H]pyridoxine, 0.2-ml aliquots of fractions are solubilized in 1 ml Hyamine hydrochloride overnight before addition of scintillation fluid and counting. When [³H]pyridoxine is used, fractions containing radioactivity and eluting in the position of unbound B₆ can be subjected to microbiologic assay to distinguish unconverted [³H]pyridoxine from [³H]pyridoxal. Pyridoxine does not have microbiologic activity in the *L. casei* assay.

Author Index

Numbers in parentheses are footnote reference numbers and indicate that an author's work is referred to although his name is not cited in the text.

Subject Index

A

Acyl lipoic acids
  photochemical preparation, 145, 146
Affinity cytochemistry
  avidin–biotin complex, 308–315
β-Alanine
  precursor of pantothenate synthetase,
    215, 218, 219
d-Allobisnorbiotin
  characterization, 387–390
  isolation, 385–387
4-Amino-5-aminomethyl-2-
    methylpyrimidine
  synthesis, 75
4-Amino-5-cyano-2-methylpyrimidine
  synthesis, 75
(4-Amino-2-methyl-5-pyrimidinylmethyl)-5-
    (2-hydroxyethyl-4-methyl)thia-
    zolethione
  synthesis, 75–76
1-Amino-D-proline
  natural antagonist of vitamin B$_6$,
    487–489
O-Amino-D-serine
  natural antagonist of vitamin B$_6$,
    492–495
Anaerobic spectrophotometric titrations
  methodology employed, 185–198
Apofatty acid synthetase
  assays, 257
  preparation from pigeon liver, 251
  preparation from rat liver, 250
Apo- and holofatty acid synthetases
  immunochemical determination, 260–262
  interconversion, 249–262
    assays for the conversion of apo- to
      holofatty acid synthetase, 257–259
    assays for the conversion of holo- to
      apofatty acid synthetase, 259–260
    separation of apo- and holo- forms of
      pigeon liver enzyme, 251–254
    separation of apo- and holo- forms of
      rat liver enzyme, 251–253
Ascorbate oxidase
  apo- and reconstituted oxidase, 37–39

  assay methods for, 30–31
  holoascorbate oxidase, 31–34
  properties, 34–37
  purification summary, 35
Ascorbic acid (*see also* Ascorbate oxidase)
  L-ascorbic acid 2-sulfate
    hydrolysis by sulfatase A, 42
    preparation of [^{35}S]-labeled compound,
      39
  determination
    animal cells, tissues, and fluids, 3-11
    ascorbic acid 2-sulfate, 15
    buffy coat content, 4, 5, 10–11
    content of adult human tissues, 5
    L-dehydroascorbic acid, 3, 15
    2,6-dichlorophenolindophenol, 6–7
    L-diketogulonic acid, 3
    2,4-dinitrophenylhydrazine, 7–8
    α,α'-dipyridyl-Fe^{2+}, 9–10
    fluorometric, 8–9
    iodine, potassium iodate, potassium
      bromate, and iodine monochloride,
      12–13
    leukocyte concentrations, 4
    liquid chromatography with ampero-
      metric detection, 15
    potentiometric titration, 14
    plasma ascorbate levels, 4
    potassium dichromate, 13
    suitability of assays for various sample
      types, 11
    urinary excretion, 4
  L-gulono-γ-lactone oxidase
    assay methods, 24
    properties, 29–30
    purification from goat liver, 27–28
    purification from rat liver, 25–27
    staining of enzyme activity on gels, 28
Ascorbic acid 2-sulfate
  determination, 15
L-Ascorbic acid 2-sulfate
  hydrolysis by sulfatase A, 42
  preparation of [^{35}S]-labeled compounds,
    39
Asparagusate dehydrogenases
  assay, 172–173